ANNUAL REVIEW OF
PLANT PHYSIOLOGY

ANNUAL REVIEW OF PLANT PHYSIOLOGY

WINSLOW R. BRIGGS, *Editor*
Carnegie Institution of Washington, Stanford, California

PAUL B. GREEN, *Associate Editor*
Stanford University

RUSSELL L. JONES, *Associate Editor*
University of California, Berkeley

VOLUME 26

1975

ANNUAL REVIEWS INC.   4139 EL CAMINO WAY   PALO ALTO, CALIFORNIA 94306

ANNUAL REVIEWS INC.
Palo Alto, California, USA

International Standard Book Number: 0-8243-0626-0
Library of Congress Catalog Card Number: A51-1660

REPRINTS

The conspicuous number aligned in the margin with the title of each article in this
volume is a key for use in ordering reprints. Available reprints are priced at the
uniform rate of $1 each postpaid. Effective January 1, 1975, the minimum acceptable
reprint order is 10 reprints and/or $10.00, prepaid. A quantity discount is available.

PRINTED AND BOUND IN THE UNITED STATES OF AMERICA

CONTENTS

ANNUAL REVIEWS INC. is a nonprofit corporation established to promote the advancement of the sciences. Beginning in 1932 with the *Annual Review of Biochemistry,* the Company has pursued as its principal function the publication of high quality, reasonably priced Annual Review volumes. The volumes are organized by Editors and Editorial Committees who invite qualified authors to contribute critical articles reviewing significant developments within each major discipline.

Annual Reviews Inc. is administered by a Board of Directors whose members serve without compensation.

Annual Reviews are published in the following sciences: Anthropology, Astronomy and Astrophysics, Biochemistry, Biophysics and Bioengineering, Earth and Planetary Sciences, Ecology and Systematics, Entomology, Fluid Mechanics, Genetics, Materials Science, Medicine, Microbiology, Nuclear Science, Pharmacology, Physical Chemistry, Physiology, Phytopathology, Plant Physiology, Psychology, and Sociology. The *Annual Review of Energy* will begin publication in 1976. In addition, two special volumes have been published by Annual Reviews Inc.: *History of Entomology* (1973) and *The Excitement and Fascination of Science* (1965).

Robert Hill

Ann. Rev. Plant Physiol. 1975. 26:1–11

DAYS OF VISUAL SPECTROSCOPY

❖7579

Robert Hill

Department of Biochemistry, University of Cambridge, Tennis Court Road, Cambridge
CB2 1QW, England

To have been asked by the editor of the *Annual Review of Plant Physiology* to prepare an introductory chapter is indeed an honor. I accepted because it might be taken as an acknowledgment of help and encouragement given to me by so many different people and also from many different plants. I choose a title which refers to the time up to the year 1945. That is, the time before the advent of photoelectric spectrum photometry which has been responsible for many major scientific advances since that date.

At Stresa in 1971 the organization by Giorgio Forti of the Second International Congress on Photosynthesis enabled many of us to celebrate the bicentenary of Joseph Priestley's discovery of reversal of animal respiration by vegetation. At that time there was some thought of having a later continuation of the history of photosynthesis after 1800. But plowing through the nineteenth century in detail would be heavy going. Even Eugene Rabinowitch hinted as much at the end of his superb historical review where he had shown the effect of the four most remarkable investigators following Priestley: Ingen-Housz, Senebier, de Saussure, and Robert Mayer. Their work had led to the overall representation of photosynthesis as we see it today in the vegetable kingdom. As given by Rabinowitch in 1945, referring to the year 1845, the equation reads:

carbon dioxide + water + light $\xrightarrow{\text{plant}}$ organic matter + oxygen + chemical energy

It happened that Dr. A. San Pietro had asked me to contribute a lecture to the Gordon Research Conference on Regulatory Mechanisms in Photosynthesis, held in August 1973. There was discussion about later publication of my talk. So, with permission, this present contribution is based on that lecture, entitled An Appreciation of Some of the Former Searches Leading to a Knowledge of Photosynthesis. The first part is concerned with influences through the years since I started as a research student in 1922; the second half indicates how, from 1935 onwards, green plants attracted much of my scientific inquisitiveness.

Chlorophyll and hemoglobin are perhaps the two most interesting coloring matters on this earth. Not so very long ago, while he was still with us, the Sage had asked me, "Who said why is grass green and blood red?" I do not know; perhaps my readers will; but this question summed up in a few words the interests I hope now to share with them.

To begin with, let us go back to 1837, when H. von Mohl described chloroplasts as discrete green bodies seen in plant cells under the microscope. He also observed the presence of starch in several kinds of plants. Later, Julius von Sachs (1832–1897) proved that the starch in chloroplasts was only formed in the light and could disappear from the chloroplasts in darkness. The starch, detected by the color with iodine, was formed very rapidly in some plants. This made it appear that starch might be the first detectable product of photosynthesis and that the whole process actually took place in a chloroplast. The classical demonstration with a stencil cut to form the word STÄRKE placed on a leaf showed how sharply defined the effect of light could be. In 1921 Schroeder partially covered a single algal cell and only the illuminated part showed starch formation. Autonomy of chloroplasts has been questioned many times, for example in 1883, when Josef Böhm showed that starch could be formed in chloroplasts if leaves were floated on sugar solutions in darkness. This led to the view that the first product of photosynthesis could be a simple sugar; besides, it can be observed that many plants do not form starch in their leaves. D. Tollenaar in 1925 made a detailed study of conversion of carbohydrates in leaves of *Nicotiana tobacum* L., which like many other solanaceous plants is a notable starch former. He found that feeding leaves with a solution of sucrose of only 0.025% gave the formation of starch at 28°. Czapek in 1911 found that for many plants about 1.0% was required. The time required for the first appearance of starch formed in darkness is measured in days, while in the light the time may be measured in minutes. The difference in time scales could be due to the slow diffusion of the carbohydrate through the cut portion of the leaf to the chloroplasts. Yet the concentration of sugar solution needed to enable the leaf of a particular plant to form starch in the dark is of the same order as that estimated for starch formation in the light. For example, leaves of the wheat *Triticum sativum* Lam. were found by Tollinaar to require feeding with a 10% solution of sucrose, and when growing in the field the leaves contain starch only under conditions allowing a rapid rate of photosynthesis. The snowdrop (*Galanthus nivalis* L.) does not form starch in the leaves and could not be induced to do so with any concentration of sucrose. The regulatory mechanisms which determine the appearance and disappearance of starch in a chloroplast would still seem to offer some challenging problems.

The study of plant physiology as a part of botany seemed to have been started by Sachs and also by Bussingault, who had a practical interest in agriculture. The influence of Sachs on the teaching of plant physiology continued long after the beginning of the present century. His *Handbuch der Experimental Physiologie*, published in Leipzig in 1865, is still regarded as a classic. The simple representation of photosynthesis as the formation of carbohydrate from CO_2 and water with elimination of oxygen stimulated theories by many of the organic chemists in the nineteenth century. In particular, J. von Liebig (1803–1873) proposed the gradual reduction of CO_2 through a series of organic acids finally to give carbohydrate. The formaldehyde hypothesis of A. von Baeyer (1835–1917), published in 1870, involved the production of formaldehyde from CO_2 as an intermediate stage in the formation of carbohydrate. This derived strong support from the discovery by A. M. Butlerow (1828–1886) that formaldehyde was condensed to a hexose sugar by means of aqueous alkalies. While Baeyer's formaldehyde hypothesis exercised a strong influence on the study of photosynthesis for over 50 years, no direct support could be found for it from experiments with living plants, except for the one fact that in a continuous run the photosynthetic quotient CO_2/O_2 is unity or very near to it.

Although little was added in chemical terms to the formulation of the production of carbohydrates from CO_2, there were many important developments in relation to biological studies throughout the last century. In 1894 T. W. Engelmann used motile bacteria under the microscope and showed that oxygen is produced by illuminating a part of the chloroplast in a living cell of *Spirogyra*. By the same method, G. Haberlandt in 1888 observed oxygen produced from an isolated chloroplast of a moss (*Funaria hygrometrica* Hedw.). This was obtained by cutting the leaf under sucrose solution so that a chloroplast could leave the cell without either osmotic or mechanical damage. In 1896 A. J. Ewart extended this method with several plants, notably with *Selaginella helvetica* (L.) Link, but *Elodea canadensis* Michx. and other phanerogams that were tried did not yield active chloroplasts. These experiments with chloroplasts probably represent the first isolation of an intracellular organelle showing some recognizable activity both inside and outside the cell. It was only later that the mitochondrion was recognized to be the structure concerned in active uptake of oxygen in intracellular respiration. We can appreciate this early work in two ways: either by putting ourselves in the position of the knowledge at the time by a great effort, or by enjoying the numerous detailed descriptions in the old literature from the advantage of present knowledge. In 1888 F. Noll described the "cavern moss" (*Schistostega osmundacea* Mohr., *S. pennata* Hook.) and showed how the lens-shaped cells of the protonema can focus the image of the limited area of light from outside onto the chloroplasts contained in them. The diffused light in the habitat of this moss may be too feeble to support the growth of any other plant. Noll's drawing of the cells of the protonema is reproduced (p. 15) in the book on photosynthesis by Walter Stiles. This book, published in 1925, was one of the first books in English devoted to this subject. It was followed in 1926 by H. A. Spoehr's monograph, published under the auspices of the American Chemical Society. The two books together give a fine critical review of the early experimental work.

From the beginning of the twentieth century the available scientific methods for the study of photosynthesis began to show continuous and rapid improvement. Experiments requiring a strong source of illumination were no longer dependent on sunlight. The methods of analysis of gases had been greatly improved, together with the methods of controlling temperature. These technical improvements were used successfully by F. F. Blackman, who carried out (in my opinion) the first *real* kinetic analysis concerning rates of photosynthesis. This was to determine the effect of varying three external factors—light intensity, carbon dioxide concentration, and temperature—one at a time, keeping the other two at constant level. He assumed that the rate was completely determined by the slowest or limiting factor (meaning a limiting *process* depending on the factor). The rate showed a linear relation at first, as the magnitude of the particular factor was increased, and then tended to become constant when one or other of the factors held at constant level became limiting. This simple conception of a limiting factor gave a plot of the rate as a uniform slope with a sharp bend to the horizontal—a situation rarely approached in practice. Usually the curve is convex to the coordinate, which refers to the variable factor, showing a uniform slope at first and then gradually reaching the horizontal region which has been referred to as "the ceiling." However, this simplified hypothesis did not obscure the importance of the experimental results. They showed that at a constant low light intensity a moderate increase in temperature did not affect the rate, while at a high light intensity the rate increased with temperature. Blackman concluded that the process of photosynthesis consisted of two parts: the light reaction, a process limited by low light intensity with the Q_{10} of unity, and the dark reaction, with a higher Q_{10} between 2 and 3 at high light intensity. However, if the temperature was raised above 40°C, the Q_{10} diminished and at about 45° was reduced to zero. His analysis showed that the dark process had the properties of enzyme reactions as it had a high Q_{10} and was destroyed above a certain range of temperature. The work of Blackman provided a basis for developments in the study of photosynthesis which have continued to the present time.

It should be emphasized that in those days the chlorophylls were the only chemical substances known to be specific for the process of photosynthesis. This made the chlorophyll molecule the key for the source of energy derived from light. In 1913 Willstätter and Stoll published their researches on the chemistry of hemin and chlorophyll from the Kaiser Wilhelm Institut für Chemie in Berlin. Willstätter had succeeded Baeyer at the Institute. A photograph of Willstätter in his laboratory has been published showing von Baeyer's wash bottle on the bench, the traditional type bottle, exactly like those we used in the 1920s. Baeyer was the famous figurehead in the rapid development in the nineteenth century of synthetic organic chemistry, culminating in his synthesis of the natural blue dye, indigo. It is said that Baeyer's success stimulated Willstätter to study the chlorophyll structure as a green pigment.

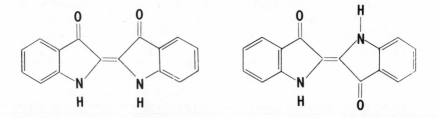

Figure 1 *(left)* Baeyer formula for indigo. *(right)* Formula for indigo established by X-ray crystal structure in 1928.

Willstätter and Stoll developed Blackman's work and also worked with leaves of land plants. The apparatus used for the measurement of photosynthesis at that time now looks formidable. The air with added CO_2 was passed over the leaf and analyzed before and after going through the leaf chamber. This meant that conditions could be maintained constant for the concentration of CO_2, but the analysis had to be very accurate as differences were small. Leaves of land plants were chosen with the object of getting variations in the internal factors, because this would entail variations in the content of chlorophyll and the enzymic factor as conceived by Blackman. Willstätter was notable for his representation of experimental results in terms of a number. His Purpurogallinzahl or PZ, which refers to the specific activity of a peroxidase, is widely known. The "Assimilationszahl" was defined as the ratio: hourly assimilated $CO_2(g)$/chlorophyll(g), while its reciprocal multiplied by 176 gave the value of the "Assimilationszeit": the time in seconds for one molecule of CO_2 to be converted by one molecule of chlorophyll in the leaf. These numbers refer to measurements of photosynthesis in approximately saturating light, with the activity expressed in terms of chlorophyll content. The values of the Assimilationszahl was found to vary greatly with the type of leaf. Some varieties of plants with yellow-green leaves gave very high values which were less affected by change in temperature than those found with leaves of the normally green variety. This showed that there could not be a constant ratio between the amount of chlorophyll and the activity of the enzymic dark process.

This division of photosynthesis into just two processes seems strange to us now, but we have to carry this idea through several more stages in following the development of the subject. Willstätter and Stoll found that yellow-green leaves which were chlorotic as a result of growth under iron deficiency gave a low photosynthetic number; this suggested that iron nutrition was important, not only for the formation of chlorophyll but also for activity of the enzymic process. The work on photosynthesis was partly carried out in Munich, and in 1917 Willstätter and Stoll collected their papers into a volume entitled: *Untersuchungen über die Assimilation der Kohlensaure,* which appeared in 1918. This book has always been a pleasure to take in hand with its fine printing, wide margins, and fascinating discussions of theory and experiment.

In 1908 Joseph Barcroft described his differential manometer which he used for the estimation of respiratory gas exchanges. This was partly derived from the constant volume apparatus for blood gas analysis originally developed by J. S. Haldane in 1902. His discovery that ferrocyanide liberated the whole of the oxygen combined with hemoglobin gave a new method for study of its physiological function. I think that the first description of the differential manometer came from botanical studies; it was given in 1902 by H. H. Dixon, who showed how it could be used to measure the respiration and photosynthesis of a seedling and moreover to prove that the plant had no use for carbon monoxide. Was this the first application of manometry to measure the rate of photosynthesis?

However, in 1908 Otto Warburg realized that a manometric method could be used for continuous observation of gas exchange on a suspension of respiring cells such as sea urchin eggs, which he had been studying at Naples. Warburg then developed his constant volume manometer and showed the advantages of using cell suspensions of a microorganism. The unicellular alga *Chlorella* was introduced by Warburg for the study of a green plant. His technique created a new epoch for experimentation with respiration and photosynthesis. With the leaf the diffusion of gas via the stomata through intercellular spaces represented an important internal factor which was operative in the analysis of Blackman. With the cell suspension the diffusion path is shorter and more easily determined.

In 1919 Warburg, following Blackman's analysis, showed that the "dark enzymic process," which he called the "Blackman reaction," was inhibited by cyanide. The "light reaction" he found to be extremely sensitive to narcotics, more sensitive than the respiration. By the use of intermittent illumination he was able to show a time separation of the light and dark processes. The brilliant development of this method in 1931–1932 by Robert Emerson and William Arnold enabled them to measure the time for completion of a dark process. This gave a most surprising result. The number of chlorophyll molecules concerned in reduction of one molecule of CO_2 was found to be between 1000 and 2000, and the time of this dark process was about 1/100th of a second. The values of the "assimilation time" given by Willstätter and Stoll for normal green leaves were usually between 10 and 30 seconds; these referred to all the chlorophyll, and division by a factor based on the result of Emerson and Arnold would make the values obtained from the leaves and from the *Chlorella* appear in reasonable agreement. But the values of the assimilation time for leaves of a golden variety were between 1.5 and 2 seconds because of the lower content of chlorophyll. So would a smaller number of chlorophyll molecules be concerned in the assimilation of one molecule of CO_2? While Emerson preferred to interpret his results in terms of an essential catalyst present in very small quantity in relation to the chlorophyll content of the algae, the idea of a photosynthetic unit was developed. This came to be thought of as representing a structure containing a definite number of chlorophyll molecules.

The widely adopted assumption that the chlorophyll molecule had a direct chemical relationship with the reduction of CO_2, leading to a subsequent production of oxygen, seems to have been a dominating idea up to the 1930s. No experiments with isolated chlorophyll and CO_2 seemed to give any positive clue to finding the mechanism in a living cell. Willstätter and Stoll in their book gave reference to 30 previous and current theories. But in 1926 a far-reaching change in outlook was on the way. By that time A. J. Kluyver and H. J. L. Donker had developed the conception of many biological processes, including photosynthesis, in terms of hydrogen transport. C. B. van Niel was developing the study of photosynthetic bacteria which he obtained in pure culture. Among the bacteria, "chemosynthetic" forms were known which could reduce CO_2 in the dark by utilization of energy derived from oxidation of components in their environment. So chlorophyll, CO_2, and light together were not fundamental for assimilation of carbon! Yet it was felt that there was nothing in this that would apply to the green plant, and besides, the properties of the chlorophyll molecule provided such an esthetic attraction. So even when van Niel announced his generalized scheme for photosynthesis,

$$CO_2 + 2H_2A \xrightarrow{\text{Light}} (CH_2O) + 2A + H_2O$$

which was based on his accurate quantitative experiments with bacteria, there was no immediate stir in the domain of plant physiology. Thus I published my contribution on the green plant in 1937 and 1939 with no reference to van Niel's work; I think he forgave me. I had found that chloroplasts from higher plants would produce oxygen in light by the reduction of substances other than CO_2. This property of isolated chloroplasts eventually emerged in the literature under the name of "the Hill reaction," indicating perhaps that none of us knew exactly how it worked. Since I have been asked occasionally how this reaction was discovered, what follows now is an attempt to record in a subjective sense some of the influences and experiments leading up to it.

In my early days I had acquired a keen interest in botany from my father and in artists' colors from my maternal grandfather, who taught art during the second half of the nineteenth century. During the first World War there was a serious shortage of dyestuffs because they were produced mainly in Germany. So during my school days the idea of reviving the traditional dyes from plants became a powerful influence. Indigo from the woad plant (*Isatis tinctoria* L.) exercised a special fascination both chemically and botanically. There was a recorded locality on a bank above the river Severn where I got some seed in 1920 and have had it growing ever since. The lower epidermis of a leaf of the woad in its first year is easy to skin off. It used to be fun to put the skinned fragments in dilute $NaHCO_3$ and to demonstrate photosynthesis by counting bubbles from the broken end of a vascular bundle. Then the gas could be collected and shown to rekindle a glowing splint.

In 1922 I was fortunate to become a research student with the active group headed by F. G. Hopkins in the biochemical laboratory in Cambridge. In those days Hopkins did not consider that plants were interesting for biochemical research. On one occasion he indicated that he thought they were rather disgusting because they had no mechanism for excretion; characteristically this was a really stimulating idea! He directed me into the study of the blood pigment, hemoglobin and its derivatives. In this field I had two notable rivals, M. L. Anson and A. E. Mirsky, in the department of physiology under the direction of Joseph Barcroft. They usually won out on most points, but I probably supplied them with their first lot of pure hemin! I think the work with blood pigments in the department of Hopkins, and later in the Molteno Institute with David Keilin, was the most important and exciting time in my scientific life. Hopkins was keen to have an explanation for the easy removal of the iron from hemoglobin (yielding the porphyrin) by dilute acid in absence of oxygen, while in oxyhemoglobin the metal had become so firmly fixed. To begin with, one had to learn to follow changes in absorption spectrum with the Zeiss direct vision spectroscope. All physiological and medical students then were taught to use this instrument for observing the reactions of blood pigments. I was led to examine the combining properties of the porphyrin with different metals. The ferrous iron porphyrin when separate from the protein (globin) would form a characteristic compound with CO both in water and when dissolved in an organic solvent, but all attempts to show the formation of a reversible compound with O_2 were fruitless. In a collaboration with H. F. Holden, we were able to split off the protein from hemoglobin reversibly. It was a great moment when we observed the two absorption bands of oxyhemoglobin with our resynthesized product, showing that the protein had retained its specific activity in directing the properties of the ferrous iron porphyrin towards molecular oxygen. Hopkins at once asked us what happened if we put the porphyrin back instead of the hematin. This produced a dramatic change in the absorption spectrum of the aqueous solution of the porphyrin. The absorption bands became very narrow and sharp, even more so than if the porphyrin was dissolved in an organic solvent.

The molecule of hemoglobin seems to have been responsible for initiating an amazing number of scientific developments. The study of the function of the hemoglobin in larvae of *Gastrophilus* had led Keilin to discover in cytochrome the key to the understanding of intracellular respiration. When J. D. Bernal (known privately as Sage) first showed us the X-ray diffraction pattern of spots from a crystal of hemoglobin, he excited all of us with wonder and anticipation. The S-shaped O_2 dissociation curve of the blood had led to the famous equation of A. V. Hill and to Gilbert Adair's determining the molecular weight of hemoglobin by his superlative skill in the construction of osmometers and in the interpretation of the measurements of osmotic pressure. The different hemoglobins provided material for molecular weight determinations with the ultracentrifuge when it was being developed by Svedberg in Sweden. Then one might have been hit by a high pressure jet of a dilute hemoglobin solution, as happened once to Hopkins when Hamilton Hartridge and F. J. W. Roughton were developing techniques for the measurement of high velocity reaction rates. R. A. Peters had succeeded in putting the chemical studies with hemoglobin on a sound basis, for in 1912 he had shown in Barcroft's laboratory that when hemoglobin was saturated with oxygen the ratio of O_2 to Fe was unity.

The biochemical department at Cambridge moved into its new building in 1923. When this was formally opened in 1924 we all had to produce some kind of display. The different metalloporphyrins made a fine lot of colors to cheer up the bench, and I had filled a large glass jar with a bright purple solution of hematoporphyrin in acid. One of the eminent guests indicated that it just looked like methylated spirit; my resulting deflation was compensated because the show of pigments led to contact with David Keilin. This soon became intimate, as indeed all Keilin's contacts through the Molteno institute were to become. I helped him in the preparation of cytochrome c from yeast. It was a baker's yeast of a Delft strain, very pleasantly aromatic, very pink, and we used to eat bits of it. It was in this yeast (which was lost during the last war) that Keilin observed for the first time the cytochrome spectrum in a microorganism. We had difficulties in preparation of cytochrome c from muscle until A. Szent-Gyorgyi showed us how to use trichloracetic acid as a start. For his work on cytochrome Keilin used the Zeiss microspectroscope. This had a right angle prism which could cover half the slit, giving a comparison spectrum so that the instrument could be used as a spectrum colorimeter. Keilin described this in his book, *The History of Cell Respiration and Cytochrome.* He emphasized the advantage of using a small dispersion spectroscope for direct observation of pigments; the absorption bands are less spread out, showing more contrast than they would with greater dispersion. Under Keilin's kind guidance I was able to use this apparatus to follow up the very important work of M. L. Anson and A. E. Mirsky on the nature of the pigment known as hemochromogen from a quantitative standpoint. The results showed that the ferrous porphyrin could form dissociable compounds with two nitrogen-containing molecules. This result, taken together with the structure Willstätter had proposed for the porphyrin, would have indicated that the iron was 6 covalent in Werner's sense and was surrounded by the nitrogen atoms forming an octohedron. But unfortunately there was a bond between the two methene carbons which joined the four pyrrol rings together. Kuster had proposed a formula in 1912 that would have been perfect for representing the hemochromogen type of structure. Willstätter and Stoll pointed out that the disadvantage of Kuster's formula was that it had the 4 nitrogens and 12 carbons forming a 16-membered ring. Adolph Baeyer's powerful influence from his "strain theory" was still operative. Yet again in the 1920s, after Hans Fischer had actually synthesized porphyrins, the results were used as evidence to support a structure he called the "indigoid," which nevertheless allowed the presence of two 8-membered rings. When Joseph Barcroft was revising his book, *The Respiratory Function of the Blood,* first published in 1913, with a second edition in 1928, he asked me for information about porphyrins. I gave him the "indigoid" structure. I have felt excessively stupid on many occasions, but when the correct formula appeared, actually before the publication of Barcroft's book, this seemed the worst. I had such a great admiration for the way Barcroft had dealt with the oxygen dissociation curve of hemoglobin; if only pages 9–11 could have been replaced! Barcroft was able to get a footnote added on page 9 indicating that Kuster's formula had been accepted.

Then I became anxious to measure oxygen dissociation curves of hemoglobin. I wanted a compact and rapid method which could be applied to quite small quantities of the blood pigment, so a spectroscopic method seemed most suitable. Malcolm Dixon very kindly loaned me the Zeiss microspectroscope he had obtained through the Royal Society. The double wedge trough for the comparison spectrum was discarded in favor of a pair of cups and plungers from an old Dubosq colorimeter. The diluted sample of hemoglobin was put in a Keilin-Thunberg vacuum tube to which a graduated pipette could be attached (see Figure 2). The method was made easy because Nature gives us a readymade standard solution of oxygen, about millimolar in terms of H equivalents. That was the only standard needed for measuring the equilibrium in dilute solutions of hemoglobin. So after removing all the oxygen from the sample in the vacuum tube, known amounts of oxygen could be added with the pipette. After the equilibrium had been established in the vacuum tube, the ratio HbO_2/Hb could be determined. The comparisons between this method and the standard procedure then in use were better than one had dared to hope for. In 1925 I had a most fascinating collaborative effort using the spectroscopic method with H. P. Wolvekamp from The Netherlands, who was working with Barcroft.

It was simple to change over and use a hemoglobin with a known affinity to determine the amount of oxygen in solution over the dissociation range of that particular hemoglobin. So it would seem interesting to put some chloroplasts in with some myoglobin, but nothing happened in the light when I had hoped to see some oxyhemoglobin appear. The only working hypothesis was a light and a dark reaction. So I decided to add the dark reaction, based on the current studies with muscle, in the form of an aqueous extract of acetone leaves, very strong and soupy. It was a very thrilling moment when I saw the spectrum of oxymyoglobin. Then later on a sad disappointment: the presence or absence of CO_2 made no difference. This was really lucky, however, because if CO_2 had "worked," I might well have got no further. It was shown that the oxygen produced in light corresponded with the reduction of a hydrogen or electron acceptor. One of the reactions catalyzed in light was the reduction of ferric potassium oxalate to the ferrous state. It seemed that the oxygen must have come from the water, thus confirming van Niel, provided that this property shown by chloroplasts was part of the process of photosynthesis.

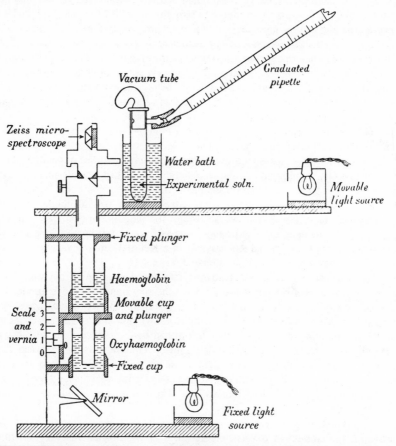

Figure 2   The Zeiss microspectroscope used to measure oxygen dissociation curves of hemoglobin.

The effort at this critical stage was enormously helped by Otto Warburg's work with *Chlorella*, which he published in 1919. Ronald Scarisbrick and I found that the production of oxygen by chloroplasts was very sensitive to urethanes. The concentrations for inhibition agreed with those Warburg had found for whole cells. Also, we found that the production of oxygen was not sensitive to cyanide. This gave evidence for assuming that the chloroplasts plus hydrogen acceptor were doing the light reaction in the sense that F. F. Blackman used. But the light curve with our isolated chloroplasts tended to a dark reaction ceiling and was sensitive to temperature at the higher light intensities. No longer then did we have to be content in photosynthesis with just a light and a dark process.

The chloroplast activity seemed to resemble the cytochrome system working the wrong way round. Keilin, by his use of the two types of inhibitor, cyanide and urethane, was led to interpret the respiratory function of cytochrome components in relation to the transfer of H or of electrons. Scarisbrick and I found that the chloroplasts had a cytochrome component very much like cytochrome c which was associated with their structure. Warburg had introduced the idea of structure bound constituents, and inhibition by urethane was used as an indication that a process was structure bound. It has always seemed remarkable to us that while in respiration the reaction with O_2 is sensitive to cyanide, in photosynthesis the process of producing O_2 does not show the same type of sensitivity. Scarisbrick had suggested a model for the photochemical production of oxygen. This was in terms of a heme protein compound like ferrihemoglobin. By the loss of three negative charges it might be brought to the state of oxyhemoglobin; the loss of a further negative charge would act like adding ferricyanide and liberate the oxygen, restoring the ferrihemoglobin. But the insensitivity of the chloroplast activity to cyanide militated against this very suggestive picture. It was an effort to be convinced that we were really on the right lines with photosynthesis. The idea of photosensitization of CO_2 by chlorophyll seemed to be so fixed in people's minds. Through Margory Stephenson we heard about the chemosynthetic organisms and about the varied activities shown by bacteria. This comparative view gave us the necessary confidence. So in 1939 Scarisbrick and I had distant visions of something like a photosynthetic chain. Then came the war and we both started different work. The description of cytochrome f was not published until 1950. In the meantime the first volume of *Photosynthesis and Related Processes* by Rabinowitch had appeared, in which he discussed the work by Scarisbrick and myself. Meanwhile, *Photosynthesis in Plants,* edited by James Franck and W. Loomis, had appeared in 1949, with a contribution on the photochemical liberation of oxygen from water by A. S. Holt and C. S. French. By this time the search for the path of CO_2 was well on the way. It was started in 1939 by Ruben, Kamen, and Hassid, with the short-lived tracer carbon ^{11}C, and continued from 1945 when supplies of ^{14}C became available. Accounts of this early work are given in the book by Franck and Loomis.

In conclusion, I must record my appreciation of having been able to witness brilliant scientific developments over the years, seemingly emanating from the colors of blood and grass, and express my gratitude for the friendship of many of the participants in this search for knowledge.

Ann. Rev. Plant Physiol. 1975. 26:13–29

PLASMODESMATA

♦7580

Anthony W. Robards

Department of Biology, University of York, Heslington, York, YO1 5DD, England

CONTENTS

INTRODUCTION

Ions, molecules, and water may move into, out of, within, and between plant cells. Movement through the extracytoplasmic compartment is said to be apoplastic, while translocation between the protoplasts of cells is termed symplastic. This concept envisages the symplasm as an intercellular continuum, a corollary of which is the need for cytoplasmic bridges between cells. It is the purpose of this article to consider current evidence relating to the structure and possible physiological functions of such connections.

Tangl (127) was the first botanist to write at length on *"offene Communicationen,"* while Strasburger introduced the term *"Plasmodesmen"* in 1901 (122). (For fuller reviews of early work see 91 and 92.) Despite other suggestions, the word plasmodesma has continued to be used to describe a protoplasmic connection. Recent physiological work has thrown increasing emphasis upon the role that plasmodesmata must play in symplastic transport, and it has therefore become topically important to consider details of plasmodesmatal distribution, structure, and variation.

DISTRIBUTION

Occurrence in Plants

Plasmodesmata have been described in angiosperms, gymnosperms, pteridophytes, bryophytes, and many algae. The situation in fungi and lichens is less clear, although earlier workers (93), as well as more recent ones (58, 126), have recorded plasmodesmata in fungal hyphae; the rare reports may be due partly to the small size (25 nm diameter) of the structures in this group (58). Plasmodesmata-like structures have also been reported in blue-green algae (14, 80). There is considerable structural variation in the plasmodesmata of both higher and lower plants, to the extent that it is sometimes difficult to distinguish a true plasmodesma from either a very thin closed tube or from an open cytoplasmic channel. The distinction must be mainly on the basis of size: cytomyctic channels such as those found between angiosperm meiocytes (63) are *not* plasmodesmata, although some authors have confusingly described them as such (18). Plasmodesmata (25.3 nm) and plasma channels (175 nm) in the pollen mother cell walls of *Lycopersicum* and *Cucurbita* have also been distinguished from each other on the basis of size (136), and similarly the large intercellular connections in *Volvox aureus* are thought of as cytoplasmic bridges rather than plasmodesmata (7). Meeuse (92) attempted a lengthy definition of plasmodesmata, but at the present time it is probably sufficiently useful and accurrate to think of them as being thin protoplasmic connections through which organelles and large macromolecules cannot normally pass (but see the special case of viruses, p. 25). While this description is far from totally satisfactory, it is impossible to produce a more specific and concise definition which would not exclude structures that clearly must be studied along with more "typical" plasmodesmata. A slightly different viewpoint is adopted by Juniper (67) in his full account of connection between plant cells.

Variation in Cells and Tissues

Plasmodesmata have been said to occur between all living plant cells (43, 91), but this is not the case. The presence or absence of plasmodesmata appears to be linked directly to the functional role that they play. They have been reported absent from the guard cells of stomata in some species (11, 129), while present in others (65, 101); absent between the zygote and surrounding cells of *Capsella,* as well as between the suspensor and embryo sac of the same species (112, 113); absent between the nucellus and embryo sac and zygote and all other cells in *Myosurus* (138); and absent between the nucellus and megagametophyte of *Zea mays* (27). They have not been found between pollen grains during their later stages of development (61, 62) or through the walls separating generative cells from each other and from surrounding tapetal cells (82). Not only are plasmodesmata absent from the walls of cells between which symplastic continuity might be assumed to be unnecessary or unacceptable, but walls between cells across which high translocatory fluxes are known to occur are often characterized by relatively high frequencies of cytoplasmic connections (Table 1). Plasmodesmata are found between adjacent cells in tissue culture systems (56, 71, 118). There are considerable technical difficulties involved in obtaining reliable estimates of plasmodesmatal frequency, but it is evident that there may be significant variations even between different walls of the same cell (Table 1). In relation to symplastic transport, it is significant that even the smallest meristematic cells have 10^3-10^5 connections with their neighbors (22), and frequencies greater than 1 million per square millimeter are common.

STRUCTURE

If the function of plasmodesmata is to be understood, then an accurate knowledge of structure is imperative. One of the most critical dimensions—the radius of the effective conducting pore—shows considerable variation in different published reports (Table 1); the situation has not been aided by confusion between the diameter of the pore through the wall and the diameter of any tubule that may be contained within that pore. In the following discussion I shall use the terms as illustrated in Figure 1. I have retained the word *desmotubule* (104, 105) for the central strand or core, as this refers to a specific component but does nothing to preclude further argument about its nature.

Optical Microscopy

The direct observation of plasmodesmata in untreated cells by optical microscopy is confined to relatively few cases; these particularly include the endosperms and cotyledons of seeds. In most other situations various stains and/or swelling agents

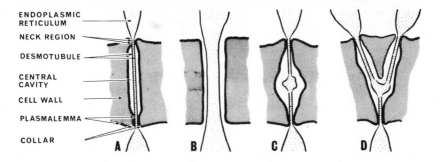

Figure 1   Using electron microscopic observations and theoretical considerations, a model has been constructed for the structure of a simple plasmodesma (105). This model envisages a thin strand of modified ER membrane (the desmotubule) running through the plasmalemma-lined plasmodesmatal canal (A). Where the constraint of the canal is less limiting (as during early stages of cell plate formation) it is possible that unmodified ER passes through the wall (B). Plasmodesmata with median nodules (C) or anastomosing desmotubules (D) may be seen as developments from the simple condition. [Figure 1 modified and adapted from Robards, 1971 (105).]

Table 1 Diameter and frequency of plasmodesmata through different plant cell walls

Species and cell type	Diameter (nm)[a]	Frequency (per μm^2)	Source
Nicotiana tabacum			
Various	500[b]	0.08 – 0.41[b]	Livingston (83)
Callus cells	300 – 500[b] 50 – 70	–	Spencer & Kimmins (118)
Viscum album			
Various	–	0.01 – 0.38[b]	Kuhla (78)
Mature cortical cells	50 – 60	0.6 – 2.4	Krull (77)
Allium cepa			
Root meristematic cells	30 – 40 100 – 200[c]	6 – 7 –	Strugger (123) Strugger (124)
Mature cortical cells	80 – 120[d]	1.5[d]	Tyree (130)
Avena sativa			
Cortical cells of mature coleoptile	60 – 100[d]	3.6[d]	Böhmer (8)
Dactylorchis fuchsii			
Archesporial cells	28	7	Heslop-Harrison (63)
Tamarix aphylla			
Walls between collecting and secreting cells of glands	80	17[d]	Tyree (130)
Zea mays			
Root cap cells			
Meristematic			
Transverse	25[e]	14.87	
Longitudinal		5.30	
Peripheral			Juniper & French (69)
Transverse	–	5.13	also Juniper &
Longitudinal		0.45	Barlow (68)
Hordeum vulgare			
Inner tangential wall of endodermal cells			
2 mm from root tip		1.19	
4 mm from root tip	60 – 70	0.69	Robards et al (106)
1 cm from root tip		0.65	
30–40 cm from root tip		0.67	
Outer tangential wall of endodermal cell			
1 cm from root tip		0.40	

Table 1   (Continued)

Species and cell type	Diameter (nm)[a]	Frequency (per μm^2)	Source
Radial wall of endodermal cell 1 cm from root tip		0.30	
Cucurbita pepo Inner tangential wall of root endodermal cells		6.2	
Outer tangential wall of root endodermal cells		10.4	Wingrave[f]
Radial wall of root endodermal cells		6.9	
Dryopteris filix-mas Root meristem primary pit fields	50 – 80	140	Burgess (12)
Older primary walls	—	10 – 20	
Polytrichum commune Leptoid end walls		16 – 20	Eschrich & Steiner (41)
Parenchyma cell cross walls		9 – 12	
Chara spp.	48 – 64	—	Pickett-Heaps (102)
Chara corallina Wall between internodal and peripheral cells	100	12	Fischer et al (45)
Wall between internodal and central cell	118	14	
Nitella translucens Wall between mature internodal cells	70	3[d]	Spanswick & Costerton (117)
Laminaria spp. Trumpet cell cross walls	60	50 – 60	Ziegler & Ruck (143)

[a] Diameter as cited. These figures will not be directly comparable because the authors have not always made it clear whether the thickness of the plasmalemma is included or not.

[b] Data derived from optical microscope observations.

[c] This larger figure obtained by "correcting" the measurement derived from electron microscopy.

[d] Calculated by Tyree (130) from the work of other authors.

[e] This measurement applies at least for the early stages of differentiation.

[f] Unpublished data obtained by Miss Sally Wingrave in the author's laboratory.

need to be used to reveal the structures, and electron microscope studies have now shown that many plasmodesmata were never seen by optical microscopy at all. Whatever their precise diameter, they always have a thickness which is close to or below the limit of resolution of the optical microscope. Cited measurements of pore diameter in the range 0.1–0.2 μm are therefore scarcely surprising. It has been suggested that the observed diameter is *smaller* than the in vivo condition (83, 84, 123, 124) because the optical microscope techniques usually involve wall swelling agents. It is significant that as recently as 1964 Livingston considered the presence of plasmodesmata so poorly substantiated that he felt it necessary to state: ". . . (this) study answers the criticism which has frequently been made in the past that these structures (plasmodesmata) are artifacts . . ." (84).

The protoplasmic nature of plasmodesmata, assumed from their first discovery, was not seriously challenged until 1930 (70). Meeuse (92) summarized the arguments up to 1957 and concluded that " . . . the accumulated evidence seems to be overwhelming and conclusive (that plasmodesmata are protoplasmic)."

Electron Microscopy

The earliest demonstrations of plasmodesmata by electron microscopy, using osmium tetroxide (15, 75, 76, 123, 124) or potassium permanganate (33, 103, 137) fixation, amply confirmed that the structures were real and continuous from cell to cell. An immediate problem was the apparent difference in size (diameter) between optical and electron microscopical observations. Those seen in the electron microscope are generally only about one tenth the diameter determined by optical microscopy. This led to the suggestion that there may be two types of plasmodesmata of different size (22, 96). There is little to substantiate this point of view, and it appears more probable that the discrepancies may be explained on the basis of diffraction effects enhanced by such features as closely adjacent strands within the thickness of the specimen, branched structures, and the possibility that the "wall tubule" is considerably wider than the protoplasmic thread itself (125).

Knowing that plasmodesmata may sometimes be "observed" by optical microscopy, it is ironic to find that lack of resolution within the specimen, when examined with an electron microscope, still precludes a generally applicable and acceptable theory for their structure. With few exceptions (17), it is thought that the plasmalemma is continuous through the canal from cell to cell. From the earliest electron micrographs it was apparent that the endoplasmic reticulum (ER) was closely associated with each end of a plasmodesma. This, together with the clear observation of strands of ER becoming trapped within the differentiating cell plate, led to the common assertion that plasmodesmata comprise a strand of ER running through the wall from cell to cell (15, 16, 49, 134); some workers adopted a less committed point of view (33, 44, 103). Briefly, the problem is this: in sectioned material the internal diameter of the neck region is approximately 20 nm (Figure 1A); this is the area where the 8 nm thick ER membranes converge on, and arguably connect with, the desmotubule; a typical thin section is 40–50 nm thick, and thus includes the whole thickness of the neck (sectioned longitudinally) together with other irrelevant information. In these circumstances, taken together with other problems of electron microscopy (105), it is virtually impossible to achieve the degree of resolution required to visualize the precise structural arrangements of the different components. With these limitations in mind, it is possible to discuss the structure of a "simple" plasmodesma, that is, one with no anastomosing arms or complicated median nodules.

The suggestion that the traversing strand is a tubule of ER is well illustrated in the work of López-Sáez et al (85), who conclude that the ER is tightly curved into a tubule so that the inner opaque layer appears as a central rod. A similar idea is propounded by Semenova & Tageeva (114), who suggest that the dark central part of the axial structure represents the two inner layers of the (ER) membrane fused together. The implications of such a structure are that the desmotubule would be nonconducting, and that any symplastic transport would have to take place between the plasmalemma and the desmotubule in the neck region. The tightness of the junction at the neck is not known, but could prove to be an important feature (20). Working from the premise that a normal bimolecular leaflet of ER membrane could not curve itself about the radius determined for a desmotubule (108), and also concerned at the apparent nonfunctional nature of the López-Sáez model, I constructed a different model from observations obtained in my own laboratory (104). This was later slightly modified (105 and Figure 1A). The most recent scheme envisages continuity of ER from cell to cell, but via a tubule which, I believe, cannot be a normal lipoprotein bilayer. Analysis of micrographs (104) and theoretical considerations have led to the proposition that the desmotubule comprises protein subunits and is in fact structurally very similar to a microtubule. This model has the virtue that continuity of the ER cavity from cell to cell is retained, and the desmotubule is a stable structure. Other authors have commented on the microtubule-like appearance of the desmotubule (e.g. 2, 60, 100), or have expressed concern at the idea that an unmodified strand of ER should pass through the canal (82). There appears to be no particular problem in the continuity of a membrane with a microtubule-like structure; there are increasing reports of such associations. Similarly, branched desmotubules impose no unacceptable demands upon credulity (Figure 1D). While this model will certainly need further modification, it does provide a structure that can be examined in the light of physiological results. One of the most relevant artifacts is shrinkage during preparation for electron microscopy. Little relevant work has been carried out on frozen cells, although it has been reported that plasmodesmata prepared by cryoultramicrotomy appear not to have a central core (110). The issue of whether all plasmodesmata always possess a desmotubule is still the subject of debate. Radiating spokes between desmotubule and plasmalemma have been described (12), and also desmotubules of "C-shaped" traverse section (28). These latter images are similar to those obtained of some mitotic microtubules (23). There are reports (79, 132) of paramural bodies (88) closely associated with the ends of the plasmodesmatal canal.

Even the structure of a "simple" plasmodesma is thus far from being unequivocally resolved. Membrane modification within the canal of this model (Figure 1A) would be a function of the space available. Where little restriction occurs, as in young plasmodesmata in the cell plate, or where the central part of the wall cavity is dilated (Figures 1B, 1C), a more normal ER membrane structure could be envisaged.

The nature of the wall through which the plasmodesmatal canal runs has been largely neglected. It has been implied that callose is always present surrounding the plasmalemma through the pore (74), but whether this is generally true remains unproved, although callose has been found associated with phloem and algal plasmodesmata (143), and also around aging plasmodesmata in a moss (41). Further, a highly resistant wall component (possibly glycoprotein) has been found close to plasmodesmata in barley aleurone cells (125).

Structural Variability

The more commonly cited deviations from a structure with a simple, straight desmotubule are those which have a median nodule (e.g. 24, 77); those with anastomosing desmotubules [these are particularly common in phloem cells (75, 97), as well as in other locations (22, 43)]; and those which do not possess a desmotubule at all (12, 95). In some cases, plasmodesmata with median nodules may become interlinked by a large cavity in the midline of the cell wall (38). The complete absence of a desmotubule (often the situation in algae) poses considerable difficulties in constructing a unified functional hypothesis, although there seems little doubt that the variation is a real one and not artifactual.

The median nodule is apparently a secondary development from a simple plasmodesma. This nodule could be important in controlling fluxes or in completely stopping them; it may also be involved in enzyme-mediated processes located within the wall cavity. This idea is supported by the reports of phosphatase activity associated with plasmodesmata (57, 107), as well as the established hydrolytic activities during sieve plate pore formation. Other variations may also occur, so that while it is first necessary to elucidate the structure and function of the simpler connections, the complete range of structural variation must be brought into consideration when attempting to construct functional hypotheses.

Phloem Types

Phloem plasmodesmata need to be considered separately because they represent a separate population of structures; they would be expected to be involved in particularly active fluxes; and some of them are the sites of initiation of sieve plate pores (36). The distribution and structure of phloem plasmodesmata are both of special interest. They have been found to connect parenchyma to parenchyma, parenchyma to companion cells, and companion cells to sieve elements (42, 98, 115, 135, 139). With few exceptions (140), their general absence between parenchyma cells and sieve elements seems to point to a specific role of companion cells in sieve tube functioning (115). The companion cell/sieve element connections are typically multibranched on the companion cell side, with a single enlarged tubule on the sieve element side (35, 139).

The plasmodesmata traversing the walls of potential sieve plates initially appear quite normal. They may develop a median nodule, and from this general area hydrolysis of the wall occurs prior to the deposition of callose to form the completed sieve plate pore (26, 36).

Phloem plasmodesmata, from their distribution and complexity, are clearly important in translocation, and presumably especially so in loading and unloading sieve elements via companion cells.

Algal Types

Two recent contributions have greatly aided our understanding of the distribution of plasmodesmata in the algae (46, 121). In general, it appears that they only occur in those algae in which cytokinesis is effected by formation of a cell plate. However, the range of structural variation is great. Brown algae have connections similar to those of higher plants, but without any central strand or desmotubule (6). The oedogonialean alga *Bulbochaete* has plasmodesmata quite unlike those found anywhere else (47): there is no desmotubule, and the structure of the pore lining is relatively complex. Another green alga, *Chara,* appears to have some plasmodesmata which contain desmotubules while others lack them (102). A desmotubule has been found in the plasmodesmata of the sea thong *Himanthalia* (5), and some sort of core exists in plasmodesmata through the cross-walls of trumpet cells in *Laminaria* (143). Connections through the walls between the nodal and internodal cells of *Nitella translucens* varied greatly in morphology, anastomoses, and median sinuses (116). The basic canal diameter was always about 50 nm internally, and "membrane systems" were reported within the pores. The algae clearly comprise an extremely diverse group!

It remains to be seen whether the reports of "microplasmodesmata" from blue-green algae (80) should be considered along with the higher plant equivalents. The very thin (25 nm) threads are extremely difficult to resolve, and it has been stated (14) that the pores through the septa do not pierce the plasma membrane, so no direct connection of the protoplasts is achieved.

FORMATION AND DEVELOPMENT

It has long been suggested that plasmodesmata are formed during cytokinesis. Electron microscopy has confirmed the trapping of ER strands within the developing cell plate, leading to the predominant suggestion that the ER constitutes the desmotubule (15, 48, 66). Tangl (127) had much earlier commented on the resemblance of groups of plasmodesmata to the mitotic spindle, a suggestion supported by some following workers (50, 51, 111) but no longer seriously considered. It has already been explained why there seems to be a necessity for trapped ER membranes to change in structure as they become confined within the very narrow plasmodesmatal canal. It also seems most unlikely that the trapping in the cell plate is an entirely random event. Plasmodesmata are frequently restricted to pit fields; they are also often found in different frequencies through different walls of the same cell (Table 1), or are sometimes completely absent. The enmeshing of ER strands must therefore be both nonrandom and to some extent controlled (as the patterns of frequency variation are often quite consistent). Alternatively, there could be a programmed loss of connections already formed through the wall—a less attractive hypothesis that has not received support from ultrastructural studies.

While it appears that plasmodesmata can develop from ER strands that traverse the cell plate of dividing cells, this postulate for their development is unacceptable as a universal rule if they are found traversing walls which have not derived directly from a cell division. There are now sufficient citations of plasmodesmata in nondivision walls to make it highly probable that intercellular connections can develop secondarily, that is, by penetrating an existing wall. There are many reports of connections between parasitic hyphal walls and host cell walls (4, 29–32, 34, 74). The early work of Hume (64) has recently been supported by ultrastructural observations (13) indicating that plasmodesmata are formed between cells originating from different species in the periclinal chimaera *Cystisus (Laburnocytisus) adami.* It is also quoted that plasmodesmata can be found between the outpushings of ray cell membranes into vessel elements (tyloses) (34). An important consequence of such citations is that some degree of coordination must be achieved by adjacent cells if there is to be any chance of the strands meeting in the midregion of the wall. A similar degree of cooperation is demonstrated in adjacent xylem cells producing aligned wall thickenings. Krull (77) proposed that plasmodesmata can divide longitudinally, but this awaits confirmation. More recently (52) it has been reported that there is seasonal variation in the frequency of plasmodesmatal threads passing through pores in the walls of onion scale epidermal cells. Although dynamic activities of plasmodesmata such as these may arouse suspicion, they should not automatically be regarded as improbable, particularly in the context of young primary walls.

Development subsequent to the formation of an intercellular connection is usually centred on the desmotubule: a median nodule may be formed; the desmotubule may become highly convoluted; it may open into the cavity and/or become attached to the plasmalemma within the cavity; it may form branches; and it may anastomose with adjacent plasmodesmata. Plasmodesmata may be modified in a number of other ways which would be expected to halt or to impede symplastic transport. Sliding growth of cells would lead to a severance of connections (34); wall material may be deposited over the ends of the pore; and less final effects, such as constriction of the plasmodesmata, may occur (99, 106).

The inevitable conclusion is that plasmodesmata are far more dynamic than most botanists have believed. They are not inert canals, but instead they are capable of changes which, it must be assumed, render them capable of carrying out their translocatory functions to a predetermined plan.

FUNCTION

With few exceptions, it has long been considered that plasmodesmata are important in the translocation of substances or the transmission of stimuli from cell to cell (55). More recently it has been suggested that they may be important for the equilibration of membrane potentials and transfer of membrane supported excitations (82). However, as recently as 1968 it has been stated that "we cannot assign any role to the plasmodesmata with any confidence" (22). The technical problems in positively demonstrating translocatory function of plasmodesmata, particularly through young, freely permeable walls, are severe. Therefore, support for believing in the involvement of plasmodesmata in symplastic transport arises from indirect circumstantial sources as well as from the less abundant (but increasing) literature reporting experiments which directly involve plasmodesmatal function.

Circumstantial Considerations

There appears to be a strong correlation between the assumed translocatory flux from cell to cell and the presence and frequency of plasmodesmata. Some cells, where other evidence suggests the need for complete isolation, have no plasmodesmata at all (p. 14); others have many connections through walls across which high flow rates would be expected; branched desmototubules are common at very active sites of solute movement. The stamen filament cells of wheat elongate at 2–3.5 mm min^{-1} and, according to Ledbetter & Porter (82), are "doubtless supplied by the abundant plasmodesmata." Connections are also present between living cells with thick walls that are considered to be impermeable (e.g. xylem ray cells and tertiary endodermal cells). If the ER constitutes the symplasm, then plant cells have the opportunity for intercellular transport of materials without the constant crossing of membranes [a situation contrasting with that in animal cells (74)].

There are therefore good a priori arguments for supposing that plasmodesmata are functional in symplastic transport. There remain difficulties to reconcile with such a role: plasmodesmata often separate cells of quite different types [e.g. tannin cells and tannin-free cells (34)], and they must therefore be functionally selective; further, individually or together, they need to be capable of sustaining bidirectional fluxes (20).

It is important to know whether plasmodesmata are, at least to some extent, open continuities from cell to cell which will allow the free interchange of ions and molecules, or whether the symplastic pathway is indeed confined to the cavity of the ER (20). Structural evidence remains equivocal, although many physiologists have worked with models depicting totally open tubes (e.g. 10, 94), a situation which can only apply to a minority of cases in higher plants.

Experimental Work

Using silver precipitation or autoradiographic techniques, a number of workers have been able to demonstrate the apparent passage of chloride ions through plasmodesmata (133, 141, 142). More recently (81, 120), chloride–precipitated silver deposits have been demonstrated, not only in plasmodesmata but also within the ER, thus supporting the idea that the ER is part of the symplastic pathway.

Arisz (1) has concluded that both longitudinal and transverse fluxes in coleoptiles take place via an energy-maintained flow through plasmodesmata. He has also shown that solutes and ions to which the cell membrane is relatively impermeable still move from cell to cell in the leaves of *Vallisneria*. Inhibition of membrane-located transport did not prevent this movement, and the view that symplastic transport takes place via cytoplasmic connections thus receives further support.

One of the few recent publications unsympathetic to the possibility that plasmodesmata carry intercellular fluxes (59) concludes that free diffusion within the desmotubules would be so restricted as to make the plasmalemma the "only means of intercellular transport." However, most recent papers support the idea of symplastic transport, whether or not this takes place through an open unrestricted pore.

The specific electrical resistance of cells of *Nitella* joined by plasmodesmata has been found to be 50 times smaller than between those cells without such connections (117), yet even this resistance was higher than would be expected if the plasmodesmata were open channels, and it was therefore assumed that there must be some restriction upon the movement of ions within the pores. Similarly (116), it has been found that the plasmodesmata constitute the path of least resistance between cells in the higher plant *Elodea*, although the resistance is again about 60 times greater than if there were completely open channels. Low resistance junctions have also been reported between cells in the developing anthers of lily (119), but Goldsmith et al were unable to demonstrate electrotonic coupling between coleoptile cells of *Avena* (54). Electrical resistance measurements, if they are to be interpreted correctly, must be made between known compartments of the cell. If, as has been suggested, the ER constitutes the symplasm, then the nature of any measured continuity will depend entirely upon where the electrode tips have lodged. In large vacuolate cells there may be a little problem, but in cells with small vacuoles there must be considerable risk that the precise point of electrode penetration will remain undetermined. Among physiologists there is a growing appreciation that ions and other solutes can be compartmented within the cytoplasm (87). Against this background it is quite acceptable that the symplast should be regarded as one cytoplasmic compartment rather than the whole of the protoplast. It has been suggested that small vacuoles or pinocytotic vesicles formed at the plasmalemma may transport ions directly, either to the vacuole or to the ER, without mixing their contents directly with the bulk cytoplasm (86, 109).

Tyree (130), examining the theory of symplastic transport "according to the thermodynamics of irreversible processes," has concluded that plasmodesmata constitute the pathway of least resistance for the diffusion of small solutes, and that diffusion will be the predominant mechanism of transport across the pores for small solutes. Certain of Tyree's assumptions relating to the structure of plasmodesmata and the viscosity of the fluid within the pore are either unclear or open to argument, but his assumptions may be considered pessimistic ones, and his conclusions therefore strongly support the idea that plasmodesmata are, in general, capable of sustaining the flow rates attributed to them. A similar treatment to that of Tyree has been carried out by Ginsburg (53), who arrived at essentially similar conclusions. Plasmodesmatal connections have been shown to contribute to a low electrical resistance between cells; calculations also clearly reveal their enhanced hydraulic conductivity compared with either cell membranes or cell walls (Table 2).

Quantitative studies of ultrastructure and ^{36}Cl movement between internodal cells of *Chara corallina* have recently been combined to determine the resistance of plasmodesmata to diffusion of chloride (45, 131). Assuming that diffusion is the rate limiting step, then the contents of the plasmodesmatal pore were estimated to be about 31 times more resistant to diffusion than water. However, cyclosis in the internodal cells may have a rate controlling effect, and it is concluded that the plasmodesmata are possibly filled only with viscous or gelled cytoplasm. In this

Table 2 Permeability of plasmodesmata compared with some other barriers to transport

Barrier	Hydraulic conductivity $(\mu m\ s^{-1}\ bar^{-1})$	Source
Plasmodesmata	13 – 1300	Clarkson, Robards & Sanderson (21)
Plasmodesmata	42	Marks (89)
Plasmodesmata	0.56 – 22	Tyree (130)
Plasmodesmata	0.003[a]	Tanton & Crowdy (128)
Biological membranes	0.37 – 270	Clarkson (19, p. 104)
Plant cells	0.001 – 0.01[b]	Briggs (9, p. 88)
Pine wood (longitudinal flow)	6000	Briggs (9, p. 88)
Pine wood (radial flow)	0.03	Briggs (9, p. 88)

[a]This may be considered an extremely low value, probably arising from an underestimate of plasmodesmatal frequency and other parameters.

[b]Briggs considers that these figures may be underestimates, although they are "unlikely to be grossly out." The permeability of higher plant cells tends to lie towards the lower end of the range cited here.

work the plasmodesmata were considered to be open pores, although the authors themselves, as well as other workers (102), have indicated that at least some of the connections in *Chara* have a further substructure. Taken together with the direct demonstration by electron cytochemical techniques of chloride movement through plasmodesmata (133, 141, 142), as well as the less direct methods of Arisz (1), the weight of evidence is now very strong for believing that this ion moves symplastically from cell to cell, even though the rate controlling processes remain unknown and the structural variability of the connections must be of great importance.

Recent work (19–21, 106) has strongly supported the supposition that plasmodesmata carry the symplastic flow across the inner tangential wall of the endodermal cells in barley roots. In particular, it has been shown that (for example) phosphate moves by this symplastic route whereas calcium does not. The calculated flow of water through the plasmodesmata was approximately three orders of magnitude greater than that across membranes from a wide variety of sources (see also Table 2). However, some surprising conclusions are reached: for example, if the flux of water across the plasmodesmata of the inner tangential endodermal wall is 9.1×10^{-5} cm^3 mm^{-2} h^{-1}, then each plasmodesma would fill and empty 1000 times per second, with a flow velocity of approximately 500 μm per second (20); despite this, the main tenet that plasmodesmata can accomodate the observed flow remains unaltered.

Viruses

If it can be shown that viruses penetrate cells through plasmodesmata, then the possibility that these connections function in the more general context of symplastic transport is greatly strengthened. There have been numerous such suggestions (3, 25, 73, 83, 90, 118), although this does not necessarily mean that the whole intact virus is transported. Early electron micrographs showed viruses within plasmodesmata (37), but the nature of the core or desmotubule remained in doubt. Most viruses found in plasmodesmata have been icosohedral particles, although rod-shaped viruses have also been reported (39, 40). Esau & Hoefert (40) conclude that beet western yellows virus in some way modifies the plasmodesmata so that they no longer possess a desmotubule (although these are present in the plasmodesmata of noninfected cells). This concept—that the presence of a virus within a cell is capable of bringing about structural changes in plasmodesmata—is now quite widely supported (25, 37, 72, 73). The micrographs of Davison (25) show tobacco ring spot virus particles enclosed within a tubule extending into the cytoplasm for some distance from the plasmodesmata. Viruses thus appear to be capable of movement from cell to cell through plasmodesmata whose structure they may directly or indirectly cause to be changed. These findings further support the idea that plasmodesmata are malleable structures, quite capable of considerable secondary modification, and not merely rigid cytoplasmic connections.

The conclusion is inevitably reached that plasmodesmata act as channels for symplastic transport and are capable of allowing the passage of molecules as large as virus particles. There seems to be ample evidence that the function of plasmodesmata can be changed or partially or totally impeded by secondary modification and/or occlusion. Numerous problems remain, including the apparent necessity for plasmodesmata to operate bidirectionally. We still know too little about these minute structures to understand them fully, and until we appreciate the nature of the simplest situations it is unlikely that progress will be made with the more complex ones.

ACKNOWLEDGMENTS

I should particularly like to thank Dr. D. T. Clarkson (A.R.C. Letcombe Laboratory), whose collaboration and helpful criticism is greatly appreciated. I am also indebted to the Science Research Council and Agricultural Research Council, who have supported some of the work referred to in this paper.

Literature Cited

1. Arisz, W. H. 1969. *Acta Bot. Neer.* 18:14–38
2. Bajer, A. 1968. *Chromosoma* 24:383–417
3. Behnke, H. D. 1966. In *Viruses of plants,* ed. A. B. R. Beemster, J. Dijksta. *Proc. Int. Conf. Plant Viruses* July 1965, 28–43
4. Bennett, C. W. 1944. *Phytopathology* 34:905–32
5. Berkaloff, C. 1963. *J. Microsc.* 2:213–28
6. Bisalputra, T. 1966. *Can. J. Bot.* 44:89–93
7. Bisalputra, T., Stein, J. R. 1966. *Can. J. Bot.* 44:1697–1702
8. Böhmer, H. 1958. *Planta* 50:461–97
9. Briggs, G. E. 1967. *Movement of Water in Plants.* Oxford:Blackwell
10. Brouwer, R. 1965. *Ann. Rev. Plant Physiol.* 16:241–66
11. Brown, W. V., Johnson, S. C. 1962. *Am. J. Bot.* 49:110–15
12. Burgess, J. 1971. *Protoplasma* 73:83–95
13. Ibid 1972. 74:449–58
14. Butler, R. D., Allsopp, A. 1972. *Arch. Mikrobiol.* 82:283–99
15. Buvat, R. 1957. *C.R.H. Acad. Sci. Ser. D* 245:198–201
16. Ibid 1960. 250:170–72
17. Chukhrii, M. G. 1971. *Izv. Akad. Nauk Mold. SSR. Ser. Biol. Khim. Nauk* 4:3–7

18. Ciobanu, I. R. 1969. *Rev. Roum. Biol. Ser. Bot.* 14:269–73
19. Clarkson, D. T. 1974. *Ion Transport and Cell Structure in Plants.* London: McGraw-Hill. 350 pp.
20. Clarkson, D. T., Robards, A. W. 1975. In *Root Structure and Function,* ed. J. Torrey, D. T. Clarkson. *Harvard Symp.* 1974. London: Academic
21. Clarkson, D. T., Robards, A. W., Sanderson, J. 1971. *Planta* 96:292–305
22. Clowes, F. A. L., Juniper, B. E. 1968. *Plant Cells.* Oxford:Blackwell
23. Cohen, W. D., Gottlieb, T. 1971. *J. Cell Sci.* 9:603–19
24. Cox, G. C. 1971. *The structure and development of cells with thickened primary walls.* PhD thesis. Oxford Univ.
25. Davison, E. M. 1969. *Virology* 37: 694–96
26. Deshpande, B. P. 1974. *Ann. Bot.* 38:151–58
27. Diboll, A. G., Larson, D. A. 1966. *Am. J. Bot.* 53:391–402
28. Dolzmann, P. 1965. *Planta* 64:76–80
29. Dörr, I. 1968. *Naturwissenschaften* 8: 396
30. Dörr, I. 1969. *Protoplasma* 67:123–37
31. Esau, K. 1948. *Bot. Rev.* 14:413–49
32. Esau, K. 1948. *Hilgardia* 18:423–82
33. Esau, K. 1963. *Am. J. Bot.* 50:495–506
34. Esau, K. 1965. *Plant Anatomy.* New York:Wiley. 2nd ed.
35. Esau, K. 1973. *Ann. Bot.* 37:625–32
36. Esau, K., Cheadle, V. I., Risley, E. B. 1962. *Bot. Gaz.* 123:233–43
37. Esau, K., Cronshaw, J., Hoefert, L. L. 1967. *J. Cell Biol.* 32:71–87
38. Esau, K., Gill, R. H. 1973. *J. Ultrastruct. Res.* 44:310–28
39. Esau, K., Hoefert, L. L. 1971. *Protoplasma* 72:255–73
40. Esau, K., Hoefert, L. L. 1972. *J. Ultrastruct. Res.* 40:556–71
41. Eschrich, W., Steiner, M. 1968. *Planta* 82:321–36
42. Evert, R. F., Murmanis, L. 1965. *Am. J. Bot.* 52:95–106
43. Fahn, A. 1967. *Plant Anatomy.* Oxford: Pergamon
44. Falk, H., Sitte, P. 1963. *Protoplasma* 57:290–303
45. Fischer, R. A., Dainty, J., Tyree, M. T. 1974. *Can. J. Bot.* 52:1209–14
46. Floyd, G. L., Stewart, K. D., Mattox, K. R. 1971. *J. Phycol.* 7:306–9
47. Fraser, T. W., Gunning, B. E. S. 1969. *Planta* 88:244–54
48. Frey-Wyssling, A., Muller, H. R. 1957. *J. Ultrastruct. Res.* 1:38–48
49. Frey-Wyssling, A., Mühlethaler, K. 1965. *Ultrastructural Plant Cytology.* Amsterdam:Elsevier
50. Gardiner, W. 1900. *Proc. Roy. Soc. B* 66:186–88
51. Gardiner, W. 1907. *Proc. Cambridge Phil. Soc.* 14:209–10
52. Genkel, P. A., Kurkova, E. B. 1971. *Fiziol. Rast.* 18:777–80
53. Ginsburg, H. 1971. *J. Theor. Biol.* 32:147–58
54. Goldsmith, M. H. M., Fernandez, H. R., Goldsmith, T. H. 1972. *Planta* 102:302–23
55. Haberlandt, G. 1914. *Physiological Plant Anatomy.* Transl. from 4th German ed. London:MacMillan
56. Haccius, B., Engel, I. 1968. *Naturwissenschaften* 55:45–46
57. Hall, J. L. 1969. *Planta* 85:105–7
58. Hawker, L. E., Gooday, M. A., Bracker, C. E. 1966. *Nature* 212:635
59. Helder, R. J., Boerma, J. 1969. *Acta Bot. Neer.* 18:99–107
60. Hepler, P. K., Jackson, W. T. 1968. *J. Cell Biol.* 38:437–46
61. Heslop-Harrison, J. 1964. *Pollen Physiology and Fertilization,* ed. H. F. Linskens
62. Heslop-Harrison, J. 1966. *Endeavour* 25:65–72
63. Heslop-Harrison, J. 1966. *Ann. Bot.* 30:221–30
64. Hume, M. 1913. *New Phytol.* 12:216–20
65. Inamdar, J. A., Patel, K. S., Patel, R. C. 1973. *Ann. Bot.* 37:657–60
66. Juniper, B. E. 1963. *J. Roy. Microsc. Soc.* 82:123–26
67. Juniper, B. E. 1975. In *Textbook of Developmental Biology,* ed. C. Graham, P. F. Wareing. Oxford:Blackwell. In press
68. Juniper, B. E., Barlow, P. W. 1969. *Planta* 89:352–60
69. Juniper, B. E., French, A. 1970. *Planta* 95:314–29
70. Jungers, V. 1930. *La Cellule* 40:7–82
71. Kassanis, B. 1967. In *Methods in Virology,* ed. K. Maramorosch, H. Koprowski, 1:537–66. New York: Academic
72. Kim, K. S., Fulton, J. P. 1973. *J. Ultrastruct. Res.* 45:328–42
73. Kitajima, E. W., Lauritis, J. A. 1969. *Virology* 37:681–85
74. Kollmann, R., Dörr, I. 1969. *Ber. Deut. Bot. Ges.* 82:415–25
75. Kollmann, R., Schumacher, W. 1962. *Planta* 58:366–86
76. Ibid 1963. 60:360–89
77. Krull, R. 1960. *Planta* 55:598–629

78. Kuhla, F. 1900. *Bot. Ztg.* 58:29–58
79. Kurkova, E. B., Vakhmistrov, D. B., Solovyev, V. A. 1974. In *Structure and Function of Primary Root Tissues,* ed. J. Kolek. Bratislava:Slovak Acad. Sci.
80. Lang, N. J., Fay, P. 1971. *Proc. Roy. Soc. B* 178:193–203
81. Läuchli, A., Kramer, D., Stelzer, R. 1974. In *Membrane Transport in Plants and Plant Organelles,* ed. U. Zimmermann, J. Dainty. Berlin:Springer
82. Ledbetter, M. C., Porter, K. R. 1970. *Introduction to the Fine Structure of Plant Cells.* Berlin:Springer
83. Livingston, L. G. 1935. *Am. J. Bot.* 22:75–87
84. Ibid 1964. 51:950–57
85. López-Sáez, J. F., Giménez-Martín, G., Risueño, M. C. 1966. *Protoplasma* 61:81–84
86. MacRobbie, E. A. C. 1969. *J. Exp. Bot.* 20:236–56
87. MacRobbie, E. A. C. 1971. *Ann. Rev. Plant Physiol.* 22:75–96
88. Marchant, R., Robards, A. W. 1968. *Ann. Bot.* 32:457–71
89. Marks, I. 1973. *Ultrastructural studies of minor veins.* PhD thesis. Queens Univ., Belfast
90. Martin, L. F., McKinney, H. H. 1938. *Science* 88:458–59
91. Meeuse, A. D. J. 1941. *Bot. Rev.* 7:249–62
92. Meeuse, A. D. J. 1957. *Protoplasmatologia* II A 1c:1–43
93. Meyer, A. 1896. *Ber. Deut. Bot. Ges.* 14:280–81
94. Minchin, F. R., Baker, D. A. 1970. *Planta* 94:16–26
95. Mueller, D. M. J. 1972. *Bryologist* 75:63–68
96. Newcomb, E. H. 1963. *Ann. Rev. Plant Physiol.* 14:43–64
97. Northcote, D. H., Wooding, F. B. P. 1966. *Proc. Roy. Soc. B* 163:524–37
98. Northcote, D. H., Wooding, F. B. P. 1968. *Sci. Progr. London* 56:35–58
99. O'Brien, T. P., Carr, D. J. 1970. *Aust. J. Biol. Sci.* 23:275–87
100. O'Brien, T. P., Thimann, K. V. 1967. *Protoplasma* 63:417–42
101. Pallas, J. E., Mollenhauer, H. H. 1972. *Science* 175:1275–76
102. Pickett-Heaps, J. D. 1967. *Aust. J. Biol. Sci.* 20:539–51
103. Porter, K. R., Machado, R. D. 1960. *J. Biophys. Biochem. Cytol.* 7:167–80
104. Robards, A. W. 1968. *Planta* 82:200–10
105. Robards, A. W. 1971. *Protoplasma* 72:315–23
106. Robards, A. W., Jackson, S. M., Clarkson, D. T., Sanderson, J. 1973. *Protoplasma* 77:291–311
107. Robards, A. W., Kidwai, P. 1969. *Planta* 87:227–38
108. Robertson, J. D. 1964. In *Cellular Membranes in Development,* ed. M. Locke, 1–81. New York/London: Academic
109. Robertson, R. N. 1968. *Protons, Electrons, Phosphorylation and Active Transport.* Cambridge Univ. Press
110. Roland, J. C. 1973. *Int. Rev. Cytol.* 36:45–92
111. Russow, E. 1883. *Sber. Ges. Naturf. Dorpat.* 6:562
112. Schulz, P., Jensen, W. A. 1968. *Am. J. Bot.* 55:807–19
113. Schulz, P., Jensen, W. A. 1969. *Protoplasma* 67:139–63
114. Semenova, G. A., Tageeva, S. V. 1972. *Dokl. Akad. Nauk SSR Ser. Biol.* 202:1427–28
115. Shih, C. Y., Currier, H. B. 1969. *Am. J. Bot.* 56:464–72
116. Spanswick, R. M. 1972. *Planta* 102:215–27
117. Spanswick, R., Costerton, J. 1967. *J. Cell Sci.* 2:451–64
118. Spencer, D. F., Kimmins, W. C. 1969. *Can. J. Bot.* 47:2049–50
119. Spitzer, N. C. 1970. *J. Cell Biol.* 45:565–75
120. Stelzer, R., Läuchli, A., Kramer, D. 1974. *Cytobiologie.* In press
121. Stewart, K. D., Mattox, K. R., Floyd, G. L. 1973. *J. Phycol.* 9:128–41
122. Strasburger, E. 1901. *Jahrb. Wiss. Bot.* 36:493–610
123. Strugger, S. 1957. *Protoplasma* 48:231–36
124. Ibid 1957. 365–67
125. Taiz, L., Jones, R. L. 1973. *Am. J. Bot.* 60:67–75
126. Takada, H., Yagi, T., Hiraoka, J. 1965. *Protoplasma* 59:494
127. Tangl, E. 1879. *Jahrb. Wiss. Bot.* 12:170–90
128. Tanton, T. W., Crowdy, S. H. 1972. *J. Exp. Bot.* 76:600–18
129. Thomson, W. W., De Journett, R. 1970. *Am. J. Bot.* 57:309–16
130. Tyree, M. T. 1970. *J. Theor. Biol.* 26:181–214
131. Tyree, M. T., Fischer, R. A., Dainty, J. 1974. *Can. J. Bot.* 52:1325–34
132. Vakhmistrov, D. B., Kurkova, E. B., Solovyev, V. A. 1972. *Fiziol. Rast.* 19:951–60
133. Van Steveninck, R. F. M., Chenoweth,

A. R. F. 1972. *Aust J. Biol. Sci.* 25:499–516
134. Wardrop, A. B. 1965. In *Cellular Ultrastructure of Woody Plants,* ed. W. A. Côté, 61–97. Syracuse Univ. Press
135. Wark, M. C. 1965. *Aust. J. Bot.* 13:185–93
136. Weiling, F. 1965. *Planta* 64:97–118
137. Whaley, W. G., Mollenhauer, H. H., Leech, J. H. 1960. *Am. J. Bot.* 47:401–49

138. Woodcock, C. L. F., Bell, P. R. 1968. *J. Ultrastruct. Res.* 22:546–63
139. Wooding, F. B. P., Northcote, D. H. 1965. *J. Cell Biol.* 24:117–28
140. Zee, S. Y., Chambers, T. L. 1968. *Aust. J. Bot.* 16:37–47
141. Ziegler, H., Lüttge, U. 1966. *Planta* 70:193–206
142. Ibid 1967. 74:1–17
143. Ziegler, H., Ruck, I. 1967. *Planta* 73:62–73

Ann. Rev. Plant Physiol. 1975. 26:31–52

CARBOHYDRATES, PROTEINS, CELL SURFACES, AND THE BIOCHEMISTRY OF PATHOGENESIS

❖7581

Peter Albersheim and Anne J. Anderson-Prouty

Department of Chemistry, University of Colorado, Boulder, Colorado 80302

CONTENTS

I. GENERAL RESISTANCE

A. Introduction

The study of plant-pathogen interactions at the molecular and physiological levels is largely an unexplored discipline which offers a tremendous opportunity for fruitful research. However, the results of the workers in this field demonstrate that fundamental advances in such areas as genetics and molecular and developmental biology will be the reward for those who investigate the sophisticated regulatory processes which have evolved in host-pathogen systems. Regulation is required in host-pathogen systems to maintain, on an evolutionary scale, a balanced state in which the pathogen is successful in living as a parasite, but not so successful that it results in extinction of its frequently obligatory host (28).

Some pathogens are able to attack a wide variety of plant species (28). Other pathogens are highly selective in their host range. Most of these specific host-pathogen species exist as a number of races, each of which is distinct from the others in its ability to attack the various varieties of its host plant species. Complex host-pathogen systems of this type have been frequently demonstrated to be gene-for-gene systems (28, 37, 38, 58, 98, 99). The rationale and significance of gene-for-gene host-pathogen systems represent a major concern of this article.

Among the myriad microorganisms, viruses, nematodes, and insects that plants are exposed to, there are relatively few to which any plant is susceptible (28, 70). Thus very effective mechanisms of general resistance enable plants to remain healthy in the face of potential attack by an enormous variety of pests. General resistance must be based on mechanisms that are effective against a large number of different species of potential pathogens and pests (28, 58, 70, 138).

B. The Role of the Cell Wall and Wall-Degrading Enzymes in Infective Processes—An Example of General Resistance

In a review written 5 years ago (7), it was concluded that pathogen-secreted polysaccharide-degrading enzymes play a fundamental role in pathogenesis. The evidence that has been obtained since confirms the importance of these enzymes. All plant pathogens have the ability to produce polysaccharide-degrading enzymes (7). These enzymes, taken together, have an ability to degrade every glycosidic linkage known to occur in the polysaccharides of primary plant cell walls (3, 6, 7). The presence of polysaccharide-degrading enzymes in infected tissues and the degradation of the host cell wall early during the infection process is additional evidence for the role of such enzymes in pathogenesis (7, 15, 30, 42, 82). So, too, are the reports that mutants of pathogens are avirulent when they are unable to produce one or more polysaccharide-degrading enzymes (7, 16). At the time that the review (7) was written, it was felt that these pathogen-secreted enzymes would likely be involved in determining the specificity of interaction between host and pathogen in gene-for-gene systems. However, we now feel that although these enzymes are involved in general pathogenesis, they are not determinants of varietal specificity.

The advances which have caused the revision in our thinking concerning the role in pathogenesis of polysaccharide-degrading enzymes include a substantial increase in our knowledge of the structures of the substrates of these enzymes, that is, the polysaccharides of the primary cell walls of plants (3, 4, 21, 137). It was proposed (7) that varietal specificity might be determined by variations in the cell wall compositions of the hosts, and that these wall differences would cause differential induction of the pathogen's polysaccharide-degrading enzymes. However, the primary cell walls of a variety of dicotyledons are very similar in that they are composed of the same monosaccharide constituents joined together by the same glycosidic linkages (3, 4, 137). It also appears that the primary cell walls of monocotyledons, although different from those of dicotyledons, are also very similar to each other (3, 4, 21). This similarity between the cell walls of different species of plants, and, of course, between different varieties of the same species (61, 87, 88; B. Nusbaum and P. Albersheim, unpublished results) directly contradicts the idea that variations in cell wall structure would control the rate at which pathogens secrete polysaccharide-degrading enzymes. Although it remains possible that quantitatively minor glycosidic constituents of the cell walls differ between plants, this possibility is becoming ever more remote.

Specific induction of the degradative enzymes of pathogens by cell wall constituents of the host is made even more unlikely by the finding that low levels of simple monosaccharides appear to be the natural inducers of the degradative enzymes of pathogens (24). These results indicate that the monosaccharides released by a pathogen's extracellular enzymes cause the pathogen to secrete greater amounts of the degradative enzymes. Since the walls of all plants are composed of the same ten or so monosaccharides (2, 3, 21, 137), there is not enough information in the monomeric constituents of walls to account for varietal specificity. Even if plants have minor differences in their cell wall glycosyl linkages, these differences would disappear upon enzymic hydrolysis. Therefore, such differences could not control polysaccharide-degrading enzyme synthesis in pathogens. Indeed, no significant difference in the pattern of enzyme secretion by fungal pathogens has been observed when cell walls isolated from susceptible or resistant plant varieties were used as a carbon source in liquid medium (62).

The fact that an endopolygalacturonase is the first wall-degrading enzyme secreted when a pathogen is grown in culture on any of the cell walls of a variety of dicotyledons (32, 62; J. M. Mullen and D. F. Bateman, personal communication) is further evidence that the walls do not have a high degree of specificity with regard to the induction of degradative enzymes. Not only are endopolygalacturonases secreted early during pathogenesis of dicotyledons, but all the dicotyledons investigated have, associated with their cell walls, proteins capable of inhibiting the action of the endopolygalacturonases secreted by plant pathogens (5, 33, 62). These inhibitors act against the endopolygalacturonases, but not against any of the other cell wall-degrading enzymes studied (5). This evidence suggests that endopolygalacturonase involvement in the degradation of dicotyledon cell walls is a general phenomenon and that plants respond to this general mechanism of pathogenesis by possessing proteins capable of blocking the action of these essential enzymes. Indeed, a highly purified endopolygalacturonase inhibitor completely negates the action of an otherwise potent mixture of wall-degrading enzymes (62). We suggest that a pathogen is incapable of attacking a plant unless the pathogen finds the proper environment to permit it to secrete sufficient amounts of endopolygalacturonase to overcome the amount of inhibitor present in the cell walls of the plant.

There is additional evidence which suggests that the endopolygalacturonase-inhibitor systems represent a general, rather than a specific, mechanism for resistance. Although there are several different endopolygalacturonase inhibitors in the cell walls of dicotyledons, each of these appears to be able to inhibit the endopolygalacturonases of a variety of unrelated plant pathogens (5, 33, 62). This was demonstrated by the ability of an apparently homogeneous protein preparation, purified from bean (*Phaseolus vulgaris*) hypocotyls, to inhibit with almost equal efficiency the endopolygalacturonases secreted by three fungal pathogens of plants: *Colletotrichum lindemuthianum, Helminthosporium maydis,* and *Aspergillus niger* (2, 33).

Colletotrichum lindemuthianum, the causal agent of anthracnose in beans, is thought to be a gene-for-gene pathogen (22). The endopolygalacturonases secreted by several races of this pathogen are indistinguishable proteins (9). Also indistinguishable from each other are the *C. lindemuthianum* endopolygalacturonase inhibitors from several varieties of beans (9). These facts, and the observation that the endopolygalacturonase inhibitors are present in the cell walls of dicotyledons even when the plants have not been challenged by a pathogen (5), support the conclusion that these inhibitors are playing a role in general rather than specific resistance.

One observation which would have suggested that the polygalacturonase inhibitors are involved in specific resistance would have been that the amount of the endopolygalacturonase inhibitor present in a plant increased as a result of an attack by an incompatible rather than by a compatible race of a pathogen. Unpublished results of the authors indicate that the amounts of endopolygalacturonase inhibitors in gene-for-gene hosts do, in fact, increase about threefold following attack by a pathogen. However, both compatible and incompatible races of a pathogen have the same stimulatory effect. Thus the results of these experiments failed to support a role in specific or varietal resistance for the endopolygalacturonase inhibitors.

Finally, it is interesting that while the inhibitors of endopolygalacturonases are present in the cell walls of dicotyledons, no such inhibitors have been detected in the cell walls of monocotyledons (unpublished results of the authors). This suggests that another degradative enzyme may be involved in initiating the breakdown of the primary cell walls of monocotyledons, but no direct information on this possibility has yet been obtained.

Another way in which plants are protected against cell wall-degrading enzymes secreted by pathogens is by modification of the primary cell walls. The changes that occur during secondary cell wall formation inhibit dramatically the ability of pathogen-produced enzymes to degrade the cell wall polymers of their hosts (2, 15, 47, 78, 129, 138). This phenomenon may be responsible for the relatively high levels of resistance displayed by the older parts of plants.

The ability of plants to produce enzymes that are able to degrade the cell walls of pathogens may represent another general defense mechanism. Plants can synthesize large amounts of lysozyme (34, 35, 44, 59, 121), chitinase (1, 44, 59, 96), and

endo-β-1,3-glucanase (1, 23, 79, 80). The chitinase of tomatoes appears to participate in the lysis of *Verticillium albo-atrum* mycelia that occurs during infection of
tomatoes by this fungus (96).

Pathogens may have evolved their own defense against the potentially lytic enzymes of their hosts. *C. lindemuthianum* has been shown to secrete a protein that
can specifically inhibit the action of such a wall-degrading enzyme present in beans,
the host of this fungal pathogen (8). The enzyme, an endo-β-1,3-glucanase is produced in large amounts in bean tissues (1). The inhibitor secreted by *C. lindemuthianum* efficiently prevents the host's enzyme from degrading the walls of
C. lindemuthianum. It is not known from the limited studies that have been conducted whether this system is involved in determining general or specific resistance.
However, comparison to the endopolygalacturonase-inhibitor systems suggests that
the endoglucanase-inhibitor system is likely to be involved in general resistance.
Presumably, an excess of endo-β-1,3-glucanase in the plant over the amount of
inhibitor secreted by the pathogen would have a deleterious effect on the virulence
of the fungus; however, if the pathogen neutralizes all of the plant's enzyme, this
defense mechanism of the plant would be circumvented.

II. VARIETAL SPECIFICITY IN HOST-PATHOGEN SYSTEMS

A. An Hypothesis to Account for Varietal Specificity in Gene-for-Gene Host-Pathogen Systems

If we accept the evidence that the cell walls of plants are not sufficiently distinct to
account for the specificity observed in gene-for-gene type host-pathogen systems,
then it might be fruitful to ask whether variations in the cell walls of microbial
pathogens could account for this phenomenon. The answer to this question is clear
and the evidence compelling. The cell walls of pathogens are indeed specific enough
to account for varietal specificity in gene-for-gene systems. The following section
will include the evidence which demonstrates the high degree of structural specificity characteristic of the cell walls of bacteria and fungi.

Although plants are resistant to the attack of most pathogens, they are susceptible
to some. Clearly, pathogens have evolved the ability to overcome the general resistance mechanisms of their hosts. However, during the evolutionary process, susceptible plants have responded to the pressure of pathogenesis by developing more
specific modes of resistance (28).

There are two types of varietal specific host-pathogen systems that have been well
characterized. One type includes those pathogens which secrete host-specific toxins
(126). Host-specific toxins are molecules capable by themselves of producing in the
host the symptoms of the disease. Varieties of the host are susceptible to such
toxin-producing pathogens when they possess receptors which interact with the
toxin molecules (126). Varieties of the host which lack such receptors are resistant.
In these types of varietal-specific host-pathogen systems, the pathogen is characterized by the existence of only a single species or race.

The second type of varietal specific host-pathogen system that has been recognized involves pathogens that exist as a number of distinct races of a single species.

Resistance in plants to this type of varietal specific pathogen is determined by dominant Mendelian genes (28, 37, 38, 58, 98, 99, 117). Each such resistance gene that a plant possesses has the potential to make the plant totally resistant to some of the races of at least one of its pathogens. However, a resistance gene is effective in protecting a plant against only those races of a pathogen which possess molecules capable of interacting with the product of the host's resistance gene. Since these molecules in the pathogen cause the pathogen to be avirulent, the genes responsible for the synthesis of these molecules are called avirulence genes rather than virulence genes. If a race of the pathogen lacks a functional avirulence gene product capable of interacting with a product of any of the resistance genes in the host, the host's resistance genes are ineffective and the pathogen can successfully parasitize the host. Many important host-pathogen systems have evolved this type of gene-for-gene interrelationship as the basis for balanced parasitism. The understanding of the molecular nature of these gene-for-gene systems is of fundamental importance to biology and is the major concern of the remainder of this article.

The interdependence of resistance and avirulence genes demonstrates that the products of the resistance genes must recognize molecules produced by the pathogen whose synthesis is controlled by avirulence genes. We believe that this recognition reaction is the key to whether a race of a gene-for-gene pathogen is virulent on a variety of its host. A positive interaction between the products of a resistance and an avirulence gene initiates a resistant response in the plant.

The products of a plant's resistance genes can be thought of as receptors for molecules whose synthesis is controlled by the avirulence genes of pathogens. We hypothesize that these receptors are proteins which recognize carbohydrates and that these proteins are likely to be located in the plasma membranes of the host. The products of the pathogen's avirulence genes are, in all likelihood, also proteins. An attractive idea is that these proteins are glycosyltransferases which participate in the synthesis of the carbohydrate containing macromolecules which we suggest are recognized by the products of the host's resistance genes. The glycosyltransferases, the products of the avirulence genes in this model, must be able to be made nonfunctional without a highly deleterious effect on the pathogen. This ability to lose the function of avirulence genes is essential, for it is the loss of the function of one or more of these genes which permits an avirulent pathogen to become virulent (27, 37, 38).

It is important to realize that the products of the avirulence genes must provide some selective advantage to the pathogen when the host lacks corresponding resistance genes, or pathogens would have lost all their avirulence genes. Indeed, it has been observed that pathogens regain the function of all their avirulence genes for which their immediate host lacks corresponding resistance genes (20, 36–38, 132, 134).

The recognition factors in pathogens, the products of the avirulence gene glycosyltransferases, are, we propose, "antigens" exposed on the surface of the pathogen. These antigens would be antigens in the classical sense if the pathogen were injected into an immunologically responsive animal (14, 77). We do not mean to imply that these antigens act immunologically when the pathogen invades its plant host. There is no evidence that plants possess a true immunological system. However, we do suggest that the pathogen's surface antigens are recognized by receptor proteins in resistant varieties of the pathogen's host and that this interaction activates the host's defenses.

It is of particular interest that highly specialized host-pathogen systems not only frequently possess gene-for-gene relationships (28, 37, 38, 58, 98, 99), but that all of the gene-for-gene host-pathogen systems studied also exhibit the same type of physiological response leading to resistance. This type of resistance is known as hypersensitive resistance (28, 58, 70, 138). Hypersensitive resistance depends upon the plant's ability to respond sufficiently quickly to the presence of a pathogen such that the pathogen cannot penetrate past the first cell or few cells of the plant with which it comes in contact. This rapid response of the plant involves the death or sacrifice of the first host cell(s) invaded, and thus the term hypersensitive response or hypersensitive resistance. These hypersensitive cells are the cells that produce molecules which either kill or at least prevent the further growth of the pathogen (13, 28, 58, 70, 78, 83, 118, 119, 138).

Molecules produced by pathogens which activate or elicit the hypersensitive response in their hosts have been called elicitors by Keen et al (65). Our hypothetical surface antigens fit this definition of an elicitor, and we shall use that term in the remainder of this manuscript.

Let us summarize. In the model we are proposing to explain the specificity of gene-for-gene host-pathogen systems, the reaction between the pathogen and the host will be an incompatible one (28) (the host will be resistant and the pathogen avirulent) when the plant has the product of at least one resistance gene capable of interacting with a product of a pathogen's avirulence gene. This interaction triggers the synthesis in the plant of molecules capable of stopping further growth of the pathogen. The reader should remember that no product of a plant's resistance gene or of a pathogen's avirulence gene in a gene-for-gene host-pathogen system has been identified or characterized. Nevertheless, these gene products clearly do function in the recognition processes that occur between plants and their pathogens. There are now a number of examples of similar biological recognition phenomena which have been examined at a molecular level. These examples will be used in the following section as models for supporting our prediction of the chemical nature of the products of the avirulence and resistance genes of the gene-for-gene host-pathogen systems. In all the examples, specificity is provided by the interaction between carbohydrate-containing molecules produced by one organism and a protein produced by the other.

B. Examples which Demonstrate that Cell Surface Recognition Phenomena are Mediated through the Interaction of Carbohydrate-Containing Macromolecules and Proteins

1. AGGLUTININS OF SEXUAL MATING TYPES IN YEAST   Sexual mating in the yeast *Hansenula wingei* occurs only between certain species, defined as species of opposite mating type (136). A macromolecule isolated from the cell surface of one

mating type will specifically agglutinate cells of the opposite mating type (128, 141). This specificity of agglutination suggests that the surface macromolecule is involved in recognition of cells of the correct mating type. The biologically active macromolecule is a glycoprotein. The carbohydrate, which accounts for from 80–90% by weight of this molecule, is composed mainly of mannosyl residues. However, the mannan of this glycoprotein is distinct from other yeast cell wall mannans that have been characterized and which are thought to have structural functions (140). It appears that the mannan-containing agglutinating glycoprotein has a polypeptide backbone to which short oligosaccharide chains are attached. The integrity of the protein-polysaccharide structure is required for the agglutination activity of the molecule. Treatment of the glycoprotein with enzymes that specifically degrade either the peptide or carbohydrate portions destroy the molecule's biological activity (141). Thus cell surface glycoproteins govern the specificity of mating types in this fungus.

2. A SPECIES SPECIFIC AGGREGATION FACTOR OF SPONGE CELLS   The involvement of cell surface glycoproteins is also demonstrated by the results of studies on sponge aggregation. When cells from a number of sponge species are dissociated and mixed, reaggregation occurs only between cells of the same species (54, 81). The specificity of aggregation has been shown to reside in factors that can be released from the surface of the sponge cells (54). These factors agglutinate cells of the donor species but fail to agglutinate, or agglutinate to a lesser extent, cells of other sponge species. Chemical analysis indicates that the agglutination factors are glycoproteins containing about 50% by weight of heteropolysaccharide (54).

3. THE DETERMINANTS OF THE A, B, O RED BLOOD CELL TYPES   The examples described for yeast and sponge indicate that these two classes of organisms produce structures on their cell surface which, although they are not essential for viability, participate in reactions of the cell that involve recognition of other cells. Although the biological activity of a glycoprotein may involve both its carbohydrate and peptide components, there are a number of examples where the specificity of such molecules is determined solely by the nature of the carbohydrate component. For example, the A, B, O red blood cell types are distinguished from each other by the nature of the terminal nonreducing glycosyl residue on the oligosaccharide portions of a plasma membrane glycoprotein (63, 133). If the oligosaccharide chains terminate in N-acetylgalactosamine, the cell is type A, whereas if this terminal glycosyl residue is galactose, the cell is type B. In the absence of either of these two sugars on the nonreducing terminus, that is, if the oligosaccharides terminate with one less glycosyl residue, the cell is type O. Thus antibodies capable of distinguishing the different blood cell types can recognize not only the absence of a sugar but also the difference between terminal galactosyl and N-acetylgalactosaminyl residues.

4. THE STRUCTURE OF BACTERIOPHAGE RECEPTORS   Another example of where a relatively minor alteration in carbohydrate structure leads to a dramatic change in recognition specificity is provided by the manner in which bacteriophage

select their hosts. It has been shown that the attachment site or receptor on the surface of Gram-negative bacteria for the tail proteins of phage is a surface component known as the O-antigen (25, 74). The O-antigens are part of the cell envelope lipopolysaccharides which consist of three covalently linked structural entities. The architectures of two of these entities, structural lipid A and the polysaccharide core, do not vary significantly between the various species of the Gram-negative bacteria, that is, they are not species specific (93). However, the O-antigen chains, which extend outward from the lipopolysaccharide core and which are exposed at the bacterial surface, are species specific (25, 76, 93, 108). The O-antigens are short polysaccharides consisting of a three or four sugar sequence that is repeated several times.

Robbins et al (76, 108) have shown that, when phage ϵ^{15} lysogenizes *Salmonella anatum*, the lysogenized bacteria possess O-antigen chains of different structure than exist on *S. anatum* not containing the prophage. Before infection, the O-antigen chains contain 6-O-acetyl-α-D-galactosyl residues, but in the lysogenized bacteria these residues are replaced by unacetylated β-D-galactosyl residues. The changes in anomeric configuration and in the loss of acetyl groups alter the antigenic nature of the bacterium to that of *Salmonella newington* (139). Moreover, the lysogenized *S. anatum* is no longer susceptible to phage ϵ^{15} infection because the altered O-antigen side chains fail to act as a receptor for this phage. However, the phage ϵ^{34}, which does not infect *S. anatum*, does infect ϵ^{15} lysogenized *S. anatum* as well as *S. newington* (52, 76, 84, 131, 139). Thus the specificity of phage infection depends upon the interaction of phage tail proteins with defined carbohydrates on the surface of their bacterial hosts.

5. A SPECIFIC SURFACE COMPONENT IS REQUIRED FOR VIRULENCE OF A PLANT PATHOGENIC BACTERIUM   The above examples demonstrate that the ability of a cell to recognize the presence of either another cell or a molecule produced by another cell involves an interaction between a carbohydrate-containing macromolecule from one cell and a protein from the other. The examples we will now discuss support the hypothesis that recognition of a pathogen by a plant involves carbohydrate-containing molecules on the surface of the pathogen as one of the determinants. The first example to be described demonstrates that a specific surface component of *Agrobacterium radiobacter* is required for virulence of this plant pathogen.

Roberts & Kerr (109) have shown that the ability of *Agrobacterium radiobacter* to induce crown gall in its host is correlated with the sensitivity of the *A. radiobacter* strain to an antibiotic, Bacteriocin 84. All virulent strains of *A. radiobacter* are susceptible to Bacteriocin 84, whereas nine out of ten avirulent strains are resistant to the antibiotic. Further, when avirulent strains are mutated to virulent strains, all of the newly virulent strains are susceptible to Bacteriocin 84. As other bacteriocins are proteins that require attachment to specific receptor sites on the cell surfaces of bacteria for their function (105), it is likely that this is also true for Bacteriocin 84. Thus it appears that a surface site on *A. radiobacter* that is responsible for the attachment of bacteriocins is required for the bacterium to be a virulent pathogen.

6. LECTINS AND THE HOST SPECIFICITY OF NITROGEN-FIXING BACTERIA
Another fine example of an interaction between specific carbohydrates on a cell surface and a protein from another cell is provided by the work of Bohlool & Schmidt (17). They have studied the specificity of the interaction of the nitrogen-fixing bacteria *Rhizobium* with legumes. Each species of *Rhizobium* will establish a symbiotic relationship with only certain species of legume. Bohlool & Schmidt attempted to correlate the binding of lectins to the surface of *Rhizobium* cells with the ability of the *Rhizobium* to establish a symbiotic relationship with that legume from which the lectin was isolated. They have found that soybean lectin combines specifically with all but three of the 25 strains of soybean nodulating bacteria that were tested. On the other hand, soybean lectin does not bind to any of 23 strains of *Rhizobium* that do not form nodules in soybeans. Lectins are proteins that bind to defined carbohydrate structures (45, 69, 75, 116). Thus the results of Bohlool & Schmidt indicate that the *Rhizobium* which nodulate soybeans have a surface distinct from the *Rhizobium* which do not nodulate soybeans. This suggests that binding between the plant's lectin, proteins, and carbohydrates on the surface of *Rhizobium* cells determine which legumes the *Rhizobium* can nodulate.

7. SPECIES SPECIFIC SURFACE ANTIGENS OF BACTERIA AND FUNGI   In examples II B 5 and 6, the nature of the interaction of the plant host with the invading bacteria appears to depend on the bacterial cell surface. These results parallel the proposal that differences between the cell surfaces of the various races of a gene-for-gene pathogen participate in determining varietal specificity in these systems. Differences of this nature are known to distinguish between closely related species of both bacteria and yeast.

The immunological properties of the surface antigens of bacteria have been used to classify the various species of a bacterial genus. The Kaufmann-White immunological classification system, which distinguishes *Salmonella* species, is correlated with the structure of the O-antigens of bacteria (77). As discussed in example II B 4, small structural differences in the O-antigens are responsible for the different antigenic responses of the bacterial species. Recently Ballou et al (14) have shown that the antigenic determinants of yeast species also reside in small structural variations in a surface polysaccharide. The antigens in this case are part of mannan-containing glycoproteins. Comparison of the structures of mannans from four yeast species show that they all possess an α-1,6-linked mannan backbone. However, there are significant differences in the length and in the glycosyl linkages and even in the composition of the side chains of these antigenic mannans.

Thus examples have been presented which demonstrate that mammalian, invertebrate, fungal, and bacterial cells can utilize the carbohydrate portions of cell surface glycoproteins or lipopolysaccharides in biological recognition phenomena. The structures of the carbohydrate portions of these surface macromolecules are varied from cell type to cell type without affecting the cells in a lethal manner. We suggest that similar variations in the carbohydrate-containing macromolecules on the surfaces of gene-for-gene pathogens act as determinants in the differential infectivity patterns observed in gene-for-gene host-pathogen systems.

8. THE MITOGENIC EFFECT OF LECTINS   In our hypothesis, host-pathogen specificity depends on the interactions of a macromolecule synthesized by one organism with a macromolecule produced by a second organism. We have considered above the type of molecules that are likely to be involved as the specificity determinants, the elicitors, in the pathogens. However, equally specific factors are required in the plant. The mechanism of resistance in a gene-for-gene host, hypersensitivity, involves a severe change in the metabolism of the plant cell that is undergoing the response. The possibility that plasma membrane proteins are involved in initiation of hypersensitivity is suggested by the last four examples to be discussed. These examples demonstrate that intracellular metabolic changes can be initiated by the interaction of plasma membrane components with extracellular molecules. In all of the examples that have been and that will be presented, the specificity of the biological phenomena being studied resides in the interaction of carbohydrates and proteins.

An excellent example of the ability of extracellular molecules to cause intracellular changes through binding to cell surfaces is the mitogenic effect caused by binding of certain lectins to the surface of lymphocytes (106). Lectins are proteins which have the property of binding strongly to molecules which possess a specific arrangement of glycosyl residues (45, 69, 116). One lectin that is mitogenic, that is, that stimulates mitosis, is Concanavalin A (106). Concanavalin A binds to molecules containing exposed α-mannosyl or α-glucosyl residues (45). Other lectins that are mitogenic require for efficient binding more precise carbohydrate structures, structures which include several specific glycosyl residues in correct spatial orientation (69, 106).

The mitogenic effect of lectins demonstrates that the binding of molecules to cell surface structures can initiate such major metabolic changes within the cell that the result is an increased rate of cell division. How the cell surface interaction is able to stimulate mitosis is unknown, but it is clear that surface interactions do alter basic cellular processes. This observation supports the hypothesis that pathogen synthesized elicitors, interacting with receptors in the plasma membranes of their hosts, could initiate a hypersensitive response.

9. THE SPECIFICITY OF COLICINS IS DETERMINED BY CELL SURFACE RECEPTORS   Other fine examples of the ability of cell surface interactions to alter dramatically the metabolism of the recipient cells come from studies on the effects and mode of action of toxins. Many toxins require the presence of a specific receptor on the plasma membrane of their target cell although their toxic effect resides in an intracellular process (90, 94, 110). The colicins are a group of toxic proteins produced by some strains of enteric bacteria which are able to kill other strains of the same bacterial species (40, 85, 90, 105, 110). There are a number of types of colicins, each of which causes death of the recipient cells by different biochemical methods (40, 85, 105). However, for all the colicins, the initial event is the binding of the colicin to a specific cell surface receptor. Purification and characterization of the receptor for colicin E3 from susceptible *Escherichia coli* cells demonstrate that the receptor molecule is a glycoprotein in which the carbohydrate portion is required

for activity (110). Sabet & Schnaitman (110) have also shown that *E. coli* cells that are resistant to colicin E3 do not possess a functional receptor in their cell membrane. Therefore, although the protein synthesizing machinery in in vitro preparations of colicin E3 resistant cells is susceptible to the colicin (18, 73), whole cells resist the colicin by lacking the requisite receptor. Like all the examples we have discussed, the specific interaction between colicins and their target cells involves a protein, the colicin, and the carbohydrate portion of a macromolecule, the glycoprotein receptor.

10. THE SPECIFICITY OF SEVERAL EUKARYOTIC TOXINS IS DETERMINED BY CELL SURFACE RECEPTORS   A number of proteins toxic to the cells of certain mammalian species have also been shown to require specific receptors on the host's cell surface before the toxin is active. These toxins include the plant proteins abrin and ricin (90–92) as well as diphtheria toxin (43, 90, 94). Diphtheria toxin is synthesized by *Corynebacterium diphtheriae*, lysogenic for strains of phage β that carry a structural gene coding for the toxin. Each of these three toxins—abrin, ricin, and diphtheria—is a single polypeptide protein which can be cleaved into two polypeptide fragments (43, 90, 94). Although neither peptide fragment is active as a toxin on whole cells, one fragment from each toxin catalytically destroys the ability to synthesize proteins in all in vitro eukaryotic systems tested (90, 94). The other polypeptide fragment of each toxin has been shown to bind to host plasma membrane preparations. Since the binding of ricin and abrin to the plasma membrane is inhibited by certain sugars (90), the receptors for these proteins are probably glycoproteins. Further, cells that are resistant to abrin, ricin, and diphtheria toxins do not bind the toxins (90). Thus, like the colicin-bacterial systems, specific cell surface receptor sites are required for the mammalian toxins to be effective.

11. A TOXIN RECEPTOR IN THE PLASMA MEMBRANE OF A PLANT   The presence of specific receptor sites on mammalian, invertebrate, fungal, and bacterial cell surfaces strongly suggests that similar specific receptors also exist on plant cell surfaces. Indeed, the presence of receptor proteins in the plasma membranes of plants has been established by the excellent work of Strobel & Steiner (122, 124–127). These plant pathologists have studied the molecular interactions of a host-specific toxin produced by *Helminthosporium sacchari*, the causal agent of eye-spot disease in sugar cane. This work is the more significant in that it is the first time that the molecules involved in determining varietal specificity have been characterized in both the plant and the pathogen.

The host-specific toxin secreted by *H. sacchari* is called helminthosporoside (2-hydroxycyclopropyl-α-D-galactopyranoside) (122). This small molecule is secreted by the pathogen when it is growing in culture on media containing sugar cane extracts as well as when it is growing on the live plant. Susceptible varieties of sugar cane have proteins in their plasma membranes which bind helminthosporoside (124). Resistant varieties of sugar cane have a cell membrane protein which is immunologically related to and which purifies in a manner similar to the binding protein from susceptible varieties. However, the protein isolated from resistant plants differs by a few amino acids from the related protein of susceptible plants

(125). Further, the protein from resistant plants fails to bind helminthosporoside (125). Since the binding protein is located in the plant's plasma membrane (127), the susceptibility of a sugar cane plant to *H. sacchari* depends on the presence of a functionally active receptor in its plasma membrane. Strobel has shown that the binding of the toxin to the membrane of a susceptible plant causes a change in the normal permeability of the plant cell (126). He suggests that the binding of helminthosporoside to the membrane protein induces a conformational change that initiates the intracellular events that occur in the susceptible response.

The results of Strobel establish that plant plasma membranes contain proteins which can recognize carbohydrate-containing molecules presented to the plasma membranes from the outside and that such binding can trigger metabolic alterations in the receptor cell. This is a direct parallel to our suggestion that it is the cell surface of pathogens which interact with the plasma membranes of their hosts to induce a hypersensitive response.

III. ELICITORS OF PHYTOALEXIN PRODUCTION

A. Introduction

The evidence from the examples presented in II B supports our hypothesis that the plasma membranes of plants contain specific receptors for antigen-like carbohydrate-containing molecules, elicitors, of pathogen origin. The examples described also support the proposal that a positive interaction between an elicitor and a membrane receptor can trigger a hypersensitive response in the cells of the host. We have proposed, too, that the elicitor molecules are likely to be on the outer surface of pathogens. Ultrastructural analysis has shown that during initial invasion by some fungal pathogens, the plasma membrane of the first host cell to come in contact with the pathogen is deeply invaginated by the penetrating hypha (31, 53, 100). In these systems, there would be direct contact between the proposed receptors located in the host cell's plasma membrane and the elicitors on the pathogen's cell surface. However, in cases where there is no direct contact between the pathogen's cell surface and a plasma membrane of the host, our hypothesis would still be viable if the pathogen's elicitors were released into the pathogen's surroundings.

The reactions in the plant which are responsible during a hypersensitive response for stopping the growth of the pathogen have not yet been determined for many host-pathogen systems. Ultrastructural studies of cells undergoing a hypersensitive response show that the walls of the plant cells surrounding the hypersensitive cells are altered (41, 57, 66, 67, 78, 129). It has been suggested that these changes are important in limiting the spread of the pathogen (41, 57, 66, 67, 78, 129). In addition, low molecular weight compounds, phytoalexins, that are able to inhibit the growth of pathogens have been isolated from the hypersensitively responding tissues of many plants (13, 28, 29, 39, 56, 60, 64, 70, 71, 83, 102, 111, 138). Since concentrations of phytoalexins that are able to halt pathogen growth are attained early in hypersensitive responding tissues, a positive role for phytoalexins in hypersensitive resistance has been implicated (13, 39, 64, 102, 111, 118).

Phytoalexins are present in only low or undetectable amounts in healthy tissues. Thus their synthesis in hypersensitive-responding cells must be a consequence of the pathogenic state. This conclusion is strengthened by the ability of molecules produced by pathogens, elicitors (65), to stimulate phytoalexin synthesis. The synthesis of phytoalexins can also be stimulated by numerous procedures that are apparently unrelated to pathogenesis. These include treatment of the plant tissues with ultraviolet light (19, 51) and freezing (101). Phytoalexin synthesis is stimulated, too, by the application to plant tissues of many chemicals such as polyamines (48, 49), antibiotics (11, 114, 115), DNA intercalating agents (50, 55), and salts of heavy metals (26, 104, 114). However, the concentrations of these reagents required to stimulate phytoalexin synthesis (10^{-3} to 10^{-5} M) are higher than those (10^{-8} to 10^{-10} M) required of the pathogen-produced elicitors.

Most of the reports of pathogen-produced elicitors refer to the presence of elicitor activity in crude preparations (26, 28, 68, 89, 104, 130). The only published report of a purification and partial characterization of an elicitor is by Cruickshank & Perrin (27). These workers isolated a small peptide, monilicolin A, from *Monilinia fructicola,* the causal agent of brown rot of soft fruits. This peptide activates phytoalexin synthesis in bean *(Phaseolus vulgaris)* tissues when applied at a concentration of about 10^{-9} M. However, beans are not a host of this fungal pathogen. In contrast monilicolin A does not stimulate the production of phytoalexins in tissues of either broadbeans *(Vicia faba)* or peas *(Pisum sativum)*, other nonhost plants of *M. fructicola* (27). As the effect of this elicitor on the hosts of *M. fructicola* has not been reported, the significance of this work awaits further experimentation.

Studies of elicitors produced by several additional pathogens will be described in the following sections. These elicitors do stimulate phytoalexin synthesis in tissues of the pathogens' hosts.

B. An Elicitor Isolated from a Fungal Pathogen of Soybeans

Keen et al (65) detected elicitors in the extracellular media of cultured *Phytophthora megasperma* var. *sojae,* the causal agent of *Phytophthora* stem and root rot in soybeans (64). Application of the extracellular media to soybean hypocotyls or cotyledons resulted in the synthesis in these tissues of the phytoalexin hydroxyphaseollin. The elicitor studied by Keen et al (65) is of small molecular weight and appeared to them to be a peptide. The unpublished results of A. Ayers, J. Ebel, and P. Albersheim have verified that the extracellular media of cultured *P. megasperma* var. *sojae* does contain a very active elicitor of hydroxyphaseollin synthesis in soybean tissues. However, Ayers et al have shown that this elicitor is a polysaccharide, containing only minor amounts of peptide, and that the elicitor is heterogeneous in size, varying from about 4,000 to 40,000 daltons. Ayers et al have also extracted an equally active elicitor from the purified mycelia walls of *P. megasperma* var. *sojae.* This elicitor is larger in size (greater than 50,000 daltons) than the culture filtrate elicitor but otherwise appears to be structurally similar.

The polysaccharide portions of highly purified preparations of the elicitors from both the extracellular media and the mycelial walls consist largely of glucan, although minor amounts of other glycosyl residues are also present. The predominant glycosyl residues of the elicitor preparation are 3-linked, 6-linked, 3,6-linked and terminal glucose. Thus the structure of these polysaccharides resembles closely the noncellulosic glucan of this pathogen's mycelial walls (142; B. Valent and P. Albersheim, unpublished results).

The purified elicitors are exceedingly active. The synthesis of hydroxyphaseollin in soybean cotyledons is stimulated by the application of an amount of elicitor equivalent to less than 10 ng of glucose (about 10^{-13} moles of elicitor). The biological activity of the elicitor depends on its detailed structure since model glucans such as laminarin [a polysaccharide containing predominantly 3- and 6-linked glucosyl residues (97)] are not active elicitors in this system. Consequently, an in-depth structural analysis of the elicitor is in progress. The findings of this study will bear on the question of whether elicitors possess antigenic determinants. If elicitors do possess such determinants, their identification may be complicated by the fact that the determinants are likely to represent quantitatively minor portions of these molecules.

C. An Elicitor Isolated from a Fungal Pathogen of Beans

Elicitors of a similar nature to those from *P. megasperma* var. *sojae* have been isolated from *Colletotrichum lindemuthianum,* a fungal pathogen of beans (*Phaseolus vulgaris*) (10). Results in our laboratory agree with those of Skipp & Deverall (119) that application of culture filtrates of *C. lindemuthianum* to the cut surface of bean hypocotyls causes the tissue to turn brown. Ultrastructural analysis of these brown tissues shows changes in the fine structure that resemble those observed in bean tissues undergoing a hypersensitive response to *C. lindemuthianum* (78). We have shown that the browning response observed with cotyledons as well as with hypocotyls is accompanied by the production of phytoalexins (10). The phytoalexins found in these tissues (10) include phaseollin (12, 13). Phaseollin has been identified by comparison with an authentic standard (the gift of Hans Van Etten of Cornell University) on thin layer chromatography as well as by combined gas chromatography-mass spectrometry. Three additional phytoalexins extracted from elicitor-treated tissues have been tentatively identified by thin layer chromatographic comparison to known standards (also the gift of Hans Van Etten). These additional phytoalexins appear to be kievitone, phaseollinisoflavan, and phaseollidin (12, 120).

The elicitor secreted by *C. lindemuthianum* has been extensively purified. The most purified preparations contain molecules ranging from 70,000 to 150,000 daltons. Less than 10^{-11} moles of the purified elicitor is required to activate phytoalexin synthesis in bean cotyledons.

The purified elicitor preparation is largely glucan in nature, but the purest preparations contain small amounts of other glycosyl residues and amino acids. An essentially identical elicitor can be extracted from purified mycelia cell walls of this fungus. The predominant glycosyl residues of the *C. lindemuthianum* elicitor preparations isolated from culture filtrate or from mycelia walls are 4-linked and 3-linked glucose. Thus this elicitor has a structure similar to the major polysaccharide component of *C. lindemuthianum* mycelia walls (B. Valent and P. Albersheim, unpublished results). The minor glycosidic components of the elicitor, which may represent its antigenic determinants, have not been characterized. As with the *P. megasperma* var. *sojae* system, model glucans, with linkages similar to those found in the *C. Lindemuthianum* elicitor, are not active in eliciting phytoalexin synthesis in the bean tissues.

D. Are Elicitors the "Antigenic" Determinants of Gene-for-Gene Pathogens?

A question of utmost importance is whether elicitors such as those isolated from *P. megasperma* var. *sojae* and *C. lindemuthianum* have the ability to activate phytoalexin synthesis in resistant varieties but not in susceptible varieties of their hosts. Keen et al (65) obtain five to tenfold greater induction of hydroxyphaseollin synthesis in hypocotyls of resistant soybeans than they do when they apply their impure elicitor preparations to hypocotyls of susceptible soybeans. Our laboratory has yet to obtain any substantial evidence of specificity with regard either to the *C. lindemuthianum* or the *P. megasperma* var. *sojae* elicitors. However, until now our laboratory has emphasized the purification and characterization of the elicitors rather than examining carefully the question of biological specificity. The question of specificity is most intriguing, but the isolation of molecules from the surface of pathogens which trigger the synthesis of phytoalexins in the host is in itself of importance. These macromolecules are effective in such minute quantities, 10^{-11} to 10^{-13} moles per cotyledon, that they undoubtedly have biological significance. In addition, the chemical nature and cellular location of the elicitors is excitingly reminiscent of the antigenic determinants of yeast and bacteria (examples II B 4–7). This encourages the thought that elicitors may represent the "antigenic" determinants of the races of gene-for-gene fungal pathogens.

IV. FURTHER CONSIDERATION OF THE HYPOTHESIS AND HOW THE GENE-FOR-GENE RELATIONSHIP MAY HAVE EVOLVED

The hypothesis put forward in this paper proposes that there exist on the surface of gene-for-gene pathogens carbohydrate-containing molecules, elicitors, that vary in structure from race to race. For a race to be virulent, the elicitor must have a structure that is not recognized by its host. The examples presented in II B suggest that the structural variations between the elicitors of each race may consist of the substitution of one glycosyl residue for another, of changes in anomeric configuration, in the addition or loss of glycosyl residues, or in the presence of other chemical moieties such as methyl ethers and acetate and phosphate esters. In order for these variations to be effective, they must control the recognition process that occurs between the host and the pathogen but must not reduce greatly the viability of the pathogen. Thus we propose that the elicitors of the races of a gene-for-gene pathogen possess a basic structure common to all of the races but which is made unique to each race by some aspect of its detailed structure.

A gene-for-gene relationship between host and pathogen has evolved as a consequence of long association and competition between the organisms. It is possible that the ancestors of present-day gene-for-gene pathogens were related to the pathogens which secrete phytotoxins. A number of the pathogen-produced phytotoxins (46, 95, 103, 107, 123, 126) are large molecular weight glycopeptides which are similar in structure and other characteristics to the elicitors described in III B and III C. The biological role of elicitors could have evolved from the development in plants of mechanisms to prevent pathogen growth, mechanisms that may have been based on attempts to recognize the potentially phytotoxic molecules. If plants evolved a mechanism for recognizing the presence of macromolecular carbohydrate-containing toxins, plants would have been well on the way to recognizing the surface of the pathogens themselves. In fact, the ability of plasma membrane proteins to recognize toxins has been established (section II b 11).

The hypothesis proposed in this paper to account for varietal specificity in a gene-for-gene system suggests that the interaction between a pathogen's surface antigen and a plasma membrane protein in the host would activate a defense mechanism of the host. An avirulent pathogen would be under selective pressure to loose the antigen that was triggering the host's defenses. In our scheme, the pathogen could achieve this goal by mutation of the avirulence gene which codes for the glycosyl transferase responsible for the synthesis of the determinant surface antigen. The phenotypic expression of this genetic loss would be a change from avirulence to virulence.

Loss of the ability to synthesize a functional avirulence gene product is associated in pathogens with a detectable loss in vigor. Field studies (20, 36–38, 132, 134) have shown that pathogens possess in a functional form all of their avirulence genes that do not correspond to any of the resistance genes in their host. Indeed, pathogens rapidly regain any avirulence genes not corresponding to resistance genes in their immediate host but whose functions were selected against because they did correspond to resistance genes in their previous host. That this is true is not surprising, for if there were no advantage to pathogens in possessing avirulence genes, pathogens would have lost all of these genes. However, if a pathogen had lost all of its avirulence genes, and if this pathogen retained sufficient vigor to infect a plant, then this pathogen would be virulent on all of the varieties of its host. This would result in an imbalance of the host-pathogen system with the eventual extinction of both the host and the pathogen.

Evolutionary selection pressure, resulting from the reduced vigor of a pathogen lacking one or more avirulence genes, would cause such a pathogen to develop another avirulence gene. In our hypothesis, this would mean the development of a second glycosyltransferase to replace the nonfunctional enzyme. The pathogen's second glycosyltransferase would result in a different surface antigen and would presumably renew the vigor of the pathogen. By this process, the pathogen species would have gained a second race distinguished from the first by the nature of its avirulence gene.

The host of a pathogen which has developed a second avirulence gene would respond by developing a second resistance gene. In our proposal, this would mean that the host would respond to the presence of the new surface antigen on the pathogen by evolving a new receptor protein in its plasma membrane. A plant could most readily achieve this by modification of the same plasma membrane protein which acted as a receptor for the first surface antigen of the pathogen. This evolutionary process would result in a new gene in a location on the host's chromosome allelic to the original gene. Each such allele would be a different resistance gene and would code for a distinct plasma membrane protein. Alternatively, the host may have achieved the same end by gene duplication followed by modification of one of the resulting genes. In this case, a second resistance gene could evolve in a location on the host's chromosome very near or next to the first resistance gene.

The repetition of such processes in the host and in the pathogen would result in today's gene-for-gene systems which have resistance genes in the host clustered at one or several neighboring loci (28, 37, 38, 98, 99, 112, 117). Since their many potential resistance genes are distributed over a few genetic loci, plants are unable to express more than a few resistance genes at one time (28). On the other hand, most of the avirulence genes of pathogens have evolved at widely diverse locations in their genome (28). This permits pathogens to have all or almost all of their avirulence genes functioning simultaneously. Since functional avirulence genes lead to increased vigor, pathogens are able to obtain all of the benefits that their avirulence genes can provide.

Elucidation of the molecular basis of the specificity exhibited by gene-for-gene pathogens is extremely important. The need to protect the world's food supply, based largely on crops with an ever-decreasing gene pool (86), is paramount. The ability of host-parasite interactions in plants to act as a model for other biological systems should not be underestimated. There are many highly important examples of organism-organism interaction which probably have a similar gene-for-gene basis, such as the symbiotic relationships exemplified by nitrogen-fixing bacteria and legumes, algae and fungi in lichens, enteric bacteria and mammals, and insects and protozoa. Of direct interest to man is the possibility that inheritance of disease resistance in mammals may be based on similar gene-for-gene relationships, genetic interactions that would be difficult to detect in animals that have immune systems.

The possibility that such genetic interactions do exist in man is supported by the examples of section II B 10 which demonstrate that toxins are selective in the eukaryotic cells that they attack. Similarly, viruses have a degree of selectivity as to which cells they infect. For example, the various types of influenza virus are specific as to which mammalian or avian cells they infect (72, 113). The attachment of the influenza virus to its host cell involves a glycoprotein on the surface of the virus called the hemagglutinin (72, 113). This hemagglutinin is the major surface antigen on the virus and is responsible for the induction in the host of neutralizing antibodies. Major structural changes in the protein component of the viral hemagglutinin have resulted in the great influenza epidemics of man (135). Although the protein portion of the hemagglutinin is coded for by the virus, the carbohydrate portion is coded for by the host cell (72, 113). Thus the carbohydrate constituents of a given strain of the virus will vary according to the host in which it is grown. With biochemical interactions of this type being demonstrable in man, it seems probable that gene-for-gene relationships similar to those of plant-pathogen systems exist in higher animals.

The evidence supporting the hypothesis described in this paper, which attempts to offer a molecular basis for the specificity of the interactions exhibited by some plants and their pathogens, is sufficient to warrant a detailed investigation of the hypothesis. A product of an avirulence gene of a pathogen must be isolated and its function elucidated. Do the products of the avirulence genes participate in the synthesis of elicitors? The resistance gene products of the hosts must be identified. Are the products of the host's resistance genes receptors for the products of a pathogen's avirulence genes? Eventually it will be necessary to learn how the interaction between these molecules stimulates the host to defend itself. Assuming that the avirulence genes control the synthesis of elicitors, there are several attractive mechanisms by which the elicitor may function. For example, elicitors may act in a fashion analogous to that of the colicins and those toxins like abrin, ricin, and diphtheria (sections II B 9 and 10), having a part of their molecules react specifically with a receptor in the membrane and a part act within the cell to cause a major metabolic alteration. Another intriguing possibility is that elicitors cause conformational changes in the plasma membranes of the host as has been suggested for the host-specific toxin helminthosporoside (section II B 11). In this proposal, the alteration in membrane structure would trigger the metabolic events leading to a hypersensitive response.

Whatever the mechanisms underlying the specificity of host-pathogen interactions, the regulatory processes which are certainly going to be discovered during investigation of these systems will be of interest to all biologists, and the results in this immediately practical discipline will be of benefit to all mankind.

ACKNOWLEDGMENTS

The authors wish to express their appreciation to Claude Renou, Joyce Albersheim, and Dora Kelling for their untiring assistance in preparation of this manuscript. The authors have been supported by the Frasch Foundation, the National Science Foundation (GB 41386), and the U.S. Atomic Energy Commission (Contract #AT(11-1)-1426).

Literature Cited

1. Abeles, F. B., Bosshart, R. P., Forrence, L. E., Habig, W. H. 1970. *Plant Physiol.* 47:129–34
2. Albersheim, P. 1965. *Plant Biochemistry*, ed. J. Bonner, J. Varner, 298–321. New York: Academic
3. Albersheim, P. 1975. *Plant Biochemistry*, ed. J. Bonner, J. Varner. New York: Academic. 2nd ed. In press
4. Albersheim, P. 1974. In *Tissue Culture and Plant Science 1974*, ed. H. E. Street. Proc. 3rd Int. Congr. Plant Tissue and Cell Culture. London: Academic
5. Albersheim, P., Anderson, A. J. 1971. *Proc. Nat. Acad. Sci. USA* 68: 1815–19
6. Albersheim, P., Bauer, W. D., Keegstra, K., Talmadge, K. W. 1973. *Biogenesis of Plant Cell Wall Polysaccharides*, 117–47. New York: Academic.
7. Albersheim, P., Jones, T. M., English, P. D. 1969. *Ann. Rev. Phytopathol.* 7:171–94
8. Albersheim, P., Valent, B. 1974. *Plant Physiol.* 53:684–87
9. Anderson, A. J., Albersheim, P. 1972. *Physiol. Plant Pathol.* 2:339–46
10. Anderson, A. J., Albersheim, P. 1973. *Phytopathol. Abstr. 0764. 2nd Int. Congr. Plant Pathol.*
11. Bailey, J. A. 1969. *Phytochemistry* 8:1393–95

12. Bailey, J. A., Burden, R. S. 1973. *Physiol. Plant Pathol.* 3:171–77
13. Bailey, J. A., Deverall, B. J. 1971. *Physiol. Plant Pathol.* 1:435–49
14. Ballou, C. E., Raschke, W. C. 1974. *Science* 184:127–34
15. Bateman, D. F., Van Etten, H. D., English, P. D., Nevins, D. J., Albersheim, P. 1969. *Plant Physiol.* 44:641–48
16. Beraha, L., Garber, E. D. 1971. *Phytopathol. Z.* 70:335–44
17. Bohlool, B. B., Schmidt, E. L. 1974. *Science* 185:269–71
18. Bowman, C. M., Sidikaro, J., Nomura, M. 1971. *Nature New Biol.* 234:133–37
19. Bridge, M. A., Klarman, W. L. 1973. *Phytopathology* 63:606–9
20. Browning, J. A., Frey, K. J. 1969. *Ann. Rev. Phytopathol.* 7:355–82
21. Burke, D., Kaufman, P., McNeil, M., Albersheim, P. 1974. *Plant Physiol.* 54:109–15
22. Charrier, A., Bannerot, H. 1970. *Ann. Phytopathol.* 2:489–506
23. Clarke, A. E., Stone, B. A. 1962. *Phytochemistry* 1:175–88
24. Cooper, R. M., Wood, R. K. S. 1973. *Nature* 246:309–11
25. Costerton, J. W., Ingram, J. M., Cheng, K. J. 1974. *Bacteriol. Rev.* 38:87–110
26. Cruickshank, I. A. M., Perrin, D. R. 1963. *Aust. J. Biol. Sci.* 16:111
27. Cruickshank, I. A. M., Perrin, D. R. 1968. *Life Sci.* 7:449–58
28. Day, P. R. 1974. *Genetics of Host-Parasite Interaction.* San Francisco: Freeman. 238 pp.
29. Deverall, B. J. 1972. *Proc. Roy. Soc. B* 181:233–46
30. Dimond, A. E. 1970. *Ann. Rev. Phytopathol.* 8:301–22
31. Ehrlich, M. A., Ehrlich, H. G. 1971. *Ann. Rev. Phytopathol.* 9:155–84
32. English, P. D., Jurale, J. B., Albersheim, P. 1971. *Plant Physiol.* 47:1–6
33. Fisher, M. L., Anderson, A. J., Albersheim, P. 1973. *Plant Physiol.* 51:489–91
34. Fleming, A. 1922. *Proc. Roy. Soc. B* 93:306
35. Fleming, A. 1932. *Proc. Roy. Soc. Med.* 26:7
36. Flor, H. H. 1953. *Phytopathology* 43:624–28
37. Flor, H. H. 1956. *Advan. Genet.* 8:29–53
38. Flor, H. H. 1971. *Ann. Rev. Phytopathol.* 9:275–96
39. Frank, J. A., Paxton, J. D. 1970. *Phytopathology* 60:315–18
40. Fredericq, P. 1957. *Ann. Rev. Microbiol.* 11:7–22
41. Friend, J., Reynolds, S. B., Aveyard, M. A. 1973. *Physiol. Plant Pathol.* 3:495–507
42. Garibaldi, A., Bateman, D. F. 1970. *Phytopathol. Mediter.* 9:136–44
43. Gill, D. M., Pappenheimer, A. M. Jr., Uchida, T. 1973. *Fed. Proc.* 32:1508–15
44. Glazer, A. N., Barel, A. O., Howard, J. B., Brown, D. M. 1969. *J. Biol. Chem.* 244:3583–89
45. Goldstein, I. J., So, L. L. 1965. *Arch. Biochem. Biophys.* 111:407
46. Goodman, R. N., Huang, J. S., Huang, Pi-Yu 1973. *Science* 183:1081–83
47. Griffey, R. T., Leach, J. G. 1965. *Phytopathology* 55:915–18
48. Hadwiger, L. A., Jafri, A., von Broembsen, S., Eddy, R. Jr. 1974. *Plant Physiol.* 53:52–63
49. Hadwiger, L. A., Schwochau, M. E. 1970. *Biochem. Biophys. Res. Commun.* 38:683–91
50. Hadwiger, L. A., Schwochau, M. E. 1971. *Plant Physiol.* 47:346–51
51. Ibid, 588–90
52. Harada, K. 1956. *Virus (Osaka)* 6:285
53. Heath, M. C., Heath, I. B. 1971. *Physiol. Plant Pathol.* 1:277–87
54. Henkart, P., Humphreys, S., Humphreys, T. 1973. *Biochemistry* 12:3045–50
55. Hess, S. L., Hadwiger, L. A. 1971. *Plant Physiol.* 48:197–202
56. Higgins, V., Millar, R. 1968. *Phytopathology* 58:1377–83
57. Hijwegen, T. 1963. *Neth. J. Plant Pathol.* 69:314–17
58. Hooker, A. L., Saxena, K. M. S. 1971. *Ann. Rev. Genet.* 5:407–24
59. Howard, J. B., Glazer, A. N. 1969. *J. Biol. Chem.* 244:1399–1409
60. Ingham, J. L. 1972. *Bot. Rev.* 38:343–424
61. Jones, T. M., Albersheim, P. 1972. *Plant Physiol.* 49:926–36
62. Jones, T. M., Anderson, A. J., Albersheim, P. 1972. *Physiol. Plant Pathol.* 2:153–166
63. Kabat, E. A. 1956. *Blood Group Substances: Their Chemistry and Immunochemistry.* New York: Academic
64. Keen, N. T. 1971. *Physiol. Plant Pathol.* 1:265–75
65. Keen, N. T., Partridge, J. E., Zaki, A. I. 1972. *Phytopathology* 62:768 (Abstr.)
66. Kitazawa, K., Tomiyana, K. 1970. *Phytopathol. Z.* 66:317–24

67. Klarman, W. L., Corbett, M. K. 1974. *Phytopathology* 64:971–75
68. Klarman, W. L., Gerdemann, J. W. 1963. *Phytopathology* 53:1317
69. Kornfeld, R., Kornfeld, S. 1970. *J. Biol. Chem.* 245:2536–45
70. Kuć, J. 1966. *Ann. Rev. Microbiol.* 20:337–70
71. Kuć, J. 1972. *Ann. Rev. Phytopathol.* 10:207–32
72. Laver, W. G. 1973. *Advances in Virus Research,* 57–103. New York: Academic
73. Levisohn, R., Konisky, J., Nomura, M. 1968. *J. Bacteriol.* 96:811
74. Lindberg, A. A., Hellerqvist, C. G. 1971. *J. Bacteriol.* 105:57–64
75. Lis, H., Sela, B. A., Sachs, L., Sharon, N. 1970. *Biochim. Biophys. Acta* 211:582–85
76. Losick, R., Robbins, P. W. 1967. *J. Mol. Biol.* 30:445–55
77. Lüderitz, O., Staub, A. M., Westphal, O. 1966. *Bacteriol. Rev.* 30:192–255
78. Mercer, P. C., Wood, R. K. S., Greenwood, A. D. 1974. *Physiol. Plant Pathol.* 4:291–306
79. Moore, A. E., Stone, B. A. 1972. *Planta* 104:93–109
80. Moore, A. E., Stone, B. A. 1972. *Virology* 50:791–98
81. Moscona, A. A. 1963. *Proc. Nat. Acad. Sci. USA* 49:742–47
82. Mullen, J. M., Bateman, D. F. 1971. *Physiol. Plant Pathol.* 1:363–73
83. Müller, K., Börger, H. 1940. *Arb. Biol. Reichsanst. Land Forstwirt. Berlin* 23:189–231
84. Nakagawa, A. 1957. *Jap. J. Bacteriol.* 12:47
85. Nomura, M. 1967. *Ann. Rev. Microbiol.* 21:257–84
86. National Research Council 1972. *Genetic Vulnerability of Major Crops.* Washington, D.C.: Nat. Acad. Sci.
87. Nevins, D., English, P., Albersheim, P. 1967. *Plant Physiol.* 42:400–6
88. Ibid 1968. 43:914–22
89. Nüesch, J. 1963. *Soc. Gen. Microbiol. Symp.* 13:335
90. Olsnes, S. 1972. *Naturwissenschaften* 59:497–502
91. Olsnes, S., Pihl, A. 1972. *FEBS Lett.* 20:327–29
92. Olsnes, S., Pihl, A. 1972. *Nature* 238:459–61
93. Osborn, M. J., Rosen, S. M., Rothfield, L., Zeleznick, L. D., Horecker, B. L. 1964. *Science* 145:783–89
94. Pappenheimer, A. M. Jr., Gill, D. M. 1973. *Science* 182:353–58
95. Patil, S. S. 1974. *Ann. Rev. Phytopathol.* 12:259–79
96. Pegg, G. F., Vessey, J. C. 1973. *Physiol. Plant Pathol.* 3:207–22
97. Percival, E. 1970. *The Carbohydrates,* ed. W. Pigman, D. Horton, A. Herp, 11B:737–65. New York: Academic
98. Person, C. 1959. *Can. J. Bot.* 37: 1101–30
99. Person, C., Sidhu, G. 1971. *IAEA (Vienna),* 31–38
100. Politis, D. J., Wheeler, H. 1973. *Physiol. Plant Pathol.* 3:465–71
101. Rahe, J. 1974. *Phytopathol. Abstr.* 196. In press
102. Rahe, J. E., Kuć, J., Chuang, Chien-Mei, Williams, E. B. 1969. *Neth. J. Plant Pathol.* 75:58–71
103. Rai, P. V., Strobel, G. A. 1969. *Phytopathology* 59:47–52
104. Rathmell, W. G., Bendall, D. S. 1971. *Physiol. Plant Pathol.* 1:351–62
105. Reeves, P. 1972. *The Bacteriocins.* Berlin, Heidelberg, New York: Springer
106. Reichart, C. F., Pan, P. M., Mathews, K. P., Goldstein, I. J. 1973. *Nature New Biol.* 242:146–48
107. Ries, S. M., Strobel, G. A. 1971. *Plant Physiol.* 49:676–84
108. Robbins, P. W., Uchida, T. 1962. *Biochemistry* 1:323
109. Roberts, W. P., Kerr, A. 1974. *Physiol. Plant Pathol.* 4:81–91
110. Sabet, S. F., Schnaitman, C. A. 1973. *J. Biol. Chem.* 248:1797–1806
111. Sato, N., Kitazawa, K., Tomiyama, K. 1971. *Physiol. Plant Pathol.* 1:289–95
112. Saxena, K. M. S., Hooker, A. L. 1968. *Proc. Nat. Acad. Sci. USA* 61:1300–5
113. Schulze, I. T. 1973. *Advances in Virus Research,* 1–55. New York: Academic
114. Schwochau, M. E., Hadwiger, L. A. 1968. *Arch. Biochem. Biophys.* 126: 731–33
115. Ibid 1969. 134:34–41
116. Sharon, N., Lis, H. 1972. *Science* 177:949–59
117. Shepherd, K. W., Mayo, G. M. E. 1972. *Science* 175:375–80
118. Skipp, R. A., Deverall, B. J. 1972. *Physiol. Plant Pathol.* 2:357–74
119. Ibid 1973. 3:299–313
120. Smith, D. A., Van Etten, H. D., Bateman, D. F. 1973. *Physiol. Plant Pathol.* 3:179–86
121. Smith, E. L., Kimmel, J. R., Brown, D. M., Thompson, E. O. P. 1955. *J. Biol. Chem.* 215:67–89
122. Steiner, G. W., Strobel, G. A. 1971. *J. Biol. Chem.* 246:4350–57

123. Strobel, G. A. 1967. *Plant Physiol.* 42:1433–41
124. Strobel, G. A. 1973. *J. Biol. Chem.* 248:1321–28
125. Strobel, G. A. 1973. *Proc. Nat. Acad. Sci. USA* 70:1693–96
126. Strobel, G. A. 1974. *Ann. Rev. Plant Physiol.* 25:541–66
127. Strobel, G. A., Hess, W. M. 1974. *Proc. Nat. Acad. Sci. USA* 71:1413–17
128. Taylor, N. W., Orton, W. L. 1967. *Arch. Biochem. Biophys.* 120:602
129. Tomiyama, K. 1967. *Ann. Rev. Phytopathol.* 1:295–324
130. Uehara, K. 1959. *Ann. Phytopathol. Soc. Jap.* 24:224
131. Uetaka, H. 1956. *Proc. Int. Genet. Symp. Tokyo,* 638
132. Van der Plank, J. E. 1968. *Disease Resistance in Plants.* New York: Academic. 206 pp.

133. Watkins, W. M. 1966. *Science* 152:172–81
134. Watson, I. A. 1970. *Ann. Rev. Phytopathol.* 8:209–30
135. Webster, R. G. 1972. *Curr. Top. Microbiol. Immunol.* 59:75–105
136. Wickerham, L. J. 1956. *C. R. Trav. Lab. Carlsberg, Ser. Physiol.* 26:423
137. Wilder, B. M., Albersheim, P. 1973. *Plant Physiol.* 51:889–93
138. Wood, R. K. S. 1972. *Proc. Roy. Soc. B* 181:213–32
139. Wright, A., Barzilai, N. 1971. *J. Bacteriol.* 105:937–39
140. Yen, P. H., Ballou, C. 1974. *Biochemistry* 13:2420–27
141. Ibid, 2428–37
142. Zevenhuizen, L. P. T. M., Bartnicki-Garcia, S. 1969. *Biochemistry* 8:1496–1502.

Ann. Rev. Plant Physiol. 1975. 26:53–72

BIOCHEMISTRY OF LEGUME SEED PROTEINS

❖7582

Adele Millerd

Division of Plant Industry, CSIRO, Canberra, A.C.T. 2601, Australia

CONTENTS

INTRODUCTION

The proteins present in seeds are of two types: metabolic proteins, both enzymatic and structural, which are concerned in normal cellular activities including the synthesis of the second type, the storage proteins. Storage proteins, together with reserves of carbohydrates or oils, are synthesized during seed development. They function following seed germination when, subsequent to hydrolytic breakdown, they provide a source of nitrogen and carbon skeletons for the developing seedling. The storage proteins occur within the cell in discrete protein bodies. The term "protein body" is used in preference to "aleurone grain" (2), reserving the latter term for granules found in aleurone layers of seeds.

Seed proteins were comprehensively reviewed by Altschul et al (2), and subsequently many aspects of seed proteins have been considered in detail (40, 46, 50,

53

68–70). The general biochemistry of protein synthesis in plant tissues has been reviewed most recently in this series by Zalik & Jones (106).

The study of seed proteins is of twofold interest to biologists. Seed proteins are an extremely, and increasingly, important component of nutrition for both humans and animals. In addition, these storage proteins are tissue-specific. When proteins were extracted from organs of the pea (*Pisum sativum*) plant and examined immunoelectrophoretically using antiserum prepared against purified storage proteins, such proteins were detected only in extracts of organs of the seed: cotyledons, epicotyls, and hypocotyl plus radicle. Extracts from all other tissues, including seed coats, were negative (Millerd & Dudman, unpublished). This indicates, as shown earlier with *Vicia faba* (35), that storage proteins are restricted to seeds. They offer, therefore, an attractive system for the study of differential gene expression. These two aspects of seed protein biochemistry interact. The limitations of seed storage proteins in the nutrition of humans and other monogastric animals are well known (46). If these limitations are to be alleviated, or removed by biological manipulation, a fundamental understanding is necessary of the controls which limit the types and amounts of protein synthesized by seeds and of the constraints which must exist in the chemistry of the storage proteins.

CHARACTERIZATION OF STORAGE PROTEINS

Clearly, if one is to understand the details of the biosynthesis of specific proteins and the controlling mechanisms involved, it is necessary to have detailed characterizations of those proteins. Such studies would ideally involve isolation and purification, definition of physicochemical behavior, determination of amino acid composition and amino acid sequence, secondary structure and subunit interactions.

There are additional reasons which make such characterizations essential. The functional requirements of seed storage proteins include a high nitrogen content, the ability to form stable protein bodies, to withstand desiccation, and to be rapidly hydrolyzed by the proteolytic enzymes of the germinating seed. Detailed chemical and physicochemical knowledge of storage proteins should indicate how the structures of such proteins are related to these roles, and should help define limits within which they may be varied. Seed proteins have evolved for the benefit, not of monogastric animals, but of plants. We do not know to what extent plant breeders may hope to alter the amino acid composition of seed proteins before seed viability will be seriously impaired.

It is generally agreed that unequivocal purification of storage proteins has proved difficult. Probably the biggest hazard has been the lack of an adequate operational definition. Legume seed storage proteins are globulins. If an investigator is isolating storage proteins from whole legume seeds rather than from protein bodies, he is looking for proteins present in large amounts, soluble in dilute salt solutions and insoluble in water. Numerous techniques of protein chemistry have been employed with varying degrees of success. A consideration of some of these purification procedures and of the problems encountered may indicate desirable future approaches.

Danielsson (24) showed that the globulin storage proteins of legume seeds were characterized in the ultracentrifuge by the presence of two peaks with respective sedimentation coefficients of about 11–12S and 7S, and that those from *P. sativum* (25) migrated as a single boundary over a wide range of pH when subjected to free-boundary electrophoresis. This indicated that storage proteins of legumes might consist of only a few molecular species. If correct, this clearly would be an advantage in detailed biochemical studies.

Much work, reviewed by Wolf (102), has been carried out on the storage proteins of soybean (*Glycine max*). However, it has been difficult to decide how many species of storage proteins were present. Hill & Breidenbach (43) characterized soybean storage proteins by sucrose density gradient sedimentation and subsequent analysis of native and urea-dissociated proteins by polyacrylamide gel electrophoresis. Three distinct sedimenting fractions (2.2S, 7.5S, and 11.8S) were obtained, and the 7.5S and 11.8S proteins accounted for up to 70% of the total seed protein. The 7.5S protein gave three major electrophoretic bands under both nondissociating and dissociating conditions; the undissociated 11.8S protein gave a single diffuse band and was dissociated to give 5 or 6 major components. The banding patterns of the dissociated 7.5S and 11.8S proteins showed there was little cross-contamination. Hill & Breidenbach's (43) data indicated that only a few protein species made up the bulk of the storage protein.

Catsimpoolas et al (22) have purified glycinin, the ∼ 11S component, by chromatography on DEAE-Sephadex. The protein was immunochemically homogeneous (21, 22). N-terminal amino acid analyses (22) and disc electrophoresis data (20) indicated that glycinin, in its monomeric form, was composed of six subunits. This finding was supported when glycinin subunits were isolated by isoelectric focusing (19). However, a higher degree of microheterogeneity was observed when a similar preparation was subjected to analytical scale isoelectric focusing (23).

Ericson & Chrispeels (32) isolated and purified two storage proteins (sedimentation coefficients 8S and 11.3S) from *Phaseolus aureus*. These proteins were isolated by chromatography on DEAE-cellulose and subjected to amino acid analysis, gel electrophoresis, and sucrose gradient determination of S values. The 11.3S protein had three subunits (mol wt 56,000, 44,000 and 16,500 daltons) and contained about 0.1% glucosamine. The 8S protein, which contained no cysteine, gave four bands on sodium dodecylsulfate (SDS) gels (mol wt 63,500, 60,000, 29,500, and 24,000) and contained 0.2% glucosamine and 1% mannose. When protein was extracted from isolated protein bodies, it was shown to be essentially only these two glyco-proteins.

The 8S protein isolated by Ericson & Chrispeels (32) corresponds to glycoprotein II described by Pusztai & Watt (83) and constitutes 35% of the total seed protein in *Ph. vulgaris* (84). Glycoprotein II was homogeneous by chemical, physical, and immunochemical criteria, contained mannose and glucosamine, had a very low cysteine content, and was composed of four subunits (83).

Blagrove & Gillespie (11) have found that the globulin storage proteins of *Lupinus angustifolius* can be satisfactorily resolved by brief electrophoresis on cellulose-acetate strips. As well as conglutin α and conglutin β (the 11.6S and 7–8S proteins described by earlier workers), another protein, conglutin γ, was demonstrated. As conglutin γ was present in protein bodies, it was considered to be a storage protein. Conglutin γ had a higher leucine, cystine, and methionine content than the α or β proteins.

Seeds were ground, extracted with hexane and then with water, and both extracts discarded. These initial steps removed much of the metabolic protein. Electrophoresis on cellulose acetate strips was then used to check the effectiveness of subsequent purification procedures. The proteins were extracted with sodium chloride, conglutin α was recovered by isoelectric precipitation, conglutin β precipitated following dialysis, and conglutin γ remained in the supernatant. The three proteins were then separately purified by ammonium sulphate fractionation.

The three lupin proteins were examined on SDS-polyacrylamide gels, both in the absence and presence of β-mercaptoethanol. Conglutin α had three or four types of noncovalently linked subunits with molecular weights in the range 55,000 to 89,000, each of which was thought to contain a disulphide bonded moiety with a molecular weight near 20,000. Conglutin γ was found to contain disulphide bonded chains of molecular weight 17,000 and 20,000, whereas the four major subunits of conglutin β which were within the molecular weight range from 30,000 to 60,000 were not covalently linked together. Conglutin β was not homogeneous since it could be separated by fractional precipitation with ammonium sulphate into a series of fractions which differed markedly in the proportion of the subunit types they contained.

The storage proteins of *Vicia* spp. and *P. sativum* have been extensively studied. These proteins were originally named by Osborne (74) legumin (sedimentation coefficient 11S) and vicilin ($\sim$ 7S). These terms are sometimes indiscriminately used, and the wisdom of their usage has been questioned (58). The immunological relationships, as demonstrated originally by Kloz & Turková (55), show clearly that these terms should be reserved for genera in the Fabeae and Trifolieae.

Legumin and vicilin have the properties of classical globulins: solubility in dilute salt solutions and isolubility in water. It is customary to make use of these properties to separate metabolic and storage protein, essentially as described by Danielsson (24). Proteins are extracted from seeds or cotyledons with buffered sodium chloride and precipitated with ammonium sulphate (70% saturation). Removal of salt by dialysis results in the precipitation of protein, which is assumed to be only legumin and vicilin. Then vicilin and legumin are separated by the insolubility of legumin at its isoelectric point, pH 4.7. Further cycles of isoelectric precipitation and dialysis may be used (3, 4, 103).

We have checked the effectiveness of this procedure using two antisera: one against total protein extracted from pea cotyledons, the other against purified pea storage proteins. When total salt-extracted protein is dialyzed, the precipitate contains legumin, vicilin, and nonstorage proteins. When legumin was subjected to several cycles of isoelectric precipitation, vicilin remained associated with it. Blagrove & Gillespie (11) also found, with lupin storage proteins, that conglutin α prepared by isoelectric precipitation contained conglutin β and could not be freed from the β component by repeated isoelectric precipitation. This implies that this type of purification has serious and incalculable quantitative limitations. It seems only too likely that when globulins, particularly if present in large amounts, precipitate out on dialysis, considerable amounts of other protein species will be trapped in the precipitate. It is customary, of course, to check this type of fractionation by electrophoresis on acrylamide gel. In the past, this technique has not given the desired degree of resolution, either with undissociated or dissociated proteins. In nondissociating gels the major bands are usually diffuse, and in dissociating gels many minor bands whose relevance cannot be evaluated are present. The author believes that when proteins are prepared by this procedure, data on subunit composition should be treated with caution. The Danielsson method may be useful in a qualitative fashion, but it cannot be used quantitatively, and this will be especially so if the investigator is following the fate of labeled amino acids being incorporated into protein; contamination of the storage protein by a small amount of metabolic protein with high specific activity would invalidate the experiment.

Bailey & Boulter (3) have characterized *V. faba* legumin and vicilin (4) prepared by a modification of the Danielsson (24) procedure. This legumin preparation moved as a single band on polyacrylamide gel electrophoresis. The amino acid composition showed the presence of large amounts of aspartic and glutamic acids and 1.26 g half-cystine per 100 g protein. Three N-terminal residues (leucine, glycine, and threonine) were found, and three subunit species (mol wt 56,000, 42,000, and 23,000) from the reduced protein were separated on SDS gels. The authors note that traces of other polymers were present "presumably due to disulphide interchange." The legumin substructure was also examined by a number of peptide mapping techniques. It was concluded that legumin contained about 140 sequences with both arginine and lysine, 14 cystine-containing sequences, and 3 methionine-containing sequences. Taken in conjunction with the amino acid analyses, the mapping experiments suggested a total sequence weight of 115,000, which was in agreement with 121,000 for the total of the 3 subunits. The ratio of the subunits displayed on SDS gels was estimated by dye-binding experiments and from the distribution of label from in vivo ^{14}C-labeled legumin. Clearly difficulty was experienced in obtaining legumin of high specific activity since that used had only 30 dpm per μg. It was suggested that the polypeptide chains of molecular weights 56,000, 42,000, and 23,000 were present in molar ratios of 1:3:6. The proposed legumin model with 10 chains would have a molecular weight of 320,000. The purified protein contained 0.1% neutral sugars.

Vicilin (4) from *V. faba* moved as a broad band on acrylamide gel, and the reduced protein gave four bands (mol wt 66,000, 60,000, 56,000, and 36,000) on SDS acrylamide gels, with traces of other polypeptides. Vicilin contained four N-terminal amino acids (leucine, threonine, serine, and lysine) in major yields, together with others in lower yields. Vicilin contained 0.5% neutral sugar.

By repeated isoelectric precipitations, Wright & Boulter (103) extracted vicilin and legumin from seeds at different stages of development. The subunit patterns of the reduced proteins were then examined by SDS gel electrophoresis. It was shown that the subunit structure of vicilin changed during development, whereas that of legumin did not. Vicilin, therefore, was not considered a single protein. During seed development, vicilin was formed prior to legumin, but the rate of synthesis of legumin was faster and in the mature seed it predominated.

Basha (7) examined legumin and vicilin from *P. sativum.* The proteins were isolated by a modification of the Danielsson (24) procedure. Legumin contained carbohydrate, 1.25% neutral sugars (glucose and mannose), and 0.1% glucosamine. This legumin preparation, with SDS and dithiothreitol, gave three major and two minor bands on acrylamide gel electrophoresis. Vicilin also contained carbohydrate (0.3% mannose and 0.2% glucosamine), and the reduced protein gave five major bands on SDS gels. When legumin and vicilin were isolated from seeds at various stages of development, both proteins showed changes in subunit structure during development.

Using *P. sativum* legumin and vicilin isolated by chromatography on DEAE-cellulose, Grant & Lawrence (38) reported four subunits for vicilin and six for legumin detectable by electrophoresis on SDS gels. All subunit fractions had two or more different N-terminal amino acids.

As part of an extensive study on purified *V. sativa* legumin, Vaintraub (95) reported that this protein was dissociated with 4M urea or guanidinium chloride to give three different subunits, A, B, and C, whose N-terminal amino acids were glycine, leucine, and threonine respectively. These subunits, each containing cysteine, were separated and purified by chromatography on DEAE-cellulose. Molecular weight determinations were made by equilibrium sedimentation and calculated from their amino acid composition. The values obtained were A: 24,300 ± 200; B: 37,000 ± 400; C: 32,600 ± 200. From earlier estimates of the molecular weight of legumin (360,000) and its dissociation into two apparently identical units, a quaternary structure of $A_6B_4C_2$ was proposed.

Purified legumin from *V. faba* was also examined by Millerd et al (60). Protein was extracted from seeds at an advanced stage of development but prior to the onset of dehydration. After fractionation with ammonium sulphate, protein was centrifuged through a sucrose gradient. Protein from the dense portion of the gradient was collected and concentrated. On acrylamide gel electrophoresis, this protein moved as a single band and was considered to be legumin on the basis of the amount present and its high molecular weight. Three subunit species from the reduced protein were separated on SDS acrylamide gels, and the molecular weights were similar to those reported by Bailey & Boulter (3) except that the molecular weight of the smallest subunit was estimated to be 18,000–20,000. Also in these experiments there were, in trace amounts, additional bands on the gels. In contrast with Bailey & Boulter (3), no cysteine was detected in this legumin preparation, indicating that the subunits were not held together by S–S bridges. The legumin preparation was used as an antigen in rabbits and then checked by immunodiffusion against the resulting antiserum. This exacting criterion showed the preparation to contain a trace of vicilin.

Using the isolation procedure described above, legumin from both *V. faba* and *P. sativum* was used as antigen. Using the increased resolving power of immunoelectrophoresis, a "third protein" of mobility intermediate between legumin and vicilin was demonstrated. This protein is present in isolated protein bodies. When peas develop under carefully controlled conditions (61), vicilin is the first storage protein detected in developing cotyledons, then legumin, and then the third protein.

Having discussed some examples of purified legume storage proteins, can we conclude that they consist of only a few species and that the proteins which have been isolated are single proteins? There is considerable evidence that much of the 11S component is a single protein. The 7–8S fraction appears heterogeneous—in some cases two proteins have been demonstrated, in others three or more. In a number of cases the subunits of the 7S proteins are not held together by covalent bonds. There can also be a significant amount of 2S storage protein.

Have any of these proteins been isolated as a single species? The answer depends on the criteria of purity used. Purity can be defined by behavior in the analytical ultracentrifuge, by immunochemical standards, by the display of a single band of undissociated protein following electrophoresis, or by the pattern of bands of the dissociated protein on acrylamide gels. There is increasing evidence for microheterogeneity. All protein preparations which have been described have shown minor or trace bands on SDS gels. At least in peas these bands do not appear to result from proteolysis since they are unaffected by prolonged incubation at 30°, or by using inhibitors of proteolysis during isolation (Higgins, personal communication). As will be discussed later, these minor bands may prove highly significant. Many storage proteins are glycoproteins, and apparent microheterogeneity may be associated with the amount or composition of the carbohydrate moiety (89).

Is it important to isolate protein species which will meet the most demanding criteria of purity? If we are to learn about control of storage protein synthesis, a fundamental question is the number of genes coding for a particular protein. Are there single copies or multiple copies of closely related genes? Detailing the number of genes for a single protein species has been achieved for hemoglobin (10, 41), for silk fibroin (93), and for keratin (51). It was shown that there was little or no reiteration of globin genes in the duck (10) or mouse (41) and an estimated one to three fibroin genes per haploid complement of DNA in the silk moth (93). In the case of feather keratin, it was estimated that there were 100–300 keratin genes in the chick genome (51). The equivalent information cannot be established until the gene product, the storage protein, has been rigorously purified and defined.

A consideration of the protein isolation procedures already discussed suggests some precautions or modifications that may be beneficial. A major difficulty in purification of storage proteins has been the lack of any assay system to check for purity during the isolation procedures. Essentially these proteins have been isolated on the basis of solubility, molecular size, and the large amounts present. Blagrove & Gillespie (11) reported electrophoresis on cellulose acetate membranes as a significant improvement in terms of speed, resolution, and potential ease of quantitation of lupin seed proteins. A very sensitive assay can be devised using the antigen-antibody technique. Purified storage proteins from soybean (21), peanuts (26), peas (61, 63), and broadbean (60) have been used as antigens.

In experiments designed to isolate pure proteins, it may be advisable to use as starting material seeds harvested prior to the onset of dehydration. We do not know if any protein breakdown occurs during seed maturation.

When mature seeds are used as the starting material, they are frequently allowed to imbibe for quite long periods (24 hr) at room temperature. There may be some protein hydrolysis during this period. Seeds may be caused to imbibe rapidly by scarifying either mechanically or by hand. Protein can be rapidly extracted from ground meal, with prior hexane treatment for seeds with a high oil content. Possible contamination by metabolic proteins can be reduced by using isolated protein bodies. Since protein bodies are associated with cell membranes, membrane proteins are likely to be extracted with the storage proteins. Possible complications caused by the association of protease activity with protein bodies from ungerminated seeds, reviewed by Ryan (86), should be remembered.

QUANTITATIVE ASPECTS OF STORAGE PROTEIN ACCUMULATION DURING SEED DEVELOPMENT

In legume seeds the cotyledons form the bulk of the seed and synthesize most of the protein. In the developing cotyledon there are two phases of growth, an initial one of intensive cell division followed by a longer period of growth by cell expansion. During expansion growth, 95 percent of the protein is synthesized (16, 60, 61, 91).

In describing the accumulation of storage proteins during seed development, there are three aspects which one would wish to describe quantitatively. 1. When does storage protein synthesis begin? 2. What proportion of total seed protein is storage protein? 3. What is the quantitative contribution of a particular species of storage protein?

When the sensitive antigen-antibody reaction is used as a method for detection, it is apparent that synthesis of storage protein begins early in seed development (60, 61). With *P. sativum* grown under controlled conditions, immunoelectrophoresis was used to determine the specific stage in seed development when vicilin and legumin first appeared (61). The growing conditions were such that seed development was rapid; for example, 8 days from flower opening, cotyledons occupied approximately 75% of the volume of the seed coat, and by 9 days there was no liquid endosperm. It was shown that the storage proteins were detectable in the cotyledons at definite and reproducible times. Vicilin was first detected 9 days from flowering when about 60% of the final cell complement was present; legumin was detected one day later when 80% of the cells were present.

Hall et al (39), using acrylamide gel electrophoresis, studied the changing protein profile during seed development in *Ph. vulgaris* and showed a continuously increasing proportion of storage proteins.

Using methods developed to characterize soybean storage proteins (43), Hill & Breidenbach (44) have described quantitative and qualitative changes during seed development. The 2.2S sedimenting proteins predominated at very early stages of development and decreased proportionately throughout maturation. The 7.5S and 11.8S components appeared to be synthesized later in maturity and in larger amounts than the 2.2S proteins. Electrophoretic studies revealed temporal differences in the accumulation of the three components of the 7.5S fraction. The 11.8S sedimenting fraction appeared throughout seed development as a homogeneous protein which accumulated in the seed with a time course similar to that of the total 7.5S protein fraction.

The patterns of protein accumulation in developing seeds of *P. sativum* and *V. faba* have received considerable attention. With one exception (60), storage proteins have been estimated by procedures based on those of Danielsson (24). It is frequently difficult to integrate much of the published data, mainly because of the great variation in growing conditions. There is a steady increase in protein content during cotyledon development and this continues longer than does starch accumulation (33, 61, 91).

Basha (7) reported a detailed investigation extending the work by Beevers & Poulson (8) on storage protein synthesis in developing cotyledons of *P. sativum* grown under controlled conditions. Globulins were synthesized up to 27 days from flowering, while albumin accumulation stopped after 24 days. At the stage of maximum protein content, the globulin fraction accounted for about 85% of the total protein and the ratio of legumin to vicilin was about 3:1. Comparisons were made (7) of the patterns of subunits on SDS gels from reduced legumin and vicilin isolated at various stages of development. Changes in subunit composition of both legumin and vicilin were observed, and it was suggested that legumin and vicilin were not synthesized as single units, but on a subunit basis. The carbohydrate content of the storage proteins was shown to change during development; labeling studies with ^{14}C-glucosamine indicated that in legumin most of the label was incorporated into the smallest subunit, while in vicilin incorporation occurred into all subunits.

Wright & Boulter (103) followed the pattern of vicilin and legumin synthesis during cotyledon development in *V. faba*. Globulin was isolated from cotyledons at various stages of development and the amount of vicilin and legumin determined. Vicilin was formed in the developing seed before legumin was detected. Legumin was synthesized at a faster rate than vicilin, and in the mature seed the ratio of legumin to vicilin was about 4:1 on a weight basis. However, as will be discussed later, the ratio of storage proteins is not a constant character even in a single cultivar. Changes in the ratios of vicilin and legumin subunits were examined by quantitating stained subunits on SDS gels. This procedure showed that the ratio of subunits one to another remained reasonably constant in legumin extracted at various stages of development. In contrast, the ratios of vicilin subunits changed during seed development, indicating, as already discussed, that vicilin was not a single protein species.

The pattern of legumin synthesis during the growth of *V. faba* cotyledons has been determined (60) by microcomplement fixation. This method provided a sensitive, specific, and quantitative assay for legumin in the presence of vicilin. A small amount of legumin was present when only 20% of the total complement of parenchymatous cells had been formed. However, when cotyledons were about 10 mm long and cell division was essentially complete, there was a sharp increase in the rate of legumin accumulation, and legumin finally accounted for about 30% of the extractable protein. Checks on the assay used indicated that a single molecular species was being measured.

The overall pattern of total protein accumulation during seed development is clear. In experiments designed to establish at what point during development storage proteins may first be detected, the answer will depend on the sensitivity of the method used. The difficulty encountered in making a clear picture of the contribution of various species of storage proteins is caused by our incomplete knowledge, as already discussed, of the number of such species involved in any particular system.

NUCLEIC ACID PROFILES DURING COTYLEDON DEVELOPMENT

RNA

The RNA content of cotyledons increases steadily during development, plateauing shortly before the maximum protein content has been reached (61, 64, 81, 88, 91, 97). This RNA is predominantly ribosomal-RNA (rRNA) (61, 81). When nucleic acids were isolated from cotyledons exposed in vivo to ^{32}P-inorganic phosphate (81) or from detached cotyledons supplied with radioactive nucleosides (64), most of the radioactivity was incorporated into rRNA, both 25S and 18S, and into a heterogenous peak comprising 5S and tRNA.

Poulson & Beevers (81) found that in developing pea cotyledons, the proportion of ribosomes present as polysomes was high (about 80%) and approximately constant up to the onset of seed maturation. During the dehydration phase, the monosome component increased.

Large amounts of rough endoplasmic reticulum are formed during cotyledon development (16, 73). Payne & Boulter's experiments (78) indicate that at the time of maximum storage protein accumulation, only membrane-bound ribosomes are synthesized.

DNA

During expansion growth in cotyledons, not only RNA but also the nuclear DNA of parenchymatous cells increases (61, 64, 88, 91). This increase beyond the diploid (2c) level begins early in some cells while other cotyledon cells are still dividing (61, 91).

It has been suggested (88, 91) that the synthesis of "extra" DNA (i.e. DNA beyond the 2c level) precedes the intensive synthesis of RNA, protein and starch. However, using pea seeds growing under controlled environmental conditions, it was observed (61) that the accumulation of chlorophyll, starch, protein, DNA, and RNA was roughly parallel. In V. faba also DNA and RNA increased simultaneously (64).

DNA isolated from V. faba cotyledons in the cell division phase of growth has been compared with DNA isolated from cotyledons undergoing expansion growth (64). A number of criteria, including buoyant density and reassociation kinetics, indicated that the DNA increase involved replication of the whole genome (endoreduplication).

The overall patterns of RNA and DNA during the period of intensive protein synthesis are clear. There is an elaboration of endoplasmic reticulum and most ribosomes are present as polysomes. In P. sativum, the nuclear DNA is endoreduplicated to a maximum of 64c. There is disagreement, however, whether DNA synthesis precedes (88, 91) or is coincident (61, 64) with RNA and storage protein synthesis.

There is a striking similarity of the cotyledon system with certain insect systems. There are numerous examples of terminally differentiated polyploid or polytene cells which synthesize large amounts of tissue-specific proteins. Such systems which have been extensively studied include the synthesis of cocoonase in the galea of the silk moth (49), the synthesis of fibroin by the posterior silk gland of *Bombyx mori* (93) and the synthesis of calliphorin by larval fat bodies of *Calliphora* (67, 82, 94).

BIOCHEMISTRY OF PROTEIN SYNTHESIS IN DEVELOPING SEEDS

Developing seeds have not been widely chosen as experimental material for studies of the detailed biochemistry of general protein synthesis. There are many metabolically active plant systems which are easier to obtain. However, it has been established that the basic reactions are the same as in all protein synthesizing systems so far examined.

The overall process of storage protein synthesis in the developing seed can be divided experimentally into three parts: transcription, translation, and deposition of storage proteins.

1. Transcription: in this process, in addition to the transcription of genes involved in basic cell metabolism, there occurs the transcription of the genes specifying the storage proteins. The genes are transcribed by RNA polymerases, and the mRNAs are transported from the nucleus to the cytoplasm where they become associated with ribosomes.

2. Translation: the process in which the information for the amino acid sequences is translated and proteins are formed. This process involves many steps from amino acid activation, the association of aminoacyl-tRNAs with ribosomes and numerous protein factors. The generalized system for protein synthesis on 70S (microbial) ribosomes has been diagrammed by Boulter (13) and by Zalik & Jones (106).

3. Deposition of storage proteins. After synthesis, storage proteins accumulate in protein bodies. These discrete deposits of protein can be isolated from mature seeds or from seeds during the later phases of development, as was first shown for peanuts (1, 29) and peas (96).

Transcription

Despite the exciting prospects for studying specific gene expression during storage protein synthesis (12), events at the level of DNA have not been extensively examined. Mehta & Spencer (personal communication) have solubilized the RNA polymerases of developing pea cotyledons and have resolved them into three distinct species by chromatography on DEAE-Sephadex. The three polymerases have distinct characteristics with regard to divalent cation requirements and reaction to inhibitors; the major component is extremely sensitive to α-amanitin.

The role of the endoreduplicated DNA in cotyledon storage cells is not known. It has been suggested (88) that in peas it is directly related to storage protein synthesis by virtue of a gene dosage effect, i.e. it is a means of making available extra copies of those cistrons involved in the synthesis of storage proteins. However, the concept of a gene dosage effect, in a major quantitative sense, is not supported in experiments described by Millerd & Spencer (61). Nuclei were isolated from pea cotyledons at various developmental stages and their endogenous RNA polymerase activity determined. It was shown that the RNA polymerase activity per unit of DNA declined sharply from the time endoreduplication commenced. This reduced activity, which was not due to RNase or to loss of enzyme, could have been caused by a major portion of the "extra" DNA being repressed. However, the template activity of isolated chromatin, assayed in the presence of nonlimiting amounts of *E. coli* RNA polymerase, was fairly constant throughout the whole period of endoreduplication. These results suggested that, although there was a proportional increase in available template during endoreduplication, there was not a proportional increase in transcriptive activity, i.e. RNA polymerase was probably limiting. In addition, in a period of cotyledon expansion growth, when there was a sixfold increase in DNA per cell, the rate of RNA increase remained constant. This implied that there was no gene dosage effect for rRNA or tRNA which accounted for most of the tissue RNA.

Translation

CELL-FREE PROTEIN SYNTHESIS AND THE CORRELATION WITH IN VIVO ACTIVITY   It has been suggested (65, 66) that protein bodies were cytoplasmic organelles possessing a separate protein synthetic mechanism independent of the characteristic cytoplasmic system. This proposal was based on experiments involving the incorporation of labeled amino acids into protein. Further investigations (100, 101) under sterile conditions have not supported this concept.

Isolated systems from developing seeds have been shown to incorporate labeled amino acids into protein (material insoluble in trichloracetic acid). In such systems there are two gross fractions: the soluble fraction which contains all the components necessary to form the aminoacyl-tRNAs and the particulate fraction containing the ribosomes, either free or membrane bound, with which are associated the mRNA molecules (natural or synthetic) and the enzymes which catalyze the synthesis of the polypeptide chain from the aminoacyl-tRNAs. As with other plant systems (13), such systems are characteristically dependent on the presence of ATP, GTP, K^+, Mg^{++}, and an ATP generating system. Amino acids, usually one radioactively labeled and 19 unlabeled, are supplied. In even the most active of such systems, the amount of protein synthesized is small compared with that in vivo. With one exception (12), in systems from developing seeds, the synthesis of specific proteins has not been studied.

Boulter and associates (79, 80) have characterized the cell-free protein synthesizing system from developing *V. faba* seeds. They isolated from cotyledons a soluble fraction and a ribosomal fraction containing both free and membrane-bound ribosomes. The system could be fractionated into its major components, the ribosomes, the tRNAs and synthetases, and recombined to give an active preparation. There was some activity due to endogenous mRNA, but a marked stimulation followed the addition of the synthetic messenger, poly U (79, 80).

The cell-free system from *V. faba* was isolated from cotyledons at intervals during development (80). With endogenous mRNA, maximal activity per milligram of RNA was observed in preparations from cotyledons 60 days from flowering, coinciding with maximal protein accumulation in vivo. This peak of activity was not simply due to increased availability of mRNA, since in the presence of poly U, the 60-day system was still the most active. It was also shown (79) in such 60-day preparations using endogenous mRNA that only membrane-bound ribosomes were involved in protein synthesis, implying that storage proteins were made on bound ribosomes.

Cell-free protein synthesizing systems from developing *P. sativum* seeds have been studied extensively by Beevers et al (8, 81, 98, 99). Beevers & Poulson (8) examined cell-free protein synthesizing systems isolated from cotyledons at intervals during development. They showed that the amino acid incorporating capacity of the ribosomal preparations was related to polysomic content. Only with the appearance of a high percentage of monosomes at the onset of seed dehydration did the preparation show a marked response to poly U, suggesting an in vivo limitation of mRNA. Supernatant fractions from cotyledons at this developmental stage also showed a reduced capacity to support in vitro amino acid incorporation.

Beevers & Poulson (8) fed [^{32}P]-orthophosphate by injection in vivo to cotyledons at three developmental stages: early, at maximum protein synthesis, and during seed maturation. After 90 min, cotyledons were harvested and the ribosomes were isolated and separated on sucrose density gradients into three fractions, one containing monosomes, one "light" polysomes, and one "heavy" polysomes. The low molecular weight RNA (putative mRNA) associated with these fractions was then resolved by acrylamide gel electrophoresis and the distribution of radioactivity determined. It was shown that certain RNA species were associated with specific ribosome fractions and that there was a consistent association of specific peaks of radioactivity with specific developmental ages. However, there was no preponderance of particular RNA species at the height of storage protein synthesis.

The biochemical mechanisms involved in the association of mRNA with ribosomes and the initiation and elongation of the polypeptide chains are extremely complex (106). Recently, in vitro protein synthesizing systems from developing cotyledons have been used with considerable success to detail the events involved in general protein synthesis (99, 100, 104, 105).

It is apparent from the studies discussed above that active in vitro protein synthesizing systems can be obtained from cotyledons synthesizing storage proteins. Intensive protein synthesis in the cotyledons is associated with marked proliferation of the rough endoplasmic reticulum, and membrane-bound ribosomes have been implicated in the synthesis of storage proteins (78, 79). However, there is as yet no data concerned specifically with synthesis of storage proteins. The isolation of mRNA by complexing through its polyadenylate fragment to poly U cellulose or poly U Sepharose is now feasible. In addition, the isolation and quantitation, by specific immunochemical precipitation, of polysomes synthesizing a particular protein is practicable. One is hopeful that advances in knowledge will be rapid.

Deposition of Storage Proteins

There are many detailed studies, using light and electron microscopy, which describe the formation of protein bodies during development of seeds, both legumes and non-legumes, e.g. *P. sativum* (6, 87), *Ph. vulgaris* (73), *Glycine max* (9), *V. faba* (16), *Ph. lunatus* (53), *Arachis hypogaea* (28), *Capsella bursa-pastoris* (28), *Gossypium hirsutum* (28, 31), *Zea mays* (52), *Ricinus communis* (92), *Sinapis alba* (85).

Two generalizations can be made: First, at the time of storage protein synthesis, there is a marked proliferation of the rough endoplasmic reticulum and, as already discussed, this may be the site of storage protein synthesis (5, 78). Second, there are many observations that, at an early stage in seed development, the large vacuoles in cells appear to be replaced by small vacuoles in which storage proteins accumulate. It has been suggested that these small vacuoles arise by dilation of the endoplasmic reticulum, but this has not been proved. How storage proteins enter such vacuoles is also unknown; no connections with the endoplasmic reticulum have been observed (16, 73, 85). That the protein in these vacuoles is storage protein was demonstrated in situ by Graham & Gunning (37) in sections of *V. faba* cotyledons using fluorescent antibodies to legumin and vicilin. Graham & Gunning (37) also found protein bodies which did not react with the fluorescent antibodies.

Although there is evidence that protein can accumulate in vesicles of the endoplasmic reticulum (6, 31), that protein granules can originate as cisternal accumulations of protein (52), and that dictyosomes (52) and Golgi vesicles (87) are associated with protein accumulation, there is no evidence that such protein is storage protein.

Since storage proteins are sequestered in vacuoles, they are separated from the cytoplasm, and the cells synthesizing them can be considered as secreting protein. It has been suggested that the process is similar to that observed in animal cells secreting protein. In certain secretory processes, e.g. synthesis and secretion of amylase by rabbit parotid gland, the sequence of events has been clearly established (18). Using biochemical analyses and autoradiography coupled with light and electron microscopy, Palade and his colleagues showed there were three phases in the formation of zymogen granules: (*a*) synthesis of the proteins on rough endoplasmic reticulum; (*b*) immediate transfer and segregation of these proteins into the cisternae of the endoplasmic reticulum; (*c*) intracellular transport through the cavities of the endoplasmic reticulum to the Golgi complex for packaging into zymogen granules.

Using a similar approach, Bailey et al (5) pulse-labeled slices of *V. faba* cotyledons with ^{3}H-leucine and followed the label by electron microscopic autoradiography. In such an experiment, a brief exposure (10 minutes in these experiments) to labeled precursor is followed by washing out of the unincorporated label by transferring the tissue to a "chase" medium. Difficulties were encountered because at the end of the labeling period a sharp break in incorporation into protein was not obtained. This indicated that the chase was not totally effective. Grain counts on sectioned material showed that, following a 10-min chase, about 67% of the silver grains were associated with the endoplasmic reticulum and 10% with the protein bodies, the remainder being associated with other organelles or with the cytoplasm. After an 80-min chase, 80% of the grains were associated with protein bodies and 20% with the endoplasmic reticulum. These results showed that there was movement of labeled material from the endoplasmic reticulum to protein bodies, but there was no indication at the ultrastructural level of the pathway involved.

Dieckert & Dieckert (28), using light and electron microscopy, have made extensive studies of protein deposition in seeds of shepherd's purse, peanut, and cotton. In their electron micrographs, protein was defined only on the basis of electron density; there was no identification of storage proteins, e.g. by the use of ferritin-labeled antibodies. However, from these studies, Dieckert & Dieckert (28) proposed that storage proteins are synthesized on the rough endoplasmic reticulum, concentrated into protein droplets in dictyosomes, and transported to the neighborhood of a protein-storage vacuole in membrane-bound vesicles. These vesicles then empty into the vacuole by a process of membrane fusion. Apart from the inherent difficulty of describing a sequence of kinetic events from a series of static pictures (71), the fact that the presence of storage protein was not specifically established would suggest there is no convincing data to support this theory.

It is apparent that there are serious gaps in our understanding of the mechanisms involved in storage protein deposition.

NUCLEIC ACID AND PROTEIN SYNTHESIS IN CULTURED COTYLEDONS

In pea plants grown under conditions of controlled environment, the synthesis of vicilin and legumin is initiated at precise and predictable times (61). It is thus possible to obtain experimental material prior to, during, and after these events. To study biochemical mechanisms involved in controlling this synthesis of storage proteins, a system more manageable than seed development on the whole plant, is desirable. It has been found that when immature pea pods, in which the seeds contained no immunoelectrophoretically detectable storage proteins, were detached and transferred to liquid culture, seeds continued to grow and synthesized both vicilin and legumin (62, 63).

It was also shown (63) that detached immature pea cotyledons with the embryonic axes removed continued to grow for several days in liquid medium containing a nitrogen source and to synthesize considerable amounts of chlorophyll, starch, DNA, RNA, and protein. Protein was synthesized from endogenous amino acids, but supplying exogenous amino acids resulted in more protein synthesis. Asparagine supplied alone could serve as a nitrogen source, showing that the cotyledons had considerable capacity for synthesis and interconversion of amino acids, as indicated by the data of Lewis & Pate (57).

When pea cotyledons were detached one day prior to the appearance of vicilin and then cultured, vicilin was readily and routinely detected one day later (63). It is clear that the genetic information for vicilin was expressed in such cultured cotyledons. A similar switching on of legumin synthesis in cultured cotyledons was not observed even in the presence of a range of growth regulators. However, cotyledons which had commenced legumin synthesis before removal from the plant continued to synthesize legumin in culture.

Since cultured pea cotyledons did initiate vicilin synthesis and continued to synthesize legumin, provided this program was set before removal from the plant, this system should be useful for studying some aspects of the control of protein synthesis.

CONSTRAINTS AND CONTROLS

It has been shown by a number of criteria that the storage proteins of many legumes are closely related. Genera in the subfamily Faboideae have been particularly studied (reviewed by Kloz 54, Boulter & Derbyshire 14).

Immunological relationships of *V. faba* legumin and vicilin with similar proteins in taxa of the Fabeae and Trifolieae have been shown (55). These observations have been extended (29a), using more stringent analysis provided by the Osserman (75) modification of immunoelectrophoresis. It was shown that proteins closely related to *V. faba* legumin were widespread. Thus there has been marked genetic conservation of the immunochemically active groups in legumin. Only *P. sativum* contained a protein immunochemically identical with vicilin of *V. faba;* the equivalent proteins of all other genera tested were immunochemically different. There is conservation of the antigenic determinants of vicilin, but not to the same degree as observed with legumin.

Boulter et al (15, 47, 48), using disc electrophoresis of undissociated and urea-dissociated storage proteins, together with fingerprints of tryptic digests, have shown similar close relationships among genera of the Fabeae, and in agreement with experiments using immunological criteria (55, 56) have shown very different patterns in members of the tribe Phaseoleae.

Although there is strong evidence for genetic conservation at the level of the whole molecule, there is also evidence for subunit polymorphism (45). The variation of the proportion of particular subunits in legumin (7) and vicilin (7, 103) has already been mentioned.

Gillespie & Blagrove (36), using electrophoresis on cellulose acetate, examined 18 cultivars of *L. angustifolius* and showed there was considerable variation in the proportions of conglutins α, β and γ; conglutin γ showed a constant mobility, whereas β, and particularly α, varied. When the subunits of the unreduced proteins were examined on SDS gels, the patterns were closely related but showed some variation in the proportion of subunits. Using the same technique, similar variations were observed with *P. sativum* (Thomson, personal communication).

Randall, Blagrove & Gillespie (personal communication) have recently shown that nutrition can exert a dramatic effect on both the amount of particular storage proteins synthesized and on the subunit composition. When *L. angustifolius* was grown under conditions of sulfur deficiency there was an almost total absence of conglutin α and γ. The subunit composition of conglutin β was dramatically altered; bands which were normally minor components were now present in major amounts.

It is apparent then that there is considerable potential for manipulation of legume storage proteins both qualitatively and quantitatively. Since the discovery (59) of the dramatic effect of the *opaque*-2 gene on the nutritional value of maize protein, the potential for improvement by changing the relative proportions of storage proteins has aroused much interest (69, 70).

The processes involved in the delivery of materials to developing seeds are potential control points. The role of transfer cells in the movement of assimilates to developing seeds has been reviewed by Pate & Gunning (77).

Pate & Flinn (76) have examined in detail carbon and nitrogen transfer to ripening seeds of *P. arvense*. Plants were pulse-labeled with ^{15}N-nitrate or ^{14}C-carbon dioxide either early or late in the life cycle and the subsequent distribution of the isotope studied. Carbon assimilated early in the life cycle had virtually no direct relevance to fruit nutrition; that fixed during reproductive development furnished the seed's requirements for this element (76). At the first blossom node, approximately two-thirds of the carbon required for the seeds was provided by pod, leaflets, and stipules at that position (34).

Assimilatory activity before flowering provided approximately one-fifth of the seed's requirements for nitrogen (76). When *P. sativum* plants were pulse-labeled with ^{15}N-nitrate, -glutamine, or -glutamic acid (57) the time course of labeling indicated that nitrogen for seeds was channeled through the vegetative organs, the stems and leaves being the main donors to the seeds.

There are two events in storage protein synthesis which are of particular interest, the initiation of storage protein synthesis and its termination prior to or at the onset of seed maturation. At present we know nothing of the mechanisms involved. Insight into the "switching off" process could be particularly valuable; conceivably it could suggest means of extending the period of protein synthesis.

It has been suggested (27) that the maternal and paternal loci for storage proteins may be differentially activated. While it is possible that switch genes of the kind postulated by Davies (27) may occur in *Pisum,* the proposal of a control mechanism appears premature. At present the globulin subunits have not been adequately assigned to parent molecules.

There have been correlative studies of the levels of growth hormones and seed development [e.g. gibberellin levels (42, 72, 90), cytokinin content (17)] but the significance, if any, in protein biosynthesis is unknown. The involvement of abscisic acid in embryogenesis and germination is discussed in another chapter in this volume (30).

At the molecular level, with the known complexity in the overall process of transcription and translation, the potential opportunities for controls and limitations are indeed awe-inspiring. At present we know nothing of the lifetimes of mRNAs for storage proteins; if these were short, it would suggest that regulation at the transcriptional level could be important. The variability of subunit composition of storage proteins, especially in relation to environmental influences such as nutrition, suggests an involvement of translational control.

CONCLUSION

A realistic assessment of the prospects of increasing the protein quality of any crop plant depends on an understanding of the genetic control of protein structure (69). Even in a relatively simple and much studied system like *Vicia* spp. or *Pisum* spp., the number of storage protein species and their exact subunit composition cannot yet be defined with confidence. Without this information, an estimate cannot be made of the number of structural genes coding for storage proteins.

Immunochemical criteria indicate that the number of storage protein species in *Vicia* or *Pisum* is small. However, there is strong evidence for heterogeneity of subunits. These observations can be reconciled by proposing, for example for vicilin, that there is a family of repeated genes. The resulting gene products are sufficiently similar to react immunochemically in identical fashion although the component subunits may differ in composition. Present knowledge is inadequate to test the validity of this proposal.

The biochemical machinery for protein synthesis in developing seeds has been well characterized. Specific knowledge is lacking concerning the synthesis of storage proteins, but the appropriate methodology is now available so that advances in this area can be expected soon. Then the even more exciting problems of control mechanisms can be approached.

ACKNOWLEDGMENT

Within the Division of Plant Industry, CSIRO, there is an interdisciplinary group interested in seed proteins. The author accepts full responsibility for the material in this review, but wishes to acknowledge that many of the ideas came from deliberations with colleagues in this group.

Literature Cited

1. Altschul, A. M., Snowden, J. E., Manchon, D. D., Dechary, J. M. 1961. *Arch. Biochem. Biophys.* 95:402-4
2. Altschul, A. M., Yatsu, L. Y., Ory, R. L., Engleman, E. M. 1966. *Ann. Rev. Plant Physiol.* 17:113-36
3. Bailey, C. J., Boulter, D. 1970. *Eur. J. Biochem.* 17:460-66
4. Bailey, C. J., Boulter, D. 1972. *Phytochemistry* 11:59-64
5. Bailey, C. J., Cobb, A., Boulter, D. 1970. *Planta* 95:103-18
6. Bain, J. M., Mercer, F. V. 1966. *Aust. J. Biol. Sci.* 19:49-67
7. Basha, S. M. M. 1974. *Protein metabolism in the cotyledons of Pisum sativum L. during seed development and germi-nation.* PhD thesis. Grad. Coll. Univ. Oklahoma, Norman. 79 pp.
8. Beevers, L., Poulson, R. 1972. *Plant Physiol.* 49:476-81
9. Bils, R. F., Howell, R. W. 1963. *Crop Sci.* 3:304-8
10. Bishop, J. O., Pemberton, R., Baglioni, C. 1972. *Nature New Biol.* 235:231-34
11. Blagrove, R. J., Gillespie, J. M. 1975. *Aust. J. Plant Physiol.* 2. In press
12. Bonner, J., Huang, R. C., Gilden, R. V. 1963. *Proc. Nat. Acad. Sci. USA* 50:893-900
13. Boulter, D. 1970. *Ann. Rev. Plant Physiol.* 21:91-114
14. Boulter, D., Derbyshire, E. 1971. See Ref. 40, 285-308

15. Boulter, D., Thurman, D. A., Derby-shire, E. 1967. *New Phytol.* 66:27–36
16. Briarty, L. G., Coult, D. A., Boulter, D. 1969. *J. Exp. Bot.* 20:358–72
17. Burrows, W. J., Carr, D. J. 1970. *Physiol. Plant.* 23:1064–70
18. Castle, J. D., Jamieson, J. D., Palade, G. E. 1972. *J. Cell Biol.* 53:290–311
19. Catsimpoolas, N. 1969. *FEBS Lett.* 4:259–61
20. Catsimpoolas, N., Campbell, T. G., Meyer, E. W. 1969. *Arch. Biochem. Biophys.* 131:577–86
21. Catsimpoolas, N., Meyer, E. W. 1968. *Arch. Biochem. Biophys.* 125:742–50
22. Catsimpoolas, N., Rogers, D. A., Circle, S. J., Meyer, E. W. 1967. *Cereal Chem.* 44:631–37
23. Catsimpoolas, N., Wang, J. 1971. *Anal. Biochem.* 44:436–44
24. Danielsson, C. E. 1949. *Biochem. J.* 44:387–400
25. Danielsson, C. E. 1950. *Acta Chem. Scand.* 4:762–71
26. Daussant, J., Neucere, N. J., Yatsu, L. Y. 1969. *Plant Physiol.* 44:471–79
27. Davies, D. R. 1973. *Nature New Biol.* 245:30–32
28. Dieckert, J. W., Dieckert, M. C. 1972. See Ref. 46, 52–85
29. Dieckert, J. W., Snowden, J. E. 1960. *Fed. Proc.* 19:126
29a. Dudman, W. F., Millerd, A. 1975. *Biochemical Systematics and Ecology.* In press
30. Dure, L. S. III 1975. *Ann. Rev. Plant Physiol.* 26:259–78
31. Engleman, E. M. 1966. *Am. J. Bot.* 53:231–37
32. Ericson, M. C., Chrispeels, M. J. 1973. *Plant Physiol.* 52:98–104
33. Flinn, A. M., Pate, J. S. 1969. *Ann. Bot.* 32:479–95
34. Flinn, A. M., Pate, J. S. 1970. *J. Exp. Bot.* 21:71–82
35. Ghetie, V., Buzilă, L. 1962. *Stud. Cercet. Biochim.* 5:541–50
36. Gillespie, J. M., Blagrove, R. J. 1975. *Aust. J. Plant Physiol.* 2. In press
37. Graham, T. A., Gunning, B. E. S. 1970. *Nature* 228:81–82
38. Grant, D. R., Lawrence, J. M. 1964. *Arch. Biochem. Biophys.* 108:552–61
39. Hall, T. C., McLeester, R. C., Bliss, F. A. 1972. *Phytochemistry* 11:647–49
40. Harborne, J. B., Boulter, D., Turner, B. L., Eds. 1971. *Chemotaxonomy of the Leguminosae.* London/New York: Academic. 612 pp.
41. Harrison, P. R. et al 1974. *J. Mol. Biol.* 84:539–54

42. Hashimoto, T., Rappaport, L. 1966. *Plant Physiol.* 41:623–28
43. Hill, J. E., Breidenbach, R. W. 1974. *Plant Physiol.* 53:742–46
44. Ibid, 747–51
45. Hynes, M. J. 1968. *Aust. J. Biol. Sci.* 21:827–29
46. Inglett, G. E., Ed. 1972. *Symposium: Seed Proteins.* Westport, Conn.: Avi. 320 pp.
47. Jackson, P., Boulter, D., Thurman, D. A. 1969. *New Phytol.* 68:25–33
48. Jackson, P., Milton, J. M., Boulter, D. 1967. *New Phytol.* 66:47–56
49. Kafatos, F. C. 1972. *Current Topics in Developmental Biology,* ed. A. A. Moscona, A. Monroy, 7:125–91. New York/ London: Academic
50. Kamra, Om P., 1971. *Z. Pflanzenzüchtg* 65:293–306
51. Kemp, D. J., Walker, I. D., Partington, G. A., Rogers, G. E. 1975. In *The Eukaryote Chromosome,* ed. R. D. Brock, W. J. Peacock. Canberra: Aust. Nat. Univ. In press
52. Khoo, U., Wolf, M. J. 1970. *Am. J. Bot.* 57:1042–50
53. Klein, S., Pollock, B. M. 1968. *Am. J. Bot.* 55:658–72
54. Kloz, J. 1971. See Ref 40, 309–65
55. Kloz, J., Turková, V. 1963. *Biol. Plant.* 5:29–40
56. Klozová, E., Kloz, J. 1972. *Biol. Plant.* 14:379–84
57. Lewis, O. A. M., Pate, J. S. 1973. *J. Exp. Bot.* 24:596–606
58. McLeester, R. C., Hall, T. C., Sun, S. M., Bliss, F. A. 1973. *Phytochemistry* 12:85–93
59. Mertz, E. T., Bates, L. S., Nelson, O. E. 1964. *Science* 145:279–80
60. Millerd, A., Simon, M., Stern, H. 1971. *Plant Physiol.* 48:419–25
61. Millerd, A., Spencer, D. 1974. *Aust. J. Plant Physiol.* 1:331–41
62. Millerd, A., Spencer, D., Dudman, W. F. 1974. In *Mechanisms of Regulation of Plant Growth,* ed. R. L. Bieleski, A. R. Ferguson, M. M. Cresswell, Bull. 12:799–803. Wellington, N. Z.: Roy. Soc. New Zealand
63. Millerd, A., Spencer, D., Dudman, W. F., Stiller, M. 1975. *Aust. J. Plant Physiol.* 2. In press
64. Millerd, A., Whitfeld, P. R. 1973. *Plant Physiol.* 51:1005–10
65. Morton, R. K., Raison, J. K. 1964. *Biochem. J.* 91:528–39
66. Morton, R. K., Raison, J. K., Smeaton, J. R. 1964. *Biochem. J.* 91:539–46

67. Munn, E. A., Feinstein, A., Greville, G. D. 1971. *Biochem. J.* 124:367–74
68. Müntz, K., Horstmann, C., Scholz, G. 1972. *Kulturpflanze* 20:277–326
69. Nelson, O. E. 1969. *Advan. Agron.* 21:171–94
70. Nelson, O. E., Burr, B. 1973. *Ann. Rev. Plant Physiol.* 24:493–518
71. O'Brien, T. P. 1972. *Bot. Rev.* 38:87–118
72. Ogawa, Y. 1963. *Plant Cell Physiol.* 4:85–94
73. Öpik, H. 1968. *J. Exp. Bot.* 19:64–76
74. Osborne, T. B., Campbell, G. F. 1896. *J. Am. Chem. Soc.* 18:583–609
75. Osserman, E. F. 1960. *J. Immunol.* 84:93–97
76. Pate, J. S., Flinn, A. M. 1973. *J. Exp. Bot.* 24:1090–99
77. Pate, J. S., Gunning, B. E. S. 1972. *Ann. Rev. Plant Physiol.* 23:173–96
78. Payne, P. I., Boulter, D. 1969. *Planta* 84:263–71
79. Payne, E. S. et al 1971. *Phytochemistry* 10:2293–98
80. Payne, E. S., Brownrigg, A., Yarwood, A., Boulter, D. 1971. *Phytochemistry* 10:2299–2303
81. Poulson, R., Beevers, L. 1973. *Biochim. Biophys. Acta* 308:381–89
82. Price, G. M. 1966. *J. Insect Physiol.* 12:731–40
83. Pusztai, A., Watt, W. B. 1970. *Biochim. Biophys. Acta* 207:413–31
84. Racusen, D., Foote, M. 1971. *Can. J. Bot.* 49:2107–11
85. Rest, J. A., Vaughan, J. G. 1972. *Planta* 105:245–62
86. Ryan, C. A. 1973. *Ann. Rev. Plant Physiol.* 24:173–96
87. Savelbergh, R., Van Parijs, R. 1971. *Arch. Int. Physiol. Biochim.* 79:1040–41
88. Scharpé, A., Van Parijs, R. 1973. *J. Exp. Bot.* 24:216–22
89. Schmid, K. 1968. *Biochemistry of Glycoproteins and Related Substances, Cystic Fibrosis II,* ed. E. Rossi, E. Stoll, 4–58. New York: Karger
90. Skene, K. G. M., Carr, D. J. 1961. *Aust. J. Biol. Sci.* 14:13–25
91. Smith, D. L. 1973. *Ann. Bot.* 37:795–804
92. Sobolev, A. M., Suvorov, V. I., Safronova, M. P., Prokof'ev, A. A. 1972. *Fiziol. Rast.* 19:1047–52
93. Suzuki, Y., Gage, L. P., Brown, D. D. 1972. *J. Mol. Biol.* 70:637–49
94. Thomson, J. A. 1973. In *The Biochemistry of Gene Expression in Higher Organisms,* ed. J. K. Pollak, J. W. Lee, 320–32. Sydney: Aust. and N. Z. Book Co.
95. Vaintraub, I. A., Nguyen-Thanh-Thien 1971. *Mol. Biol.* (Eng. transl.) 5:59–68
96. Varner, J. E., Schidlovsky, G. 1963. *Plant Physiol.* 38:139–44
97. Walbot, V. 1973. *New Phytol.* 72:479–83
98. Wells, G. N., Beevers, L. 1973. *Plant Sci. Lett.* 1:281–86
99. Wells, G. N., Beevers, L. 1974. *Biochem. J.* 139:61–69
100. Wheeler, C. T., Boulter, D. 1966. *Biochem. J.* 100:53p
101. Wilson, C. M. 1966. *Plant Physiol.* 41:325–27
102. Wolf, W. J. 1970. *J. Agr. Food Chem.* 18:969–76
103. Wright, D. J., Boulter, D. 1972. *Planta* 105:60–65
104. Yarwood, A., Boulter, D., Yarwood, J. N. 1971. *Biochem. Biophys. Res. Commun.* 44:353–61
105. Yarwood, A., Payne, E. S., Yarwood, J. N., Boulter, D. 1971. *Phytochemistry* 10:2305–11
106. Zalik, S., Jones, B. L. 1973. *Ann. Rev. Plant Physiol.* 24:47–68

Ann. Rev. Plant Physiol. 1975. 26:73–100

ASSIMILATORY NITRATE–NITRITE REDUCTION [1]

❖7583

E. J. Hewitt

Long Ashton Research Station, University of Bristol, Long Ashton, Bristol, BS18, 9AF, England

CONTENTS

[1]Abbreviations used: BV, MV,BV·, MV· [oxidized and reduced (free-radical) forms respectively of benzyl or methyl viologens]; DCMU [3-(3',4'-dichlorophenyl)-1,1-dimethylurea]; DNP [2,4-dinitrophenol].

INTRODUCTION

This subject has developed intensively in several contexts in the six years since the last excellent review in this series (26). Other specialist or general reviews appearing since that time (27, 114, 115, 140, 252) supplement the present work in different directions. I have attempted to deal comparatively with the three main subjects, since underlying principles and interesting developments tend to be shared or contrasted over a wide phylogenetic range. I have excluded dissimilatory systems and referred to bacteria in limited relevant or comparative contexts. I express regret to friends and authors for deliberate or accidental omissions, for errors or inaccuracies, and for the condensed style.

NITRATE REDUCTASE

Enzyme Systems

Well-known systems have been thoroughly reviewed (26, 27, 115, 122, 140), but flavin and iron and sulfide components, constitution, and assembly merit further attention. In *Chlorella vulgaris* (Berlin strain) no flavin presence or need was initially detected in pure NADH-dependent enzymes which contained an essential cytochrome b_{557} carrier (308, 309, 338) as in *N. crassa* (91, 92) and probably 1 atom molybdenum (306). The *C. fusca* (syn. *pyrenoidosa*) enzyme responded to added FAD (K_m 2.5 μM) (286) for activity or stability (12, 13 100, 166, 260, 356), especially after gel elution. The absorbance at 420 nm (354, 356) may indicate a heme component as in *C. vulgaris*. Direct comparison of *Chlorella* preparations appeared to confirm these differences (339), but FAD is now identified as present in pure *C. vulgaris* enzyme as well as heme.[2] Loss of activity from spinach was restored (260), or activity was not influenced (B. A. Notton, P. Brownell, and E. J. Hewitt, unpublished) by FAD after agarose elution or in sucrose gradients, but was preserved in barley by FAD when sedimented in sucrose (346). Responses for wheat embryo (315) and sugar cane (206) required flavin additions which may fulfill a stabilizing role (155, 260, 356). Nitrate reductase of *Torulopsis nitratophila* was stimulated twelvefold specifically by FAD with NAD(P)H (270). Nitrate reductase of *Aspergillus nidulans* (68) was found to be a classical NADPH-flavomolybdo-protein with 1 mol FAD in mol wt 197,000 and 1 atom molybdenum (71) but devoid of heme components which were spectral contaminants of earlier preparations, though probably a *b*-type cytochrome persisted in other preparations (199a) but was not functionally identified. In *N. crassa* the more readily dissociable FAD is a carrier from NADPH to the cytochrome b_{557} component which links the NADPH dehydrogenase moiety to the molybdoprotein component in one crystallizable complex of 228,000 mol wt from wild strain (91, 92), but molybdenum might be one or two atoms per molecule. The 6.8S (160,000 daltons) $FADH_2$-$MV^{\cdot}$ nitrate reductase moiety of *N. crassa nit3* mutant (see below) also retained the cytochrome b_{557}

[2]L. P. Solomonson, G. H. Lorimer, R. L. Hall, R. Borchers, and J. L. Bailey; submitted to *J. Biol. Chem.*

(11a), although the induced *nit1* mutant also appeared to provide this component (185, 226) which may be constitutive but repressible.

Horseradish peroxidase and other peroxidases (but not iron or other hemes) were found in comparison with nitrate reductase to reduce nitrate to nitrite rapidly when diethyldithiocarbamate and sulfite with or without persulfate were used as the electron donor system (142, 250, 251). Removal of heme eliminated both peroxidase and nitrate reductase, which were restored by reconstitution of apoprotein and crystalline heme.

Independent evidence (disputed) for molybdenum or iron components comes from radioactive isotope incorporation into purified proteins (13, 179, 232, 234, 336). Nitrate reductase of *Ankistrodesmus braunii* was dependent for formation on iron, which was concentrated in purified enzyme (359) by contrast with *C. fusca* (13). For spinach, discordant mol wt values of 230,000 (237) or 500,000 (12, 260) were obtained on agarose, and dimeric behavior may occur. Values of 19S (600,000) for wheat (10), 500,000 for *C. fusca* (12, 354), 8S (about 230,000) for barley (346), and 160,000 for maize (104) indicate the probable range in green plants. Molybdenum analyses are essential to interpret catalytic mechanisms (see 89, 115) and minimal weights. Reaction mechanisms may be either ordered (ping-pong bi-bi) (1, 74) or random (199a), suggesting differences in structure or cofactors.

Bacterial nitrate reductases of two types (see 252) are differentiated by reduction of nitrate and chlorate (A), or by nitrate only but inhibited by chlorate (B). The A-type enzymes have been purified from *Micrococcus denitrificans, M. halodenitrificans,* and *E. coli* (K12 strain) (82–84, 272). They indicate a new class of nitrate reductases foreshadowed by the *E. coli* protein terminal moiety with a mol wt of 10^6, which was reported to have 1 molybdenum and 40 iron atoms (325). The *Micrococcus*-soluble brown enzymes (mol wt 160,000) contained 1 atom molybdenum, 2 of nonheme iron, and 4 of acid labile sulfide in *M. halodenitrificans* (272), or 4 to 8 atoms iron and 10 labile sulfide in *M. denitrificans* (82), from which molybdenum (1 atom?) was easily lost. In the *E. coli* protein (83) there were 20 atoms each of nonheme iron and labile sulfide and 1 or 2 atoms molybdenum in mol wt 320,000. Flavins were absent but could function like MV· and BV· donors, whereas NAD(P)H was inactive. Light absorption increased below 600 nm with a broad shoulder around 410 nm, producing a brown color which was partly bleached by dithionite. The iron sulfide and molybdenum components were both functional in EPR spectra (84).

Nitrate reductase from *Azotobacter chroococcum* mol wt 100,000 (101) in which dependence on molybdenum was shown by tungsten antagonism also functioned only with MV· or BV· or $FMNH_2$ or $FADH_2$, but ferredoxin might have served in an enzyme from *Cl. perfringens* (50a). The *Azotobacter* enzyme was activated by cyanate and unaffected by azide, which are usually powerful competitive inhibitors (260, 270, 309, 337), but nitrate reductase of *Achromobacter fisheri* was not inhibited by cyanide with BV· (277), possibly because of reaction conditions (13). Nitrate reductase from *Thioparus denitrificans* functioned only with sulfite as a possible physiological donor or with MV·, BV·, or dithionite (1). In *Thiobacillus denitrifi-*

cans, sulfite oxidation to sulfate was linked to nitrate or oxygen as oxidants in a membrane-bound system (7), but the purified enzyme utilized only ferricyanide or cytochrome carriers (8, 9). Liver sulfite oxidase is also a molybdenum enzyme, but it cannot utilize nitrate (51–53). EPR signals for Mo^V in sulfite oxidase were reported for wheat germ and for *Thiobacillus thioparus* (163) where nitrate could not act as an oxidant by contrast with *Thioparus denitrificans.* The reversible reduction of carbon dioxide to formate in *Cl. pasteurianum* by ferredoxin–CO_2 oxidoreductase (328, 329) depended on molybdenum at 1 nM during growth and was prevented by tungsten. The system thus resembled the formate-nitrate dehydrogenase and hydrogen lyase systems of *E. coli,* but there was no selenium requirement that was absolute for *E. coli* (187, 254, 328).

Constitution and Assembly of Nitrate Reductase

COMPONENTS In fungi (92, 310, 311), algae (194, 195), and higher plants (232, 239, 291) whole nitrate reductase shows three functions: (*a*) NAD(P)H nitrate reduction; (*b*) NAD(P)H dehydrogenase with cytochrome *c,* DCPIP, ferricyanide, or tetrazolium acceptors; and (*c*) MV·, BV·, $FMNH_2$, $FADH_2$, and rarely ferredoxin (50a, 107, 108) donor-nitrate reductase (see 26, 115). Bacteria have analogous systems (see 115). Multiple genetic loci control the synthesis of the separate polypeptide components of these systems in bacteria (see 276, 297) and in fungi (17, 56–58, 225, 247, 310, 312, 314), and heterokaryon recombination readily occurs. Another NAD(P)H dehydrogenase component, defined here as 3.7–4.5S, is nitrate-inducible and related as a subunit to the complete nitrate reductase (defined here as 7.8–8S). A 10–11S dehydrogenase occurs independently (225). In algae and higher plants, where nitrate-defective mutants are less common, no genetic separation of the moieties has been reported, but as the 3.8–4.5S component or 8S nitrate reductase may vary independently in plants and algae (232, 233, 336, 346), separate genes may be involved. The polypeptides may be spliced during biosynthesis (14) or naturally covalently fused (see 6), and detergent (340) or hydrolytic (311) procedures used to separate the wild-type bacterial or fungal complex into components may be useful. In *N. crassa* the MV·, BV·, and $FMNH_2$ activity was substantially increased both in *nit3* mutant extracts, which lack all NADPH activities over that found in wild-type mycelia (225), and when the NADPH system was denatured by a mercurial or by heat (92), as was also shown for *Cyanidium caldarium* (262, 265, 266). In *Chlorella* similar heating allowed Mo^V to donate directly (13), and loss of the NAD(P)H cytochrome *c* reductase moiety may expose the molybdenum site more accessibly, but this was not found for spinach (R. W. Jones, and E. J. Hewitt, unpublished). Loss of NADPH but persistence of BV· and MV· activity under semianaerobic conditions (230, 342) in *N. crassa* wild strain suggests that whereas *nit3* mycelia cannot grow with nitrate, anaerobiosis may induce a physiological donor for the MV· $FMNH_2$ system when the NADPH system is physiologically repressed.

ASSEMBLY IN VITRO AND IN VIVO The nitrate-induced 4.5S NADPH cytochrome *c* reductase of the *nit1 N. crassa* mutant which contained the cytochrome

b_{557} (185, 226) complemented in vitro with the $FMNH_2$ BV·, MV· nitrate reductase of the *nit3* mutant, where it was constitutive, to form wild-type NADPH nitrate reductase (243). Analogous complementation was shown for *E. coli* chlorate-resistant *ChlA* and *ChlB* mutants, lacking nitrate reductase, under strict conditions (22) with altered K_m for nitrate. Multiple deletion mutant proteins complemented when mixed in appropriate combinations (205).

Genetic evidence from *A. nidulans* (17, 57, 247, 283) indicated that *nia* structural genes controlled synthesis of NADPH cytochrome *c* reductase protein components and a complex locus *cnx* controlled production of an additional factor required for the manifestation of both xanthine dehydrogenase and nitrate reductase. As these proteins are different, production of a common cofactor based on molybdenum was deduced. This labile *cnx* cofactor structure was sought in diverse molybdoproteins and was found to be dissociated by treatment at pH 2–3 from xanthine, aldehyde, and sulfite oxidases, Fe–Mo nitrogenase proteins, and plant, bacterial, and *N. crassa* wild-type or *nit3* (MV·) nitrate reductases (169, 183, 184, 226, 227). The dissociated structure was then used to reconstitute in vitro typical wild-type nitrate reductase by incubation with extracts obtained from *nit1* mutants. These contained only nitrate-induced NADPH cytochrome *c* reductases from which the 4.5S component was depleted after the reconstitution (227) as well as after complementation by the *nit3* MV· nitrate reductase (225). Ionic molybdenum and numerous amino acid molybdenum complexes were inactive (227). The nitrogenase MoFe proteins yielded appreciable amounts of *cnx* component with less or no acid treatments (227). Nitrate reductase-defective complementing mutants *ChlA* and *ChlB* of *E. coli* provided the dissociable cofactor from *B* but not from *A* (201). The *cnx* component which had a mol wt of possibly 1000 (226) was transferred by dialysis from *Rhodospirillum rubrum* after tryptic digestion (170). The *cnx* component was thermolabile (50°C, 5 min) (184, 227) and was stabilized by ionic molybdate (184); probably it was stabilized when protein-bound, as shown by activation effects of heating on MV· nitrate reductase (above). The dialyzable cofactor was obtained from several bacteria (171). In vitro incorporation of ^{99}Mo into the cofactor was obtained, and enhanced activity occurred if molybdate (10 mM) was added immediately before incubation with induced *nit1* extracts (183, 184). A polypeptide about 1000 mol wt containing molybdenum was isolated from the MoFe component of nitrogenase of *A. vinelandii* (88) and was dissociated by 0.5 M NaCl at pH 8.5–9. It had an EPR signal at approximately $g = 2.0$.

Nitrate reductase is not reactivated by molybdate in cell-free extracts of molybdenum-deficient cells (2, 124, 322). The "kinetics" of nitrate reductase formation in vivo in cauliflower leaves induced by nitrate or molybdenum were similar (2) and were correspondingly inhibited by inhibitors of protein synthesis when measured at hourly intervals (3). In *Chlorella* over a single 4-hr period (336), cycloheximide inhibited by 50% the increase of enzyme activity induced by molybdenum; and in cauliflower, puromycin inhibited tungsten incorporation over 80% during 16 hr (236). Whereas de novo protein synthesis was proved by buoyant density measurements for induction by nitrate (353), responses to molybdenum were oppositely interpreted as mainly dependent on de novo protein synthesis (3, 124, 236) or mainly on an in vivo reaction with existing apoprotein (336). In molybdenum-deficient cell suspensions of Paul's Scarlet Rose (121), grown with urea and induced by pre-incubation with nitrate, the addition of molybdate caused steady enzyme formation without the lag period that occurred when the two treatments were reversed. When cycloheximide or puromycin was given after nitrate and before molybdate, about 30% of normal enzyme was formed in 2 hr and was then stabilized by cycloheximide but soon declined with puromycin. When nitrate and molybdate were reversed, enzyme formation was totally inhibited. In molybdenum-deficient spinach grown with nitrate, serological studies (231) indicated that cross-reacting protein, possibly apoprotein, was only 2% of normal cell nitrate reductase, but it rose to 20% in plants given additional ammonium sulfate to supplement nitrate. In *N. crassa,* nitrate-induced wild-type mycelia grown with ammonium but without molybdenum contained over 90% of the 7.8S NADPH cytochrome *c* reductase but only 30% of the normal NADPH nitrate reductase (319, 322). Aggregation of the 4.5S component and production of 7.8S apoprotein were thus not dependent on molybedenum, unlike the effects of the *cnx* component on proportions of 8S and 4.5S proteins in *nit1* and reconstituted systems (226, 227). In *Azotobacter* (223a) and *Klebsiella* (42a), molybdenum deficiency resulted in 95% loss of serologically reactive component I nitrogenase apoprotein although available nitrogen was provided.

In *A. nidulans* (69) some *cnx* mutants which were unable to utilize molybdenum lacked nitrate reductase but produced the 7.8S nitrate-inducible cytochrome *c* reductase component, and aggregation of the 4.5S subunits was not dependent on molybdenum. In other *cnx* mutants (70, 200) the proportions of the 7.8S and 4.5S components varied greatly, possibly because of differential stability (200). Heterokaryons were obtained between *nia* structural and *cnx* cofactor mutants but in vitro complementation of wild-type nitrate reductase was not obtained (70), unlike *N. crassa* (225). *A. nidulans* produced some mutants having nitrate-inducible 7.8S MV·-nitrate reductase which lacked both NADPH nitrate reductase and cytochrome *c* reductase (70). The 4.5S NADPH cytochrome *c* reductase in other mutants was thought to be the FAD-bearing component (71) which was more readily lost than from the complete wild-type enzyme. The 7.8S MV·-nitrate reductase must have contained molybdenum and the cofactor uniting the subunits but may have lacked flavin. Another mutant, *niaD8,* produced 7.8S NADPH cytochrome *c* reductase but lacked both NADPH and MV· nitrate reductase, although it had the same molybdenum content as wild-type enzymes (71). The *cnxE14* mutant produced 7.8S NADPH cytochrome *c* reductase but lacked NADPH nitrate reductase and was defective in molybdenum (71), although it still had 30% of wild-level nitrate-inducible MV· nitrate reductase (70). It seemed that a normal molybdenum level was not essenfial for uniting 4.5S subunits. When *cnxE* mutants lacking nitrate reductase were grown on 33 or 330 mM molybdate instead of 3.3 μM, some repair of nitrate reductase resulted with a tenfold increase in K_m for nitrate (17), but other *cnx* mutants were not repairable. The *cnxE* mutants, when nitrate-induced, also produced 15% of wild-type MV·-nitrate reductase (70) and thus resembled *N. crassa nit3* mutants except in being inducible. The *cnxE* mutants may have an exposed weakly binding molybdenum site. The *ChlD* nitrate reductase-defective mutant of *E. coli* was repaired but not saturated by 0.1 mM molybdate compared with 1 μM normally (94). The *ChlA* and *ChlB* complementing mutants above (201) were not individually repairable by excess molybdenum (94). Mutants of *Pseudomonas aeruginosa,* lacking nitrate reductase, were repaired by 50 μM molybdate (106).

STRUCTURAL MODELS   The model for nitrate reductase in *A. nidulans* (57, 200) predicts that two 4.5S flavin-bearing (71) cytochrome *c* reductase subunits are united into 7.8S wild-type NADPH nitrate reductase by a cofactor subunit coded at the *cnx* locus, which is regarded as a molybdenum-binding protein of 10,000–20,000 mol wt and carries the electron transferring site common to xanthine dehydrogenase and nitrate reductase (285) and probably other enzymes. The cofactor subunit may lack molybdenum but it can still unite 4.5S components, thus giving 7.8S cytochrome *c* reductase only. In some mutants where a defective cofactor subunit is produced, excess molybedenum may repair nitrate reductase but with impaired substrate affinity. The united 7.8S protein may lose flavin as 1 mole remains after isolation (68). Stability of the apoprotein (121, 231) or the molybdenum-binding or other components (200) may vary, and enzymes specifically degrading other enzymes (147, 152) such as nitrate reductase (343, 344), or which attack apoprotein as in pyridoxal enzymes (156) may determine net levels of the components in different species. The inducer may influence nucleotide specificity for the NAD(P)H moiety (292, 293), possibly by irrelevant protein changes.

Several genes may control molybdenum utilization (57, 200, 285), i.e. possibly by molybdenum uptake or insertion into the cofactor (Mo permease and transferase), by synthesis of the 1000 mol wt peptide and the 10,000–20,000 protein subunit, and by unidentified factors as in *niaD8* (71) or repaired *cnxE* (17) which could influence nitrate binding or orientation to molybdenum reflected also in widely different natural K_m values (see 115). After first reacting in the *trans* configuration, nitrate coordination, studied in a new model oxomolybdenum complex (89), needs to be bidentate with only one oxygen atom coordinated and *cis* to the oxo group on mononuclear molybdenum already as Mo^V in a hydrophobic environment which can then act as a single electron donor to produce nitrogen IV (di)oxide. This is believed to disproportionate very rapidly to yield nitrite and nitrate. Further slow reduction could yield NO^+ at the oxidation III level of nitrite. Molybdenum analyses favor a mononuclear mechanism, but transformations based on Mo^{IV} or Mo^V may be visualized in this context (see 115). EPR measurements can be decisive in deciding between the alternatives.

TUNGSTEN AND VANADIUM ANALOGS   Tungsten depressed nitrate reductase activity in barley (346) and tobacco cells (112) and induced excessive NADH cytochrome *c* reductase activity in 3.7S and 8S components and excessive nitrate accumulation in barley. Similar differential effects on nitrate reductase and NADH dehydrogenase appeared in *Chlorella* (336). This was shown to result from in vivo replacement of molybdenum by tungsten in nitrate reductase protein in spinach (233, 235, 237), *Chlorella* (245), and probably *N. crassa* (322) to produce an inactive (8S) nitrate reductase analog which retained NAD(P)H dehydrogenase function.

Molybdenum deficiency in spinach caused appearance of multiple electrophoretically separate NADH dehydrogenases which were decreased by application of tungsten in vivo, but total dehydrogenase activity was similar in normal, molybdenum-deficient or tungsten-treated plants (233). Tungsten was held firmly in the spinach protein and could not be exchanged for molybdenum in vitro (234), whereas exchange occurred in vivo without protein synthesis in tobacco cells (111) or *N. crassa* (313). The tungsten binding was less stable than molybdenum in *Chlorella* (245) and possibly spinach (233) proteins during electrophoresis. The *N. crassa* tungsten analog obtained by in vivo incorporation was less stable to heat or acid than the molybdoprotein (185). A tungsten analog of nitrogenase MoFe protein from *A. vinelandii* (30, 223a) lacked activity and bound tungsten less stably than molybdenum (30), which could displace tungsten in vivo without protein synthesis. By contrast, *Klebsiella* produced only 5% of serologically reactive component I protein as tungsten analog (42a), which was either highly unstable or not formed. Unlike the *Azotobacter* situation, molybdenum deficiency or tungsten substitution also caused absence of component II. Production of the tungsten analog in cauliflower was inhibited by puromycin but not by L-azetidine-2-carboxylic acid (231, 236), showing that although protein synthesis was somehow involved, altered tertiary structure (85) assumed to have prevented nitrate reductase formation (124) did not prevent metal incorporation. Tungsten inhibited incorporation of molybdenum into xanthine and sulfite oxidases but did not replace it to a corresponding extent in preformed apoprotein (see 147a). Although tungsten antagonized the role of molybdenum in nitrate-dependent formation of formate (DCPIP acceptor) dehydrogenase-nitrate reductase of *E. coli* (76, 254), it enhanced or replaced molybdenum in the production of formate (NADP or MV acceptor) dehydrogenase by *Cl. thermoaceticum* (11) or *Cl. formicoaceticum* (10a) with enhanced activity.

In *A. vinelandii* during prolonged growth vanadium either substituted for molybdenum in vivo in a nitrogenase protein having apparently modified activities (47) or possibly stabilized apoprotein which enhanced molybdenum utilization (29, 208) but did not produce a component I analog during 3 hr in molybdenum-deficient cells (223a) in contrast with tungsten. Vanadium was not incorporated in vivo as an analog of spinach nitrate reductase NADH dehydrogenase (235) or by *Cl. thermoaceticum* (11). Vanadium was incorporated by *N. crassa* with much weaker stability into a protein analog having no nitrate reductase (185), but its NADPH cytochrome *c* reductase activity was not reported. Vanadium inhibited nitrate reductase in wheat embryo (315) and in tomato (46) where inhibition was reversed by EDTA and may be unrelated to molybdenum.

NITRITE REDUCTASE

Plants and Green Algae

Since the discovery of BV·-dependent nitrite reductase in green plants (60, 61, 103), the suggested role of ferredoxin (224) was shown wherever tested for chlorophyllous tissues (33, 117, 119, 133, 139, 151, 197, 243, 244, 258), with K_m values of 5–10 μM (33, 107, 108, 119, 258) or 70 μM (128). Several highly purified preparations have been described (48, 49, 119, 128, 133, 136, 141, 294, 355). Increases in specific activity of 950 to 1000 for spinach (48, 119) and 1200 to 1300 for marrow (119, 133, 136) appear maximal, and yields may be between 5 and 20% (48, 49, 355). Best specific activity from marrow (136) (60–85 μmole nitrite reduced min^{-1} mg^{-1} at 27°C with MV') gives a turnover rate of 5.3 X 10^3 mole (3.2 X 10^4 e$^-$) min^{-1} mole $^{-1}$. Polyacrylamide gel electrophoresis revealed twin isoenzymes from *Chlorella* (355), but these were not seen in marrow (119, 136). Spinach preparations showed single (128, 141) or possibly twin components (119), but sodium dodecyl sulfate produced subunits of 37,000 daltons (141), compared with 72,000 found by sedimentation or 60,000 by gel exclusion, in agreement with other values between 61,000 and 70,000 found by either method for spinach (48, 118, 128), marrow (49, 118, 119, 133), *A. cylindrica* (108), *Chlorella* (355), or *Dunaliella tertiolecta* (97). Two isoenzymes (63,000–67,000) in maize scutellum were separated on DEAE-cellulose (64, 131) or by electrophoresis (D. P. Hucklesby, unpublished; 334a). One was common to chlorophyllous, etiolated, and nonchlorophyllous cells, but properties were similar. In barley (40), the enzyme purified 60 times appeared to be similar for leaves and roots. In situ identification of nitrite reductase in polyacrylamide gels is most useful (132, 334a).

Effects of iron deficiency (50, 61, 166), cyanide, and inhibition by carbon monoxide not reversed for marrow (120, 134) or reversed by light for *Chlorella* (354, 355) suggested the role of iron, although *o*-phenanthroline, α,α-dipyridyl, 8-hydroxyquinoline, and EDTA had insignificant effects. Ironically, the inhibitory effects of bathophenanthroline and bathocuproine sulfonates (134) appear to be artifacts of the BV' assay. Iron content was proved by ^{59}Fe incorporation into nitrite reductase of highly purified marrow (136) or *Chlorella* (13) enzyme. Isotopic exchange occurred at 0.8% per month at 3°C (136). Direct analyses showed two atoms iron/ mole in *Chlorella* (355), spinach (48), calabash (49), and marrow preparations (136). Acid-labile sulfide (2.7–3.6 atoms/mole) occurs (135, 136, 355). Labile sulfide may be easily lost (215), and the ratio 2Fe:4S as in nitrate reductase of *Micrococcus* (272) is possible. Plant and algal nitrite reductases are red brown and show similar light absorption. In addition to the peak at about 278 nm and a shoulder at 293 nm probably due to tryptophan, there are major and lesser peaks (relative to 278 nm) at 380 (0.28) and 570 nm (0.12) (48) or 388 (0.25) and 573 nm (0.05) (128) for spinach; at 384 (0.34), 573 (0.085), and 635 nm (< 0.02) with faint shoulders at 530 and 560 nm for *Chlorella* (355); at 384 (0.43) and 572 nm (0.13) with weak shoulders or peaks at 540 and 700 nm for marrow (136); and at 370 (0.58) and 580 nm (0.11) for calabash (49). On reduction of marrow enzyme with minimal dithionite (135, 136), the peak at 384 nm shifted to 392 nm and the 572 nm peak to 590 nm, while the absorbence at 700 nm was bleached. Autoxidation restored the peaks at 700 and 540 nm but not at 384 nm. With *Chlorella* enzyme (355), dithionite caused loss of the 573 nm peak, production of peaks at 555 and 585 nm, and loss of the weak 635 and 692 nm maxima. Autoxidation or nitrite restored the 573 nm peak but not the absorbence beyond 600 nm. Changes at 384 nm were not recorded. Dithionite, however, does not serve effectively as a direct donor for either enzyme, but does so in crude preparations (40). All agree that flavins are absent on the evidence of absorption spectra. Inhibition after short treatment of marrow enzyme by mersalyl was reversible by GSH, but prolonged treatment inhibited irreversibly with immediate bleaching at 384 nm and slow bleaching at 570 nm (D. P. Hucklesby and E. J. Hewitt, unpublished).

Plant and algal enzymes are practically specific for single electron donors—ferredoxin and MV· or BV· radicals with certain exceptions—and do not utilize NAD(P)H directly (61, 97, 103, 107, 108, 119, 133, 258, 279, 355, 357). Reduced flavins are generally ineffective (107, 108, 119, 151, 258, 355, 357). However, in marrow preparations $FMNH_2$ reduced by NADPH and the flavoprotein sometimes produced 5–30% of the activity found with BV· or ferredoxin reduced by the same system (119, 133). This behavior was confirmed with *Dunaliella tertiolecta* (97). The *Chlorella* enzyme functioned with illuminated chloroplasts and flavodoxin (355), a low molecular weight flavoprotein produced in response to iron deficiency (358). The *Anacystis nidulans* enzyme functioned with phytoflavin obtained therefrom when reduced by chloroplasts but not by dithionite (38), although this served with ferredoxin. Inconsistent activity with $FMNH_2$ might result from the presence in partially purified preparations of enzyme-bound ferredoxin or of a residual protein carrier. Flavoprotein carriers can yield reduced flavin semiquinone couples of relatively low potential (-0.46 V) (77) which do not readily equilibrate with the high potential oxidized flavin semiquinone couple (347) and are effectively one-electron donors. In solubilized systems from roots (39, 40, 210, 211) or maize scutellum (64, 131), MV· or ferredoxin, which is probably absent from the tissues, functioned efficiently, but the unknown electron donating system might be present in acetone precipitates (40). Uncoupling reagents inhibited nitrite reduction by *Ankistrodesmus* (4, 130, 167), which was dependent on high energy phosphate supply in light, and in darkness (63a, 164) unless hydrogenase (E^0, -0.42 V) was utilized (165) or light intensity was saturating (164). In vivo nitrite reduction in barley aleurone was inhibited by DNP or anaerobically (80). In barley roots a pyruvate (E_0' -0.7 V)-dependent particulate nitrite reductase functioned when ATP was added (39) and DNP was inhibitory in vitro (40).

The K_m values for nitrite are usually between 100 and 300 μM in leaf enzymes (119, 141, 151, 258) and may be below 40 μM (61, 131) or exceptionally 2 mM in root and leaf of barley (40) when obtained with BV· or MV· or with dithionite-reduced ferredoxin, whereas values are about 10 times higher for hydroxylamine (see 115). When ferredoxin was reduced by chloroplast grana and light, the apparent K_m for nitrite was often much smaller (119) and varied in relation to the "energy input" in terms of light intensity or chlorophyll concentration, and nitrite was sometimes sharply inhibitory at concentrations which were tolerated in other experiments or conditions (73, 119, 279). These results may be explained by inhibition by nitrite of the reversible reduction and re-oxidation of ferredoxin in the presence of illuminated grana and nitrite, especially when "energy input" was weak or pH was below 7.5 (119, 293).

Bacteria, Yeasts, and Fungi

In *Azotobacter chroococcum* (335) the enzyme appeared to be a simple polypeptide, mol wt 67,000, and like the plant enzymes reduced nitrite quantitatively to ammonia but differed in having a dissociable FAD requirement, in utilizing essentially NADH, in functioning with dithionite as a direct donor, and in the low K_m for nitrite (5 μM). Use of ferredoxin was not reported. A homogeneous preparation from *Achromobacter fischeri,* mol wt 95,000 (138, 256, 257), reduced nitrite (K_m 50 μM) directly to ammonia with a turnover number of 1.43 X 10^4 mole mol^{-1} min^{-1} at 30°C or 1.2 X 10^4 for 80,000 daltons (138a), eight times faster than hydroxylamine (K_m 5 mM). Reduction of this, as in plants (119), was inhibited by nitrite but not vice versa, and sulfite was not reduced or inhibitory. Inhibition by carbon monoxide was light reversible when $FMNH_2$ alone or NADPH were used with a flavoprotein carrier but not when BV· alone was the donor, possibly because of absorption of the actinic light. Although dithionite reduced the autoxidizable cytochrome c_3 component (two hemes/mole protein, which comprised two equal subunits) (138a), and this was reoxidized by nitrite or hydroxylamine, dithionite was as ineffective for plants as a donor for the enzyme. This resembled the cytochrome c_3 nitrite reductase from *Desulfovibrio desulfuricans* (see 256, 257). The c-type heme does not confrom with the siroheme class of proteins described below, considered to be the prosthetic group in plant and bacterial systems, which also show significant similarities in amino acid composition (138b) including histidine, of possible mechanistic significance. Although ammonia was the enzymic product, the separated heme when reduced by dithionite yielded nitric oxide (316). A homogeneous sulfite and hydroxylamine reductase from *A. nidulans* (349) utilized MV· or BV· but not NAD(P)H. Nitrite reduction was not reported. The protein sedimented at 4.2S ($\sim$120,000 daltons) and had absorption maxima at 384 and 585 nm as in plants. Yeast *(S. cerevissiae)* and *E. coli* contain sulfite reductases of great complexity, which also reduce nitrite and hydroxylamine, but nitrite reduction in vivo in *E. coli* probably depends physiologically on the separate NADH- and nitrite-specific but sulfite-inhibited nitrite reductase (54, 55, 159). The complex yeast and *E. coli* enzymes (mol wt 350,000 to 670,000) each contained one to four mole of FAD and FMN. They contained up to six nonheme iron and three labile sulfide units in yeast (348, 350–352), but heme iron was said to be absent (351). In *E. coli* there was 12 to 20 iron atoms and 12 to 15 labile sulfide units (298, 299, 301). A similar complex system occurred in *Salmonella typhimurium* (300). These proteins absorbed strongly at 386 and 587 nm, and at 714 nm in *E. coli* in addition to the flavin peak at 455 nm, and reacted with carbon monoxide (see below).

Prosthetic Group and Possible Mechanisms

The *E. coli* chromophore was extracted by acid acetone and was identified as an iron tetrahydro iso-bacteriochlorin with eight carboxyl side chains (221). The prosthetic groups of *E. coli* sulfite reductase and the assimilatory and dissimilatory sulfite reductases (desulfoviridin and desulforubidin) of two species of *Desulfovibrio* are probably identical with the bacteriochlorin above (220, 223) in spite of major differences in light absorption, behavior with carbon monoxide, molecular weight classes, and product stoichiometry (182, 221, 223). Whereas acid acetone extraction of the *E. coli* assimilatory and the *Desulfatomaculum nigrificans* dissimilatory sulfite reductase proteins yielded the iron porphyrin siroheme, the *Desulfovibrio gigas* desulfoviridin yielded the metal-free porphyrin sirohydrochlorin (220).

The similarities between plant sulfite reductase (18) and nitrite reductases were noted (220, 355), and a common prosthetic chromophore was suggested for plant nitrite and bacterial sulfite reductases (220). Preparations of spinach nitrite reductase and *E. coli* sulfite reductase treated with acid acetone were found to yield an identical iron-containing porphyrin identified as siroheme or iso-bacteriochlorin by several criteria (222). This was considered to be uniquely capable of mediating the six-electron transfer reactions of both the nitrite and sulfite reductases, whose specificity is dependent on the proteins and may occur in *N. crassa* (178a).

Chlorella nitrite reductase yielded a red pigment in acid butanol (355) and was inhibited by carbon monoxide, but no spectral changes were reported and neither pyridine nor cyanide indicated heme components. The marrow enzyme or ferredoxin, when treated with ferrozine at pH 5 or below, both rapidly produced large increases in absorbance at 560 nm (D. P Hucklesby and E. J. Hewitt, unpublished work). A possible resemblance of nitrite reductases to hydrogenase systems is supported by hydrogenase from *Desulfovibrio vulgaris* (186). This had a shoulder peak at about 380 nm, which was bleached by hydrogen or dithionite to yield a small peak at 408 nm. The dimeric mol wt was 60,000, three or four atoms of iron and labile sulfide were present, and molar absorbance of about 4×10^3 at the 400 nm region was characteristic of several sulfite/nitrite reductases. Inhibition of *Cl. pasteurianum* hydrogenase by carbon monoxide was reversible by light under correct conditions (330) unlike those used with the marrow enzyme (134).

Several nitrite reductase systems appear to have critical redox requirements below a certain potential (perhaps −0.32 V) first noted for the marrow enzyme with BV· (61) and confirmed for *D. tertiolecta* (97). This concept was supported by work with ferredoxin (119, 133), with marrow, and confirmed with *A. cylindrica*, where activity with ferredoxin ceased when NADPH was only partially reoxidized (107, 108). Rate constants for reduction by one-electron donors may be determined by their redox potential differences (259), and for nitrite reductase an enzyme-bound intermediate step may have an unfavorable potential comparable, for example, to possibly NO to NOH (−0.3 V estimated) (180), although free nitric óxide is not reduced (253). The role of ATP in nitrite reduction in nonchlorophyllous and some other systems above suggests: (a) the possible operation of a membrane system or protein which separates conformationally stored energy from an electron flux generated in the aqueous phase in a nonspecific manner (345), whereas the solubilized enzyme in vitro utilizes nonspecific low potential one e^- donors; or (b) the production of a specific phosphorylated low potential reductant such as phosphoenolpyruvate, able to donate electrons to an unknown cofactor which reacts only weakly with dithionite (40).

No nitrite reductase tested from the chlorophyllous plants can reduce sulfite (108, 133, 151, 355) by contrast with microorganisms (above). Highly purified preparations vary in their ability to reduce hydroxylamine. In marrow and spinach, the ratio of nitrite to hydroxylamine reductase with ferredoxin or MV· ranged between 6:1 and 40:1 (119, 133, 258), and was 6:1 for *Chlorella* (355), 25:1 for *Anabaena* (108), and 8:1 for *A. Fisheri* (256). In recent marrow preparations (136; and unpublished work) the ratio exceeded 100:1. Other hydroxylamine-reducing proteins (118, 119,

133) were deemed to be eliminated in these experiments. Purified nitrite reductase from marrow or spinach or reduced ferredoxin were unable to reduce hyponitrite (133, 253) or nitric oxide (253), although this may be a product of nonenzymic nitrite reduction (133). These results, the kinetics for hydroxylamine or nitrite alone or together (33, 119, 133), and full yields of ammonia produced from nitrite often by pure enzymes (33, 61, 103, 108, 119, 128, 131, 151, 258, 279, 335, 355) show that fascile notions based on the Meyer & Schultze sequence and still cited with unwarranted authority derived from insufficient evidence obtained in early studies with fungi (229) are not relevant to higher plants, some algae, bacteria, or *N. crassa* (178a).

Nitrite reductase may nevertheless have the dual activity reported above and envisaged in earlier schemes (119, 160). It is difficult to visualize how nitrite can be converted to ammonia via bound intermediates without the penultimate stage corresponding to a hydroxamate, oxime, or other derivative of hydroxylamine (115, 119), but once combined with nitrite or exposed to it before extraction, the protein may be conformed in such a way that exchange between bound and free hydroxylamine does not occur (160) and nitrite would then severely inhibit free hydroxylamine reduction (119, 256). Covalent substrate binding is possible (28), and covalent mechanisms based either on a histidine nitrosamine structure (199) followed by hydroxamate and hydrazone structures or on an iron-sulfide nitrito bridge structure (115) followed by nitrosyl, nitroxyl, and hydroximate steps are independently suggested. Nitrozation at a histidine residue (199) would have to occur either by HONO after local protonation or by NO_2^- at neutral pH for which a mechanism was described involving aldehyde, or possibly carbonyl catalyzed imminium ion formation (157). A carbonyl group in siroheme might provide this requisite. Purified sufite-nitrite-hydroxylamine reductase (desulfoviridin) of *D. vulgaris* was able to reduce trimethylamine *N*-oxide (177), but tertiary amine *N*-oxide reductase of *E. coli* was considered to be separate from nitrite reductase (277a), so *N*-oxide or nitroxide reduction expected in the above schemes is experimentally observed, possibly for a nitrite reductase enzyme.

REGULATORY AND PHYSIOLOGICAL ASPECTS OF NITRATE AND NITRITE REDUCTION

Locations of Activity and Synthesis

Nitrite reductase is present in chloroplasts (33, 65, 269, 323) which maintained nitrite reduction in light at 0.1–0.2 μmole/min/mg chlorophyll without any additions when suitably prepared (73, 202, 213, 214). Chlorophyll fluorescence in algae was quenched by nitrite but not by nitrate (168), showing the close photochemical relationship for nitrite reduction. Results obtained by improved methods (214) refuted the supposed location in microbodies (189, 191). The location of both ferredoxin and ferredoxin NADP reductase on the exterior of the thylakoid membranes (see 332) makes it probable that nitrite reductase is similarly located, but other leaf cytoplasmic locations seem possible (98, 282), and particulate locations are indicated for nonchlorophyllous cells (39, 40, 66, 131, 210, 211), possibly in proplastids. The location of nitrate reductase in leaves is controversial, and cell grinding media may be critical. In sucrose-phosphate Mg^{2+}/K^+ media (65, 66) a cytoplasmic location was indicated for spinach or tobacco leaves and wheat roots, after removal of an inhibitor in the tobacco cytoplasmic fraction and by avoiding adsorption. Other studies on tobacco and barley using Honda's medium showed co-attachment of nitrate reductase, glycolate oxidase, and catalase to particles like microbodies in linear gradients, but solubilization of nitrate reductase and glycolate oxidase was caused by discontinuous gradients or hard acceleration (190, 190a, 275). Gibberellic acid or red light enhanced and far-red light impaired the particulate attachment of both enzymes (190). Transfer to nitrate-free media caused solubilization of glycolic oxidase, which remained attached when the tungsten analog of nitrate reductase was formed. A membrane association (see 42, 63) might explain both the participation of phytochrome control above and in triggered induction of the enzyme in etiolated pea seedlings (148, 149) and operation of a circadian rhythm in wheat (334). Location on or within the outer chloroplast membrane was thought possible on the basis of inhibitory effects of ADP (75) or isolation procedures (see 26, 27, 269) which indicate particulate binding in barley roots (211). The loss of ascorbic acid caused by molybdenum deficiency (116) is tentatively ascribed to superoxide-peroxide destruction (5). The similarity between probable superoxide damage to chloroplasts by pyridylium herbicides (67, 105) and cytological effects of molybdenum deficiency which was substantially reversed by tungsten (R. J. Fido, C. S. Gundry, E. J. Hewitt, and B. A. Notton, unpublished) suggests that superoxide peroxidation of chloroplast and tonoplast membranes may be induced in molybdenum-deficient plants by the abnormal multiple NADH dehydrogenases (233) which may have effects like NADPH cytochrome c reductase in lipid peroxidation (34, 249).

The sites of biosynthesis may differ from the ultimate location of nitrate and nitrite reductases as deduced with some inconsistency from effects of chloramphenicols, lincomycin, and cyclohexamide. It appeared that nitrite reductase synthesis involved 70S (chloroplast) ribosomes in maize (289) and 70S (280) or 70S and 80S (cytoplasmic) ribosomes (281) in rice. Both nitrate and nitrite reductases appeared to be synthesized on 80S ribosomes in *Lemna minor* (317). In beans (303) the induction of both enzymes was inhibited by cycloheximide but not by chloramphenicol in illuminated green leaves. In etiolated leaves, induction of both enzymes was inhibited by both antimetabolites when applied on illumination but not if used 24 hr later. A role of chloroplasts in addition to a cytoplasmic site of synthesis was inferred. Induction by nitro compounds including chloramphenicol in light or dark with modified properties for NAD(P)H (175, 292, 293) complicates the subject. Induction of nitrate reductase in maize (289) and resynthesis in wheat or barley following impaired synthesis during water stress (41, 218, 240, 255) were inhibited by cycloheximide although not by chloramphenicol, but some role of chloroplasts in synthesis or in ultimate location of nitrate reductase was still inferred (255, 280, 282, 303). Compounds which inhibit oxygen evolution may inhibit activity in vivo (213) or induction of nitrate reductase (123, 280) or enhance both its activity and nitrate content in tissues (see 19, 261), possibly depending on critical concentrations (280). The requirement for carbon dioxide for induction of nitrate reductase (153, 280) may be related to these suggested roles of chloroplasts in terms of redox control (280) or the production of ATP (20).

Permeases and Pools

Formation, activation, and stability of nitrate reductase independently depend on nitrate access controlled by permeases and sequestrating pools. A nitrate-induced permease was revealed in tobacco XD cells (110), where threonine inhibited nitrate uptake except by a threonine-tolerant strain, whereas tungsten did not inhibit uptake by contrast with reduction. In nitrate-free cells, induction occurred rapidly at low cell nitrate concentrations but ceased soon after removal of the external nitrate supply, although cell nitrate concentration remained high (111), thus differentiating between pools for storage and induction. A nitrate-induced permease was present in sycamore cells (172). Comparable pools were identified in pawpaw fruit (209), in bean leaves by use of $^{15}NO_3^-$ (207), and in cotton (35). In tobacco XD cells and barley aleurone (81) a metabolic nitrate pool was easily exhausted, but was replenished from a storage pool when cells were exposed to monohydroxy alcohols, DNP or pyrazole. Nitrate uptake capacity of maize seedlings was promoted by nitrate, and inhibitory effects of cyclohexamide, puromycin, or actinomycin D on this behavior indicated the role of a nitrate-inducible permease (143). Ammonium did not inhibit nitrate uptake by L. minor (318). A study of ^{15}N and ^{14}N fluxes in rye grass roots suggested that passive influx and efflux, aerobically dependent active transport, and recycling of nitrate in layers adjacent to each side of the plasmalemma occurred simultaneously (217). The net active uptake was stimulated by Mg^{2+} ions (217) and by increasing charge of the accompanying cations (23). Whereas external NO_3^- and Cl^- ions did not compete (304), high internal concentrations were reciprocally inhibitory in barley and carrot (59, 304). Photosynthetically produced malate was supposed to be decarboxylated in roots, yielding bicarbonate ions which exchanged for nitrate that was reduced in leaves and promoted malate formation (31).

Penicillium chrysogenum nitrate permease was nitrate induced and was inactivated by ammonium, glutamine, or asparagine. Cycloheximide prevented inactivation by ammonium, and both inactivation and decay of the nitrate permease were more rapid than for the reductase (95). Nitrate and nitrite permease systems of N. crassa wild strain and nitrate-defective mutants were both nitrate and nitrite induced (287, 288). Neither was repressed by ammonium, but nitrate permease was repressed by most amino acids. Ammonium and nitrite inhibited the nitrate permease noncompetitively. It had a half-life of 3 hr and was stabilized by cycloheximide. In A. nidulans, external (but not internal) ammonium was considered to regulate both permease and reductase systems by a membrane-bound glutamate dehydrogenase (248). This interpretation was disputed (15), but mutant deletion of NADP glutamate dehydrogenase also abolished repression of nitrate reductase by ammonium (16).

Induction, Repression, and Degradation

The distinctions between inactivation in vivo and repression, or between reactivation and derepression or induction must be based on time scales and evidence for de novo synthesis (see 26, 27) and turnover (see 137).

In tobacco XD callus cells, nitrate-induced de novo synthesis and simultaneous decay were proved by buoyant density separations (353). Nitrate reductase decay in higher plants shows first-order kinetics in dark or absence of nitrate ($t^{\frac{1}{2}}$, 1–24 hr) (19, 111, 137, 238, 291, 331, 334, 353) and is more rapid than nitrite reductase decay (158). A specific degrading enzyme was found in maize roots (343, 344) and in rice roots where only the NADH component is attacked (151a). Enzymic breakdown was inferred in darkness in barley (331) or without nitrate in fungi (188, 313, 320) as tungstate accelerated loss and cycloheximide stabilized the existing enzyme. The mRNA for nitrate reductase in maize roots ($t^{\frac{1}{2}}$, 0.3 hr) was much less stable than the enzyme (238). In contrast with 6-methyl purine, actinomycin D often failed to inhibit mRNA activity (78, 162, 174, 238, 268), but in Paul's Scarlet Rose cells (121) it was severely inhibitory.

The role of nitrate in induction is well established in plants (see 2, 3, 26, 27, 114), but product induction was described in bean seed cotyledons (192), in pea roots (102, 278), and in etiolated barley leaves (154) where nitrate appeared inactive. Product induction, which is usually associated with mixed dissimilatory and assimilatory paths (see 284), obviously is not expected in plants for nitrate assimilation. Glycollate acted synergistically in barley with nitrate in the dark (154, 273, 274), possibly to produce nitrite. Induction by aromatic nitro compounds including chloramphenicol having 5^{+} oxidation states might be expected, as found for rice (175, 292, 293), but the altered specificity in favor of NADPH, which extended to cytochrome *c* reductase (293), may be a nonspecific effect of nitro compounds on –SH or –NH$_2$ groups, possibly in incompletely conformed polypeptides, and effects on mixed NAD(P)H specificity of other enzymes or on NADPH phosphatase (344a) should be investigated.

The light requirement for induction is complex and multiple (see 26, 27, 114). Phytochrome (148) or other rapid photochemical processes may be involved (149). In barley or fenugreek (*T. foenum-graecum*) cytokinins or gibberellic acid GA$_3$ (193, 268, 274) substituted for the light requirement with nitrate, but in rice (86) GA$_3$ was more effective than kinetin and light was still needed when roots were excised. In *Agrostemma githago,* fenugreek, and *Cucumis sativus,* cytokinins substantially replaced or enhanced the effect of nitrate as an inducer (37, 161, 174, 246). Cytokinins appeared especially to derepress the formation of other microbody enzymes, namely glycolate oxidase and hydroxypyruvate reductase (36). This supports the proposed location of nitrate reductase on microbodies (190, 275). The enzyme formed in cucumber or *A. githago* by cytokinin induction (127, 162) or treatment with succinic acid-2, 2-dimethylhydrazide (B.9) (176) resulted from de novo synthesis. Instead of induction, the hormones may derepress the formation of a constitutive enzyme.

Repression by ammonium is not widely reported for higher plants but was reported for barley roots (305) and *Lemna minor* (150, 241), where arginine may be the actual repressor (318). In soybean cell cultures the enzyme formation appeared to depend on production from ammonium of a growth enhancing factor which was synergistic with 2,4-dichlorophenoxy acetate, but ammonium itself independently repressed formation of the enzyme (24, 25, 237a). Nitrate reductase and cytochrome *c* reductase activity were repressed by ammonium in various green algae (113, 263, 308, 326, 327) during growth and were constitutive and derepressed by aspartate (113), glutamate (263), or nitrogen deficiency (267, 324, 336), but activation of a degrading enzyme by ammonium or a product or labilization to attack remain possibilities (241). In *N. crassa,* differential time/inhibitor studies (312, 313, 320, 321) showed that ammonium repressed NADPH and MV· nitrate reductase formation and accelerated its degradation in wild strain. The first reaction was antagonized by nitrate, which also stabilized the enzyme against breakdown, as concluded from serological equivalence with specific activity (313). Unlike the behavior in algae where NADH dehydrogenase persisted (196, 198) during short-term ammonium inactivation of NADPH and MV· nitrate reductase, this loss was paralleled by loss of the NADPH cytochrome *c* reductase in *N. crassa* wild strain (320) over similar periods of 2 hr. Ammonium or nitrogen deficiency did not inactivate the MV· system in the *nit3* mutant having only MV· nitrate reductase, and correspondingly in the *nit1* mutant having only NADPH cytochrome *c* reductase this was not inactivated (320). In wild strain, nitrate was required for formation of NADPH cytochrome *c* reductase and NADPH and MV· nitrate reductase, but in the *nit3* mutant, MV· nitrate reductase was constitutive. Thus the NADPH cytochrome *c* reductase moiety conferred repressibility on formation and susceptibility to breakdown and was regulatory (See 72 and below) in two respects (312, 313, 320).

Regulatory Patterns

DEHYDROGENASE FUNCTION   The NAD(P)H dehydrogenase function common to fungal, algal, and plant nitrate reductases may have a regulatory role. Ammonium caused reversible inactivation in vivo of NADH and $FMNH_2$ nitrate reductase from *C. fusca,* but NADH cytochrome *c* dehydrogenase was not inactivated in the same short period by contrast with slower repression. Reactivation of nitrate reductase occurred in vitro in 24 hr at 0°C unless pretreated with Sephadex G-25 (198). The *C. vulgaris* enzyme extracted from nitrate-grown cells was almost inactive for nitrate reduction, but NADH dehydrogenase was fully active (309, 338). Crude nitrate reductase was spontaneously reactivated slowly and more rapidly by nitrate with phosphate at pH 6 in air (338). The inactive *C. vulgaris* enzyme (145, 307) and the *C. fusca* enzyme which had been inactivated by incubation with NADH (216, 337) were both immediately reactivated by ferricyanide which is an oxidant for the dehydrogenase site and will compete for the electron supply to nitrate (309).

Both *Chlorella* enzymes were inhibited by cyanide for nitrate reduction but not for dehydrogenase activity. Inhibition was severe, irreversible, and noncompetitive when cyanide was added with NADH or before nitrate, but competitive with nitrate, less severe (and reversible by dialysis) when added after nitrate and before (or without) NADH (309, 337). Reversible inhibition of spinach enzyme showed that molybdenum was not removed (234) and inhibition increased progressively when cyanide was added during enzyme turnover (A. R. J. Eaglesham and E. J. Hewitt, unpublished), or was irreversible and noncompetitive when added with NADH (260). The irreversible, noncompetitive inhibition by cyanide was eliminated by ferricyanide (307, 337) and converted to reversible and competitive with nitrate in *Chlorella* and was also reversed in *N. crassa* wild-strain nitrate reductase (90) where cytochrome *c* also served. Cyanide inhibited the $FMNH_2$ (MV·) nitrate reductase in *C. fusca,* also irreversibly when the enzyme was preincubated with dithionite although this was not an effective donor alone (337). In *C. vulgaris* (306) inactivation was supposed to involve NADH and cyanide simultaneously. The reaction between enzyme and cyanide was 1:1, suggesting 1 atom Mo/mole and endogenous cyanide were said to be present in inactive enzyme systems. Reversible inactivation of one or both of *Chlorella* enzymes as shown for NADH above was 1. produced by NADPH (96, 306, 307), which was ineffective as a donor for nitrate reduction; 2. enhanced by addition of ADP (203) or at high pH indicating a pK_a for possibly a thiol group (216); 3. produced by addition of a thiol or sulfite which may rupture disulfide bonds (96); and 4. enhanced by the presence of endogenous components having mol wt about 10^3–10^4 and possibly containing nonheme iron which promoted the effects of NAD(P)H (146, 307). Inactivation was reversed by ferricyanide. Nitrite, which is not an oxidant but competed with nitrate (309), mercurials, flavins, e.g. the flavoprotein ferredoxin NADP oxidoreductase (17 μM), and *N*-ethylmaleimide, which inhibited NADH dehydrogenase activity, all prevented inactivation (216,306) which was reversed by $NAD(P)^+$ and the endogenous compounds (146) in a pH-dependent system. Carbon monoxide inhibited spontaneous aerobic reactivation of the inactive crude, but not purified, *C. vulgaris* enzyme (145). The *Nitrobacter* enzyme showed comparable effects (112a). Prolonged incubation of spinach nitrate reductase with NADH (the specific donor) or NADPH caused inactivation of both the NADH and the $FMNH_2$ nitrate reductase but not the dehydrogenase (242), and both nitrate reductase systems were reactivated by ferricyanide unless the NADH dehydrogenase was first denatured. In rice, however, NADH activated the enzyme (87), possibly by protecting it from an inactivating enzyme in roots (151a).

Two physiological regulation systems based on the dehydrogenase site have been suggested. One was based on the reversible inactivation by endogenous cyanide and a reductant and reactivation by endogenous dehydrogenase oxidants (93a, 306). In sorghum, cyanogenic glycosides released enough cyanide during extraction to inactivate nitrate reductase which was protected by nickel (204). Physiological cyanide release might inactivate in such species and an oxidant such as cytochrome *c* and its oxidase could reactivate, as indicated from effects of carbon monoxide in *C. vulgaris* (145). The other system (194) was based on over-reduction of molybdenum in a redox system controlled by relative rates of photosynthetic phosphorylation of ADP and the uncoupling effects of ammonium (see below) on the ratios of

NAD(P)H–NAD(P)⁺. This mechanism is possible in *Chlamydomonas reinhardii* (198) where NADH and $FMNH_2$ nitrate reductase was reversibly inactivated in vivo by ammonium or by arsenate but the dehydrogenase was not altered (113, 196). Light, carbon dioxide, and ability to evolve oxygen were all needed for inactivation by either ammonium or arsenate, which was supposed to cause uncoupling, and a fall in cell redox potential which was achieved also by anaerobic conditions. Rapid reversible inactivation occurred in dark or light with *Chlorella* (129). Ammonium inactivation also occurred in *Cyanidium caldarium* (262, 265, 267) but light was not required here for ammonium inactivation (264). Whereas uncoupled phosphorylation is not equivalent to exogenous ADP in terms of NADH/NAD⁺, this could be equivalent in terms of the adenylate energy charge concept (21). The active dehydrogenase site was supposed to be essential for reversible control in *Chlorella* (194, 306) and in spinach (242). However, in *T. nitratophila* (270, 271) no NAD(P)H system was detected but inactivation by cyanide or by pre-incubation with MV· before nitrate were both reversed by ferricyanide. The participation of the cytochrome b_{557} in *C. vulgaris* was obscure, as it was reduced by all inactivating reductants (307) which were not all donors for nitrate. As *N. crassa* wild strain behaved like *Chlorella*, it would be useful to test effects of ferricyanide on the cyanide inhibition of the *nit3* mutant (above) lacking both the NADPH cytochrome *c* reductase and cytochrome b_{557} moieties, especially as cytochrome *c* reductase was suspected to have a physiological role which could not be identified (72) but see above. It was assumed (306, 337) for both *Chlorella* systems that irreversible non-competitive inactivation by cyanide involved a reduction state of Mo below V to IV or III but this can only be decided by EPR. It is uncertain whether the reoxidation of molybdenum by ferricyanide occurs by direct reaction (see 337) or by reverse electron flow to the dehydrogenase site as the integrity or absence of this indicate conflicting situations (see 194, 242 270, 271, 306). Ferricyanide was thought alternatively to destroy excess NADH or other reductants (90). Some (NADPH, dithionite) were not direct donors for nitrate reduction, so an additional regulatory site may be involved, but dithionite or NADPH and transhydrogenase could both maintain supplies of NADH if enzyme bound.

KINETIC MECHANISMS   Nitrate reduction in vivo measured as ^{18}O reduction to water attained only one sixth the extracted activity in barley aleurone (79), suggesting inhibitory regulation. Anaerobic conditions or antimycin A increased nitrite appearance. This was attributed to competitive nitrate respiration, but could result from impaired nitrite reduction caused by lack of ATP (see above). Competition by dehydrogenase site oxidants would not be influenced by antimycin A. Endogenous inhibitors found in tobacco (65), ADP (74, 75, 228, 341), carbamyl phosphate (219, 290, 302, 337), or cyanate (219, 309, 337) or cyanide (above) may all be significant in vivo. ADP showed mixed inhibition of spinach nitrate reductase which was transformed to competitive with NADH in the presence of glutathione concentrations (75) similar to those produced by chloroplasts in light (126). Effective ADP concentrations were unphysiologically high, but NADH may not be saturating (see 75) and a glutathione-mediated switch mechanism was envisaged (75). Phosphate activates nitrate reductases (see 115) but usually only about 30 to 100% in vitro. In barley aleurone (79) phosphate activated by ninefold in vivo, so it is probable that other reactions, possibly generation of NADH by glyceraldehyde-3-phosphate (173) or glycolysis (341), were also influenced to produce synergistic effects.

Regulation at the level of nitrite by reciprocal mechanisms in chloroplasts was postulated (114, 119), and inhibition by nitrite of photosynthesis was observed in spinach chloroplasts (98, 99), but inhibition of oxygen evolution by nitrite when $NADP^+$ was a Hill reagent (99) was an artifact and nitrite did not prevent reduction of $NADP^+$ (100, 114, 119). However, $NADP^+$ and illuminated chloroplasts (295) or $NADP^+$ added to grana with ferredoxin in photosystem I (32, 119) totally inhibited nitrite reduction until $NADP^+$ was reduced. Nitrite, especially below pH 7, severely inhibited the reversible reduction of ferredoxin by photosystem I in vitro (253). Reduced ferredoxin (or $MV^.$) and a protein factor were essential for light activation of chloroplast fructose diphosphate phosphatase (43, 45) which was previously thought to be inhibited by nitrite (probably as HONO) in *Chlorella* in vivo (125). As this system was held to be of key importance in sequential activation of ribulose diphosphate carboxylase and ADP glucose pyrophosphorylase (44), nitrite might have inhibitory (regulatory) effects at this stage. Large differences in K_m for ferredoxin for the $NADP^+$ reductase (0.3 μM) (296) and for nitrite reductase (10 μM) (33, 119) might account for preferential NADP reduction and might apply also to the activation of the phosphatase, but only if depletion conditions for reduced ferredoxin were established. Under conditions of a minimal $NADP^+/NADPH$ ratio, nitrite reduction would be maximal, and nitrite could serve as a Hill oxidant for coupled phosphorylation and oxygen evolution (244). The continuous generation of nitrite outside the chloroplast might accelerate phosphorylation by its entry as a weak anion (pK_a 3.5) (144) which was then removed by reduction. The ratio of 1.5 mole O_2/mole nitrite (244) did not hold for leaf discs or isolated whole chloroplasts without supplements (212, 213), where nitrite promoted oxygen evolution only when present in a critical concentration (99, 333). Uncoupling reagents did not inhibit nitrite reduction (212, 213, 333), which was possibly enhanced by ammonium (213), and DCMU had relatively little effect on this process compared with inhibition of carbon dioxide fixation in leaf discs or of nitrate reduction and oxygen evolution by chloroplasts (213). Ascorbate, which is abundant in chloroplasts (93), may be an alternative physiological electron donor for nitrite reduction mediated by photosystem 1 instead of OH^- ions and photosystem 2.

Rapid carbon dioxide reduction by isolated chloroplasts is matched by the rate of nitrate or nitrite incorporation into amino acids (202). The rapid conversion of ammonia to glutamate is not consistent with the unfavorable kinetics for glutamic dehydrogenase (see 181, 213, 214), but this can now be adequately accounted for by ATP-dependent glutamine synthetase of chloroplasts (214, 239a) with a low K_m for ammonia and a new ferredoxin-dependent glutamate synthase which transfers the amide group of glutamine to α-ketoglutarate (181) so that glutamate fulfills a catalytic role. Production of ammonium in excess of its assimilation might cause uncoupling of phosphorylation for which the K_i is 200–600 μM (62, 178). Consequent deficiency of ATP might limit glutamine synthetase activity, thereby promoting further ammonium accumulation, and also result in decreasing the export of dihydroxyacetone phosphate (109) to yield glyceraldehyde-3-phosphate for nitrate reduction (173), although glycolate may be an alternative (273). ADP might be additionally inhibitory for nitrate reductase (above), especially if synergistic with a product arising from the inhibition of fructose diphosphate phosphatase. It is necessary in this scheme to postulate that progressive uncoupling predicted from the ATP requirement in glutamine synthetase could be halted or reversed when nitrite production and reduction had both ceased. This condition might be achieved by slow operation of the separate NADPH glutamate dehydrogenase in chloroplasts and by a high affinity for ATP by glutamine synthetase. The steady state concentrations of ATP and ammonium might then determine the relative rates of carbon dioxide fixation and incorporation of nitrate into glutamate, and explain the well-known situation that nitrate may accumulate whereas nitrite and ammonium do not (see 114).

ACKNOWLEDGMENTS

I appreciate the benefits from discussions with my colleagues, Dr. D. P. Hucklesby and Dr. B. A. Notton, and the support of the United Kingdom Agricultural Research Council for work in this field. I especially thank Miss M. Banwell for her excellent assistance in checking references and Mrs. M. Griffin, Miss J. L. Ogborne, and Miss J. A. Pemberton for their intelligent performance of the tedious task of typing.

Literature Cited

1. Adams, C. A., Warnes, G. M., Nicholas, D. J. D. 1971. *Biochim. Biophys. Acta* 235:398–406
2. Afridi, M. M. R., Hewitt, E. J. 1964. *J. Exp. Bot.* 15:251–71
3. Ibid 1965. 16:628–45
4. Ahmad, J., Morris, I. 1967. *Arch. Mikrobiol.* 56:219–24
5. Allen, J. F., Hall, D. O. 1973. *Biochem. Biophys. Res. Commun.* 52:856–62
6. Aloj, S., Bruni, C. B., Edelhoch, H., Rechler, M. M. 1973. *J. Biol. Chem.* 248:5880–86
7. Aminuddin, M., Nicholas, D. J. D. 1973. *Biochim. Biophys. Acta* 325:81–93
8. Aminuddin, M., Nicholas, D. J. D. 1974. *J. Gen. Microbiol.* 82:103–13
9. Ibid, 115–23
10. Anacker, W. F., Stoy, V. 1958. *Biochem. Z.* 330:141–59
10a. Andreesen, J. R., Ghazzawi, E. E., Gottschalk, G. 1974. *Arch. Mikrobiol.* 96:103–18
11. Andreesen, J. R., Ljungdahl, L. G. 1973. *J. Bacteriol.* 116:867–73
11a. Antoine, A. D. 1974. *Biochemistry* 13:2289–94
12. Aparicio-Alonso, P. J. 1971. *Mecanismo molecular de la reduccion enzimatica del nitrato a nitrito en plantas.* PhD thesis. Anal. Univ. Hisp. Sevilla No. 13, Spain. 94 pp.
13. Aparicio, P. J. et al 1971. *Phytochemistry* 10:1487–95
14. Apte, B. N., Zipser, D. 1973. *Proc. Nat. Acad. Sci. USA* 70:2969–73
15. Arst, H. N., Cove, D. J. 1973. *Mol. Gen. Genet.* 126:111–41
16. Arst, H. N., MacDonald, D. W. 1973. *Mol. Gen. Genet.* 122:261–65
17. Arst, H. N., MacDonald, D. W., Cove, D. J. 1970. *Mol. Gen. Genet.* 108:129–45
18. Asada, K., Tamura, G., Bandurski, R. S. 1969. *J. Biol. Chem.* 244:4904–15
19. Aslam, M., Huffaker, R. C. 1973. *Physiol. Plant.* 28:400–4
20. Aslam, M., Huffaker, R. C., Travis, R. L. 1973. *Plant Physiol.* 52:137–41
21. Atkinson, D. E., Walton, G. M. 1967. *J. Biol. Chem.* 242:3239–41
22. Azoulay, E., Puig, J. 1968. *Biochem. Biophys. Res. Commun.* 33:1019–24

23. Bassioni, N. H. 1973. *Agrochimica* 17:341–46
24. Bayley, J. M., King, J., Gamborg, O. L. 1972. *Planta* 105:15–24
25. Ibid, 25–32
26. Beevers, L., Hageman, R. H. 1969. *Ann. Rev. Plant Physiol.* 20:495–522
27. Beevers, L., Hageman, R. H. 1972. In *Photophysiology,* ed. A. C. Giese, 7:85–113. New York/London: Academic
28. Bell, R. M., Koshland, D. E. 1971. *Science* 172:1253–56
29. Benemann, J. R., McKenna, C. E., Lie, R. F., Traylor, T. G., Kamen, M. D. 1972. *Biochim. Biophys. Acta* 264:25–38
30. Benemann, J. R., Smith, G. M., Kostel, P. J., McKenna, C. E. 1973. *FEBS Lett.* 29:219–21
31. Ben Zioni, A., Vaadia, Y., Lips, S. H. 1971. *Physiol. Plant.* 24:288–90
32. Betts, G. F. 1965. *Ferredoxin in nitrite and hydroxylamine reductase systems from higher plants.* PhD thesis. Bristol Univ., England. 151 pp.
33. Betts, G. F., Hewitt, E. J. 1966. *Nature* 210:1327–29
34. Bidlack, W. R., Okita, R. T., Hochstein, P. 1973. *Biochem. Biophys. Res. Commun.* 53:459–65
35. Bilal, I. M., Rains, D. W. 1973. *Physiol. Plant.* 28:237–43
36. Boer, J. de, Feierabend, J. 1974. *Z. Pflanzenphysiol.* 71:261–70
37. Borriss, H. 1967. *Wiss. Z. Univ. Rostock Math.-Naturwiss. Reihe* 16:629–39
38. Bothe, H. 1969. *Progr. Photosyn. Res.* 3:1483–91
39. Bourne, W. F., Miflin, B. J. 1970. *Biochem. Biophys. Res. Commun.* 40:1305–10
40. Bourne, W. F., Miflin, B. J. 1973. *Planta* 111:47–56
41. Boyer, J. S. 1973. *Phytopathology* 63:466–72
42. Briggs, W. R., Rice, H. V. 1972. *Ann. Rev. Plant Physiol.* 23:293–334
42a. Brill, W. J., Steiner, A. L., Shah, V. K. 1974. *J. Bacteriol.* 118:986–89
43. Buchanan, B. B., Kalberer, P. P., Arnon, D. I. 1967. *Biochem. Biophys. Res. Commun.* 29:74–79
44. Buchanan, B. B., Schürmann, P. 1973. *Curr. Top. Cell. Regul.* 7:1–20
45. Buchanan, B. B., Schurmann, P., Kalberer, P. P. 1971. *J. Biol. Chem.* 246:5952–59
46. Buczek, J. 1973. *Acta Soc. Bot. Pol.* 42:223–32
47. Burns, R. C., Fuchsman, W. H., Hardy, R. W. F. 1971. *Biochem. Biophys. Res. Commun.* 42:353–58

48. Cardenas, J., Barea, J. L., Rivas, J., Moreno, C. G. 1972. *FEBS Lett.* 23:131–35
49. Cardenas, J., Rivas, J., Barea, J. L. 1972. *Publ. Rev. Real Acad. Cienc. Exact Fis. Nat. Madrid* 66:565–77
50. Cardenas, J., Rivas, J., Paneque, A., Losada, M. 1972. *Arch. Mikrobiol.* 81:260–63
50a. Chiba, S., Ishimoto, M. 1973. *J. Biochem.* 73:1315–18
51. Cohen, H. J., Fridovich, I. 1971. *J. Biol. Chem.* 246:359–66
52. Ibid, 367–73
53. Cohen, H. J., Fridovich, I., Rajagopalan, K. V. 1971. *J. Biol. Chem.* 246:374–82
54. Cole, J. A. 1968. *Biochim. Biophys. Acta* 162:356–58
55. Cole, J. A., Ward, F. B. 1973. *J. Gen. Microbiol.* 76:21–29
56. Cove, D. J. 1970. *Proc. Roy. Soc. B* 176:267–75
57. Cove, D. J., Arst, H. N., Scazzocchio, C. 1975. *Bacteriol. Rev.* In press
58. Cove, D. J., Pateman, J. A. 1963. *Nature* 198:262–63
59. Cram, W. J. 1973. *J. Exp. Bot.* 24:328–41
60. Cresswell, C. F., Hageman, R. H., Hewitt, E. J. 1962. *Biochem. J.* 83:38p–39p
61. Cresswell, C. F., Hageman, R. H., Hewitt, E. J., Hucklesby, D. P. 1965. *Biochem. J.* 94:40–53
62. Crofts, A. R. 1966. *Biochem. Biophys. Res. Commun.* 24:127–34
63. Cumming, B. G., Wagner, E. 1968. *Ann. Rev. Plant Physiol.* 19:381–416
63a. Czygan, F. C. 1963. *Planta* 60:225–42
64. Dalling, M. J., Hucklesby, D. P., Hageman, R. H. 1973. *Plant Physiol.* 51:481–84
65. Dalling, M. J., Tolbert, N. E., Hageman, R. H. 1972. *Biochim. Biophys. Acta* 283:505–12
66. Dalling, M. J., Tolbert, N. E., Hageman, R. H. 1972. *Biochim. Biophys. Acta* 283:513–19
67. Dodge, A. D. 1971. *Endeavour* 30:130–35
68. Downey, R. J. 1971. *J. Bacteriol.* 105:759–68
69. Downey, R. J. 1972. *Am. Soc. Microbiol. Ann. Meet. Abstr.* No. P236, p. 175
70. Downey, R. J. 1973. *Microbios* 7:53–60
71. Downey, R. J. 1973. *Biochem. Biophys. Res. Commun.* 50:920–25
72. Downey, R. J., Cove, D. J. 1971. *J. Bacteriol.* 106:1047–49

73. Eaglesham, A. R. J., Hewitt, E. J. 1970. *Rep. Long Ashton Res. Sta. 1969,* 52–53
74. Eaglesham, A. R. J., Hewitt, E. J. 1971. *Biochem. J.* 122:18–19p
75. Eaglesham, A. R. J., Hewitt, E. J. 1971. *FEBS Lett.* 16:315–18
76. Enoch, H. G., Lester, R. L. 1972. *J. Bacteriol.* 110:1032–40
77. Entsch, B., Smillie, R. M. 1972. *Arch. Biochem. Biophys.* 151:378–86
78. Ferrari, T. E., Varner, J. E. 1969. *Plant Physiol.* 44:85–88
79. Ferrari, T. E., Varner, J. E. 1970. *Proc. Nat. Acad. Sci. USA* 65:729–36
80. Ferrari, T. E., Varner, J. E. 1971. *Plant Physiol.* 47:790–94
81. Ferrari, T. E., Yoder, O. C., Filner, P. 1973. *Plant Physiol.* 51:423–31
82. Forget, P. 1971. *Eur. J. Biochem.* 18:442–50
83. Ibid 1974. 42:325–32
84. Forget, P., Dervartanian, D. V. 1972. *Biochim. Biophys. Acta* 256:600–6
85. Fowden, L., Lewis, D., Tristram, H. 1967. *Advan. Enzymol.* 29:89–163
86. Gandhi, A. P., Naik, M. S. 1974. *FEBS Lett.* 40:343–45
87. Gandhi, A. P., Sawhney, S. K., Naik, M. S. 1973. *Biochem. Biophys. Res. Commun.* 55:291–96
88. Ganelin, V. L., L'vov, N. P., Sergeev, N. S., Shaposhnikov, G. L., Kretovich, V. L. 1972. *Dokl. Akad. Nauk SSSR* 206:1236–38
89. Garner, C. D., Hyde, M. R., Mabbs, F. E., Routledge, V. J. 1974. *Nature* 252:579–80
90. Garrett, R. H., Greenbaum, P. 1973. *Biochim. Biophys. Acta* 302:24–32
91. Garrett, R. H., Nason, A. 1967. *Proc. Nat. Acad. Sci. USA* 58:1603–10
92. Garrett, R. H., Nason, A. 1969. *J. Biol. Chem.* 244:2870–82
93. Gerhardt, B. 1964. *Planta* 61:101–29
93a. Gewitz, H. S., Lorimer, G. H., Solomonson, L. P., Vennesland, B. 1974. *Nature* 249:79–81
94. Glaser, J. H., DeMoss, J. A. 1971. *J. Bacteriol.* 108:854–60
95. Goldsmith, J., Livoni, J. P., Norberg, C. L., Segel, I. H. 1973. *Plant Physiol.* 52:362–67
96. Gomez-Moreno, C., Palacian, E. 1974. *Arch. Biochem. Biophys.* 160:269–73
97. Grant, B. R. 1970. *Plant Cell Physiol.* 11:55–64
98. Grant, B. R., Atkins, C. A., Canvin, D. T. 1970. *Planta* 94:60–72
99. Grant, B. R., Canvin, D. T. 1970. *Planta* 95:227–46
100. Grant, B. R., La Belle, R., Mangat, B. S. 1972. *Planta* 106:181–84
101. Guerrero, M. G., Vega, J. M., Leadbetter, E., Losada, M. 1973. *Arch. Mikrobiol.* 91:287–304
102. Hadacova, V., Sahulka, J. 1973. *Biol. Plant.* 15:346–49
103. Hageman, R. H., Cresswell, C. F., Hewitt, E. J. 1962. *Nature* 193:247–50
104. Hageman, R. H., Hucklesby, D. P. 1971. *Methods Enzymol.* 23A:419–503
105. Harris, N., Dodge, A. D. 1972. *Planta* 104:201–9
106. Hartingsveldt, J. van, Stouthamer, A. H. 1972. *Antonie van Leeuwenhoek* 38:447
107. Hattori, A., Myers, J. 1967. *Plant Cell Physiol.* 8:327–37
108. Hattori, A., Uesugi, I. 1968. *Plant Cell Physiol.* 9:689–99
109. Heber, U. 1974. *Ann. Rev. Plant Physiol.* 25:393–421
110. Heimer, Y. M., Filner, P. 1970. *Biochim. Biophys. Acta* 215:152–65
111. Ibid 1971. 230:362–72
112. Heimer, Y. M., Wray, J. L., Filner, P. 1969. *Plant Physiol.* 44:1197–99
112a. Herrera, J., Nicholas, D. J. D. 1974. *Biochim. Biophys. Acta* 368:54–60
113. Herrera, J., Paneque, A., Maldonado, J. M., Barea, J. L., Losada, M. 1972. *Biochem. Biophys. Res. Commun.* 48:996–1003
114. Hewitt, E. J. 1970. In *Nitrogen Nutrition of the Plant,* ed. E. A. Kirkby, 78–103. Univ. Leeds
115. Hewitt, E. J. 1974. *MTP Int. Rev. Sci. Biochem. Ser.* 1, 11:199–245
116. Hewitt, E. J., Agarwala, S. C., Jones, E. W. 1950. *Nature* 166:1119–20
117. Hewitt, E. J., Betts, G. F. 1963. *Biochem. J.* 89:20p
118. Hewitt, E. J., Hucklesby, D. P. 1966. *Biochem. Biophys. Res. Commun.* 25:689–93
119. Hewitt, E. J., Hucklesby, D. P., Betts, G. F. 1968. In *Recent Aspects of Nitrogen Metabolism in Plants,* ed. E. J. Hewitt, C. V. Cutting, 47–81. New York/London: Academic, 1968
120. Hewitt, E. J., Hucklesby, D. P., James, D. M. 1969. *Rep. Long Ashton Res. Sta. 1968,* 33–34
121. Hewitt, E. J., Jones, R. W., Abbott, A. J., Best, G. R. 1974. *3rd Int. Congr. Plant Tissue Cell Culture, Leicester.* Abstr. 131
122. Hewitt, E. J., Nicholas, D. J. D. 1964. In *Modern Methods of Plant Analysis,* ed. H. F. Linskens, B. D. Sanwal, M. V. Tracey, 7:67–122. Berlin: Springer

123. Hewitt, E. J., Notton, B. A. 1966. *Biochem. J.* 101:39–40C
124. Hewitt, E. J., Notton, B. A. 1967. *Phytochemistry* 6:1329–35
125. Hiller, R. G., Bassham, J. A. 1965. *Biochim. Biophys. Acta* 109:607–10
126. Hirose, S., Yamashita, K., Shibata, K. 1971. *Plant Cell Physiol.* 12:775–78
127. Hirschberg, K., Hubner, G., Borriss, H. 1972. *Planta* 108:333–37
128. Ho, C. H., Tamura, G. 1973. *Agr. Biol. Chem.* 37:37–44
129. Hodler, M., Morgenthaler, J. J., Eichenberger, W., Grob, E. C. 1972. *FEBS Lett.* 28:19–21
130. Hofmann, A. 1972. *Planta* 102:72–84
131. Hucklesby, D. P., Dalling, M. J., Hageman, R. H. 1972. *Planta* 104:220–33
132. Hucklesby, D. P., Hageman, R. H. 1973. *Anal. Biochem.* 56:591–92
133. Hucklesby, D. P., Hewitt, E. J. 1970. *Biochem. J.* 119:615–27
134. Hucklesby, D. P., Hewitt, E. J., James, D. M. 1970. *Biochem. J.* 117:30p
135. Hucklesby, D. P., James, D. M., Banwell, M. J., Hewitt, E. J. 1974 *Rep. Long Ashton Res. Sta 1973,* 70–71
136. Hucklesby, D. P., James, D. M., Hewitt, E. J. 1974. *Biochem. Soc. London Trans.* 2:436–37
137. Huffaker, R. C., Peterson, L. W. 1974. *Ann. Rev. Plant Physiol.* 25:363–92
138. Husain, M., Sadana, J. C. 1972. *Anal. Biochem.* 45:316–19
138a. Husain, M., Sadana, J. C. 1974. *Eur. J. Biochem.* 42:283–89
138b. Husain, M., Sadana, J. C. 1974. *Arch. Biochem. Biophys.* 163:21–28
139. Huzisige, H., Sato, K., Tanaka, K., Hayashida, T. 1963. *Plant Cell Physiol.* 4:307–22
140. Ida, S. 1974. *Bull. Res. Inst. Food Sci. Kyoto Univ.* 37:60–78
141. Ida, S., Morita, Y. 1973. *Plant Cell Physiol.* 14:661–71
142. Ivanova, N. N., Peive, Y. V. 1973. *FEBS Lett.* 31:229–32
143. Jackson, W. A., Flesher, D., Hageman, R. H. 1973. *Plant Physiol.* 51:120–27
144. Jagendorf, A. T., Uribe, E. 1966. *Brookhaven Symp. Biol.* 19: Energy Conversion by the Photosynthetic Apparatus, 215–45
145. Jetschmann, K., Solomonson, L. P., Vennesland, B. 1972. *Biochim. Biophys. Acta* 275:276–78
146. Jetschmann, K., Vennesland, B. 1973. *Abstr. Int. Biochem. Congr. Stockholm*
147. John, P. C. L., Thurston, C. F., Syrett, P. J. 1970. *Biochem. J.* 119:913–19
147a. Johnson, J. L., Wand, W. R., Cohen, H. J., Rajagopalan, K. V. 1974. *J. Biol. Chem.* 249:5056–61
148. Jones, R. W., Sheard, R. W. 1972. *Nature* 238:221–22
149. Jones, R. W., Sheard, R. W. 1973. *Can. J. Bot.* 51:27–35
150. Joy, K. W. 1969. *Plant Physiol.* 44:849–53
151. Joy, K. W., Hageman, R. H. 1966. *Biochem. J.* 100:263–73
151a. Kavam, S. K., Gandhi, A. P., Sawhney, S. K., Naik, M. S. 1974. *Biochim. Biophys. Acta* 350:162–70
152. Kahl, G., Stegemann, H. 1973. *FEBS Lett.* 32:325–29
153. Kannangara, C. G., Woolhouse, H. W. 1967. *New Phytol.* 66:553–61
154. Kaplan, D., Roth-Bejerano, N., Lips, S. H. 1974. *Eur. J. Biochem.* 49:393–98
155. Kaplan, F., Setlow, P., Kaplan, N. O. 1969. *Arch. Biochem. Biophys.* 132:91–98
156. Katunuma, N., Kominami, E., Kominami, S. 1971. *Biochem. Biophys. Res. Commun.* 45:70–75
157. Keefer, L. K., Roller, P. P. 1973. *Science* 181:1245–46
158. Kelker, H. C., Filner, P. 1971. *Biochim. Biophys. Acta* 252:69–82
159. Kemp, J. D., Atkinson, D. E. 1966. *J. Bacteriol.* 92:628–34
160. Kemp, J. D., Atkinson, D. E., Ehret, A., Lazzarini, R. A. 1963. *J. Biol. Chem.* 238:3466–71
161. Kende, H., Hahn, H., Kays, S. E. 1971. *Plant Physiol.* 48:702–6
162. Kende, H., Shen, T. C. 1972. *Biochim. Biophys. Acta* 286:118–25
163. Kessler, D. L., Rajagopalan, K. V. 1972. *J. Biol. Chem.* 247:6566–73
164. Kessler, E. 1957. *Planta* 49:505–23
165. Kessler, E. 1957. *Arch. Mikrobiol.* 27:166–81
166. Kessler, E., Czygan, F. C. 1968. *Arch. Mikrobiol.* 60:282–84
167. Kessler, E., Hofmann, A., Zumft, W. G. 1970. *Arch. Mikrobiol.* 72:23–26
168. Kessler, E., Zumft, W. G. 1973. *Planta* 111:41–46
169. Ketchum, P. A., Cambier, H. Y., Frazier, W. A., Madansky, C. H., Nason, A. 1970. *Proc. Nat. Acad. Sci. USA* 66:1016–23
170. Ketchum, P. A., Sevilla, C. L. 1973. *J. Bacteriol.* 116:600–9
171. Ketchum, P. A., Swarin, R. S. 1973. *Biochem. Biophys. Res. Commun.* 52:1450–56
172. King, J. 1974. *3rd Int. Congr. Plant Tissue Cell Culture, Leicester.* Abstr. 133

173. Klepper, L., Flesher, D., Hageman, R. H. 1971. *Plant Physiol.* 48:580–90
174. Knypl, J. S. 1973. *Z. Pflanzenphysiol.* 70:1–11
175. Knypl, J. S. 1973. *Planta* 114:311–21
176. Knypl, J. S. 1974. *Z. Pflanzenphysiol.* 71:37–48
177. Kobayashi, K., Seri, Y., Ishimoto, M. 1974. *J. Biochem.* 75:519–29
178. Krogmann, D. W., Jagendorf, A. T., Avron, M. 1959. *Plant Physiol.* 34:272–77
178a. Lafferty, M. A., Garrett, R. H. 1974. *J. Biol. Chem.* 249:7555–67
179. Lam, Y., Nicholas, D. J. D. 1969. *Biochim. Biophys. Acta* 178:225–34
180. Latimer, W. 1952. *The Oxidation States of the Elements and Their Potentials in Aqueous Solutions.* New York: Prentice Hall
181. Lea, P. J., Miflin, B. J. 1974. *Nature* 251:614–16
182. Lee, J. P., LeGall, J., Peck, H. D. 1973. *J. Bacteriol.* 115:529–42
183. Lee, K. Y., Pan, S. S., Erickson, R. H., Nason, A. 1973. *Fed. Proc.* 32:629. Abstr. 2335
184. Lee, K. Y., Pan, S. S., Erickson, R. H., Nason, A. 1974. *J. Biol. Chem.* 249:3941–52
185. Lee, K. Y., Erickson, R. H., Pan, S. S., Jones, G., May, F., Nason, A. 1974. *J. Biol. Chem.* 249:3953–59
186. LeGall, J., Dervartanian, D. V., Spilker, E., Lee, J. P., Peck, H. D. 1971. *Biochim. Biophys. Acta* 234:525–30
187. Lester, R. L., DeMoss, J. A. 1971. *J. Bacteriol.* 105:1006–14
188. Lewis, C. M., Fincham, J. R. S. 1970. *J. Bacteriol.* 103:55–61
189. Lips, S. H. 1972. *Proc. 2nd Int. Congr. Photosyn. Stressa, Italy, 1971,* ed. G. Forti, M. Avron, A. Melandri, 2241–49. The Hague: Junk
190. Lips, S. H. 1974. *8th Int. Conf. Plant Growth Substances, Tokyo, 1973*
190a. Lips, S. H. *Plant Physiol.* In press
191. Lips, S. H., Avissar, Y. 1972. *Eur. J. Biochem.* 29:20–24
192. Lips, S. H., Kaplan, D., Roth-Bejerano, N. 1973. *Eur. J. Biochem.* 37:589–92
193. Lips, S. H., Roth-Bejerano, N. 1969. *Science* 166:109–10
194. Losada, M. 1973. *3rd Int. Symp. Metab. Interconversion of Enzymes, Seattle*
195. Losada, M., Aparicio, P. J., Paneque, A. 1969. *Proc. Symp. Progr. Photosyn. Res. Tübingen* 3:1504–9
196. Losada, M., Herrera, J., Maldonado, J. M., Paneque, A. 1973. *Plant Sci. Lett.* 1:31–37
197. Losada, M., Paneque, A., Ramirez, J. M., Del Campo, F. F. 1963. *Biochem. Biophys. Res. Commun.* 10:298–303
198. Losada, M. et al 1970. *Biochem. Biophys. Res. Commun.* 38:1009–15
199. Loussaert, D., Hageman, R. H. 1974. *Plant Physiol.* Suppl. p. 65, Abstr. 367
199a. MacDonald, D. W., Coddington, A. 1974. *Eur. J. Biochem.* 46:169–78
200. MacDonald, D. W., Cove, D. J., Coddington, A. 1974. *Mol. Gen. Genet.* 128:187–99
201. MacGregor, C. H., Schnaitman, C. A. 1972. *J. Bacteriol.* 112:388–91
202. Magalhaes, A. C., Neyra, C. A., Hageman, R. H. 1974. *Plant Physiol.* 53:411–15
203. Maldonado, J. M., Herrera, J., Paneque, A., Losada, M. 1973. *Biochem. Biophys. Res. Commun.* 51:27–33
204. Maranville, J. W. 1970. *Plant Physiol.* 45:591–93
205. Marcot, J., Azoulay, E. 1971. *FEBS Lett.* 13:137–39
206. Maretzki, A., de la Cruz, A. 1967. *Plant Cell Physiol.* 8:605–11
207. Martin, P. 1973. *Z. Pflanzenphysiol.* 70:158–65
208. McKenna, C. E., Benemann, J. R., Traylor, T. G. 1970. *Biochem. Biophys. Res. Commun.* 41:1501–8
209. Menary, R. C., Jones, R. H. 1972. *Aust. J. Biol. Sci.* 25:531–42
210. Miflin, B. J. 1968. See Ref. 119, 85–88
211. Miflin, B. J. 1970. *Rev. Roum. Biochim.* 1:53–60
212. Miflin, B. J. 1972. *Planta* 105:225–33
213. Ibid 1974. 116:187–96
214. Miflin, B. J. 1974. *Plant Physiol.* 54:550–55
215. Mitsui, A., San Pietro, A. 1973. *Plant Sci. Lett.* 1:157–63
216. Moreno, C. G., Aparicio, P. J., Palacian, E., Losada, M. 1972. *FEBS Lett.* 26:11–14
217. Morgan, M. A., Volk, R. J., Jackson, W. A. 1973. *Plant Physiol.* 51:267–72
218. Morilla, C. A., Boyer, J. S., Hageman, R. H. 1973. *Plant Physiol.* 51:817–24
219. Morris, I., Syrett, P. J. 1963. *Biochim. Biophys. Acta* 77:649–50
220. Murphy, M. J., Siegel, L. M. 1973. *J. Biol. Chem.* 248:6911–19
221. Murphy, M. J., Siegel, L. M., Kamin, H., Rosenthal, D. 1973. *J. Biol. Chem.* 248:2801–14
222. Murphy, M. J., Siegel, L. M., Tove, S. R., Kamin, H. 1974. *Proc. Nat. Acad. Sci. USA* 71:612–16

223. Murphy, M. J. et al 1973. *Biochem. Biophys. Res. Commun.* 54:82–88
223a. Nagatani, H. H., Brill, W. J. 1974. *Biochim. Biophys. Acta* 362:160–66
224. Nason, A. 1962. *Bacteriol. Rev.* 26:16–41
225. Nason, A., Antoine, A. D., Ketchum, P. A., Frazier, W. A., Lee, D. K. 1970. *Proc. Nat. Acad. Sci. USA* 65:137–44
226. Nason, A., Lee, K. Y., Pan, S. S., Erickson, R. H. 1973. *First International Conference on Chemistry and Uses of Molybdenum*, ed. P. C. H. Mitchell, 233–39. London: Climax Molybdenum
227. Nason, A. et al 1971. *Proc. Nat. Acad. Sci. USA* 68:3242–46
228. Nelson, N., Ilan, I. 1969. *Plant Cell Physiol.* 10:143–48
229. Nicholas, D. J. D. 1957. *Ann. Bot. London N.S.* 21:587–98
230. Nicholas, D. J. D., Wilson, P. J. 1964. *Biochim. Biophys. Acta* 86:466–76
231. Notton, B. A., Graf, L., Hewitt, E. J., Povey, R. C. 1974. *Biochim. Biophys. Acta* 364:45–58
232. Notton, B. A., Hewitt, E. J. 1971. *Plant Cell Physiol.* 12:465–77
233. Notton, B. A., Hewitt, E. J. 1971. *Biochem. Biophys. Res. Commun.* 44:702–10
234. Notton, B. A., Hewitt, E. J. 1971. *FEBS Lett.* 18:19–22
235. Notton, B. A., Hewitt, E. J. 1972. *Biochim. Biophys. Acta* 275:355–57
236. Notton, B. A., Hewitt, E. J. 1973. See Ref. 226, 228–32
237. Notton, B. A., Hewitt, E. J., Fielding, A. H. 1972. *Phytochemistry* 11:2447–49
237a. Oaks, A. 1974. *Biochim. Biophys. Acta* 372:112–16
238. Oaks, A., Wallace, W., Stevens, D. 1972. *Plant Physiol.* 50:649–54
239. Oji, Y., Izawa, G. 1969. *Plant Cell Physiol.* 10:743–49
239a. O'Neal, D., Joy, K. W. 1973. *Nature New Biol.* 246 (150):61–62
240. Onwueme, I. C., Laude, H. M., Huffaker, R. C. 1971. *Crop Sci.* 11:195–200
241. Orebamjo, T. O., Stewart, G. R. 1974. *Planta* 117:1–10
242. Palacian, E., de la Rosa, F., Castillo, F., Gomez-Moreno, C. 1974. *Arch. Biochem. Biophys.* 161:441–47
243. Paneque, A., Del Campo, F. F., Losada, M. 1963. *Nature* 198:90–91
244. Paneque, A., Ramirez, J. M., Del Campo, F. F., Losada, M. 1964. *J. Biol. Chem.* 239:1737–41
245. Paneque, A. et al 1972. *Plant Cell Physiol.* 13:175–78

246. Parkash, V. 1972. *Planta* 102:372–73
247. Pateman, J. A., Cove, D. J., Rever, B. M., Roberts, D. B. 1964. *Nature* 201:58–60
248. Pateman, J. A., Kinghorn, J. R., Dunn, E., Forbes, E. 1973. *J. Bacteriol.* 114:943–50
249. Pederson, T. C., Aust, S. D. 1973. *Biochem. Biophys. Res. Commun.* 52:1071–78
250. Peive, Y. V., Ivanova, N. N. 1970. *Dokl. Akad. Nauk SSSR* 195:1456–59
251. Peive, Y. V., Ivanova, N. N., Drobysheva, N. I. 1972. *Fiziol. Rast.* 19:340–47
252. Pichinoty, F. 1973. *Bull. Inst. Pasteur* 71:317–95
253. Pickard, M., Hewitt, E. J. 1973. *Rep. Long Ashton Res. Sta. 1972*, 70–71
254. Pinsent, J. 1954. *Biochem. J.* 57:10–16
255. Plaut, Z. 1974. *Physiol. Plant.* 30:212–47
256. Prakash, O., Sadana, J. C. 1972. *Arch. Biochem. Biophys.* 148:614–32
257. Prakash, O., Sadana, J. C. 1973. *Can. J. Microbiol.* 19:15–25
258. Ramirez, J. M., Del Campo, F. F., Paneque, A., Losada, M. 1966. *Biochim. Biophys. Acta* 118:58–71
259. Rao, P. S., Hayon, E. 1973. *Nature* 243:344–46
260. Relimpio, A. M., Aparicio, P. J., Paneque, A., Losada, M. 1971. *FEBS Lett.* 17:226–30
261. Ries, S. K., Wert, V. 1972. *Weed Sci.* 20:569–72
262. Rigano, C. 1971. *Arch. Mikrobiol.* 76:265–76
263. Rigano, C., Aliotta, G., Violante, U. 1974. *Plant Sci. Lett.* 2:277–81
264. Rigano, C., Aliotta, G., Violante, U. 1974. *Arch. Mikrobiol.* 99:81–90
265. Rigano, C., Violante, U. 1972. *Biochem. Biophys. Res. Commun.* 47:372–79
266. Rigano, C., Violante, U. 1972. *Biochim. Biophys. Acta* 256:524–32
267. Rigano, C., Violante, U. 1973. *Arch. Mikrobiol.* 90:27–33
268. Rijven, A. H. G. C., Parkash, V. 1971. *Plant Physiol.* 47:59–64
269. Ritenour, G. L., Joy, K. W., Bunning, J., Hageman, R. H. 1967. *Plant Physiol.* 42:233–37
270. Rivas, J., Guerrero, M. G., Paneque, A., Losada, M. 1973. *Plant Sci. Lett.* 1:105–13
271. Rivas, J., Tortolero, M., Paneque, A. 1974. *Plant Sci. Lett.* 2:283–88
272. Rosso, J. P., Forget, P., Pichinoty, F. 1973. *Biochim. Biophys. Acta* 321:443–55

273. Roth-Bejerano, N., Lips, S. H. 1973. *Isr. J. Bot.* 22:1–7
274. Roth-Bejerano, N., Lips, S. H. 1973. *New Phytol.* 72:253–57
275. Roth-Bejerano, N., Lips, S. H. 1975. *Plant Physiol.* In press
276. Ruiz-Herrera, J., Showe, M. K., De Moss, J. A. 1969. *J. Bacteriol.* 97:1291–97
277. Sadana, J. C., McElroy, W. D. 1957. *Arch. Biochem. Biophys.* 67:16–34
277a. Sagai, M., Ishimoto, M. 1973. *J. Biochem.* 73:843–59
278. Sahulka, J. 1973. *Biol.Plant.* 15:298–301
279. Sanderson, G. W., Cocking, E. C. 1964. *Plant Physiol.* 39:423–31
280. Sawhney, S. K., Naik, M. S. 1972. *Biochem. J.* 130:475–85
281. Sawhney, S. K., Naik, M. S. 1973. *Biochem. Biophys. Res. Commun.* 51:67–73
282. Sawhney, S. K., Prakash, V., Naik, M. S. 1972. *FEBS Lett.* 22:200–2
283. Scazzocchio, C. 1973. See Ref 226, 240–42
284. Scazzocchio, C. 1973. *Mol. Gen. Genet.* 125:147–55
285. Scazzocchio, C., Holl, F. B., Foguelman, A. I. 1973. *Eur. J. Biochem.* 36:428–45
286. Schloemer, R. H., Garrett, R. H. 1973. *Plant Physiol.* 51:591–93
287. Schloemer, R. H., Garrett, R. H. 1974. *J. Bacteriol.* 118:259–69
288. Ibid, 270–74
289. Schrader, L. E., Beevers, L., Hageman, R. H. 1967. *Biochem. Biophys. Res. Commun.* 26:14–17
290. Schrader, L. E., Hageman, R. H. 1967. *Plant Physiol.* 42:1750–56
291. Schrader, L. E., Ritenour, G. L., Eilrich, G. L., Hageman, R. H. 1968. *Plant Physiol.* 43:930–40
292. Shen, T. C. 1972. *Plant Physiol.* 49:546–49
293. Shen, T. C. 1972. *Planta* 108:21–28
294. Shimizo, J., Tamura, G. 1974. *J. Biochem.* 75:999–1005
295. Shin, M., Oda, Y. 1966. *Plant Cell Physiol.* 7:643–50
296. Shin, M., Tagawa, K., Arnon, D. I. 1963. *Biochem. Z.* 338:84–96
297. Showe, M. K., De Moss, J. A. 1968. *J. Bacteriol.* 95:1305–13
298. Siegel, L. M., Kamin, H. 1967. *Flavins and Flavoproteins,* 2nd Conf., ed. K. Yagi, 15–40. Tokyo, Baltimore, Manchester: Univ. Tokyo Press and Univ. Park Press. 1968.
299. Siegel, L. M., Kamin, H. 1971. *Methods Enzymol.* 17:539–45
300. Siegel, L. M., Kamin, H., Rueger, D. C., Presswood, R. P., Gibson, Q. H. 1969. *Flavins and Flavoproteins,* ed. H. Kamin, 523–54. 3rd Int. Symp. Durham, N. C. Baltimore and London: Univ. Park Press and Butterworth's
301. Siegel, L. M., Murphy, M. J., Kamin, H. 1973. *J. Biol. Chem.* 248:251–64
302. Sims, A. P., Folkes, B. F., Bussey, A. H. 1968. See Ref. 119, 91–114
303. Sluiters-Scholten, C. M. T. 1973. *Planta* 113:229–40
304. Smith, F. A. 1973. *New Phytol.* 72:769–82
305. Smith, F. W., Thompson, J. F. 1971. *Plant Physiol.* 48:219–23
306. Solomonson, L. P. 1974. *Biochim. Biophys. Acta* 334:297–308
307. Solomonson, L. P., Jetschmann, K., Vennesland, B. 1973. *Biochim. Biophys. Acta* 309:32–43
308. Solomonson, L. P., Vennesland, B. 1972. *Plant Physiol.* 50:421–24
309. Solomonson, L. P., Vennesland, B. 1972. *Biochim. Biophys. Acta* 267:544–57
310. Sorger, G. J. 1963. *Biochem. Biophys. Res. Commun.* 12:395–401
311. Sorger, G. J. 1966. *Biochim. Biophys. Acta* 118:484–94
312. Sorger, G. J., Davies, J. 1973. *Biochem. J.* 134:673–85
313. Sorger, G. J., Debanne, M. T., Davies, J. 1974. *Biochem. J.* 140:395–403
314. Sorger, G. J., Giles, N. H. 1965. *Genetics* 52:777–88
315. Spencer, D. 1959. *Aust. J. Biol. Sci.* 12:181–91
316. Spilker, E., LeGall, J., Dervartanian, D. V. 1972. *Am. Soc. Microbiol. Ann. Meet.,* Abstr. No. P233, p. 174
317. Stewart, G. R. 1972. *J. Exp. Bot.* 23:171–83
318. Stewart, G. R. 1972. *Symp. Biol. Hung.* 13:127–35
319. Subramanian, K. N., Sarma, P. S. 1969. *Indian J. Biochem.* 6:235–36
320. Subramanian, K. N., Sorger, G. J. 1972. *J. Bacteriol.* 110:538–46
321. Ibid, 547–53
322. Subramanian, K. N., Sorger, G. J. 1972. *Biochim. Biophys. Acta* 256:533–43
323. Swader, J. A., Stocking, C. R. 1971. *Plant Physiol.* 47:189–91
324. Syrett, P. J., Hipkin, C. R. 1973. *Planta* 111:57–64
325. Taniguchi, S. 1961. *Z. Allg. Mikrobiol.* 1:341–75

326. Thacker, A., Syrett, P. J. 1972. *New Phytol.* 71:423–33
327. Ibid, 435–41
328. Thauer, R. K., Fuchs, G., Jungermann, K. 1974. *J. Bacteriol.* 118:758–60
329. Thauer, R. K., Fuchs, G., Schnitker, U., Jungermann, K. 1973. *FEBS Lett.* 38:45–48
330. Thauer, R. K., Kaüfer, B., Zähringer, M., Jungermann, K. 1974. *Eur. J. Biochem.* 42:447–52
331. Travis, R. L., Jordan, W. R., Huffaker, R. C. 1969. *Plant Physiol.* 44:1150–56
332. Trebst, A. 1974. *Ann. Rev. Plant Physiol.* 25:423–58
333. Ullrich, W. R. 1974. *Planta* 116:143–52
334. Upcroft, J. A., Done, J. 1972. *FEBS Lett.* 21:142–44
334a. Upcroft, J. A., Done, J. 1974. *J. Exp. Bot.* 25:503–8
335. Vega, J. M., Guerrero, M. G., Leadbetter, E., Losada, M. 1973. *Biochem. J.* 133:701–8
336. Vega, J. M., Herrera, J., Aparicio, P. J., Paneque, A., Losada, M. 1971. *Plant Physiol.* 48:294–99
337. Vega, J. M., Herrera, J., Relimpio, A. M., Aparicio, P. J. 1972. *Physiol. Veg.* 10:637–52
338. Vennesland, B., Jetschmann, C. 1971. *Biochim. Biophys. Acta* 227:554–64
339. Vennesland, B., Solomonson, L. P. 1972. *Plant Physiol.* 49:1029–31
340. Villarreal-Moguel, E. I., Ibarra, V., Ruiz-Herrera, J., Gitler, C. 1973. *J. Bacteriol.* 113:1264–67
341. Vunkova, R. V., Vaklinova, S. G. 1972. *C. R. Acad. Bulg. Sci.* 25:677–80

342. Walker, G. C., Nicholas, D. J. D. 1961. *Nature* 189:141–42
343. Wallace, W. 1973. *Plant Physiol.* 52:197–201
344. Wallace, W. 1974. *Biochim. Biophys. Acta* 341:265–76
344a. Wells, G. N., Hageman, R. H. 1974. *Plant Physiol.* 54:136–41
345. Williams, R. J. P. 1972. *J. Bioenerg.* 3:81–93
346. Wray, J. L., Filner, P. 1970. *Biochem. J.* 119:715–25
347. Yoch, D. C. 1972. *Biochem. Biophys. Res. Commun.* 49:335–42
348. Yoshimoto, A., Naiki, N., Sato, R. 1971. *Methods Enzymol.* 17B:520–28
349. Yoshimoto, A., Nakamura, T., Sato, R. 1967. *J. Biochem.* 62:756–66
350. Yoshimoto, A., Sato, R. 1968. *Biochim. Biophys. Acta* 153:555–75
351. Ibid, 576–88
352. Ibid 1970. 220:190–205
353. Zielke, H. R., Filner, P. 1971. *J. Biol. Chem.* 246:1772–79
354. Zumft, W. G. 1970. *Ber. Deut. Bot. Ges.* 53:221–28
355. Zumft, W. G. 1972. *Biochim. Biophys. Acta* 276:363–75
356. Zumft, W. G., Aparicio, P. J., Paneque, A., Losada, M. 1970. *FEBS Lett.* 9:157–60
357. Zumft, W. G., Paneque, A., Aparicio, P. J., Losada, M. 1969. *Biochem. Biophys. Res. Commun.* 36:980–86
358. Zumft, W. G., Spiller, H. 1971. *Biochem. Biophys. Res. Commun.* 45:112–18
359. Zumft, W. G., Spiller, H., Yeboah-Smith, I. 1972. *Planta* 102:228–36

Ann. Rev. Plant Physiol. 1975. 26:101–15

TREE PHOTOSYNTHESIS ❖7584

M. Schaedle
Department of Forest Botany and Pathology, State University of New York, College of Environmental Science and Forestry, Syracuse, New York 13210

CONTENTS

INTRODUCTION

Many excellent discussions have been published on various aspects of tree photosynthesis (10, 15, 37, 96, 101, 109, 110, 122, 123, 125, 130, 141, 144, 156, 158, 159, 163, 184, 214, 233, 254). This review will emphasize the in vivo and in vitro performances of chloroplast systems of woody plants. Only a few references to nonarborescent species are included in the bibliography. Omitted are major ecophysiological topics of adaptation to light, moisture, temperature, and nutrient conditions as well as genetics and translocation of products.

Trees can be defined as perennial plants with upright stems. Such an arbitrary morphological grouping must include species from a wide variety of plant families. Such a broad group of organisms would be expected to vary in the significant details of the photosynthetic reaction pathways. Lack of detailed information on photosynthetic processes in any one tree species, however, prevents one from developing a more specific and therefore more satisfactory discussion of photosynthesis in trees.

IN VITRO STUDIES

Isolation and Performance of Cell-Free Systems

The isolation of cell-free preparations of organelles and enzymes from tree tissues for biochemical studies is a prerequisite to the more comprehensive, quantitative understanding of photosynthesis. With the development of new isolation techniques (9, 94, 99, 231, 232), several laboratories have isolated chloroplast preparations which attained rates of cell-free CO_2 fixation equivalent to if not higher than those of intact tissues. This provides assurance that phenomena studied with isolated chloroplasts are physiologically similar to those of the intact leaf. However, nearly all these studies have utilized leaves of agricultural crop plants.

In vitro studies with tree tissues are made difficult by the presence of inhibitors that inactivate the membrane-bound components and enzymes of biological systems. In intact plants most phenolics, terpenes, tannins, and mucilages are stored in concentrated form in membrane-bound vacuoles (16, 36, 78). When the tissue is cut or ground, these substances are released and mixed with the remaining cell contents, initiating fairly rapid deactivation processes. To eliminate or reduce the danger of inactivation from inhibitors, combinations of antioxidants and precipitation/adsorption agents have been included in the reaction mixtures (40, 69, 130, 136, 174, 199, 232, 242).

No safe strategy for the isolation of chloroplasts from tree tissues is presently available. The recipes reported to date are not universal in their applications and will have to be modified to suit the tissues studied. Several general procedural considerations seem promising. First, young tissues usually contain lower concentrations of inhibitors and are softer; therefore, they are usually more successfully macerated. Thus isolation should be attempted initially with younger tissues. However, in leaves full photosynthetic capacity is achieved at or near full leaf expansion, and therefore very young leaves may not always be satisfactory (49, 115, 206). Second, it could be advantageous to vacuum infiltrate the tissue with a special isolation medium containing higher concentrations of protective agents before cutting or grinding to deactivate the inhibitors before the moment of cell rupture (83). Third, activity of the preparations can also be increased by using reaction media that are more suitable for the process studied and differ from the isolation medium by lower concentrations or complete absence of protective agents (40, 83, 99, 171, 232). A problem aggravating the isolation of cell-free preparations from tree tissues is the toughness of the cells and tissues. Vigorous methods of grinding to free organelles and enzymes are required and these tend to fragment plastids and also to deactivate enzymes (232).

Isolation of chloroplast fragments and chloroplasts from tree tissues has been achieved in a number of systems. Hill reaction activity or oxygen evolution has been assayed in many tree species (42, 43, 50, 170–172, 174, 193, 222, 240). Hill reaction activity in some materials was effected by use of simple buffered media. However, in most cases active preparations could be obtained only by including antioxidants and absorption/precipitation agents in the reaction medium. Oku and co-workers (171) have achieved NADP reduction and cyclic photophosphorylation in a pine system by use of 10% polyethylene glycol in the isolation medium. No polythylene glycol was added to the washing medium and reaction mixture. Chloroplasts capable of cyclic and noncyclic photophosphorylation and low rates of carbon dioxide fixation have been isolated from *Populus deltoides* (83, 84). The isolation medium contained 1.5% polyethylene glycol and isoascorbate, but polyethylene glycol was omitted from the reaction mixture. Vacuum infiltration with the isolation medium prior to grinding was found to increase reaction rates by 300% (83). However, chloroplast fragments from *Pinus pinae* cotyledons reduced NADP in the light without the addition of protective agents (240). This suggests that cotyledons are low in enzyme inhibitors.

Practically none of the enzymes or enzyme activities involved directly in the photosynthetic process have been isolated from or assayed in tree tissues. The few exceptions are: the ferredoxin-NADP reductase from cotyledons of *Pinus pinae* (59); ferredoxin from several gymnosperms (241); starch synthesizing enzymes from *Vitis vinifera* (54); RuDP-carboxylase activity from coca (13) and *Populus deltoides* (50); glycolate oxidase, glyoxylate reductase, PEP-carboxylase activities from *Citrus limon* (188); and glycolate oxidase from several deciduous tree species (52). Reported enzyme activities peripheral to the process of photosynthesis include adenosine triphosphatase from *Pinus silvestris* (21) and peroxidases from cottonwood (13, 73) and from elm hybrids (58). The enzyme chlorophyllase, possibly involved both in the addition and hydrolysis of the phytol chain to chlorophyllides, was originally studied by Willstatter & Stoll (236). The enzyme is of special interest since it could be involved in thylakoid synthesis and in senescence (138, 149).

Early Products of CO₂ Fixation

The photosynthetic carbon dioxide fixation pathway has not been studied in any detail in trees. Nishida (167), using exposure periods of 30 sec, reported that the early intermediates in *Acer trifidium* were typical of C_3 plants, and he noted the occurrence of label in alanine. In *Aegiceras majus* (mangrove), after 10 sec most of the label was found in aspartate (98), suggesting the existence of an active PEP-carboxylase. After a 1-min exposure of *Abies grandis* to $^{14}CO_2$ and 21% O_2, nearly 30–40% of the label was found in phosphoglyceric acid (PGA) and 50% in sugars and starch. Increase of the oxygen concentration to 100% reduced the label in PGA and increased the proportion found in glycine and serine (153).

Exposure of *Pinus silvestris* and tea for 30 min to $^{14}CO_2$ (243, 244) leads to considerable incorporation into quinic and shikimic acid, phenolic compounds, and also sugars, pigments, and amino acids. With *Rhus typhina* after 10 min in $^{14}CO_2$, 20% of the label was found in tannins and tannin precursors, while the majority of the CO_2 was incorporated into carbohydrates (196). With grape leaves Kriedman

(115) found after 20-min exposures that most of the label was either in sugars or the ethanol insoluble fraction. In young leaves 19% of the label was in amino and organic acids, but with increased age less activity was found in this fraction. With longer exposures, as would be expected, the ^{14}C label appears in all compounds studied, with the predominant fraction being found in sugars and sugar polymers (72, 169, 195, 196, 205). This lack of data on the photosynthetic carbon metabolism in trees limits the interpretation of physiological and ecological measurements and is especially disturbing in view of the recent emphasis on the diversity of photosynthetic carbon metabolism (24, 120).

IN VIVO STUDIES

Photosynthesis of True Leaves

Different tissues of trees contribute in varying degrees to their organic nutrient supply. This includes leaves, the major producer, as well as cotyledons, bark, fruit, and flower parts. The literature on leaf photosynthesis is voluminous and several aspects have been reviewed recently (37, 101, 109, 122, 158, 233). Therefore, this section is limited to topics reflecting on the performance of the chloroplast system. In the literature on tree photosynthesis, one repeatedly encounters remarks concerning the low effectiveness of the photosynthetic system of tree leaves. Reference is made to the predominantly low net photosynthetic rates reported over the last 80 years (37, 158). However, most of these measurements were not made with the demonstration of the maximum net photosynthetic rates of tree leaves as the primary objective. As can be seen from the data in Table 1, tree leaves under certain conditions are capable of net photosynthetic rates similar to leaves of cultivated crops.

With the exception of mangroves (98), all trees studied up to now can be tentatively classified as C_3 plants. This implies the absence of a highly photosynthetically active vascular bundle sheath in the leaf tissue, the presence of photorespiration, the predominance of carboxydismutase as the major photosynthetic CO_2 fixation enzyme, and a minimum CO_2 compensation point of above 10 ppm CO_2 (24, 251, 252).

Table 1 Selected net photosynthetic rates of tree foliage

Species	Net Photosynthesis mg CO_2		Reference
	dm^2/hr	g dry wt/hr	
Pinus ellioti	19		228
Eucalyptus regans	28		116
McIntosh apple	30–14		82
Alnus rubra	17	30	117
Pinus silvestris		35	250
Picea abies		20	250
Bancroft apple	25		152
Cotyledons		25–35	210
Cecropia pelbata	26.0		216

This, however, does not necessarily mean that trees should have low photosynthetic rates, since some C_3 plants have very high net CO_2 fixation (56, 150, 230). Especially under conditions of low temperature, C_3 plants have higher rates of net assimilation than C_4 plants (23).

DEVELOPMENT OF THE PHOTOSYNTHETIC SYSTEM   As a consequence of the cyclic nature of leaf life, there is a similarity in the patterns of leaf development regardless of species. At the beginning of leaf expansion, the net photosynthetic rates of leaves are negative or very low (49, 111, 115, 135). Especially in conifers, the rate of respiration may exceed the rate of photosynthesis for periods of several weeks (118, 135, 165). However, leaves are not homogeneous and contain cells of different developmental age (90, 97, 126, 128). Even in the early stages of leaf development, some of the leaf cells will be able to export surplus photosynthate despite the fact that the leaf as a whole may have a negative CO_2 balance. In addition, chloroplast structure changes with distance from the leaf epidermis, possibly as a consequence of different light conditions (207, 208).

Rates of net photosynthesis increase usually until full leaf expansion (14, 32, 49, 86, 107, 111, 115, 127, 135, 168, 186, 189). During this period the chlorophyll content and the activity of key photosynthetic enzymes increases (66, 71, 84, 111, 131, 160, 168). As Dickman (50) has shown with *Populus deltoides,* the Hill reaction and carboxydismutase activities increase rapidly during the formative stages of leaf development. The chloroplast in such young leaves usually have well developed grana (10, 38, 53, 84).

Following the completion of leaf expansion, a more or less steady state condition of photosynthetic performance persists for a duration of 10 to 40 days depending on species, time of year, and environmental conditions. Subsequently, with increasing leaf age, the net photosynthetic rates decline (49, 57, 86, 115, 154, 168, 186). Associated with this decline is a reduction in the chlorophyll (10, 19, 66, 71, 86, 211, 212), protein (10, 180, 211, 213, 237), and nucleic acid (104, 212) content of the leaf. The CO_2 fixation and phosphorylation performance of the chloroplast also declines (83). Chloroplast grana structure becomes diffuse as the number of thylakoid membranes increase and they appear thinner (10, 35, 53, 84, 154). The thinning of the membranes could be the result of a reduction of the chlorophyll and protein content in the membranes. Such membranes could also be functionally defective (10). In late stages of senescence, hydrolytic enzyme activity increases significantly (148, 212), and the internal structure of the chloroplast may become completely disorganized.

In those conifers that retain their needles for 2 or more years, the photosynthetic rates of the needles decline from year to year (41, 61, 118, 238).

PHOTORESPIRATION   Photorespiration can be defined as the oxidation of carbon compounds to CO_2 in the light, induced by the activity of the photosynthetic system. It could include the direct oxidation of organic acids by the two photosystems (252), the production of higher concentrations of oxidizable substrates (251), increase in the level of substrates available for mitochondrial oxidation, as well as the oxygen dependent oxidation of ribulose diphosphate by RuDP-carboxylase (4, 11). At least three simultaneous processes occur in a tissue in light: mitochondrial respiration, photorespiration, and photosynthesis. Photorespiration must, therefore, be measured indirectly, and as a consequence, the magnitude of photorespiration is always somewhat in doubt (60, 70, 203).

The occurrence of photorespiration has been reported in *Pseudotsuga menziesii* (33), different *Populus* species and clones (143), *Liriodendron tuliperifera* and *Fraxinus americana* (46), *Picea glauca* (181, 182), *Coffea arabica* (47), *Picea sitchensis* (44, 142), *Pinus silvestris* (247), *Abies grandis* (153), and different citrus species (188). Glagoleva (67) did not observe any photorespiration in desert species. Photorespiration tended to increase with oxygen concentration (44, 45, 142, 182, 188) light intensity (33, 181) and leaf age (188).

Since photorespiration is a function of leaf age, developmental conditions, temperature, CO_2 concentration light intensities, etc, one expects the specific values reported in the literature to be highly variable and of limited universal significance (60, 70, 252). In studies with tree tissues this problem is further complicated by the use of twigs rather than individual leaves or needle fascicles. The nonleaf parts contribute respiratory CO_2 to the system, thus obscuring the actual metabolic relationships of the leaves (246).

CO_2 COMPENSATION POINT   The CO_2 compensation point represents the CO_2 concentration at which the total CO_2 fixation is equal to the total CO_2 evolution (252). It is, therefore, a highly complex quantity representing the balance between light plus dark CO_2 fixation and CO_2 evolution by dark plus light respiration. Changes in photosynthetic rates and changes in photorespiration as well as measurement conditions and tissue development (55, 203, 251) will have effects on the magnitude of the CO_2 compensation point. However, since compensation point measurements are steady state and are relatively easy to perform, they do represent a useful index of plant performance.

All reported CO_2 compensation points in tree tissues are above 10 ppm, as would be expected from C_3 plants (55). Values generally are not as excessively high as reported by some investigators (162, 226). Some representative values are: *Picea sitchensis* 40–50 ppm (142), mangrove branches 75 ppm (161), citrus leaves 21–83 ppm (112), and *Pinus silvestris* seedlings 65 ppm (246). Increasing the temperature of lemon foliage from 6 to 40°C increased in the CO_2 compensation point from 21 to 83 ppm (112). Similar patterns of temperature dependence have been found in *Coffea arabica* (81), Douglas fir (33), *Picea sitchensis* (164), oil palm (151), and several *Populus* clones (144). In lemon, rapid dessication raised the compensation point to 200–350 ppm (112).

Leaf age also strongly affects the CO_2 compensation point. Generally very young leaves have very high (306 ppm) compensation values that become smaller (50 ppm) as the leaf develops (111). This is presumed to result from an increase in photosynthetic rates and a decline of respiration.

With increasing light intensity the CO_2 compensation point declines (44, 106, 246), but may reach a plateau at higher light intensities (246), possibly as the result of rising rates of photorespiration. Since so many factors affect the rates of photosynthesis, dark respiration, and photorespiration, the interpretive use of the compensation point requires not only a precise definition of the measurement conditions and tissue development but an explanation of the physiological meaning of the selected experimental parameters.

LIGHT COMPENSATION POINT   The light compensation point is the light intensity at which the rate of CO_2 release is equal to CO_2 uptake at constant CO_2 concentrations and some physiological temperature. It is a useful index of the ability of plants to maintain net photosynthetic rates at low light intensities (80, 132).

Characteristically, shade-adapted foliage of trees usually has a much lower compensation point than light-adapted foliage (30, 63, 77, 133, 134, 200, 217), reflecting low dark respiration rate (63, 134, 179, 200, 248) and good rates of photosynthesis at low light intensities (91, 134, 178, 179, 200, 234). Light compensation points tend to increase with increasing temperature as a consequence of a larger temperature coefficient for CO_2 release than for assimilation (80, 121).

Considering the many sources of variability, interpretive use of the light compensation points also requires a careful definition of measurement conditions as well as of tissue development.

Photosynthesis of Plant Parts

Many tree tissues and cells form chlorophyll and develop chloroplasts. The presence of chlorophyll pigments has been demonstrated in bark (214, 215); primary and secondary wood (75, 76, 114, 209, 219, 235, 245); the outer layer of fruit (17, 105, 113, 119, 223, 229); cotyledons (147, 166, 177); reproductive organs (51, 74); and buds (68, 227). In some of these tissues the presence of chloroplasts has been confirmed by electron microscopy. However, it should be noted that the presence of chlorophyll and chloroplasts does not necessarily imply that the cells photosynthesize. Induction of chlorophyll formation may occur at light intensities below those needed for photosynthesis and may require only short periods of illumination. Presence of chlorophyll cannot, therefore, be considered as evidence for the occurrence of photosynthesis.

CORTICULAR PHOTOSYNTHESIS   The existence of chlorophyll containing cells in the bark tissues of woody plants was observed over 80 years ago (201). Chlorophyll is found in the cortex of most stems, especially in younger stems and twigs (3, 31, 75, 76, 108, 114, 173, 175, 176, 192–194, 198, 209, 219–221, 235). This layer has been shown to contain chloroplasts with fully developed grana thylakoids and starch grains (2, 114, 220, 221, 225).

The first direct experimental evidence for the occurrence of corticular photosynthesis in trees was provided by Larsen in 1936 (124). Since then, many investigators, using a variety of species, have shown the corticular chlorophyllous layer to be involved in carbon dioxide fixation or oxygen evolution (1–3, 8, 92, 102, 103, 114, 124, 157, 175, 176, 192, 193, 218–221).

Bark as a photosynthetic tissue differs from leaves and needles in several respects. The chlorophyllous layer is covered by the periderm or outer bark, a tissue whose degree of development and permeability to light varies from species to species with the age of the tissue (219) and with environmental conditions (29, 175, 194). Measurements with *Populus deltoides* and *Fraxinus oranus* suggest that up to 15% of the light in the range between 400–700 nm is transmitted through the outer bark (204, 218, 219). Older shoots of grape vines with a heavy outer bark have a light transmission of only 2% (114). The light transmission of the cell layers covering the chlorophyllous bark will by necessity be a limiting factor in the ability of the bark to photosynthesize.

Bark tissues may vary greatly in their gas permeability, depending on species and development (87, 88, 214). Bark diffusion resistance of relevance to photosynthetic gas exchange is not known, but appears to be high as indicated by the low rates of H_2O loss (65, 103, 155) and the high CO_2 concentrations found in stem tissue. In stems of the current year's growth, a small number of stomata are present (20, 155, 198, 219). This could result in somewhat higher gas permeabilities; however, no measurements of gas permeability of young stems have been reported.

The CO_2 content of stems during the growing season has been found to be between 1 and 26%, with an approximately proportional decrease in the oxygen concentration. During the winter months the gas composition of the stem is much closer to atmospheric (39, 88, 93, 145, 204). The high summer concentrations suggest that in most species the periderm may represent a zone of high diffusion resistance and externally added CO_2 may not enter rapidly (87). Thus the chlorophyllous cells of the stem probably utilize primarily internal respiratory CO_2 as a substrate for carbon dioxide fixation (192, 249).

Because of the possibly high periderm diffusion resistance, quantitative measurements of bark photosynthesis are uncertain since externally added labeled $^{14}CO_2$ may not enter the stem rapidly enough to permit the assumption that the external and internal specific activities are the same. Infrared CO_2 measurements of bark photosynthesis are also made uncertain by the possible lack of equilibration across the periderm and the suggested preferential use of internal carbon dioxide.

If one places a stem in an assimilation chamber in the field, one is also confronted with the problem of the heterogeneity of the system. The bark of the north side of a tree is different structurally and physiologically from that of the south side (29, 175, 194, 218). Since the tree is opaque, radiation conditions vary continuously as one proceeds around the stem. This is reflected in the difference in the development of chloroplasts on the north and south sides (218; I. Ames, personal communications). In addition to differences in thylakoid structure, shade side chloroplasts are nearly free of starch, whereas sun side chloroplasts contain substantial amounts of starch.

The difference in radiation regime is also reflected in gradients of surface temperature as well as internal temperatures (48, 79, 85, 95, 187). Temperature differences of up to 20° can be measured with thermocouples between the sun and shade sides, with the shaded side remaining near ambient. This difference varies continuously as a function of light intensity, the incident angle of light, the reflectance of the stem, and a myriad of factors affecting the efficiency of molecular heat transfer between stem and air (48, 79, 85, 95, 187).

Considering the many variables affecting the rate of bark photosynthesis, one can predict that the contribution of bark photosynthesis to the carbohydrate supply of woody plants will vary greatly. On a whole tree basis this contribution was estimated to be up to 5% (191) for *Populus deltoides*.

More data is available on the rate of bark photosynthesis as related to the rate of dark respiration of the stem tissue (8, 102, 103, 108, 114, 124, 157, 192). Gross photosynthesis, when expressed as percent of dark respiration, varies from 10 to 90% of the latter. Occurrence of actual net photosynthesis has been reported for *Cercidium floridum* stem tissue (1) and by tissue slices of winter bark (176). With younger branches under optimum conditions most of the values for gross photosynthesis fall between 30% and 50% of dark respiration. Numerically this represents a highly significant saving in carbohydrate resources of the stem.

One can speculate that in the case of woody plants growing under desert conditions, with small leaves that are shed rapidly under dry conditions, the contribution of stem photosynthesis to plant growth and survival is highly significant (1–3, 159). Similarly, in northern climates where leafless periods are long, bark photosynthesis could be physiologically significant to the survival of *Populus deltoides* and related species. However, during periods of low temperature actual rates of bark photosynthesis are very low (108). Even so, it is quite possible that bark photosynthesis may be of considerable importance to the normal physiological function of the bark tissue and that a consideration of only its contribution to the overall carbohydrate budget of a tree may be incorrect.

COTYLEDONS   It has been known since the end of the last century that cotyledons of young tree seedlings contain chlorophyll (25–28, 147, 197). Electron micrographs show cotyledons of woody plants to contain chloroplasts with developed grana and stroma thylakoids as well as starch grains (5, 100, 166). In the case of some conifers that form chloroplasts (5, 100) and chlorophyll in the dark (28, 34, 197), illumination increases chlorophyll content and grana structure. Cotyledons developed in light contain 3 to 18 μg chlorophyll per milligram dry weight (26, 147, 177) and can attain net CO_2 fixation rates of 25 mg CO_2 per gram dry weight per hour (210). Cotyledon photosynthesis can exceed dark respiration by a factor of 3 to 8 (191, 210). This suggests that cotyledons contribute significantly to the carbohydrate supply of the seedling before the emergence of first leaves (147, 190). However, considerable species differences should be expected, considering the reported variation in structure and composition of cotyledons (139, 140).

FRUIT   Many unripe fruits have a green coloration as the result of chlorophyll in the outer or peel tissues (18, 22, 119, 129, 146, 229, 236).

During early stages of development, the fruit epidermis contains stomata (6, 7) and well-developed chloroplasts (12, 62, 224). Both chlorophyll content (236) and grana development decline with the aging and ripening of fruit. The chloroplasts in some fruit are finally modified into chromoplasts (185, 224). It is not known whether the stomata observed in the young fruit tissues are able to open or close nor whether they are a conduit for gas exchange. Rapid cutin deposition should increase the diffusion resistance of the fruit with increasing age (6, 89, 183). This possible low gas permeability of the fruit rind may interfere seriously with attempts to estimate CO_2 fixation rates with externally fed $^{14}CO_2$ and complicate the interpretation of gas exchange measurements, as was discussed earlier. That green fruit can photosynthesize was clearly demonstrated by Willstatter & Stoll in 1918 (236). However, quantitative aspects of peel photosynthesis are still uncertain. When expressed on a weight basis, the chlorophyll content of a young fruit is 1/5 to 1/10 that of a leaf. However, since most fruit are globular, they have geometrically a maximum possible volume to surface ratio. This would reduce both the chlorophyll content and the photosynthetic rates to artificially low values if the results are expressed on a weight basis. Estimates of the amounts of chlorophyll per surface area fall within the lower range of leaf chlorophyll content (223). Most rates of photosynthesis are also expressed on a weight or per fruit basis with few exceptions (18, 223, 236). On an area basis photosynthesis of the fruit of orange, lemon, and pears were in the range of approximately 2–3 mg $CO_2/dm^2/hr$. The photosynthetic performance declines with fruit age (17, 64, 113, 223, 229, 236). Since in all fruit materials studied respiration exceeded photosynthesis, the data represent light-induced reduction of CO_2 release. Kriedman (113) showed that DCMU eliminated the effect of light on CO_2 release, providing evidence for its photosynthetic nature. In a young green fruit photosynthetic CO_2 fixation is of sufficient magnitude to conserve 20 to 80% of the CO_2 released by dark respiration (17, 105, 223). Some fruit show, however, very low rates of photosynthesis (113, 236) even during their period of optimum performance. The presently available data are not sufficiently detailed to assess the significance of fruit photosynthesis to the fruit biomass and fruit quality. However, if one considers the magnitude of the carbon conservation reported, this process analogous to photosynthesis in the flower parts of agricultural crops (137, 239) may be of considerable importance to the fruit bearing of trees.

MISCELLANEOUS   Scattered information with regard to the presence and function of a photosynthetic system is available on a few other tree tissues. Green strobili of *Pinus resinosa* are capable of CO_2 fixation at roughly half the rate of dark respiration (51). Chlorophyll has been found in: the hypocotyl of *Pinus nigra* (74); cones of *P. silvestris* (74); the megagametophyte and the embryo of *Gingko biloba* (74); and the megagametophyte of *Pinus jeffreyi* (197). Weakly developed chloroplasts have been observed in the embryo of *Tilia platyphyllas* (253) and apple petioles (202). Many tree species have been found to contain chlorophyll in their bud tissues (68, 227). Whereas the contribution of these tissues to the overall production of carbohydrates by trees is probably small, the local significance of the photosynthetic activity could be large. Insufficient data are presently available to decide this issue.

CONCLUDING REMARKS

Despite the economic and ecological importance of trees, the presently available biochemical, ultrastructural, and organizational data describing tree photosynthesis is limited and scattered. As a consequence, the interpretation of in vivo experiments must rely on comparative analysis using data from agricultural crops, wild herbaceous species, and algae. Significant diversity of the photosynthetic system of different plant species has been reported recently. Therefore, one must guard against the indiscriminate application of comparative analysis. Current work with C_3 and C_4 plants has begun to focus attention upon the importance of: (a) the organization of the conducting tissue in relation to the tissues and cells involved in photosynthesis; (b) the existence of and significance of differential distribution of photosynthetic enzymes in the chloroplast and cytoplasm; (d) the importance of internal recirculation of CO_2 on the net photosynthetic performance of plants; and (c) the value to the plant of the functional differentiation of photosynthetic tissues as illustrated by C_4 plant (24, 120). Very little such information is presently available for trees.

No doubt some generalizations are essential for operational conceptualization and will remain valid as long as they correctly reflect the molecular similarity of the photosynthetic process and the cyclic nature of plant or leaf life. However, the goal of the physiological understanding of tree photosynthesis will be to obtain sufficient detailed knowledge of the systems involved in photosynthesis so that the structural, molecular, and kinetic details can be evaluated quantitatively with regard to their significance for the photosynthetic performance of a leaf. Such quantitative data will then form the basis for genetic design of plants to optimize net photosynthesis for optimum productivity under a variety of soil and climatic conditions (137, 156, 239).

ACKNOWLEDGMENTS

I would like to thank Dr. H. B. Tepper and Mr. Stewart Cameron for helpful suggestions and criticism of this manuscript.

Literature Cited

1. Adams, M. S., Strain, B. R. 1968. *Oecol. Plant.* 3:285–97
2. Adams, M. S., Strain, B. R. 1969. *Photosynthetica* 3:55–62
3. Adams, M. S., Strain, B. R., Ting, I. P. 1967. *Plant Physiol.* 42:1797–99
4. Andrews, T. J., Lorimer, G. H., Tolbert, N. E. 1973. *Biochemistry* 12:11–18
5. Anikushin, N. F. 1971. *Bot. Zh.* 56:1687–89
6. Albrigo, L. G. 1972. *J. Am. Soc. Hort. Sci.* 97:220–23
7. Ibid, 761–65
8. Audus, L. J. 1947. *Ann. Bot. London* 11:165–201
9. Avron, M., Gibbs, M. 1974. *Plant Physiol.* 53:140–43
10. Baddeley, M. S. 1971. *Ecology of Leaf Surface Microorganisms,* ed. T. E. Preece, C. H. Dickinson, 415–29. New York: Academic
11. Bahr, J. T., Jensen, R. G. 1974. *Biochem. Biophys. Res. Commun.* 57: 1180–85
12. Bain, J. M., Mercer, F. V. 1964. *Aust. J. Biol. Sci.* 17:78–85
13. Baker, N. R., Hardwick, K. 1973. *New Phytol.* 72:1315–24
14. Barua, D. N. 1964. *J. Agr. Sci.* 63: 265–71
15. Baumgartner, A. 1969. *Photosynthetica* 3:127–49
16. Baur, P. S., Walkinshaw, C. H. 1974. *Can. J. Bot.* 52:615–19
17. Bean, R. C., Porter, G. G., Barr, B. K. 1963. *Plant Physiol.* 38:285–90
18. Bean, R. C., Todd, G. W. 1960. *Plant Physiol.* 35:425–29
19. Benecke, U. 1972. *Angew. Bot.* 46: 117–35
20. Benson, C. A. 1969. Anatomy of developing internodes of *Fraxinus americana.* PhD thesis. SUNY Coll. Environ. Sci. Forest., Syracuse, NY
21. Bervaes, J. C. A. M., Kylin, A. 1972. *Physiol. Plant.* 27:178–81
22. Biale, J. B., Young, R. E. 1962. *Endeavor* 19:164–74
23. Bjorkman, O., Berry, J. 1973. *Sci. Am.* 229:80–93
24. Black, C. C. 1973. *Ann. Rev. Plant Physiol.* 24:253–86
25. Bogdanovic, M. 1968. *Zemljiste Biljka* 17:2–7
26. Bogdanovic, M. 1973. *Physiol. Plant.* 29:17–18
27. Ibid, 19–21

28. Bogorad, L. 1950. *Bot. Gaz.* 111:221–41
29. Borger, G. A., Kozlowski, T. T. 1972. *Can. J. Forest Res.* 2:190–97
30. Boysen Jensen, P. 1932. *Die Stoffproduction der Pflanzen.* Jena, Germany: Fisher
31. Bray, J. R. 1960. *Can. J. Bot.* 38:313–33
32. Briggs, G. E. 1920. *Proc. Roy. Soc. London Ser. B* 91:249–67
33. Brix, H. 1968. *Plant Physiol.* 43:389–93
34. Burgenstein, A. 1900. *Ber. Deut. Bot. Ges.* 18:168–84
35. Buttrose, M. S., Hale, C. R. 1971. *Planta* 101:166–70
36. Campbell, R. 1972. *Ann. Bot. London* 36:711–20
37. Carter, M. C. 1972. In *Net Carbon Dioxide Assimilation in Higher Plants,* ed. C. C. Black. Raleigh, N. C.: Cotton Inc.
38. Chabot, J. F., Chabot, B. F. 1974. *Protoplasma* 79:349–58
39. Chase, W. W. 1934. *Univ. Minn. Agr. Exp. Sta. Tech. Bull.* 99, 5–51
40. Chernov, I. A., Krainova, N. N. 1971. *Sov. Plant Physiol.* 18:382–85
41. Clark, J. 1961. *Photosynthesis and Respiration in White Spruce and Balsam Fir.* State Univ. Coll. Forest., Syracuse, NY Tech. Bull. 85
42. Clendenning, K. A., Brown, T. E., Waldov, E. E. 1956. *Physiol. Plant.* 9: 519–32
43. Clendenning, K. A., Gorham, P. R. 1950. *Can. J. Res.* 28:114–39
44. Cornic, G., Jarvis, P. G. 1972. *Photosynthetica* 6:225–39
45. Decker, J. P. Wien, J. D. 1958. *J. Sol. Energy Sci. Eng.* 2:39–41
46. Decker, J. P. 1955. *Plant Physiol.* 30:82–84
47. Decker, J. P., Tio, M. A. 1959. *J. Agr. Univ. P.R.* 47:50–55
48. Derby, R. W., Gates, D. M. 1966. *Am. J. Bot.* 53:580–87
49. Dickman, D. I. 1971. *Bot. Gaz.* 132: 253–59
50. Dickman, D. I. 1971. *Plant Physiol.* 48:143–45
51. Dickman, D. I., Kozlowski, T. T. 1970. *Life Sci.* 9:549–52
52. Dietrich, W. E., Rose, J. R. 1974. *50th Ann. Meet. Am. Soc. Plant Physiol.* No. 350. Cornell Univ., Ithaca, NY
53. Dodge, J. D. 1970. *Ann. Bot. London* 34:817–24
54. Downtown, W. J. S., Hawker, J. S. 1973. *Phytochemistry* 12:1557–63
55. Downtown, W. J. S., Treguna, E. B. 1968. *Can. J. Bot.* 46:207–15
56. Evans, L. T., Dunstone, R. L. 1970. *Aust. J. Biol. Sci.* 23:725–41
57. Farukawa, A. 1973. *J. Jap. Forest. Soc.* 55:119–23
58. Feret, P. P. 1972. *Can. J. Forest. Res.* 2:264–70
59. Firenzuoli, A. M., Ramponi, G., Vanni, P., Zanobini, A. 1968. *Life Sci.* 7: 905–13
60. Fock, H. 1970. *Biol. Zentralbl.* 89: 545–72
61. Freeland, R. O. 1952. *Plant Physiol.* 27:685–90
62. Frey-Wyssling, A., Kreutzer, E. 1958. *J. Ultrastruct. Res.* 1:397–411
63. Geis, J. W., Tortorelli, R. L., Boggess, W. R. 1971. *Oecologia* 7:276–89
64. Geisler, G., Radler, F. 1963. *Ber. Deut. Bot. Ges.* 76:112–19
65. Geurten, I. 1950. *Forstwiss. Zentralbl.* 69:704–43
66. Ghosh, S. P. 1973. *J. Hort. Sci.* 48:1–9
67. Glagoleva, T. A., Reinus, R. M., Gedemov, T. G., Mokronosov, A. T., Zalensky, O. V. 1972. *Bot. Zh.* 57:1097–1107
68. Godnev, T. N., Terentev, M. V. 1952. *Dokl. Akad. Nauk USSR* 33:481–84
69. Goldstein, J. L., Swain, T. 1965. *Phytochemistry* 4:185–92
70. Goldworthy, A. 1970. *Bot. Rev.* 36: 321–40
71. Goodwin, T. W. 1958. *Biochem. J.* 68:503–11
72. Goral, I. 1973. *Acta Soc. Bot. Pol.* 12:555:65
73. Gordon, J. C. 1971. *Plant Physiol.* 47:595–99
74. Grill, R., Sprint, C. J. P. 1972. *Planta* 108:203–13
75. Gundersen, A. K. 1954. *Nature* 174: 87–88
76. Gundersen, A. K., Friis, J. 1956. *Bot. Tidsskr.* 53:60–66
77. Harder, R. 1923. *Ber. Deut. Bot. Ges.* 41:194–98
78. Harris, W. M. 1971. *Can. J. Bot.* 49:1107–9
79. Harvey, R. B. 1923. *Ecology* 4:261–65
80. Heath, O. V. S. 1969. *The Physiological Aspects of Photosynthesis.* Stanford Univ. Press
81. Heath, O. V. S., Orchard, B. 1957. *Nature* 180:180–81
82. Heinicke, A. J., Hoffman, M. B. 1933. *Cornell Univ. Agr. Exp. Sta. Bull. 577.* Ithaca, NY
83. Hernandez-Gil, R., Schaedle, M. 1972. *Plant Physiol.* 50:375–79
84. Ibid 1973. 51:245–49
85. Herrington, L. P. 1969. *Yale Univ. Sch. Forest. Bull. 73*
86. Hoffmann, P. 1962. *Flora* 152:622–54

87. Hook, D. D., Brown, C. L. 1972. *Bot. Gaz.* 133:304–10
88. Hook, D. D., Brown, C. L., Wetmore, R. H. 1972. *Bot. Gaz.* 133:443–54
89. Horrocks, R. L. 1964. *Nature* 203:547
90. Isebrands, J. G., Larson, P. R. 1973. *Am. J. Bot.* 60:199–208
91. Ivanov, L. A., Orlov, I. M. 1931. *Bot. Zh.* 16:139–57
92. Jankiewicz, L. S., Antoszewski, R., Klimovicz, E. 1967. *Biol. Plant. (Praha)* 9:116–21
93. Jensen, K. F. 1969. *Forest Sci.* 15:246–51
94. Jensen, R. G., Bassham, J. A. 1966. *Proc. Nat. Acad. Sci. USA* 56:1095–1101
95. Jensen, R. E., Savage, E. F., Hayden, R. A. 1970. *J. Am. Soc. Hort. Sci.* 95:286–92
96. Jeremias, K. 1964. *Uber die jahresperiodisch bedingten Veranderungen der Ablagerungsformen der Kohlenhydrate in vegetative Pflanzenteilen.* Jena, Germany: Fischer
97. Jones, H., Eagles, J. E. 1962. *Ann. Bot. London* 26:505–10
98. Joshi, G. V., Karekar, M. D., Gowda, C. A., Bhosale, L. 1974. *Photosynthetica* 8:51–52
99. Kalberger, P. P., Buchanan, B. B., Arnon, D. I. 1967. *Proc. Nat. Acad. Sci. USA* 57:1542–49
100. Kawamatu, S. 1967. *Bot. Mag.* (Tokyo) 80:233–40
101. Keller, T. 1972. *Photosynthetica* 6:197–206
102. Ibid 1973. 7:320–24
103. Keller, T., Beda-Puta, H. 1973. *Schweiz. Z. Forstw.* 124:433–41
104. Kessler, B., Engelberg, N. 1962. *Biochim. Biophys. Acta* 55:70–82
105. Kidd, F., West, C. 1947. *New Phytol.* 46:274–75
106. Koch, W. 1969. *Flora* 158B:402–28
107. Koch, W., Keller, Th. 1962. *Ber. Deut. Bot. Ges.* 74:64–74
108. Konovalov, I. N., Michaelova, E. N. 1957. *Izd. Acad. Nauk USSR* 248–56
109. Kozlowski, T. T., Keller, T. 1966. *Bot. Rev.* 32:294–382
110. Kramer, P. J., Kozlowski, T. T. 1960. *Physiology of Trees.* New York: McGraw-Hill
111. Kriedman, P. E. 1968. *Vitis* 7:213–20
112. Kriedman, P. E. 1968. *Aust. J. Biol. Sci.* 21:895–905
113. Ibid, 907–16
114. Kriedemann, P. E., Buttrose, M. S. 1971. *Photosynthetica* 5:22–27
115. Kriedemann, P. E., Kliewer, W. M., Harris, J. M. 1970. *Vitis* 9:97–104
116. Kriedemann, P. E., Neales, T. F., Ashton, D. H. 1964. *Aust. J. Biol. Sci.* 17:591–600
117. Kruger, K. W., Ruth, R. H. 1969. *Can. J. Bot.* 47:519–27
118. Kunstle, E. 1972. *Angew. Bot.* 46:49–58
119. Kursanov, A. L. 1934. *Planta* 22:240–50
120. Laetsch, W. M. 1974. *Ann. Rev. Plant Physiol.* 25:27–52
121. Larcher, W. 1961. *Planta* 56:575–606
122. Larcher, W. 1969. *Photosynthetica* 3:150–66
123. Ibid, 167–98
124. Larsen, P. 1936. *Forest. Forsoegsv Den.* 14:13–52
125. Larson, P. R. 1972. *Proc. Symp. Isotopes Radiat. Soil Plant Relat. Int. At. Energy Agr. Vienna,* 277–300
126. Larson, P. R., Dickson, R. E. 1973. *Planta* 111:95–112
127. Larson, P. R., Gordon, J. C. 1969. *Am. J. Bot.* 56:1058–66
128. Larson, P. R., Isebrands, J. G., Dickson, R. E. 1972. *Planta* 107:301–14
129. Lewis, L. N., Coggins, C. W., Garber, M. J. 1964. *Proc. Am. Soc. Hort. Sci.* 84:177–80
130. Leyton, L. 1972. *Proc. Symp. Isotopes Radiat. Soil Plant Relat. Int. At. Energy Agr. Vienna,* 263–42
131. Lichtenthaler, H. K. 1971. *Z. Naturforsch.* 26b:832–42
132. Lieth, H. 1960. *Planta* 54:530–54
133. Ibid, 555–76
134. Loach, K. 1967. *New Phytol.* 66:607–21
135. Loach, K., Little, C. H. A. 1973. *Can. J. Bot.* 51:1161–68
136. Loomis, W. D., Battaile, J. 1966. *Phytochemistry* 5:423–38
137. Loomis, R. S., Williams, W. A., Hall, A. E. 1971. *Ann. Rev. Plant Physiol.* 22:431–68
138. Looney, N. E., Patterson, M. E. 1967. *Nature* 214:1245–46
139. Lovell, P. H., Moore, K. G. 1970. *J. Exp. Bot.* 21:1017–30
140. Ibid 1971. 22:153–62
141. Luckwill, L. C., Cutting, C. V., Eds. 1970. *Physiology of Tree Crops.* London: Academic
142. Ludlow, M. M., Jarvis, P. G. 1971. *J. Appl. Ecol.* 8:925–53
143. Luukkanen, O. 1972. *Silva Fenn.* 6:63–89
144. Luukkanen, O., Kozlowski, T. T. 1972. *Silvae Genet.* 21:220–29
145. MacDougal, D. T., Working, E. B. 1933. *Carnegie Inst. Wash. Publ.* 441
146. MacKinney, G. 1961. *The Orange, Its Biochemistry and Physiology,* ed. W. B.

Sinclair, 302–33. Berkeley: Univ. California Press
147. Marshall, P. E., Kozlowski, T. T. 1974. *Can. J. Bot.* 52:239–45
148. Martin, D., Thimann, K. V. 1972. *Plant Physiol.* 49:64–71
149. McFeeter, R. F., Chichester, C. C., Whitaker, J. R. 1971. *Plant Physiol.* 47:609–18
150. McNaughton, S. J., Fullem, L. W. 1970. *Plant Physiol.* 45:703–7
151. Meidner, H. 1961. *J. Exp. Bot.* 36: 409–13
152. Mika, A., Antoszewski, R. 1972. *Photosynthetica* 6:381–86
153. Mileszewski, D., Levanty, Z. 1972. *Z. Pflanzenphysiol.* 67:305–10
154. Miloszvljevic, M., Nikolic, D. 1974. *Vitis* 12:306–15
155. Mokeava, E. A. 1971. *Bot. Zh.* 56: 1693–97
156. Monsi, M., Uchijima, Z., Oikawa, T. 1973. *Ann. Rev. Ecol. Syst.* 4:301–27
157. Mooney, H. A. 1972. *Bot. Rev.* 38: 455–69
158. Mooney, H. A. 1972. *Ann. Rev. Ecol. Syst.* 3:315–46
159. Mooney, H. A., Strain, B. R. 1964. *Madrono* 17:230–33
160. Moore, K. G. 1965. *Ann. Bot. London* 29:433–44
161. Moore, R. T., Miller, P. C., Ehleringer, J., Lawrence, W. 1973. *Photosynthetica* 7:387–94
162. Moss, D. N. 1962. *Nature* 193:587–88
163. Neales, T. F., Incoll, L. D. 1968. *Bot. Rev.* 34:107–25
164. Neilson, R. E., Ludlow, M. M., Jarvis, P. G. 1972. *J. Appl. Ecol.* 9:721–45
165. Neuwirth, G. 1959. *Biol. Zentralbl.* 78:559–84
166. Nikolic, D., Bogdanovic, M. 1972. *Protoplasma* 75:205–13
167. Nishida, K. 1962. *Physiol. Plant.* 15: 47–58
168. Nixon, R. W., Wedding, R. T. 1956. *Proc. Am. Soc. Hort. Sci.* 67:265–69
169. Nizcolek, S., Kaczkowski, J., Zelawski, W. 1961. *Bull. Acad. Sci. Pol.* 17:363–67
170. Oku, T., Kawahara, H., Tomita, G. 1971. *Plant Cell Physiol.* 12:556–66
171. Oku, T., Sugahara, K., Tomita, G. 1974. *Plant Cell Physiol.* 15:175–78
172. Oku, T., Tomita, G. 1971. *Photosynthetica* 5:23–31
173. Ovington, J. D., Lawrence, D. B. 1967. *Ecology* 48:515–24
174. Pavlova, I. W. 1972. *Sov. Plant Physiol.* 19:743–49
175. Pearson, L. C., Lawrence, D. B. 1958. *Am. J. Bot.* 45:383–87
176. Perry, T. O. 1971. *Forest Sci.* 17:41–43
177. Pinfield, N. J., Stobart, A. K. 1972. *Planta* 104:134–45
178. Pisek, A., Tranquillini, N. 1954. *Flora* 141:237–70
179. Pisek, A., Winkler, E. 1959. *Planta* 53:532–50
180. Plaisted, P. H. 1958. *Contrib. Boyce Thompson Inst.* 19:245–54
181. Poskuta, J. 1968. *Physiol. Plant.* 21: 1129–36
182. Poskuta, J., Nelson, C. D., Krotkov, G. 1967. *Plant Physiol.* 42:1187–90
183. Possingham, J. V., Chambers, T. C., Radler, F., Grancarevic, M. 1967. *Aust. J. Biol. Sci.* 20:1149–53
184. Priestley, C. A. 1962. *Carbohydrate Resources Within the Perennial Plant.* Commonwealth Agr. Bur. Tech. Commun. 27. Bucks, England: Farnham Royal
185. Rhodes, M. J. C., Wooltorton, L. S. C. 1967. *Phytochemistry* 6:1–12
186. Richardson, S. D. 1957. *Acta Bot. Neer.* 6:445–57
187. Sakai, A. 1966. *Physiol. Plant.* 19: 105–14
188. Salin, M. L., Homann, P. H. 1971. *Plant Physiol.* 48:193–96
189. Sanderson, G. W., Sivapalan, K. 1966. *Tea Quart.* 37:11–26
190. Sasaki, S., Kozlowski, T. T. 1968. *Can. J. Bot.* 46:1173–83
191. Sasaki, S., Kozlowski, T. T. 1970. *New Phytol.* 69:493–500
192. Schaedle, M., Foote, K. C. 1971. *Forest Sci.* 17:308–13
193. Schaedle, M., Iannaccone, P., Foote, K. C. 1968. *Forest Sci.* 4:222–23
194. Schenk, W. 1952. *Planta* 41:290–310
195. Schilling, N., Dittrich, P., Kandler, A. 1971. *Ber. Deut. Bot. Ges.* 84:457–63
196. Schilling, N., Ferguson, J. A., Kandler, O. 1973. *Ber. Deut. Bot. Ges.* 86:393–401
197. Schmidt, A. 1924. *Bot. Arch.* 5:260–82
198. Schneider, H. 1955. *Am. J.Bot.* 42:893–905
199. Schneider, V., Heber, U. W. 1970. *Planta* 94:134–39
200. Schulze, E. D. 1970. *Flora* 159:177–232
201. Scott, D. G. 1907. *Ann. Bot.* 21:437–39
202. Seryczynska, H., Kaminski, M., Zavadzka, B. 1971. *Bull. Acad. Sci. Pol.* 19:759–63
203. Sestak, Z., Catsky, J., Jarvis, P. G., Eds. 1971. *Plant Photosynthetic Production: Manual of Methods.* The Hague: Junk

204. Shepard, R. K. 1970. *Some Aspects of Bark Photosynthesis in Bigtooth Aspen (Populus grandidentata) and Trembling Aspen (Populus tremuloides).* PhD thesis. Univ. Michigan, Ann Arbor
205. Shiroya, T., Slankis, V., Krotkov, G., Nelson, C. D. 1962. *Can. J. Bot.* 40: 669–75
206. Singh, B. N. 1935. *Ann. Bot.* 49:291–307
207. Skene, D. S. 1974. *Proc. Roy. Soc. London Ser. B* 186:75–78
208. Slater, C. H. W., Beakbane, A. B. 1973. *East Malling Res. Sta. Rep.* 1972, p. 66
209. Sokolov, S. R. 1953. *Bot. Zh.* 38:661–68
210. Sorensen, F. C., Ferrell, W. K. 1973. *Can. J. Bot.* 51:689–98
211. Specht-Jürgensen, I. 1967. *Flora* 157A: 426–53
212. Spencer, P. W., Titus, J. S. 1972. *Plant Physiol.* 49:746–50
213. Ibid 1973. 51:89–92
214. Srivastava, L. M. 1964. *Int. Rev. Forest. Res.* 1:203–77
215. Stalfelt, M. G. 1960. *Encycl. Plant Physiol.* 5/2:1–7
216. Stephens, G. R., Waggoner, P. E. 1970. *Biol. Sci.* 20:1050–59
217. Stocker, O. 1935. *Planta* 24:402–45
218. Strain, B. R., Johnson, P. L. 1963. *Ecology* 44:581–84
219. Szujko-Lacza, J., Fekete, G., Faludi-Daniel, A. 1970. *Acta Bot. Sci. Hung.* 16:393–404
220. Szujko-Lacza, J., Fokovan, J. M., Horwath, G., Fekete, G., Faludi-Daniel, A. 1971. *Acta Agron. Budapest* 20:247–60
221. Szujko-Lacza, J., Rakovan, J. N., Fekete, G., Horvath, G. 1972. *Acta Agron. Budapest* 21:41–56
222. Tairbekov, M. G., Starzecki, W. W. 1970. *Sov. Plant Physiol.* 17:573–78
223. Todd, G. W., Bean, R. C., Propst, B. 1961. *Plant Physiol.* 36:69–73
224. Thomson, W. W. 1966. *Bot. Gaz.* 127:133–39
225. Thomson, W. W., Platt, K. 1973. *New Phytol.* 72:791–97
226. Townsend, A. M., Hanover, J. W., Barner, B. V. 1972. *Silvae Genet.* 21: 133–39
227. Tverkina, N. D. 1970. *Sov. Plant Physiol.* 17:685–88
228. van der Driessche, R. 1973. *Aust. J. Forest.* 36:125–37
229. van der Meer, Q. P., Wassink, E. C. 1962. *Meded. Landbouwhogesch. Wageningen* 62:1–9
230. Van Steveninck, M. E., Goldney, D. C., van Steveninck, R. F. M. 1972. *Z. Pflanzenphysiol.* 67:155–60
231. Walker, D. A. 1965. *Plant Physiol.* 40:1157–61
232. Walker, D. A. 1971. *Methods Enzymol.* 23A:211–20
233. Wardlaw, I. F. 1968. *Bot. Rev.* 34:79–105
234. Wassink, E. C., Richardson, S. D. 1956. *Acta Bot. Neer.* 5:247–56
235. Wiebe, H. H., Al-Saadi, H. A., Kimball, S. L. 1974. *Am. J. Bot.* 61:444–49
236. Willstatter, R., Stoll, A. 1918. *Untersuchungen uber die Assimilation der Kohlensaure.* Berlin: Springer
237. Wolf, F. T. 1956. *Am. J. Bot.* 43: 714–18
238. Wright, R. D., Mooney, H. A. 1965. *Am. Midl. Natur.* 73:257–84
239. Yoshida, S. 1972. *Ann. Rev. Plant Physiol.* 23:437–64
240. Zanobini, A., Vanni, P., Firenzuoli, A. M., Ramponi, G. 1968. *Phytochemistry* 7:1297–98
241. Zanobini, A., Vanni, P., Mastronuzzi, E., Firenzuoli, A. M., Ramponi, G. 1967. *Phytochemistry* 6:1633–35
242. Zanobini, A., Vanni, P., Mastronuzzi, E., Firenzuoli, A. M., Ramponi, G. 1967. *Experientia* 23:1015–16
243. Zaprometov, M. N., Bukhlaeva, V. Ya. 1967. *Sov. Plant Physiol.* 14:167–76
244. Ibid 1970. 17:227–31
245. Zavalisina, S. F. 1951. *Dokl. Akad. Nauk. USSR* 78:137–39
246. Zelawski, W. 1967. *Acta Soc. Bot. Pol.* 36:713–23
247. Zelawski, W. 1967. *Bull. Acad. Sci. Pol.* 15:565–71
248. Zelawski, W., Kinelska, G., Lotocki, A. 1968. *Acta Soc. Bot. Pol.* 37:505–18
249. Zelawski, W., Riech, F. P., Stanley, R. G. 1970. *Can. J. Bot.* 48:1351–54
250. Zelawski, W., Szaniawski, R., Dybczynski, W., Piechurowski, A. 1973. *Photosynthetica* 7:351–57
251. Zelitch, I. 1971. *Photosynthesis, Photorespiration and Plant Productivity.* New York: Academic
252. Zelitch, I. 1973. *Curr. Advan. Plant Sci.* 3:44–54
253. Zhukova, G. Y. 1972. *Bot. Zh.* 57: 290–98
254. Zimmermann, M. H., Brown, C. L. 1971. *Trees Structure and Function.* New York: Springer

Ann. Rev. Plant Physiol. 1975. 26:117–26

THERMOGENIC RESPIRATION IN AROIDS

♦7585

Bastiaan J. D. Meeuse

Department of Botany, University of Washington, Seattle, Washington 98195

CONTENTS

INTRODUCTION

For reasons which will soon become obvious, this review—at first glance polemic and too inclusive—pays more than the customary attention to biological matters. It deals with thermogenicity in plants, a property which, according to the available evidence, is connected with "cyanide-resistant respiration," a type of cellular respiration insensitive to inhibition by terminal inhibitors such as cyanide, azide, and carbon monoxide (CO), and by inhibitors such as antimycin A and HOQNO, which act between b and c-type cytochromes (18, 62, 63, 73, 80; cf 105, 106). The definition is not absolute, since degrees of cyanide resistance varying from 0 to 100%, and even stimulation by cyanide (100), have been found in plant tissues, pollen grains, and mitochondria (18, 19, 23, 25, 46, 70, 82, 86, 145, 146, 201), often as a function of plant or organ development (131, 200). The insensitivity resides in the mitochondria (19, 64, 65, 150, 188, 205, 206). When isolated from resistant tissues, they turn out to contain a "dual pathway" for respiratory electron transfer: the classical, cyanide-sensitive electron transport system which is coupled to phosphorylation, and a cyanide-insensitive pathway which branches from the classical one on the substrate side of cytochrome c and is phosphorylative to a much lesser extent (19, 64, 65, 150, 188, 205, 206).

The alternate pathway is specifically inhibited by iron-complexing agents such as hydroxamic acids, α, α'-dipyridyl, K-thiocyanate, and 8-hydroxyquinoline (19, 25, 46, 170). A ferrosulfoprotein with an apparent K_m for O_2 lower than that of a flavoprotein oxidase is probably (but not certainly) involved (83). Temporary replacement of a functioning classical electron transport chain by the alternate one may lead to a lowering of the ATP level. This form of "uncoupling" should be distinguished clearly from what may occur in the classical chain under the influence of uncouplers such as 2,4-dinitrophenol (74, 204) or fatty acids (12). The latter type will be referred to as "endogenous uncoupling." Both may lead to an increase in the rate of heat production (103, 151, 204, 205).

PHOSPHORYLATIVE EVENTS

Experimentally it was found that improved isolation methods for plant mito-chondria (47, 48, 79, 81) applied to various plant materials, including arum lilies, can lead to preparations which show respiratory control with tight coupling and ADP:O ratios similar to, but slightly lower than, those found with mammalian mitochondria (37, 49, 54, 133, 199, 203). The three sites of phosphorylation sug-gested for plant mitochondria agree well with those in mammalian mitochondria (37, 188), with site II between b and c-type cytochromes and site III between cytochrome a and a_3. In mitochondria from aroid appendices, all three sites can be operational (112, 204), resulting in ADP:O ratios as high as 2.7 for malate in *Sauromatum* (204). However, on the day of flowering the *Sauromatum* mito-chondria are uncoupled (204), either endogenously or because the alternate pathway becomes operational. The low phosphorylative efficiency (75%) of mung bean mito-chondria has been ascribed to the latter factor. Phosphorylative site I is retained (13, 14, 150, 188, 205, 206).

TEMPERATURE RISE IN FLOWERS AND INFLORESCENCES

Thermogenicity is especially obvious in floral organs (56) and has been studied most often in the water lily *Victoria* (45, 59, 95, 189) and in arum lilies such as *Arum italicum* (104, 132, 149), *A. maculatum* (24, 58, 66, 86–88, 104, 111, 112, 152, 167, 168, 172–175), *Sauromatum guttatum* (29–31, 38–40, 75, 76, 90, 135, 138, 140, 141, 147, 176, 192–195), *Symplocarpus foetidus* (13, 35, 53, 61, 97, 98, 186–188, 196), *Typhonium divaricatum* (168, 169), *Alocasia pubera* (190), *Schizocasia portei* (52), and *Philodendron selloum* (143). The temperature difference with the environment may reach a value of 22° C in *Colocasia odora* and *Schizocasia portei,* and at least as much in *Philodendron selloum* and *Symplocarpus foetidus.* In biochemical cir-cles, there has been a remarkable lack of precision in referring to the particular species and organ involved, e.g. "skunk cabbage," supposed to be identical with *Symplocarpus foetidus,* could refer just as well to *Lysichitum americanum* (Western skunk cabbage) which, however, has never been investigated biochemically.

The central column or spadix of *Symplocarpus* is completely covered with small hermaphroditic flowers which warm up, whereas in *Sauromatum* and *Arum* there has been a differentiation leading to separate staminate and pistillate flowers as well as a special, sterile, club or finger-like organ—the appendix or osmophore (198)—which abundantly produces heat (and smell). In *Philodendron selloum,* incomplete staminate flowers are involved. Frequently overlooked also is the fact that the number of temperature maxima within an inflorescence varies from species to species (124), from two in *Arum, Philodendron,* and *Sauromatum* to five in *Colocasia odora.* In forms possessing an appendix, the first maximum (produced right there, and the only one studied biochemically so far) is by far the highest. Since the heat serves as a "volatilizer" for the odoriferous compounds—often amines (176) or indole (38–40)—that attract the pollinators, and since in these forms obligatory cross-pollination is combined with proterogyny (55, 96, 102, 134, 153, 191, 198), it is no surprise that this maximum precedes the shedding of pollen by many hours (96).

HISTORY

Explicit formulation of the dual pathway concept goes back to 1932, when Okunuki (145, 146) discovered the cyanide and CO insensitivity of *Lilium auratum* pollen. Light-reversal of the partial CO inhibition in certain pollens sometimes led to respiration values exceeding those of untreated material—the first indication that the action of cyanide and CO may involve more than a simple replacement of the blocked classical pathway by another (cf 70).

Thermogenesis in *Arum* was discovered by Lamarck in 1778 (104). Garreau (58) demonstrated the close relationship between heat development and oxygen consumption here. van Herk (192) ascribed the cyanide insensitivity of the respiration of the *Sauromatum* appendix to the absence of the cytochrome/cytochrome oxidase system and the presence of an autoxidizable flavoprotein. His ideas were essentially adopted, in the case of *Arum maculatum,* by James & Beevers (85, 86). However, the presence of the classical electron transfer system can be demonstrated easily in both cases, and at present the dual pathway concept seems the most suitable for explaining the cyanide-resistant respiration in *Arum, Sauromatum,* and *Symplocarpus* (172, 173, 207, 208). The various alternatives have been discussed ably by Bendall & Bonner (19). In the meantime, cyanide-resistant respiration had been demonstrated in storage tissues of potato, sweet potato, carrot, and Jerusalem artichoke (6–9, 27, 42, 50, 63–65, 68, 69, 101, 115–123, 126, 144, 155, 158, 161–164, 197). It is now known to play a role also in the so-called climacteric respiration of fruit (180, 181), in roots (91–94), in mung bean seedlings (25, 79, 81, 82, 182–185), and in various microorganisms (67, 107–110, 171). Several reviews of cyanide-insensitive respiration in plants and the dual pathway concept are now available (18, 25, 80).

The brilliant contributions made by the workers at the Johnson Foundation (13, 19, 25, 36, 37, 49, 53, 105, 113, 170, 180–188) in "sequencing" the electron carriers in mitochondria have been acknowledged by Ikuma (80), but new evidence requires modification of his scheme (p. 429). In 1974, Storey (personal communication) pulsed anaerobic CO-saturated *Symplocarpus* mitochondria with O_2 and looked at the kinetics of the carriers oxidized mainly by the alternate pathway. Rapid kinetics were observed for ubiquinone (UQ, midpoint potential $+70$ mV) and a portion of the flavoprotein component, indicating that a nonfluorescent flavoprotein F_{ma}, with a midpoint potential of about 20 mV in *Symplocarpus* and 40 mV in mung bean mitochondria, is the link between the classical and the alternate pathway. The more highly oxidized F_{ma} is, the less well it functions as electron donor to the alternate oxidase. The redox states of UQ and F_{ma} in state 4 and 3 are such that the plant mitochondrion can shift most of its electron transport through the cytochrome chain in state 3 and through the alternate pathway in state 4.

STRUCTURAL AND DEVELOPMENTAL FEATURES

The importance of cellular organization in fruit ripening has been stressed by Solomos & Laties (179). During the development of the appendix of *Arum* and *Sauromatum*, when important permeability changes occur (51), there is an enlargement of individual mitochondria with a corresponding increase in the number of cristae per mitochondrion (22, 175). Microbodies or peroxisomes (21, 149) do not seem to be involved in the metabolic flare-up. Nitrogen metabolism (20, 57) is especially intense on the day of flowering and afterwards. The notion that the amines which attract the pollinators are formed through decarboxylation of amino acids (156, 174) has recently been challenged (71, 72); amino acid/aldehyde transamination may be the preferred pathway for biosynthesis of primary aliphatic amines in flowering plants. In *Sauromatum*, the respiratory CO_2 produced by the appendix shows a marked decrease in ^{13}C on flowering day (204; cf 84). This may reflect a greater metabolic participation of fatty acids, important because these may act as uncouplers (12).

The increased cyanide-insensitive respiration of slices of storage tissue is accompanied by an increase in the number of mitochondria, presumably through fission of preexisting mitochondria (123) without a corresponding increase in cell number (8, 9, 122, 144). The newly formed mitochondria are heavier than the preexisting ones, probably because they contain more NADH dehydrogenase (64) and nonheme iron oxidase and are more resistant to cyanide (144, 164). Their aerobic biogenesis is, understandably, accompanied by the synthesis of new RNA species and proteins (11, 28, 42, 50, 121, 197), while various metabolic activities such as the uptake of phosphate, sulfate, and glucose (6, 64, 69, 100, 117, 126, 158), operation of the pentose phosphate pathway (6), and glycolysis-fed Krebs cycle activities (3–5, 117, 118, 158) are also boosted. In potato slices, certain changes occur in the NAD/NADP ratio (27), important because plant mitochondria have been said to lack NADP (78, 80, 113).

FUNCTION OF THE ALTERNATE PATHWAY

Upon injury, many plants release HCN from cyanogenic glycosides, but the latter are not in evidence where the alternate pathway is the most obvious (*Arum, Symplocarpus, Lilium auratum* pollen). Free CO, although present in concentrations up to 12% in the internal cavity of the brown alga *Nereocystis* (114, 157), is rare in plants. Confrontation of plant tissues with cyanide or CO thus is largely a laboratory event—hardly something Nature could have selected for. Therefore, no student of evolution can easily accept the idea that the alternate pathway has become fairly common in plants when it will become functional only after exposure of tissues to cyanide or CO (188).

Of course, the situation in isolated mitochondria may not always reflect the situation in intact tissues. When plants produce the heat that will volatilize the odoriferous principles that attract the pollinators, the survival value of the alternate pathway seems obvious. In *Symplocarpus,* the long-lasting heat production guarantees development of the inflorescence and pollination, even at subfreezing environmental temperatures (97, 98). Associating the alternate pathway with thermogenesis remains legitimate, for although it is true that its operation does not in all cases lead to a drop in the ATP level (which may even rise somewhat), it is obvious that a greatly increased flow of respiratory electrons is needed to maintain such ATP constancy when the number of phosphorylative sites is reduced from three to one (188, 206). Crudely phrased, more "fuel" has to be "burned" to obtain the same amount of ATP as in the classical situation, and this constitutes thermogenicity.

For *Arum* mitochondria, Passam & Palmer (150) recently have supplemented this circumstantial evidence with direct experimental data. The rate of oxidation of ascorbate plus tetramethylphenylene diamine (TMPD), which enters the cytochrome chain at cytochrome c, in this case is somewhat lower than that of malate or succinate, in contrast to the situation in mitochondria from rat liver, Jerusalem artichoke, and mung bean, where the effect of ascorbate plus TMPD far surpasses that of the other two electron donors. In the absence of cyanide, cytochrome oxidase therefore may not always act as the major terminal oxidase in *Arum* appendix mitochondria. Since heat development in the appendix of *Arum* and *Sauromatum* is confined to less than 12 hr, meticulous attention to the particular developmental stage of the appendices used is clearly imperative. In 1974, Lance (112) reached a similar conclusion on the basis of ADP:O ratios. As the inflorescence of *Arum* develops, the efficiency of oxidative phosphorylation decreases due to increased participation of the alternate pathway, endogenous uncoupling, and activity of a mitochondrial ATPase (cf 29).

In bean hypocotyl mitochondria, where thermogenicity is not at issue, the alternate pathway comes into play only when the ADP level is low enough to limit the cytochrome pathway rate (13). It may be required either to increase the flux through the citric acid cycle or to increase the oxidation of cytoplasmic NADH (48) in the absence of a phosphate acceptor (81). The alternate pathway probably modulates a balance between the availability of reducing equivalents and that of high energy adenylates (3–5, 63, 80, 100). The latter affect various enzymes allosterically (10, 90, 159), are obligatory for the functioning of succinyl-CoA synthetase in plants (148), and strongly influence the fate of malate (210), which has been receiving increased attention (77, 128–130).

CONTROL OF THE ALTERNATE PATHWAY: RELATIONSHIP WITH CLIMACTERIC FRUIT RESPIRATION AND ETHYLENE

In slices of storage tissues, oxygen plays an important role in the development of cyanide-resistant respiration (64, 115; cf 33). Volatile aldehydes may act as the trigger (115, 116). In the brown adipose tissue of mammals, thermogenicity—probably based on endogenous uncoupling—is under the control of hormones and external temperature (15, 41, 43, 44, 89, 125, 177, 178, 202). The environmental temperature rise in the daytime has also been invoked to help account for the climacteric events in *Arum;* unfortunately, they do not start until late in the afternoon. In actuality, thermogenicity in arum lilies is controlled primarily by the light/dark regime (135, 136, 139, 167; cf 59), and secondarily by hormonal influences (30, 139, 193, 194). Exposure of *Sauromatum* inflorescences, kept in constant light, to a single 6-hr "dark shot" leads to a metabolic peak 40–45 hr after the beginning of the shot (31). The regime probably leads to production in the staminate flower primordia of a triggering hormone ("calorigen"), which can be shown to be present in the appendix about 22 hr before the heating starts. Injection of the extracted hormone into appendices amputated 2 days before the expected metabolic explosion leads to heating and smell production after a lag time of about a day (30, 193, 194). Chen & Meeuse (38, 40), concentrating on the production of the easily demonstrable compound indole under the influence of the hormone, have designed a bioassay for calorigen and have purified two active principles (calorigen I and II) with its aid. Both are low-molecular compounds which have now been characterized chemically to a considerable extent.

At the intracellular level, Bahr & Bonner (13) have shown the complete independence in vitro of the alternate path from ATP and ADP. For isolated *Symplocarpus* mitochondria, they have suggested (14) that the distribution of respiratory electrons over the two pathways is regulated by an equilibrium mechanism of two postulated carriers possessing E_0' values of such magnitude that they ensure full reduction of the component connected to the cytochrome oxidase, while the carrier feeding electrons into the alternate path is completely or partially oxidized. It is difficult to see, however, how such a system could lead to the nearly complete suppression of the classical pathway under certain circumstances.

Ethylene, like cyanide, often stimulates respiration (1, 2). In postharvest fruit respiration (160), the exact role of ethylene is hard to evaluate because of the multiplicity of events (32, 103, 142, 209). However, for intact avocados and for potatoes (where ripening is not at issue) Solomos & Laties (180, 181; cf 33, 154) could show that ethylene and HCN (gas) produce identical responses in glycolysis and respiration. The presence of the cyanide-resistant path, and not necessarily "ripening," seems to be the prerequisite for ethylene to stimulate respiration. Indeed, tissues stimulated by ethylene are also stimulated by cyanide (154), while conversely ethylene has no effect on plant materials strongly inhibited by cyanide (180). Both agents are thought to divert electrons actively from the classical respiratory chain to the alternate path, an ability which they probably share with calorigen (137). The mechanism giving rise to the increase in glycolysis which accompanies the respiratory boost is not clear. In some other instances of glycolytic boosts (no matter what the cause), phosphofructokinase (PFK) and/or pyruvate kinase (PK) have been implicated (16, 17, 26, 34, 60, 75, 76, 99, 127, 155). In banana, preclimacteric PFK displays toward its substrate a negative cooperativity (165, 166) which is partially abolished at the start of the climacteric; this amounts to an activation of the rate-limiting enzyme PFK.

EPILOGUE AND PROGNOSIS

In 1973, Dizengremel et al (46) found that cyanide-resistant and cyanide-sensitive mitochondria contain about the same proportion of ferrosulfoproteins with nonheme iron and labile sulfur. They interpreted this to mean that the alternate pathway is common but is operational in some cases only. They use the observation that cyanide resistance can be induced easily by certain experimental treatments (64, 107) as further evidence that the "switching on" of the alternate pathway does not depend on a de novo large-scale synthesis of ferrosulfoproteins, but rather on a control mechanism that makes enzymatic proteins already present in the mitochondrial membrane more accessible to oxygen (cf 84, 204). The problem is how to reconcile this concept with the proved production of new and heavy mitochondria in storage tissue slices (8, 9, 122, 123, 144). The first order of the day will be to solve this controversy. Identification of the "second oxidase" is essential. It is also imperative to elucidate the chemical nature of calorigen and to compare its triggering action with that exerted by ethylene, CO, and cyanide. Details of respiratory electron transfer mechanisms and phosphorylative sites remain to be worked out in several cases.

Literature Cited

1. Abeles, F. B. 1972. *Ann. Rev. Plant Physiol.* 23:259–92
2. Abeles, F. B. 1973. *Ethylene in Plant Biology.* New York: Academic. 302 pp.
3. Adams, P. B. 1970. *Plant Physiol.* 45:495–99
4. Ibid, 500–3
5. Adams, P. B., Rowan, K. S. 1970. *Plant Physiol.* 45:490–94
6. ap Rees, T., Beevers, H. 1960. *Plant Physiol.* 35:839–47
7. ap Rees, T., Bryant, J. A. 1971. *Phytochemistry* 10:1183–90
8. Asahi, T., Honda, Y., Uritani, I. 1966. *Arch. Biochem. Biophys.* 113:498–99
9. Asahi, T., Majima, R. 1969. *Plant Cell Physiol.* 10:317–23
10. Atkinson, D. E. 1966. *Ann. Rev. Biochem.* 35:85–124
11. Bacon, J. S. D., MacDonald, I. R., Knight, A. H. 1965. *Biochem. J.* 94:175–82
12. Baddeley, M. S., Hanson, J. B. 1967. *Plant Physiol.* 42:1702–10
13. Bahr, J. T., Bonner, W. D. Jr. 1973. *J. Biol. Chem.* 248:3441–45
14. Ibid, 3446–50
15. Ball, E. C., Jungas, R. L. 1961. *Proc. Nat. Acad. Sci. USA* 47:932–41
16. Barker, J., Khan, M. A. A., Solomos, T. 1967. *New Phytol.* 66:577–96
17. Barker, J., Solomos, T. 1962. *Nature* 169:189–91
18. Beevers, H. 1961. *Respiratory Metabolism in Plants.* Evanston, Ill. & White Plains, N.Y.: Row-Peterson. 232 pp.
19. Bendall, D. S., Bonner, W. D. Jr. 1971. *Plant Physiol.* 47:236–45
20. Berger, C. 1970. *Z. Pflanzenphysiol.* 62:259–69
21. Berger, C., Gerhardt, B. 1971. *Planta* 96:326–38
22. Berger, C. Schnepf, E. 1970. *Protoplasma* 69:237–51
23. Bonner, W. D. Jr. 1965. *Plant Biochemistry,* ed. J. Bonner, J. E. Varner, 89–123. New York/London: Academic. 1054 pp.
24. Bonner, W. D. Jr., Bendall, D. S. 1968. *Biochem. J.* 109:47P
25. Bonner, W. D. Jr., Christensen, E. L., Bahr, J. T. 1972. *Biochemistry and Biophysics of Mitochondrial Membranes,* ed. G. F. Azzone, 113–19. New York/London: Academic
26. Bourne, D. T., Ranson, S. L. 1965. *Plant Physiol.* 40:1178–90
27. Brinkman, F. G., van der Plas, L. H. W., Verleut, J. D. 1973. *Z. Pflanzenphysiol.* 68:364–72
28. Bryant, J. A., ap Rees, T. 1971. *Phytochemistry* 10:1191–97
29. Buggeln, R. G., Meeuse, B. J. D. 1967. *Proc. Kon. Ned. Akad. Wetensch. Ser. C.* 70:515–25
30. Buggeln, R. G., Meeuse, B. J. D. 1971. *Can. J. Bot.* 49:1373–77
31. Buggeln, R. G., Meeuse, B. J. D., Klima, J. R. 1971. *Can. J. Bot.* 49:1025–31
32. Burg, S. P., Burg, E. A. 1967. *Plant Physiol.* 42:144–52
33. Burton, W. G. 1950. *New Phytol.* 49:121–34
34. Chalmers, D. J., Rowan, K. S. 1971. *Plant Physiol.* 48:235–40

35. Chance, B., Bonner, W. D. Jr. 1965. *Plant Physiol.* 40:1198–1204
36. Chance, B., Bonner, W. D. Jr., Storey, B. T. 1968. *Ann. Rev. Plant Physiol.* 19:295–320
37. Chance, B., Williams, G. R. 1956. *Advan. Enzymol.* 17:65
38. Chen, J., Meeuse, B. J. D. 1971. *Am. J. Bot.* 58:478
39. Chen, J., Meeuse, B. J. D. 1971. *Acta Bot. Neer.* 20:627–35
40. Chen, J., Meeuse, B. J. D. 1972. *Plant Cell Physiol.* 13:831–41
41. Christiansen, E. N., Pedersen, J. I., Grav, H. J. 1969. *Nature* 222:857–60
42. Click, R. E., Hackett, D. P. 1963. *Proc. Nat. Acad. Sci. USA* 50:243–50
43. Dawkins, M. J. R., Hull, D. 1964. *J. Physiol.* 172:216–38
44. Dawkins, M. J. R., Hull, D. 1965. *Sci. Am.* 213:62–67
45. Decker, J. S. 1936. See Ref. 189, pp. 479, 496
46. Dizengremel, P., Chauveau, M., Lance, C. 1973. *C. R. Acad. Sci. Paris Ser. D* 277:239–242
47. Douce, R., Christensen, E. L., Bonner, W. D. Jr. 1972. *Biochim. Biophys. Acta* 275:148–60
48. Douce, R., Mannella, C. A., Bonner, W. D. Jr. 1973. *Biochim. Biophys. Acta* 292:105–16
49. Dutton, P. L., Storey, B. T. 1971. *Plant Physiol.* 47:282–88
50. Edelman, J., Hall, M. A. 1965. *Biochem. J.* 95:403–10
51. Eilam, Y. 1965. *J. Exp. Bot.* 16:614–27
52. El-Din, S. M. 1968. *Naturwissenschaften* 12:658–59
53. Erecinska, M., Storey, B. T. 1970. *Plant Physiol.* 46:618–24
54. Estabrook, R. W. 1961. *J. Biol. Chem.* 236:3051–57
55. Faegri, K., van der Pijl, L. 1971. *The Principles of Pollination Ecology.* Oxford: Pergamon. 291 pp. 2nd rev. ed.
56. Fischer, H. 1960. *Encycl. Plant Physiol.* 2:520–35
57. Fischer, H., Specht-Jürgensen, I., Fleck-Gerndt, G. 1972. *Beitr. Biol. Pflanz.* 48:243–53
58. Garreau, M. 1851. *Ann. Sci. Nat. Bot. (Paris) 3e Ser.* 16:250–56
59. Gessner, F. 1960. *Planta* 54:453–65
60. Ghosh, A., Chance, B. 1964. *Biochem. Biophys. Res. Commun.* 16:174–81
61. Hackett, D. P. 1956. *Plant Physiol.* 31: suppl. XL
62. Hackett, D. P. 1959. *Ann. Rev. Plant Physiol.* 10:113–46
63. Hackett, D. P. 1963. *Control Mechanisms in Respiration and Fermentation,* ed. B. Wright, 105–27. New York: Ronald
64. Hackett, D. P., Haas, D. W., Griffiths, S. K., Niederpruem, D. J. 1960. *Plant Physiol.* 35:8–19
65. Hackett, D. P., Rice, B., Schmid, C. 1960. *J. Biol. Chem.* 235:2140–44
66. Hackett, D. P., Simon, E. W. 1954. *Nature* 173:162–63
67. Hall, D. O., Greenawalt, J. W. 1964. *Biochem. Biophys. Res. Commun.* 17:565–69
68. Hanebuth, W. F., Chasson, R. M. 1972. *Plant Physiol.* 49:857–59
69. Hanebuth, W. F., Chasson, R. M., Pittman, D. 1974. *Physiol. Plant.* 30:273–78
70. Hanes, C. S., Barker, J. 1931. *Proc. Roy. Soc. B* 108:95–118
71. Hartmann, T., Dönges, D., Steiner, M. 1972. *Z. Pflanzenphysiol.* 67:404–17
72. Hartmann, T., Ilert, H.-I., Steiner, M. 1972. *Z. Pflanzenphysiol.* 68:11–18
73. Hartree, E. F. 1957. *Advan. Enzymol.* 18:1–64
74. Hess, C. M., Meeuse, B. J. D. 1967. *Acta Bot. Neer.* 16:188–96
75. Hess, C. M., Meeuse, B. J. D. 1968. *Proc. Kon. Ned. Akad. Wetensch. Ser. C* 71:443–55
76. Ibid, 456–71
77. Hobson, G. E. 1970. *Phytochemistry* 9:2257–63
78. Ikuma, H. 1967. *Science* 158:529
79. Ikuma, H. 1970. *Plant Physiol.* 45:773–81
80. Ikuma, H. 1972. *Ann. Rev. Plant Physiol.* 23:419–36
81. Ikuma, H., Bonner, W. D. Jr. 1967. *Plant Physiol.* 42:67–75
82. Ibid, 1535–44
83. Ikuma, H., Schindler, F. D., Bonner, W. D. Jr. 1964. *Plant Physiol.* 39: suppl. LX
84. Jacobson, B. S., Laties, G. G., Smith, B. N., Epstein, S., Laties, B. 1970. *Biochim. Biophys. Acta* 216:295–304
85. James, W. O. 1953. *Plant Respiration.* Oxford: University Press. 282 pp.
86. James, W. O., Beevers, H. 1950. *New Phytol.* 49:353–74
87. James, W. O., Elliott, D. C. 1955. *Nature* 175:89
88. James, W. O., Elliott, D. C. 1955. *New Phytol.* 57:230–34
89. Joel, C. D. 1965. *Adipose Tissue (Handb. Physiol. Sect. 5),* ed. A. E. Renold, G. F. Cahill Jr., 59–86. Washington, D.C.: Am. Physiol. Soc.

90. Johnson, T. F., Meeuse, B. J. D. 1972. *Proc. Kon. Ned. Akad. Wetensch. Ser C* 74:1–19
91. Kano, H., Kumazawa, K. 1972. *Plant Cell Physiol.* 13:237–44
92. Ibid 1973. 14:673–80
93. Kano, H., Kumazawa, K., Mitsui, S. 1969. *J. Sci. Soil Animal Fertilizers Jap.* 40:473–78
94. Ibid 1970. 41:213–17
95. Knoch, E. 1899. *Bibliog. Bot.* 9, 47:1–60
96. Knoll, F. 1926. *Abh. Zool.-Bot. Ges. Wien* 12:379–482
97. Knutson, R. M. 1972. *Am. Midl. Natur.* 88:251–54
98. Knutson, R. M. 1974. *Science* 186: 746–47
99. Kohr, M. J., Beevers, H. 1971. *Plant Physiol.* 47:48–52
100. Kolattukudy, P. E., Reed, D. J. 1966. *Plant Physiol.* 41:661–69
101. Kozuka, Y., Uritani, I. 1973. *Plant Cell Physiol.* 14:193–96
102. Kugler, H. 1970. *Einführung in die Blütenökologie.* Stuttgart: G. Fischer. 345 pp. 2nd ed.
103. Lacher, J. R., Amador, A., Snow, K. 1966. *Plant Physiol.* 41:1435–38
104. Lamarck, J. B. de 1778. *Flore Française* 3:538
105. Lambowitz, A. M., Bonner, W. D. Jr. 1973. *Biochem. Biophys. Res. Commun.* 52:703–11
106. Lambowitz, A. M., Bonner, W. D. Jr. 1973. *Plant Physiol.* 51: suppl. 10
107. Lambowitz, A. M., Slayman, C. W. 1971. *J. Bacteriol.* 108:1087–96
108. Lambowitz, A. M., Slayman, C. W., Slayman, C. L. 1972. *J. Biol. Chem.* 247:1536–45
109. Lambowitz, A. M., Smith, E. W., Slayman, C. W. 1972. *J. Biol. Chem.* 247:4850–58
110. Ibid, 4859–65
111. Lance, C. 1972. *Ann. Sci. Nat. Bot. 12e Ser.* 13:477–95
112. Lance, C. 1974. *Plant Sci. Lett.* 2:165–71
113. Lance, C., Bonner, W. D. Jr. 1968. *Plant Physiol.* 43:756–66
114. Langdon, S. 1916. *Publ. Puget Sound Biol. Sta.* 1:237–46
115. Laties, G. G. 1962. *Plant Physiol.* 37:679–90
116. Laties, G. G. 1963. See Ref. 63, 129–55
117. Laties, G. G. 1964. *Plant Physiol.* 39:391–97
118. Ibid, 654–63
119. Laties, G. G. 1967. *Aust. J. Sci.* 30:193–203
120. Laties, G. G., Hoelle, C. 1965. *Plant Physiol.* 40:757–64
121. Leaver, C. J., Key, J. L. 1967. *Proc. Nat. Acad. Sci. USA* 57:1338–44
122. Lee, S. G., Chasson, R. M. 1966. *Physiol. Plant.* 19:194–98
123. Ibid, 199–206
124. Leick, E. 1915. *Ber. Deut. Bot. Ges.* 33:518–536
125. Lindberg, O., Ed. 1970. *Brown Adipose Tissue.* New York: Elsevier
126. Loughman, B. C. 1960. *Plant Physiol.* 35:418–24
127. Lynen, F. 1963. See Ref. 63, 289–306
128. Macrae, A. R. 1971. *Phytochemistry* 10:1453–58
129. Ibid, 2343–47
130. Macrae, A. R., Moorhouse, R. 1970. *Eur. J. Biochem.* 16:96–102
131. Marsh, P. B., Goddard, D. R. 1939. *Am. J. Bot.* 26:724–28
132. Matile, P. 1958. *Ber. Schweiz. Bot. Ges.* 68:295–306
133. Matlib, M. A., Kirkwood, R. C., Smith, J. E. 1971. *J. Exp. Bot.* 22:291–303
134. Meeuse, B. J. D. 1961. *The Story of Pollination.* New York: Ronald. 243 pp.
135. Meeuse, B. J. D. 1966. *Sci. Am.* 215: 80–88
136. Meeuse, B. J. D. 1968. *Atomes* 256: 428–36
137. Meeuse, B. J. D. 1972. *What's New in Plant Physiology,* ed. G. J. Fritz, 4(2): 1–4
138. Meeuse, B. J. D., Amundson, R. G. 1973. *Plant Physiol.* 51: suppl. 48
139. Meeuse, B. J. D., Buggeln, R. G. 1969. *Acta Bot. Neer.* 18:159–72
140. Meeuse, B. J. D., Buggeln, R. G., Summers, S. N. 1969. *Abstr. 11th Int. Bot. Congr. Seattle,* 144
141. Meeuse, B. J. D., Chen, J., Johnson, T. F. 1971. *Plant Physiol.* 47: suppl. 27
142. Millerd, A., Bonner, J., Biale, J. B. 1953. *Plant Physiol.* 28:521–31
143. Nagy, K. A., Odell, D. K., Seymour, R. S. 1972. *Science* 178:1195–97
144. Nakano, M., Asahi, T. 1970. *Plant Cell Physiol.* 11:499–502
145. Okunuki, K. 1932. *Bot. Mag. (Tokyo)* 47:45–62
146. Okunuki, K. 1939. *Acta Phytochim.* 11:27–64
147. Olason, D. M. 1967. *Changes in cofactor levels in the flowering sequence of some arum lilies.* MSc. thesis. Univ. Washington, Seattle
148. Palmer, J. M., Wedding, R. T. 1966. *Biochim. Biophys. Acta* 113:167–74
149. Parish, R. W. 1972. *Z. Pflanzenphysiol.* 67:430–42

150. Passam, H. C., Palmer, J. M. 1972. *J. Exp. Bot.* 23:366–74
151. Poe, M., Estabrook, R. W. 1968. *Arch. Biochem. Biophys.* 126:320–30
152. Prime, C. T. 1960. *Lords and Ladies.* London: Collins. 241 pp.
153. Proctor, M., Yeo, P. 1973. *The Pollination of Flowers.* London: Collins. 418 pp.
154. Reid, M. S., Pratt, H. K. 1972. *Plant Physiol.* 49:252–55
155. Ricardo, C. P. P., ap Rees, T. 1972. *Phytochemistry* 11:623–26
156. Richardson, M. 1966. *Phytochemistry* 5:23–30
157. Rigg, G. B., Swain, L. A. 1941. *Plant Physiol.* 16:361–71
158. Romberger, J. A., Norton, G. 1961. *Plant Physiol.* 44:311–12
159. Rowan, K. S. 1966. *Int. Rev. Cytol.* 19:301–91
160. Sacher, J. A. 1973. *Ann. Rev. Plant Physiol.* 24:197–224
161. Sakano, K., Asahi, T. 1969. *Agr. Biol. Chem.* 33:1433–39
162. Sakano, K., Asahi, T. 1971. *Plant Cell Physiol.* 12:417–26
163. Ibid, 427–36
164. Sakano, K., Asahi, T., Uritani, I. 1968. *Plant Cell Physiol.* 9:49–60
165. Salminen, S. O., Young, R. E. 1974. *Nature* 247:389–91
166. Salminen, S. O., Young, R. E. 1974. *Plant Physiol.* In press
167. Schmucker, T. 1925. *Flora* 118:460–75
168. Schnepf, E. 1965. *Planta* 66:374–76
169. Schnepf, E., Czygan, F. C. 1966. *Z. Pflanzenphysiol.* 54:345–55
170. Schonbaum, G. R., Bonner, W. D. Jr., Storey, B. T., Bahr, J. T. 1971. *Plant Physiol.* 47:124–28
171. Sharpless, T. K., Butow, R. A. 1970. *J. Biol. Chem.* 245:58–70
172. Simon, E. W. 1957. *J. Exp. Bot.* 8:20–35
173. Ibid 1959. 10:125–33
174. Ibid 1962. 13:1–4
175. Simon, E. W., Chapman, J. A. 1961. *J. Exp. Bot.* 12:414–20
176. Smith, B. N., Meeuse, B. J. D. 1966. *Plant Physiol.* 41:343–47
177. Smith, R. E., Horwitz, B. A. 1969. *Physiol. Rev.* 49:330–425
178. Smith, R. E., Roberts, J. C. 1964. *Am. J. Physiol.* 206:143–48
179. Solomos, T., Laties, G. G. 1973. *Nature* 245:390–92
180. Solomos, T., Laties, G. G. 1974. *Science* 54:506–11
181. Solomos, T., Laties, G. G. 1974. *Plant Physiol.* In press
182. Storey, B. T. 1970. *Plant Physiol.* 45:447–54
183. Ibid, 46:13–20
184. Ibid, 625–30
185. Storey, B. T. 1971. *Fed. Proc.* 30:1189
186. Storey, B. T. 1971. *Plant Physiol.* 48:493–97
187. Storey, B. T., Bahr, J. T. 1969. *Plant Physiol.* 44:115–25
188. Ibid, 126–34
189. Valla, J. J., Cirino, D. R. 1972. *Darwiniana* 17:477–98
190. van der Pijl, L. 1933. *Trop. Natuur* 22:210–214
191. van der Pijl, L. 1937. *Rec. Trav. Bot. Neer.* 34:157–67
192. van Herk, A. W. H. 1937. *Rec. Trav. Bot. Neer.* 34:69–156
193. van Herk, A. W. H. 1937. *Proc. Kon. Ned. Akad. Wetensch.* 40:607–14
194. Ibid, 709–19
195. van Herk, A. W. H., Badenhuizen, N. P. 1934. *Proc. Kon. Ned. Akad. Wetensch.* 37:99–105
196. Van Norman, R. W. 1955. *Plant Physiol.* 30: suppl. 29
197. Vaughan, D., MacDonald, I. R. 1967. *Plant Physiol.* 42:456–58
198. Vogel, S. 1963. *Akad. Wiss. Lit. Mainz, Abh. Math.-Naturwiss. Kl.* 1962:605–763
199. Wakiyama, S., Ogura, Y. 1970. *Plant Cell Physiol.* 11:835–48
200. Wedding, R. T., McCready, C. C., Harley, J. L. 1973. *New Phytol.* 72:1–13
201. Ibid, 15–26
202. Whittow, C. G., Ed. 1973. *Comparative-Physiology of Thermoregulation, Vol. 3: Special Aspects of Thermoregulation.* New York: Academic. 278 pp.
203. Wilson, R. H., Hanson, J. B. 1969. *Plant Physiol.* 44:1335–41
204. Wilson, R. H., Smith, B. N. 1971. *Z. Pflanzenphysiol.* 65:124–29
205. Wilson, S. B. 1970. *Biochem. J.* 116: 20
206. Wilson, S. B. 1970. *Biochim. Biophys. Acta* 223:383–87
207. Yocum, C. S., Hackett, D. P. 1955. *Plant Physiol.* 30: suppl. 30
208. Ibid 1957. 32:186–91
209. Young, R. E., Biale, J. B. 1967. *Plant Physiol.* 42:1357–62
210. Zimmerman, E. J., Ikuma, H. 1970. *Plant Physiol.* 46: suppl. 37

Ann. Rev. Plant Physiol. 1975. 26:127–58

CHLOROPHYLL-PROTEINS: ❖7586
LIGHT-HARVESTING AND
REACTION CENTER COMPONENTS
OF PLANTS[1]

J. Philip Thornber

Department of Biology and Molecular Biology Institute, University of California, Los Angeles, California 90024

CONTENTS

[1]Abbreviations used: SDS (sodium dodecyl sulfate); SDBS (sodium dodecyl benzene sulfonate).

INTRODUCTION

The chemical environment of chlorophyll in a photosynthetic organism has intrigued researchers for several decades. It has long been recognized that the conversion of light energy into chemical energy in photosynthetic organisms will be fully understood only when the molecular architecture describing the relationship of chlorophyll molecules to each other and to other substances in the membrane is known. Biochemists, influenced by knowledge of other porphyrin-containing respiratory proteins and enzymes, initially favored the idea that chlorophyll is conjugated with a single protein species. Now it is apparent that most, and perhaps all, of the chlorophyll in an organism is complexed to several proteins possessing distinctive compositions. Because of the inherent difficulties encountered with water-insoluble proteins and the fewer number of researchers in this field, advances on this class of proteins have not kept pace with those on other important proteins. Very few reviews devoted solely to plant chlorophyll-proteins are available (51, 92, 135). Much of the history of this subject prior to 1960 has been effectively summarized by Kupke & French (92). Some historical details that are particularly pertinent to this review will, however, be included here.

Smith & Pickels (139) obtained the first evidence that chlorophyll was complexed with protein in a component of the size expected for a fully molecularly dispersed protein. In Smith's earlier investigation (138) digitonin, bile salts, and deoxycholate were observed to clarify aqueous opalescent extracts of spinach or *Aspidistra* chloroplast lamellar fragments in which the chlorophyll-proteins were not in true solution; these detergents, however, removed chlorophyll from its conjugation with protein. A nonpigmented boundary of 13.5S was seen upon ultracentrifugation of the detergent extracts. Since this was the only protein observed, it was concluded that the chlorophylls had been split from this component by the detergent; but in retrospect this protein may well have been coupling factor (70) and therefore did not represent a chlorophyll-containing component in vivo. Smith & Pickels (139) observed that in sodium dodecyl sulfate (SDS) extracts of chloroplasts the chlorophyll sedimented together with a protein boundary of 2.6S and thus provided the first indication that a chlorophyll-protein complex existed. This observation has withstood the test of time and will be discussed further later in this article. However, Smith & Pickels (139) observed no difference in the behavior of chlorophylls *a* or *b* or the carotenoids in the SDS extracts, and they suggested that the fundamental chlorophyll-protein unit was one that contained three chlorophyll *a* and one chlorophyll *b* molecules.

Analysis of anionic detergent extracts of chloroplast lamellae was extended during the next two decades (33, 72, 81, 163). Chiba (33) obtained the first evidence for the multiplicity of pigmented complexes upon free electrophoresis of SDS-neutral detergent extracts; a green component was resolved from a yellow-green zone. The possibility that the chlorophylls were located in more than one molecular environment in the membrane was further strengthened in 1964 when evidence was presented by two groups (21, 30) that a chlorophyll a-enriched fraction was more readily extracted by detergent solutions from chloroplast lamellae than was chlorophyll a and b-containing material. Sironval et al (137) further substantiated this possibility by fractionating two chlorophyll-proteins by electrophoresis of SDS extracts in agar gels. One chlorophyll-protein had a chlorophyll a:b ratio of 4–5/1, the other of about 1.8/1. The advent of polyacrylamide-gel electrophoresis permitted a finer resolution of the chlorophyll-proteins in anionic detergent extracts and, as a result, a rapid proliferation of data on this class of proteins ensued; this information will be elaborated in the next section.

There were other interesting advances on chlorophyll-proteins prior to 1965 that did not employ anionic detergents as their solubilizing agent. Chlorophyll-protein crystals were obtained by Takashima (144; see also 32) following treatment of clover extracts with 50% α-picoline, but later it was clearly demonstrated that these crystals were not composed of naturally occurring leaf constituents (7, 89, 113, 147). Kahn and co-workers (77–79) used DEAE-cellulose chromatography of 1% Triton extracts to fractionate a chlorophyll-protein containing only chlorophyll a. Unfortunately, no characterization of the protein moiety was possible. The relationship of this fraction to the chlorophyll-proteins now known to occur in chloroplasts merits reinvestigation. In addition, Allen et al (6) obtained a chlorophyll-protein from *Chlorella* with a chlorophyll a:b of about 1.

The research prior to 1965 supported the idea that chlorophyll was located in more than one environment; however, some of the evidence presented to substantiate the chlorophyll composition, the singular nature of the protein composition, or whether proteins actually were present with the pigments was not compelling in some cases. At the close of this period the laboratories of N. K. Boardman, L. P. Vernon, and J. S. C. Wessels were actively engaged in the separation of photosynthetic membranes into photosystem I and photosystem II fractions. Since these fractions are not composed solely of a single chlorophyll-protein species (10, 84, 137, 149), and because their nature has been reviewed elsewhere (20, 27), these fractions will be mentioned only when it is pertinent to the chlorophyll-protein being discussed.

POLYACRYLAMIDE GEL ELECTROPHORESIS OF ANIONIC DETERGENT EXTRACTS

In 1966 concurrent publications by two groups (114, 152) provided substantial evidence for the presence of two major chlorophyll-proteins in higher plant chloroplasts: Ogawa et al (114) used SDS at a ratio of 125 moles per mole of chlorophyll to achieve complete solubilization of photosynthetic membranes. SDBS was used by Thornber et al (149, 152) at a much lower concentration (SDBS: chlorophyll = 2.5:1, w/w) necessitating five successive extractions to dissolve well-washed chloroplast lamellae fully. Essentially all of the chlorophyll and protein in the SDS or SDBS extracts entered the polyacrylamide gel columns upon electrophoresis (114, 149). Three chlorophyll-containing zones were resolved in detergent-containing polyacrylamide gels at pH 8.9 or 10.3. The three pigmented zones were termed Component (114) or Complex (149) I, II, or III, corresponding to the increasing order of their electrophoretic mobility. After separation, each zone was extracted from the gel support either by further electrophoresis of the component into free solution (114), or by careful dissection of discs containing each component, followed by extrusion through a nylon mesh, and removal of the pigmented component from the gel particles by filtration through Sephadex (149). Components or Complexes I and II were shown to be chlorophyll-proteins, whereas it was concluded that Zone III was detergent-complexed free pigment since the zone contained no protein. Ogawa et al (114) estimated that Components I and II accounted for some 20% and 50–60%, respectively, of the chlorophyll in the starting extract. With gradual increase of the molar ratio of SDS to chlorophyll used to obtain the detergent extracts (125/1 to 400/1), the content of Component I approached a constant value of 12%, whereas the content of Component II continued to decrease. These observations as well as those on the re-electrophoresis of Complexes I and II (16, 149) supported the notion that the free pigment zone (III) was derived entirely from pigment associated in the plant with the two chlorophyll-proteins. In a later section (CONTENT OF CHLOROPHYLL-PROTEINS) it will be pointed out that this may not necessarily be the case; for even now, the free pigment zone remains an enigma. Thornber et al (149) calculated that 28% of the protein in the starting SDBS extract was contained in Complex I, whereas 49% was present in Complex II, thereby demonstrating that the majority of the protein in chloroplast lamellae is involved in the organization of chlorophylls.

The component of lowest electrophoretic mobility (I) had a chlorophyll $a:b$ ratio of greater than 7, which was reflected in the bluish-green color of this zone in the gel. In some preparations chlorophyll b was absent from this chlorophyll-protein (149). Zone II of yellow-green color contained most of the chlorophyll b in the starting extract and had a chlorophyll $a:b = 1.1 - 1.6$ (149) or 1.8 (114). The amino acid composition of the two chlorophyll-proteins (Table 1) showed that both contained a high proportion of nonpolar amino acid residues as expected for membrane proteins (153). Although the two analyses were similar, obvious differences in their content of lys, his, pro, glu, ile were noted. The amino acid and pigment compositions of these two complexes permitted the conclusion that they represented two distinct membrane entities. The sedimentation coefficients of the two chlorophyll-proteins were 9S for Complex I and 2–3S for Complex II (153). In both cases the pigment was observed to sediment with the protein boundary, thereby further substantiating the chlorophyll-protein nature of the two complexes. Based on the sedimentation coefficient (2–3S) and the large proportion of chlorophyll-protein II, it is very likely that this component is analogous to the complex described in the

Table 1 Amino acid composition (mole %) of chlorophyll-proteins and structural protein[a]

	P700-Chlorophyll a-Protein		Light-Harvesting Chlorophyll a/b-Protein		Lamellar Protein	
	Spinach, beet (153)	Phormidium luridum (148)	Spinach, beet (153)	Chlamydomonas[a]	Whole lamellae (76)	Structural protein (103)
Asp	8.4	8.4	9.3	8.8	8.8	8.6
Thr	5.2	5.9	5.3	5.0	4.7	5.5
Ser	5.5	5.9	4.3	3.6	5.7	7.7
Pro	4.2	4.7	7.4	7.2	5.9	5.9
Glu	7.8	6.5	9.2	9.1	9.2	10.4
Gly	11.0	10.0	13.0	12.5	10.5	10.9
Ala	9.8	10.0	10.6	11.2	9.6	10.4
Val	6.3	6.2	6.7	4.4	6.5	6.7
Cys	0.3	0.9	0.5	0.6	—	—
Met	1.4	1.9	1.6	1.7	1.7	1.4
Ile	6.5	6.2	4.5	4.4	5.3	4.5
Leu	11.4	11.2	9.8	11.5	11.0	9.4
Phe	6.8	6.2	5.8	6.8	3.8	4.2
Tyr	2.7	3.1	2.6	3.3	6.5	1.8
Lys	2.9	3.4	5.8	5.3	5.5	6.5
His	5.9	4.0	1.2	1.5	1.4	1.7
Arg	3.6	3.4	3.0	3.2	4.2	4.4
Try	1.1	1.9	—	—	—	—

[a]Numbers in parentheses indicate references. Data for Chlamydomonas are that of K. S. Kan and J. P. Thornber, unpublished.

initial studies on chlorophyll-proteins (72, 139, 163). Ogawa et al (114) inferred the function of the two chlorophyll-proteins from the analogy between their chlorophyll a:b ratios with those of photosystem I and II fractions isolated by Boardman & Anderson (21). Thornber et al (149) and Sironval and co-workers (137) demonstrated directly that SDS-treated digitonin photosystem I and II particles were enriched in Complexes I and II, respectively. It was thus concluded that chlorophyll-proteins I and II represented smaller subunits of photochemical systems I and II, respectively. No photochemical activities were detected in either chlorophyll-protein at this time. Each of these chlorophyll-proteins will now be considered separately.

THE P700-CHLOROPHYLL a-PROTEIN

Studies since 1967 on Component or Complex I have indicated that a more appropriate name for this chloroplast constituent is the P700-chlorophyll a-protein. Bailey & Kreutz (16) were the first to observe light-induced absorbance changes due to P700 in Complex I, albeit such signals were smaller than in other P700-enriched fractions. Subsequently, Dietrich & Thornber (40), using blue-green algae, discovered that the Complex I zone in polyacrylamide gel columns was highly enriched in P700 (chlorophyll/P700 = 50/1). The reader should be aware that this pigment-protein has also been termed the photosystem I chlorophyll-protein, chlorophyll-protein or pigment-protein complex I, CPI, or HAI (e.g. 4, 14, 39, 54, 60, 65, 90). Later in this section the relationship of this chlorophyll-protein to functional photosystem I particles will be clarified.

Isolation

The original gel electrophoretic procedure was a laborious and possibly degradative method in which only a single protein purification step was used to provide small yields of the complex. A more satisfactory technique was later developed (148) that enabled this chlorophyll-protein to be isolated in larger yields by application of semiconventional protein chromatographic techniques to SDS extracts of a blue-green alga. Detergent extracts of washed algal membranes were precipitated by ammonium sulfate, the precipitate was mixed with Celite, and the slurry poured into a column. A methanol-ammonium sulfate solution eluted the free pigment; thereafter a SDS solution removed all the remaining chlorophyll-containing material which was then chromatographed on hydroxylapatite. The resulting chlorophyll-protein fraction was passed through Sephadex to yield a homogeneous and stable preparation of the chlorophyll-protein (40, 148). The homology of this isolated chlorophyll-protein to the higher plant Complex I was quite obvious from their compositions and sizes (148). However, two perplexing differences were noted (151) between their characteristics: the red wavelength maximum of the algal component was 677 nm, whereas that of the higher plant constituent was 671–673 nm (certainly a component representing the heart of photosystem I should have a wavelength maximum greater than 673 nm); secondly, the P700 content varied substantially between the two preparations. Kung & Thornber (90) attempted to resolve these discrepancies by applying fractionation techniques developed for isolation of the blue-green algal chlorophyll-protein to SDS extracts of higher plant chloroplasts. Although the chlorophyll-protein was isolated, its spectral characteristics and P700 content were those of the electrophoretically isolated higher plant material rather than those of the algal complex. The matter was resolved by Shiozawa et al (136), who found that a fraction containing the chlorophyll-protein with the desired characteristics could be obtained when non-ionic detergent (preferably Triton X-100) extracts of higher plant chloroplasts were chromatographed on hydroxylapatite. Addition of SDS to this Triton-isolated P700-chlorophyll *a*-protein shifted the red wavelength maximum to a lower wavelength (671–673nm) and caused a loss of detectable P700 activity. These observations and others demonstrated that the electrophoretically prepared Complex I is an SDS-altered form of the P700-chlorophyll *a*-protein. Brown et al (28, 29) have shown that this chlorophyll-protein complex is altered by SDS in all eukaryotic organisms examined, whereas the prokaryotic blue-green algal complex is stable to this detergent (136). A molecular explanation for this variation in stability to SDS of obviously homologous protein complexes has been postulated (136) to be due to the presence only in the blue-green algal complex of a disulfide bridge that prevents the detergent from unwinding the polypeptide chain.

The Triton-P700-chlorophyll *a*-protein preparations contain, in addition to the chlorophyll-protein, cytochromes f and b_6 plus probably other proteins and low molecular weight components that are not connected with the chlorophyll-protein. Therefore, the biochemical composition of the chlorophyll-protein complex summarized below is based on analysis of either Complex or Component I of higher plants or the P700-chlorophyll a-protein of blue-green algae; only where appropriate (spectral characteristics and function) is data on the Shiozawa et al (136) preparation included in the summary. The particular virtues of this Triton-prepared complex, besides the fact that it has explained the variation in the spectrum and P700 content of different preparations of the chlorophyll-protein, is that it provides a more rapid technique for obtaining the heart of photosystem I in a spectrally purer form than is currently available. The virtue of the SDS or SDBS electrophoretic isolation procedure is that it provides a rapid method for monitoring the presence and content of this chlorophyll-protein in anionic detergent lamellar extracts in addition to yielding the chlorophyll-protein in a homogeneous but altered state. For more complete studies of the composition, three-dimensional structure, and the orientation of pigments within the protein framework of this complex, the blue-green algal P700-chlorophyll a-protein will have to be used, although the Shiozawa et al (136) preparation may prove suitable after further fractionation.

Composition and Size

Chlorophyll a and β-carotene account for essentially all of the pigment in the complex; the molar ratio of chlorophyll a to β-carotene is 20-30/1 (114, 136, 148, 149). The chlorophyll $a:b$ ratio of the higher plant complex determined by conventional spectrophotometric procedures is sufficiently high ($>$ 8/1) (114, 136, 149) that the equations are unreliable for an accurate estimation of chlorophyll b. Other evidence has indicated that chlorophyll b is almost certainly not a constituent of this chlorophyll-protein: thin layer chromatography shows the absence of chlorophyll b (136, 149); the millimolar extinction coefficients at various wavelengths for the algal complex (40), which cannot contain chlorophyll b, closely agree with those of the higher plant material (136), and re-electrophoresis of Complex I shows that any minor amounts of chlorophyll b present can be removed completely to provide a stable component of constant composition (16). Furthermore, it is unlikely, on a theoretical basis, that this ubiquitous plant complex (29) would contain a component (chlorophyll b), the distribution of which is limited to a few classes of plants.

   A quinone which is not plastoquinone, but may be naphthoquinone or tocopherylquinone, is present in the blue-green algal chlorophyll-protein (40); quinones are also present in Component I according to Ogawa et al (114). Galacto and phospholipids are present in the complex in trace amounts (40, 153), but since their content and nature varied from preparation to preparation, it was thought (40) that they represented membrane lipids which are displaced in most cases by SDS during the isolation of the complex rather than specific lipid associations. Flavins and pteridines, which have been proposed to be the primary electron acceptor for P700, are absent from the P700-chlorophyll a-protein (40). Dietrich & Thornber (40) did not unequivocally eliminate the possibility that sufficient iron may be present in the P700-chlorophyll a-protein to function as this acceptor. In light of the demonstrated photochemical activity of this complex, it is very probable that the primary electron acceptor of photosystem I, P430 (80), is present; thus it should be worthwhile to determine whether the 8000 dalton iron-sulfur protein proposed to be the primary electron acceptor of photosystem I (102) is present in the complex.

The amino acid composition of the complex is given in Table 1. There is virtually no difference between the composition of the eukaryotic and prokaryotic proteins except for their cysteine and histidine content. It has been proposed (118, 148) that the high proportion of nonpolar residues explains the water-insolubility of this chlorophyll-protein. The molecular weight of the native chlorophyll-protein has been determined by calibrated polyacrylamide gel electrophoresis to be about 110,000 daltons (42,90; S. Pastryk and J. A. Shiozawa, in preparation). The ratio of chlorophyll/protein of both the higher plant and blue-green algal chlorophyll-protein is 14 moles chlorophyll/110,000 g complex (148, 153). The chlorophyll/P700 in the complex is 40–45/1 (40,136) when a differential extinction coefficient of 64 mM^{-1}cm^{-1} is used for P700 (66).

Dietrich & Thornber (40) pointed out that, on the basis of the size and the chlorophyll/P700 and chlorophyll/protein ratios of the complex, not every 110,000 dalton unit could contain P700. They proposed that two slightly different types of chlorophyll-protein molecules are present in the homogeneous preparation. Adaptation of their notions to the current characteristics of the complex would mean that one-third of the molecules contain P700 together with 14 light-harvesting chlorophyll molecules, and two-thirds would only contain 14 light-harvesting chlorophyll molecules. An aggregate of both types would constitute the heart of photosystem I. Since it was found to be impossible to resolve two different chlorophyll-protein types using the finest protein separation techniques—the chlorophyll/P700 ratio in the leading and trailing edges and in the middle of the zone in SDS-gels was constant (40)—then it was concluded (40) that only slight differences could exist between the two protein species. It may well be that a P700-reaction center protein, possibly analogous to the bacterial reaction center-protein (34, 133), is attached to one of every three identical chlorophyll-protein molecules that contain about 14 chlorophyll a molecules and have a molecular weight of 110,000 daltons. Addition of such a small component might not increase the size sufficiently to resolve it from the other type of molecules in this molecular weight range on SDS-polyacrylamide gel electrophoresis.

Occurrence in the Plant Kingdom

The P700-chlorophyll a-protein or its SDS-altered form (Complex I) has been observed in all plants examined that contain P700. It has been found in chloroplasts of mono- and dicotyledons (tobacco, spinach, bean, soybeans, pea, *Antirrhinum*, peanut, barley, oat, and maize) (2, 5, 14, 49, 60, 90, 99, 114, 136, 149, 150, 153); of gymnosperms (pine, cedar, and *Ginkgo*) (29; R. S. Alberte and P. McClure, in preparation); of green algae (*Chlamydomonas, Dunaliella, Scenedesmus*) (28, 29, 54; K. S. Kan and J. P. Thornber, in preparation); of *Euglena* (28, 29, 49); of yellow-green, brown, red (29, 151), and blue-green algae (40, 148). Thus Brown et al (29) have proposed that this chlorophyll-protein is ubiquitous in the plant kingdom. The complex represents 4–30% of the total chlorophyll in all organisms examined (29); higher plant chloroplasts have 10–18% of their chlorophyll in this chlorophyll-protein (2, 29, 49, 90, 136, 150). Interestingly, the only plant photosynthetic membranes that do not contain the chlorophyll-protein are those of mutants which lack or may lack P700 (54, 60) or those of chlorotic (iron-deficient) plants (99). Gregory et al (54) and Herrmann (60) have demonstrated the absence of Complex I in *Scenedesmus* mutant 8 and *Antirrhinum majus* mutant en:alba-1, respectively. The algal mutant shows no light-induced absorbance changes of P700, while the other mutant is impaired in photosystem I activity. It also seems probable that a P700-lacking mutant of *Chlamydomonas* (ac-80) (50) may not contain Complex I. It remains to be substantiated whether these mutants lack a specific P700-reaction center-protein or whether both it and the 110,000 dalton light-harvesting chlorophyll-protein existing as a complex (40) are absent. A mutation that eliminates synthesis of a P700-reaction center-protein could render the 110,000 dalton units more susceptible to dissociation by SDS such that they electrophorese elsewhere on gels, or it may sympathetically switch off synthesis of the light-harvesting chlorophyll-protein units, as would be predicted from the observation (5) that insertion of P700 into developing photosynthetic membranes precedes that of its closely associated light-harvesting chlorophyll-protein molecules. In relation to these points, it would be expected that complete elimination of the entire chlorophyll-protein complex should change substantially the chlorophyll *a:b* ratio. No difference in this ratio was observed in the *Scenedesmus* mutant (54). Further work on these mutants could be very important, not only to gain an understanding of the precise consequence of the mutation but also to gain an elucidation of the structure of the complex.

Absorption and Fluorescence Spectra

Most of the absorption peaks in the visible spectrum of the complex are due to chlorophyll *a* (40, 136, 148), with the exception of a shoulder at about 490 nm on the major Soret peak (437 nm) which has been attributed to β-carotene present in the complex (136, 148). This characteristic spectrum is a useful criterion of spectral purity of preparations, and provides a rapid monitoring system for confirming the presence of the chlorophyll-protein in SDS-gels. The millimolar absorptivity based on chlorophyll of the 677 nm peak is 60 cm^{-1} for the blue-green algal complex (148) and 59 cm^{-1} for the higher plant component (136). This decrease in the absorptivity value (74 mM^{-1}cm^{-1}) of chlorophyll *a* in organic solvents is explained by the presence of several spectral forms in the complex. Analysis of liquid nitrogen spectra has shown that the following spectral forms of chlorophyll *a* are present: C*a* 662, 669, 677, and 686 nm (28, 29, 136). The chlorophyll-protein is particularly enriched in C*a* 686 compared with the concentration of this spectral form in intact chloroplast lamellae (47). Such an enrichment would be expected for a component at the heart of photosystem I. The blue-green algal complex, but not the higher plant component, shows the presence of an even longer wavelength spectral form at 710 nm (29, 40, 46). This spectral form has been shown (40) not to be due to absorbance of P700 as suggested by Butler (31).

The emission spectrum of only the blue-green algal complex has been examined to date. Room temperature spectra show a maximum emission at 682 nm with the presence of quite minor bands at 692, 720, and 740 nm (108). At 77°K the ratio of the intensity of fluorescence at longer (720–740 nm) to shorter (680–685 nm) wavelengths increased to 0.9 in dilute samples (108); fluorescence at 695 nm was also observed at 77°K. Mohanty et al (108) demonstrated that earlier published emission spectra of this complex (40), which showed a longer/shorter wavelength emission ratio of eight, had been considerably distorted by the reabsorption of the 685 nm-fluorescence.

Function

The chlorophyll-protein represents a three to twelvefold (depending on the organism) enrichment of P700 (29, 136); thus the complex is thought (4, 5) to represent the heart but not the whole of photosystem I. Both the higher plant and blue-green algal complex show light-induced bleaching and dark recovery of P700 (40, 136); however, the kinetics of these spectral changes differ greatly between the two preparations. The higher plant preparation is apparently much more effective in cycling electrons from P700 through electron acceptors and returning them to P$^+$700, whereas the algal chlorophyll-protein requires the addition of electron donors and acceptors or of PMS to obtain the same rapid kinetics (136). It appears as though SDS interrupts in some way the cyclic electron flow associated with the algal complex. In this respect, it is interesting to recall that the Triton-prepared higher plant chlorophyll-protein contains cytochromes f and b_6 (136). Thus it may be possible to delineate the electron carriers involved in the cyclic electron transfer reactions around photosystem I by contrasting the composition of the chlorophyll-proteins from the two sources and by measurement of rapid spectral changes in the higher plant material. It has been suggested (40) that the as yet unidentified quinone may participate in these cyclic reactions, giving rise to some of the light-induced changes in the ultraviolet spectral region. The quantum requirement for P700 photooxidation in the algal complex to which sodium ascorbate and methyl viologen had been added was measured to be 10 ± 2 (40); the analogous measurement on the eukaryotic complex has yet to be made.

The P700 absorbance changes in the red and blue spectral regions of both complexes are identical to those published elsewhere (66, 88). The red wavelength of maximum bleaching is 697 nm in the chlorophyll-protein prepared from all sources (29, 40, 136). In addition, an increase in absorbance around 825 nm is observed; this change in absorbance has been correlated to P$^+$700 (24, 71). The close stoichiometric equivalence of P700 and β-carotene may indicate that a small fraction of this carotenoid may function to protect the reaction center from photochemical damage (136).

The chlorophyll-protein represents the most enriched form in which the reaction center of photosystem I has been obtained, although there have been unconfirmed reports of chlorophyll/P700 ratios as low as 16/1 (132). How then may the reaction center be isolated devoid of light-harvesting chlorophyll as has been accomplished for some purple photosynthetic bacteria (34, 133)? It appears that the best hopes reside in (*a*) the discovery of a mutant that contains P700 but lacks the 110,000 dalton unit; (*b*) obtaining the required material from tissues at very early stages of greening, as suggested by Alberte et al (5); or (*c*) application of other gentle dissociating or chemical modification techniques which will further dissociate the presently available P700-enriched fractions but will not alter their photochemical characteristics. To date, however, no success has been achieved by the last possibility.

Relationship to Other Photosystem I Fractions

The P700-chlorophyll a-protein must be very closely related to the HP700 fraction of Vernon and co-workers (157, 158). Both components have very similar spectra, chlorophyll/P700 ratios, functions, and compositions. However, there are more peptides in HP700 preparations than just those of the chlorophyll-protein (84, 85). HP700 is essentially devoid of β-carotene, thereby demonstrating that the β-carotene in the P700-chlorophyll a-protein is not essential for P700 activity but not eliminating its possible role in photoprotection. In contrast, HP700 contains chlorophyll b (chlorophyll a:b = 4/1); however, it appears as though this pigment is not an integral part of the heart of photosystem I since it is absent from the P700-chlorophyll a-protein. It would be of interest to treat the P700-chlorophyll a-protein in the same manner that yielded HP700 from Triton subfraction (TSF) 1 (157, 158) to see just how enriched P700 could be obtained in the resulting material.

Photosystem I particles prepared by digitonin or by physical processes (20, 27) have been shown to be rich in Complex I (13, 28, 29, 137, 149). It can be deduced that additional pigment-protein(s) must be present in such fractions to account for (a) the chlorophyll b present and (b) the fact that about half of the lamellar chlorophyll is contained in such particles (20), whereas Complex I represents only 10–18% of the total chlorophyll in the intact membranes. The presence of small amounts of the light-harvesting chlorophyll a/b-protein (Complex II) has been demonstrated in photosystem I particles prepared with digitonin (13, 137, 149) or by physical methods (28, 29).

THE LIGHT-HARVESTING CHLOROPHYLL a/b-PROTEIN

Thornber & Highkin (150) proposed the name "light-harvesting chlorophyll a/b-protein" for the pigment-protein that has been variously referred to as Component II (114), Complex II (149, 152), chlorophyll-protein or pigment-protein Complex II (13, 14, 49, 60, 64, 65, 99), photosystem II chlorophyll-protein (90, 91), CPII (2, 55), or HA II (39). The new name was thought to be more descriptive and appropriate since the pigment-protein may not be connected solely with photosystem II. In addition, higher plants can live photosynthetically in the complete absence of the complex (14, 64, 150), and hence the pigment-protein can have no photochemical capacity essential for photosynthesis; it is most likely that this component functions in a light-harvesting role, as will be discussed later in this section.

Isolation

The complex was first obtained in homogeneous form after electrophoresis of anionic detergent extracts in polyacrylamide gels. The pigment-protein was extracted from the gel by those methods described for the P700-chlorophyll a-protein (114, 149). It has been pointed out (16, 114, 149) that the composition of the complex may vary depending on the detergent/chlorophyll ratio used to make the extracts and the length of time of electrophoresis of the extracts. A new isolation procedure was developed (90, 151) to overcome some of these difficulties; hydroxylapatite chromatography of SDS extracts again proved successful. In essence, the method involves the removal of the P700-chlorophyll a-protein entity from the column prior to elution with 0.4 M sodium phosphate-0.05% SDS-1 mM MgCl$_2$ of a fraction containing the desired chlorophyll a/b-protein. Rechromatography and ammonium sulfate fractionation of this eluate yields a homogeneous preparation of the light-harvesting chlorophyll a/b-protein (90). Argyroudi-Akoyunoglou et al (14) have used a modification of this method to prepare both chlorophyll-proteins discussed so far; however, the homogeneity of the protein in their modified procedure has never been investigated.

Polyacrylamide gel electrophoresis of anionic detergent extracts of chloroplasts or fractions thereof permits a rapid method for monitoring the presence and content of the complex. With certain reservations about the pigment content, much data on the composition of this chlorophyll-protein can be, and has been, obtained from the polyacrylamide gel fractionated-material, although analysis of chromatographically purified material is preferred for such data.

Composition and Size

It is generally agreed that there are equimolar quantities of chlorophyll a and b in the complex (2, 13, 90, 150, 151). Analysis of electrophoretically isolated material (114, 149) initially indicated a slightly greater chlorophyll $a:b$ ratio, but more recently the ratio has been quoted as 1.0 for material obtained either by electrophoresis (2, 4, 150) or by chromatography (13, 90, 151; K. S. Kan and J. P. Thornber, unpublished data). Every carotenoid present in the chloroplast is contained in this chlorophyll-protein, although not in the same stoichiometric ratio. Lutein and β-carotene represent the major portion of the carotenoid (114, 153); also, a large proportion of the neoxanthin present in an organism is associated with the chlorophyll-protein (114, 150, 153). The chlorophyll/carotenoid ratio varies from 3-7/1 (mole/mole) (114, 153). Traces of galactolipids and phospholipids occur in preparations of this complex (114, 153); however, the amount is so small that it is likely that this lipid is not a constituent of the chlorophyll-protein, but represents membrane lipids which have not been completely displaced by the detergent.

The pigments in preparations of the chlorophyll-protein sediment with a protein boundary of 2.3-3.1S (90, 153; K. S. Kan and J. P. Thornber, unpublished data). The molecular weight of the native pigment-protein of higher plants and green algae has been determined on calibrated SDS-polyacrylamide gels. Values ranging from 27,000-35,000 daltons have been reported (42, 65, 90; K. S. Kan and S. Pastryk, unpublished data). Kung et al (91) calculated a molecular weight for the protein of about 30,000 daltons based on the number of peptides produced by tryptic digestion of the chlorophyll-protein in relation to its amino acid composition (Table I). Extrapolation of these peptide data indicates that the complex is composed of one or at the most two *different* polypeptide chains. Many data have been presented (see section on membrane polypeptides) that the peptide(s) correlated to this pigment-protein is larger than 21,000 daltons. However, if two different chains are present, then one of them must have a molecular weight less than half (i.e. 15,000 daltons) that determined for the complex. Further, the molecular weight of the apoprotein (complex less chlorophyll produced by incubation in SDBS) has a size only 3000 daltons lower than the pigment-protein (49). Therefore, it is unlikely that more than one polypeptide chain is present in the light-harvesting chlorophyll a/b-protein.

The chlorophyll/protein ratio in the electrophoretically prepared complex was determined to be slightly greater than two moles chlorophyll/30,000 g protein (153). Since this complex accounts for at least 40% of the total lamellar chlorophyll, and since the intact membranes contain 230 moles of chlorophyll/9.3 $\times$ 10^5 g protein (119), such a ratio cannot be rationalized. A much more reasonable value of 6 moles chlorophyll/28,000 g protein has been determined recently for chromatographically isolated material from *Chlamydomonas* (K. S. Kan and J. P. Thornber, unpublished data). The older value may reflect removal of pigment during the electrophoretic isolation procedure, or the presence of colorless proteins in the preparation, rather than a difference between species. The newer value permits one or more carotenoid molecules per protein molecule.

Occurrence in the Plant Kingdom

It is very likely that the complex occurs in all chlorophyll b-containing plants, in which it is present as aggregates (multiples of the 30,000 dalton unit). Hiller et al (65) have demonstrated a dimer (mol wt 69,000) as well as the monomer of the complex upon electrophoresis at 4° of SDBS extracts obtained with a low detergent/chlorophyll ratio—conditions which would favor the stability of aggregates. Herrmann & Meister (62) have also observed what are probably still larger aggregates in addition to the monomer and dimer (zones 2a–d) of this chlorophyll-protein in detergent extracts obtained at very low SDS/chlorophyll ratios.

The complex has been observed in angiosperms: tobacco, spinach, beans, peas, peanuts, cotton, soybean, *Antirrhinum,* oat, barley, and maize (2, 4, 13, 62, 64, 65, 90, 99, 114, 150, 153); in gymnosperms: cedar, pine, and *Ginkgo* (29; R. S. Alberte and P. McClure, unpublished data); in green algae: *Chlamydomonas, Dunaliella,* and *Scenedesmus* (28, 29, 54; K. S. Kan and J. P. Thornber, unpublished data); and in *Euglena* (49); however, Brown et al (29) are less certain of its presence in the last organism. This chlorophyll-protein is by far the major pigment-protein in higher plants and green algae since it accounts for 40–60% of the total chlorophyll (28, 29, 49, 136, 150); *Euglena* have 4% of their chlorophyll in this complex (49). The lamellar content of this pigment-protein is more variable than that of the P700-chlorophyll a-protein (49); this variation will be discussed in detail in the section on content of chlorophyll-proteins in photosynthetic membranes. It suffices to say here that the only known photosynthetically grown higher plant or green alga that does not contain this chlorophyll-protein is the chlorophyll b-less barley mutant (10, 49, 150).

Absorption and Fluorescence Spectra

Room temperature spectra of the chlorophyll-protein show a double peak for the red wavelength maxima of chlorophylls a and b at 672 and 653 nm, respectively (13, 14, 90, 151). Two Soret peaks at 437 and 470 nm corresponding to those of chlorophylls a and b, and a shoulder on the longer wavelength side of the Soret peaks, probably contributed by carotenoids, are observed. The millimolar extinction coefficients for chlorophylls a and b in the complex are both about 30 cm^{-1} at their respective red wavelength maxima—values which are compatible with the presence of equimolar quantities of chlorophylls a and b and of different spectral forms of chlorophyll a. Several spectral forms of chlorophyll have been detected by analyses of liquid nitrogen spectra (28, 29) which show the presence of Cb650, Ca662, 670, 677, and 684; Ca684 is reduced and Cb650 is enriched in the complex in comparison to the proportions of these spectral forms present in whole chloroplasts (47). None of the minor long wavelength forms are present in the chlorophyll-protein (29).

The complex exhibits a single fluorescence emission peak at 685 nm (R. J. Strasser and M. Kitajima, unpublished data), and shows a high efficiency for energy transfer from chlorophyll b to chlorophyll a (J. P. Thornber and J. M. Olson, unpublished data). Further examination, particularly of fluorescence depolarization, could be instructive in a determination of the orientation of the pigment molecules within the protein.

Function

Because this chlorophyll-protein contains such a large proportion of the chlorophyll in some plants, and because a plant lacking this component (10, 49, 150) is photosynthetically competent (22), it has been deduced (150) that this pigment-protein is the major light-harvesting component in the photosynthetic membranes of higher plants and green algae, and that it does not perform any photochemical function required for the operation of the Z scheme. This view is supported by the data of Herrmann (61), who showed that a mutant defective in photosystem II activity still contains the chlorophyll-protein, and by that of Hiller et al (64), who observed that during greening of etiolated bean plants with flashed light photosystem II activity appears whereas the chlorophyll-protein does not. Genge et al (49) have pointed out that the amount of the complex in a plant is related to the chlorophyll b content. This correlation has been recently put on a more quantitative basis. Thornber and co-workers (3, 29) have determined that all the chlorophyll b (together with an equivalent amount of chlorophyll a) in SDS lamellar extracts is located in this chlorophyll-protein. This observation permits one to estimate the percentage of the total chlorophyll in a plant or fractions thereof associated with this chlorophyll-protein from careful measurements of the chlorophyll a:b ratio alone. Application of this relationship to data on the change in the chlorophyll a:b ratio that occurs in response to nitrogen feeding (18) suggests an additional function for this chlorophyll-protein as a store of protein-nitrogen in a plant in a manner analogous to phycocyanin in blue-green algae (155). It is also likely that other growth parameters are correlated with the lamellar content of this chlorophyll-protein (see section on content of chlorophyll-proteins in photosynthetic membranes).

Light energy absorbed by chlorophyll b in a plant is efficiently transferred to chlorophyll a (41). Presumably this transfer is occurring predominantly, and perhaps entirely, within the light-harvesting chlorophyll a/b-protein. And, since chlorophyll b feeds its absorbed energy mainly to the trap of photosystem II, the chlorophyll-protein therefore functions as the major light-harvesting antenna for that photosystem. However, since more than 50% of the chlorophyll in some plants is associated with this chlorophyll-protein, then some energy absorbed by this chlorophyll-protein must be fed to P700 to maintain an even distribution of energy between the two photosystems. The author favors an organization of the pigments between the two photosystems as described by Seeley (134). It can be easily envisaged that the chlorophylls in the sets 17–28 of Figure 1 of Seeley (134) could be entirely contributed by the chlorophylls of this chlorophyll-protein; such a location for the chlorophyll-protein would fit all the known data (particularly on the spectral forms) for the complex. Furthermore, if this chlorophyll-protein is located only in sets 17–28, and if the number of chlorophylls in these sets is flexible, then it can be rationalized how the considerable variation that occurs in the content of this chlorophyll-protein need not affect the even distribution of energy between the photosystems. In the author's view such an organization of pigments is more appealing than the separate package concept (20, 112) in which a smaller portion of the light-harvesting chlorophyll a/b-protein molecules would be attached to a structure that is photosystem I and a larger portion to one that is photosystem II.

Argyroudi-Akoyunoglou & Akoyunoglou (13) have proposed that another function of this chlorophyll-protein is in the formation of grana. Exactly how the chlorphyll-protein was involved in lamellar stacking was not stated; however, one infers that they believe the complex initiates stacking. Anderson & Levine (10) and Thornber & Highkin (150) have posed the possibility that this chlorophyll-protein may cause appressed lamellae to remain in contact with each other. All these postulations arise from observations that photosynthetic membranes exhibiting little or no stacking (e.g. bundle sheath chloroplasts, etiolated beans exposed to flashed light, some higher plant and green algal mutants) contain little or no light-harvesting chlorophyll a/b-protein (4, 10, 14, 49, 64, 150; see also 25). This pigment-protein cannot be directly or solely responsible for grana formation since it is not present in the chlorophyll b-less barley mutant which possesses some grana. Therefore, at present it appears that the relationship between grana and the proportion of the major pigment-protein complex is that the grana stacks are the principal location of the complex (28, 29) and that these stacks may be changed in size to accommodate the quantity of the chlorophyll-protein present. This view is supported by the observations that grana stacks in shade plants, which have low chlorophyll a:b ratios (8, 53, 123) and thus contain greater proportions of the chlorophyll-protein, are higher than in sun plants (8, 52).

There are no data available on the function of the carotenoids in this chlorophyll-protein, but interestingly, these pigments occur in a characteristic, stoichiometric ratio.

Relationship to Other Photosystem II Fractions

Photosystem II preparations (20, 27), with the exceptions of the TSF 2a fraction of Vernon et al (157) and the photosystem II reaction center fraction of Wessels et al (161) and Ke et al (82), have low chlorophyll a:b ratios (1.2–2.3/1). In light of what has been said previously, it is not surprising therefore that photosystem II fractions contain large proportions of the light-harvesting chlorophyll a/b protein (14, 137, 149).

OTHER CHLOROPHYLL-PROTEINS IN PLANTS

Detergent-Soluble Complexes

The two known chlorophyll-proteins (or aggregates thereof) account for at most 75% of the chlorophyll in higher plant and green algal chloroplast lamellae and for much less in most other plants (29). The organization of those chlorophylls and carotenoids that are not located in these two chlorophyll-proteins is essentially unknown. In a wide variety of plants, such chlorophyll is generally found in the free pigment zone [Component or Complex III (114, 149)] at the completion of electrophoresis (28, 29).

Some reports on the occurrence of other chlorophyll-proteins electrophoresing in SDS-polyacrylamide gels between the two major chlorophyll-proteins have appeared (49, 56, 62, 65, 149). The presence of these other components in gels is more obvious when low detergent/chlorophyll ratios are used to dissociate photosynthetic membranes. Hiller et al (65) demonstrated that the dimer of the light-harvesting chlorophyll *a/b*-protein, when present, is located in this region. The pigmented zones (Ia and IIb, c, d) observed by Herrmann & Meister (62) are probably also polymers of either the P700-chlorophyll *a*-protein or the light-harvesting chlorophyll *a/b*-protein despite the slight differences between the spectra of these zones and those of their monomers. Furthermore, zone Ia is not present in the en:alba-1 plastome mutant of *Antirrhinum* (60), which lacks zone I (P700-chlorophyll *a*-protein).

Genge et al (49) provide some of the only evidence for the existence of a third, distinct chlorophyll-protein in higher plants and *Euglena*. SDBS extracts of chloroplasts of the chlorophyll *b*-less barley mutant exhibit a chlorophyll-containing band which electrophoreses between the P700-chlorophyll *a*-protein component and the position in which the other major chlorophyll-protein would be located if it were present. This band is very likely to be a chlorophyll *a*-protein of rather limited stability to SDBS since it can only be observed in electrophoretic runs made at 4°, and is scarcely discernible at 26° (49). Genge et al (49) have also reported the presence of this component in detergent extracts of *Euglena* and maize in which it represents a minimum of 5% and 3% of the total chlorophyll, respectively. Since most other investigators use SDS which is not as soluble as SDBS at 4°, it is unlikely that they would have detected this pigment-protein. A fraction which may be related to this chlorophyll-protein was obtained by Brown et al (28, 29) from hydroxylapatite chromatography of Triton X-100 extracts of *Euglena;* this fraction exhibits the unusual characteristic of a narrow band width 684 nm-peak in its 77°K spectrum. Furthermore, they (29) have also observed an unstable chlorophyll-containing component which initially electrophoreses in the location of a 45,000 dalton component in SDS extracts of the major algal classes that do not contain chlorophyll *b*. Eventually the pigments are released from this component and move into the free pigment region. This component may correspond to the third chlorophyll-protein described by Genge et al (49). In view of these recent developments it now appears quite probable that much of the free pigment arises from a third chlorophyll *a*-containing protein (49) rather than from dissociation of the two known chlorophyll-proteins as previously thought (16, 90, 114, 149). If a SDS-labile chlorophyll *a*-protein runs close to the light-harvesting chlorophyll *a/b*-protein on SDS gel electrophoresis, and if it dissociates during the separation of the major pigment-protein zones, it would explain the observation described by Kung & Thronber (90) that the chlorophyll *a:b* ratio of the light-harvesting chlorophyll *a/b*-protein zone decreases during electrophoresis, as well as some other reports (2, 114, 149) of chlorophyll *a:b* ratios greater than one for this chlorophyll-protein after separation on polyacrylamide gel electrophoresis. Furthermore, if the presence of this postu-lated third chlorophyll-protein is confirmed, and if it represents a major portion of the pigments in photosynthetic membranes, then this component would most likely function in a light-harvesting capacity predominantly in photosystem I since the reaction center of photosystem II is abundantly equipped with the light-harvesting chlorophyll *a/b*-protein. With respect to the trap of photosystem II, no evidence has been presented so far that locates the trap on SDS-polyacrylamide gels. One can only conclude that the pigments from this reaction center are present in the free pigment zone in a photochemically inactive state.

Guignery et al (56) have reported that two protochlorophyllide-peptides of 29,000 and 21,000 daltons in etiolated maize plantlets are photoconverted to two chloro-phyll-proteins of the same size during chloroplast development and are still detect-able after 48 hr of greening. However, neither of these two chlorophyll-proteins corresponded to the light-harvesting chlorophyll *a/b*-protein. Further characteriza-tion and determination of the concentration of these two pigment-protein complexes in mature plants should be of value in attempts to account completely for the organization of chlorophyll in vivo. It can be anticipated that information on the chlorophyll-protein nature of additional light-harvesting chlorophyll *a*-proteins in higher plants and green algae, of the photosystem II reaction center, and of much of the pigments in plants other than those containing chlorophyll *b* will be resolved upon more complete studies of the dissociation of photosynthetic membranes either by those techniques already in use or by other existing techniques (e.g. chaotrophic agents, chemical modifications, etc).

Water-Soluble Complexes

A chlorophyll-protein (CP668) representing a very small fraction of the total chloro-phyll of *Chenopodium album* was first isolated by Yakushiji et al (164). This component occurs in some, but not all, members of the Chenopodiaceae, Amaran-thaceae, and Cruciferae (141, 142). It is prepared by extraction of such plants with aqueous buffers in dim light, and obtained in homogeneous form by ammonium sulfate fractionation and Amberlite CG-50 chromatography (164). Terpstra (145) modified this procedure to include Carbowax fractionation. The complex is remark-ably stable since its spectrum and phototransformability are largely unaltered by boiling for 25 min (115–117).

The size of the complex has been reported (110, 141) as 78,000 (*Chenopodium*), 40,000 (*Lepidium*), and 93,000 (*Brassica*) daltons. Sedimentation coefficients of 2.7 and 4.7S have been reported (141). The chlorophyll-protein contains 6–8 chloro-phyll *a* and one chlorophyll *b* molecules per protein moiety (143, 164, 165). Carotenoids are not present. The component is termed CP668 (*Chenopodium*), CP674 (*Brassica*), or CP662 (*Lepidium*) in relation to its red wavelength maximum. French et al (48) have observed that some of these chlorophyll-proteins contain spectral forms of chlorophylls *a* or *b* that are not usually found in higher plants. The fluorescence spectrum (maxima at 672–683 nm and 730–740 nm) indicates an efficient energy transfer from chlorophyll *b* to *a* within the protein molecule (110).

The characteristic feature of these complexes is that illumination converts the red peak of chlorophyll to longer wavelengths (743 nm) (141, 164) though some preparations do not show full photoconversion (141). The photoconversion phenomenon is reversible and requires oxidizing or reducing conditions to shift to longer or shorter wavelengths, respectively (141). Evidence has been presented that a protein conformation change occurs during photoconversion (57). The apo-protein can be reconstituted with chlorophylls *a* and *b* to yield the original photoconvertible CP668, 674, or 662 (111, 165). Regardless of the proportion of chlorophylls *a* and *b* in the reconstituting medium, the product has a spectrum identical with the original material, whereas reconstitution with chlorophyll *b* alone or reconstitution of CP743 (the photoconverted complex) yields a photoinactive component.

No physiological significance has been attributed to this minor chlorophyll-protein complex, although in view of its low concentration in a plant, the chlorophyll-protein may represent an enzyme involved in chlorophyll synthesis or in transport of chlorophyll within the plant cell. Further, there is no obvious significance in the distribution of the chlorophyll-protein in the plant kingdom. It might not be unreasonable that those plants containing a natural detergent are the ones from which this component can be isolated. It is interesting that the isolated chlorophyll-protein (CP668) and therefore the form probably occurring in vivo is the 668 nm form and not the phototransformed 743 nm form; perhaps the conditions in vivo are sufficiently reducing to favor the 668 nm form. The relationship, if any, of these water-soluble chlorophyll-proteins to (*a*) the spinach protein factor of Terpstra (146), which is a colorless protein prepared in the same manner as CP668 and which catalyzes the photooxidation of added chlorophylls, and to (*b*) other observations on transformation of 668 nm forms to 740 nm (cf further Aghion 1) remains to be investigated.

A water-soluble peridinin-chlorophyll *a*-protein has been isolated from several, but not all, dinoflagellates; *Gonyaulax polyedra, Glenodinium* sp., *Ampidinium loefleri,* and *Cachonina niei* in particular provide good yields of the pigment-protein (58, 59, 122). The carotenochlorophyll-protein is purified from clarified aqueous extracts of broken cells by Sephadex and ion-exchange chromatography. This red complex accounts for 20–40% of the chlorophyll (58, 122) and most of the peridinin (122) in these organisms. The pigment-protein has a molecular weight of 35,000–38,000 (58, 59, 122) and may be composed of two identical polypeptide chains (59, 122). The *Cachonina niei* complex contains six peridinin and two chlorophyll *a* molecules per protein of 35,500 daltons (59), while the *Glenodinium* sp. component contains slightly more peridinin (peridinin/chlorophyll = 4) (122). Both types of pigments are released from the protein by heating, proteolytic digestion or treating with organic solvents (58), or by treatment with SDS (B. L. Prézelin, F. T. Haxo, and H. W. Siegelman, unpublished data).

Fluorescence excitation spectra of the complex indicate efficient energy transfer from peridinin to chlorophyll a (122). Comparison of the in vivo absorption spectrum and the action spectrum for photosynthetic oxygen evolution strongly suggests that the complex functions in the antennae of photosystem II.

CONTENT OF CHLOROPHYLL-PROTEINS IN PHOTOSYNTHETIC MEMBRANES

The P700-chlorophyll a-protein and the light-harvesting chlorophyll a/b-protein account for 10–18% and 40–60%, respectively, of the chlorophyll in higher plant and green algal membranes (28, 29, 49, 114, 136, 150). In *Euglena* much less (6–13%) of the chlorophyll is represented by these two components (29, 49). In other eukaryotic algae and in blue-green algae 4–18% and about 30% of the chlorophyll, respectively, can be accounted for in the P700-chlorophyll a-protein (29). The remaining chlorophyll in these organisms occurs in as yet unsubstantiated components.

Genge et al (49) observed that maize mesophyll cell chloroplasts had a similar composition to other higher plant chloroplasts, whereas bundle sheath chloroplasts with a high chlorophyll $a:b$ ratio contained 50% of their chlorophyll in the P700-chlorophyll a-protein but only 5% in the other major complex. This latter value, as well as the value obtained by them for mesophyll cells, is lower than anticipated from the chlorophyll $a:b$ ratios. This may indicate that the light-harvesting chlorophyll a/b-protein is less stable to SDBS than SDS, since SDS was used to establish the relationship between the chlorophyll $a:b$ ratio and the percentage of chlorophyll present in this complex (3, 28, 29). Bailey et al (15) also observed a decreased content of the light-harvesting chlorophyll a/b-protein in bundle sheath cells, but it appears that a lower proportion of the chlorophyll than measured by Genge et al (49) is present in the other complex.

Brown et al (28, 29) have contrasted the composition of grana and stroma lamellar fractions (73, 131), and in relationship to the chlorophyll $a:b$ ratios, they found a predominance (43% compared to 12%) of the chlorophyll in the stroma fraction in the P700-chlorophyll a-protein while a predominance (54% compared to 12%) of the other chlorophyll-protein was present in the grana lamellae. This quantitative analysis does not agree with the proposal that stroma lamellae structures (primary thylakoids) contain the P700-chlorophyll a-protein whereas the grana contain the chlorophyll b-rich complex (14).

Brown et al (28, 29) have examined the effect of light intensity on the composition of jack bean, soybean, and pine chloroplast membranes. Growth under low light compared to that under naturally occurring intensities decreased the chlorophyll $a:b$ ratio but increased the chlorophyll/P700 ratio and the proportion of the light-harvesting chlorophyll a/b-protein present in the leaves. These findings agree with the general observation that plants grown in lower light intensities have lower chlorophyll $a:b$ ratios (8, 53, 123).

The lamellar content of the major light-harvesting complex has been studied in response to salinity conditions during the growth of the green alga *Dunaliella* and the marsh grass *Spartina* (28, 29; D. J. Longstreth, personal communication).

Lowering the salinity reduced the lamellar content of the complex in these halophytes. Low level water stress has also been shown to reduce the rate of formation of the light-harvesting chlorophyll a/b-protein (R. S. Alberte, E. L. Fiscus, and A. W. Naylor, unpublished data). In addition, it has been found that chlorotic (iron deficient) plants have greatly reduced amounts of this chlorophyll-protein while the P700-chlorophyll a-protein is not present (99, 100). From data on the chlorophyll $a:b$ ratios of plants grown under different environmental conditions, it can be inferred that other stress conditions also alter the lamellar content of the major light-harvesting component (e.g. availability of nitrogen as discussed in this context previously). Thus it appears that the plasticity of this major chlorophyll a/b-protein is a highly adaptive feature of the photosynthetic apparatus.

Several mutants of higher plants showing large variations in the content of this chlorophyll-protein compared to the wild types have been reported: the chlorophyll b-less barley mutant totally lacks the chlorophyll-protein (10, 49, 150), a virescent pea mutant (63) has a much reduced content (10, 49), and certain temperature-sensitive nuclear mutants of soybean and cotton vary in the percentage of their total chlorophyll associated with this complex (2).

BIOSYNTHESIS OF THE CHLOROPHYLL-PROTEIN COMPLEXES

Alberte et al (4) examined the time of appearance of the two known major chlorophyll-proteins in greening jack bean leaves. The light-harvesting chlorophyll a/b-protein appeared first (after 2 hr), and thereafter (after 6 hr) the P700-chlorophyll a-protein was seen. The appearance of the former chlorophyll-protein was correlated with a decrease in the chlorophyll $a:b$ ratio while the appearance of the latter complex corresponded to the first detection of light-induced absorbance changes of P700 in the tissue (4). Guignery et al (56) have also found this same order of appearance for the chlorophyll-proteins in greening *Zea mays*. If a third major chlorophyll a-containing protein does exist in vivo, the evidence for which has been summarized in this article, then its appearance during greening could be inferred from the presence of chlorophyll in the free pigment zone. In this case, such a component is present after 2 hr of greening (4), and thus its presence could explain the very high chlorophyll $a:b$ ratio during the first two hours of chloroplast development. This is of course a very tenuous extrapolation of the data since any intermediate in the synthesis of the known chlorophyll-proteins could also be dissociated by detergent. Whether such extrapolation of the data is valid requires reinvestigation of the system using a monitoring technique in which the stability of the postulated complex is improved.

Alberte et al (5) have posed the question whether a reaction center of a photosystem is inserted into the membrane before or after assembly of its antenna chlorophyll. In the case of photosystem I it appears quite definite that insertion of the reaction center (P700) precedes deposition of its closely associated light-harvesting chlorophyll-protein (5).

The appearance of chlorophyll-proteins has also been examined during greening of etiolated beans in flashed light (2 min light, 98 or 118 min dark). Plants greened in this way contain chlorophyll *a* only (13, 14, 64, 126), and in some cases show (64), while in other cases do not show (126, 140), photosystem II activity. Argyroudi-Akoyunoglou et al (14) and Hiller et al (64) observed that of the two major chlorophyll-proteins only the P700-chlorophyll *a*-protein is present in these flashed leaves; shortening of the dark period leads to chlorophyll *b* formation and the appearance of the chlorophyll *a/b*-protein (64).

Site of Coding and Synthesis

The en:alba-1 mutant of *Antirrhinum majus* does not synthesize the P700-chlorophyll *a*-protein (60). Since this plant is reported to be a plastome mutant, the site of coding of this component is construed to be in the chloroplast (60); however, more rigorous proof is required to demonstrate that the lesion resides in chloroplast DNA. Furthermore, if the chlorophyll-protein complex does contain two slightly different moieties, as explained previously (see composition of the complex), the expression of the mutation may affect either entity, allowing the same observation to be made. The mode of inheritance of the primary structure of the light-harvesting chlorophyll *a/b*-protein into interspecific, reciprocal hybrids was used by Kung et al (91) to deduce that nuclear DNA codes for the protein.

Machold & Aurich (101) investigated the site of synthesis of the chlorophyll-proteins by studying the incorporation of labeled amino acids into the two complexes during antibiotic inhibition of protein synthesis. With the proviso that antibiotic inhibition of chlorophyll synthesis will almost certainly affect synthesis of the protein moieties of any chlorophyll-protein complex (98, 100), it was concluded (101) that the light-harvesting chlorophyll *a/b*-protein is synthesized on cytoplasmic ribosomes. The other major chlorophyll-protein, or at least a main component of it, was thought to be synthesized on chloroplast ribosomes, whereas a subunit of it may be synthesized in the cytoplasm. Eaglesham & Ellis (42) have determined more directly that neither major chlorophyll-protein is synthesized by isolated, intact pea chloroplasts. They conclude the chlorophyll-proteins or a necessary part of them are made on cytoplasmic ribosomes, but point out that since chloroplasts containing functional photosystems were studied, less mature ones might conceivably synthesize one or both of these complexes.

Many of the points in this section will be taken up in the following section in which similar studies on the chloroplast membrane polypeptides are described.

CHLOROPLAST MEMBRANE POLYPEPTIDES

Studies have been made on the nature of the polypeptides present in chloroplast membranes from which aqueously soluble proteins, pigments, and lipids have been removed. Certain aspects of these investigations are particularly pertinent to chlorophyll-protein research and thus will be considered here.

The water-insoluble protein residue after buffer and solvent extraction of chloroplast membranes was first examined in detail by Menke & Jordan (105). The yield of what they called lamellar structural protein was approximately half the weight of the chloroplast with carbohydrate (156) representing 3–9% of the weight of the protein fraction. N- and C-terminal amino acid analysis showed that this fraction contained more than ten polypeptide chains (105), the amino acid composition of which was reported later (159). This lamellar structural protein was first obtained in aqueous solution by formylation of the proteins (160). Subsequently, Biggins & Park (19) and Lockshin & Burris (97) observed that upon treatment of lamellar suspensions (97) or SDS-dissociated lamellae (19) with n-butanol the aqueous phase contained all the protein in a water-soluble form, whereas the organic phase contained the lipids and pigments. Since then many groups have obtained the pigment-extracted, lamellar protein fraction in solution by treatment with SDS (9, 37, 43, 67, 75, 76, 84, 93, 106, 109). The membrane protein has also been obtained in an aqueously soluble form by addition of phenol-formic acid-water to Triton X-100 (pH 3.7)-extracted membranes (11), or by treatment with acidic ethanol followed by elution of the proteins from hydroxylapatite columns (104). Ultracentrifugation of the solubilized protein exhibits a boundary of 2.2–2.9S (19, 76, 93, 106, 109, 129). A review of the behavior of such extracts in the ultracentriguge can be found in Rogers et al (129).

Criddle & Park (37, 38) fractionated this so-called lamellar structural protein. Acetone powders of spinach chloroplast lamellar membranes were solubilized in cholate-deoxycholate or SDS-urea and fractionated by ammonium sulfate. A component was obtained which exhibits a sedimentation coefficient of 2.2S (38) and a molecular weight of 23,000 daltons. Further, only one N-terminal amino acid residue was present, unlike the heterogeneous whole membrane fraction of Menke & Jordan (106). The fraction represented 24% or more of the protein in the chloroplast membrane. As Criddle (37) pointed out, this component meets the criteria for true structural protein as defined for the mitochondrion in that it was a single protein species of molecular weight 23,000 and functioned primarily as a major organization and stabilization component of the membrane. The protein was found to bind many low molecular weight membrane constituents; in particular, one mole of chlorophyll was bound per 26,000–31,000 g protein. However, this may have been a nonspecific binding of pigment since binding of chlorophyll to several nonplant proteins has been observed (162).

In a reinvestigation of this structural protein component of the membrane, Mani & Zalik (103) showed that material prepared from wheat or bean chloroplasts in an identical manner to Criddle & Park (38), was a heterogeneous mixture of polypeptides. Unfortunately, no N-terminal amino acid data were reported to substantiate the correspondence of their material in this regard to Criddle's (37) structural protein preparations. Nevertheless, it appears most likely that chloroplast lamellae do not contain a single major protein species that functions solely in a structural capacity as previously thought.

The amino acid composition of the whole lamellae (17, 76, 97, 129) and of the fractionated structural protein (103) has been reinvestigated. The analysis of structural protein is included in Table 1 for comparison with that of the chlorophyll-proteins.

The structural protein research has been described here in some detail because of its probable relationship to the chlorophyll-proteins. A large proportion (50%) of the chloroplast lamellar protein is contributed by the light-harvesting chlorophyll

a/b-protein (149), and the characteristics (size, sedimentation coefficient, and amino acid composition) of lamellar or structural protein are very similar to those of the major pigment-protein; thus it is very likely that structural protein and even whole lamellar protein, although heterogeneous, are largely composed of the protein of the chlorophyll a/b-protein complex.

Despite the homogeneity observed in the ultracentrifuge for solutions in which the whole chloroplast membrane is dissolved, the original observation (106) of the multiplicity of polypeptides in the chloroplast membrane has been confirmed by many groups. Polyacrylamide gel electrophoresis of solubilized chloroplast membranes either before or after lipid extraction has revealed the presence of as many as 21 protein zones (9, 11, 26, 35, 42, 43, 56, 62, 67, 75, 76, 84, 93, 95, 101, 104, 106, 107, 124, 129, 150); N-terminal amino acid analyses corroborate this complexity (23, 36). However, those groups (11, 26, 104) that have dissolved their membrane preparations in solvents other than SDS observe fewer zones on polyacrylamide gel electrophoresis. Apel & Schweiger (11) and McEvoy & Lynn (104) find a maximum of five polypeptides associated with the membrane when treated in this way. Both of these groups demonstrate that the polypeptides present in SDS-dissociated EDTA extracts, which are enriched in coupling factor subunits, account for many of the peptides present in their membrane preparations. In this respect, Ridley et al (128) have demonstrated how difficult it is to remove completely by successive aqueous washes of chloroplast membranes all those proteins that are generally classified among the aqueously soluble protein of the chloroplast. It is therefore probable that some of the bands observed in the gels of other investigators are not truly membrane polypeptides even though the preparations have been extracted with EDTA in most instances.

Studies of the polypeptide composition of pigment-extracted chloroplast membranes are further complicated by the fact that the function of most of the zones in the gel cannot be unequivocally ascertained. The function of an electrophoretically separated polypeptide is postulated by extrapolating its molecular weight (i.e. its electrophoretic mobility) to that of some known constituent of the intact membrane (84, 86). This is difficult to do with certainty when so many peptides are present in the gel columns; Eaglesham & Ellis (42) have highlighted this problem in their study. Ultimately it is hoped that standard protein chemistry techniques will be applied to each polypeptide in the gel and that these data will be correlated to that of isolated proteins of known function so that the identity of a zone in the acrylamide gel columns can be substantiated.

Workers in this area all agree that the chloroplast membrane after lipid extraction contains a major polypeptide of about 25,000 daltons (9–11, 35, 43, 67, 84–86, 93–95, 104, 124, 150). Some 36–40% of the protein stain in a gel is associated with this component (67, 94). Furthermore, this major component is generally agreed to be the protein moiety of the light-harvesting chlorophyll a/b-protein. This has been interpreted not only from its size, but also from the occasional presence of small amounts of pigment in this zone, and from the observation that those organisms lacking or having greatly reduced amounts of the chlorophyll-protein also lack or have decreased quantities of the major polypeptide component (9, 10, 84, 85, 93, 94,

124). Furthermore, this zone is highly enriched in digitonin, Triton, or French cell-prepared photosystem II fractions which possess large quantities of the chlorophyll-protein (9, 74, 84, 95, 124, 127).

Some groups observe that the major 25,000 dalton zone is composed of two polypeptides that differ in molecular weight by 2000–3000 daltons (9, 67, 84, 85, 95). It has been interpreted that both polypeptides are most likely correlated with the same membrane constituent, i.e. the light-harvesting chlorophyll a/b-protein, since blue-green algae and those plant mutants lacking or deficient in the chlorophyll-protein show decreased amounts of *both* polypeptides in the major zone (9, 10, 85). These observations would indicate that the light-harvesting chlorophyll a/b-protein is composed of two different polypeptide chains. However, this does not agree with subunit structure of the isolated complex (see section on the light-harvesting chlorophyll a/b-protein). It is important to note that the behavior of the two polypeptides is not always consistent with the notion that they, in fact, represent two distinct subunits of the pigment-protein. The proportion between the two varies significantly between different species, and furthermore it can be influenced by acetone extraction of the membranes (9, 10, 124). In this respect it may be significant that Lagoutte & Duranton (93) have observed that chloroform-methanol extraction of membranes yields a residue that is completely soluble in SDS, while acetone-extracted material is not completely soluble. Furthermore, those groups using high concentrations of organic acids together with phenol (11) or ethanol (104) to extract and dissociate the chloroplast membrane find that the major membrane protein (25,000 daltons) contains only a single polypeptide chain. If the dual nature of a single polypeptide chain in the polyacrylamide gels is confirmed to be an artifact of the extraction procedure, then the presence of other artifactual zones will have to be carefully examined. A slight difference in lipid content or in the extent of folding of the polypeptide may explain the duality of the major membrane protein as well as Hoober's (68) observation that only the smaller polypeptide occurs in an aqueous soluble form in the cell during chloroplast biogenesis.

The function of many of the other polypeptides in SDS-polyacrylamide gels of lipid-extracted chloroplast membranes has not been designated with the same degree of confidence as that of the 25,000 dalton component. Certain polypeptides (termed Group I polypeptides in some reports) in the gel patterns have been associated with photosystem I, where others (Group II) have been correlated to photosystem II (9, 10, 74, 84, 95, 124, 127). The major Group I polypeptides have molecular weights of 50,000–70,000 whereas those of Group II range between 25,000–30,000 daltons. Levine and co-workers (9, 10, 96) and Remy (125) have presented evidence that some, or perhaps all, of the Group II polypeptides provide a membrane with its ability to form stacks. The involvement in grana formation of the light-harvesting chlorophyll a/b-protein, the polypeptides of which are among those of Group II, was discussed previously. The function of this pigment-protein need not necessarily initiate grana stacks as proposed (13) since Group II contains several polypeptides in addition to those attributed to the complex. Since Group I polypeptides are present in isolated photosystem I fractions, the subunits of the P700-chlorophyll a-protein must occur among the group I polypeptides. Several investigators (10, 84,

85, 93, 124) have suggested that one or more 60,000 dalton polypeptides, which usually represented the second major polypeptide of the chloroplast membrane, may be derived from the chlorophyll-protein. Unfortunately, since such a size is the same as that of the major subunit of the coupling factor (104), and since there is no corroborating evidence similar to that which exists for the major polypeptide's identity, conclusive proof of the subunit structure of the P700-chlorophyll a-protein awaits a more direct determination. The polypeptide present in the photosystem II reaction center-enriched fraction (TSF 2a) of Vernon et al (158) has been studied (84, 85). A 44,000 dalton component is highly enriched in this fraction. This polypeptide may be that of the photosystem II reaction center, or since the polypeptide is absent from preparations enriched in the other two chlorophyll-proteins but is present in a chlorophyll a-enriched fraction (84, 85), it is also possible that it may correspond to the protein moiety of the proposed third chlorophyll a-protein.

Site of Coding and Synthesis

Although the function of the polypeptides observed in solubilized lipid-extracted chloroplast lamellae may be equivocal, certain data on the synthesis of some of the polypeptides correlate well with the previously described observations on the biosynthesis of the chlorophyll-proteins. For instance, it has been reported (12, 43, 67–69) that the major polypeptide is synthesized on cytoplasmic ribosomes, and hence it is most likely that nuclear DNA codes for this protein. Also, it has been observed (35, 44, 67–69, 94, 125, 126) that there is a rapid increase in the content of this major polypeptide in greening tissues of higher plants and *Chlamydomonas reinhardi* mutant y-1. Light is required to stimulate the synthesis (44, 45, 56, 68), which has been correlated to chlorophyll accumulation (44). All these observations correspond well with those on the biosynthesis of the light-harvesting chlorophyll a/b-protein (see section on biosynthesis of chlorophyll-proteins). The site of synthesis of the other major chlorophyll-protein cannot be delineated from the evidence presented to date on biosynthesis of membrane polypeptides. This is mainly because the polypeptide(s) postulated to be derived from the pigment-protein have not yet been studied in the same detail afforded the major polypeptide. It can be said with general agreement of the researchers in this area (12, 35, 42, 43, 67) that production of photosynthetically competent membranes requires components synthesized on cytoplasmic as well as chloroplastic ribosomes.

There is considerable debate on whether the polypeptides of the two complexes preexist in the etioplast. Hoober (67) maintains that most, and possibly all, membrane polypeptides are only synthesized during biogenesis of the membrane structure. Supportive evidence has been produced by Eytan & Ohad (43), who have observed differences in the polypeptide pattern between etioplasts and chloroplasts, and by others (35) who believe that de novo synthesis of at least some of the membrane polypeptides occurs. Other groups (13, 94, 125, 126) contend that all polypeptides are present prior to greening, and they particularly emphasize the preexistence of the protein moieties of the two major chlorophyll-proteins. However, the method of detection used by one group (13) is equivocal. Stepwise elution of protein by a certain phosphate concentration from hydroxylapatite columns is a poor criterion for establishing the presence of a specific pigment-less polypeptide, particularly when the purity of the eluates has never been determined. Hiller and co-workers (64) initially supported the notion that pigment is added during greening to proteins already present in the etioplast; however, they later (49) found that they could not establish with certainty the preexistence of the apoprotein of the light-harvesting chlorophyll a/b-protein. Remy (125) and Collet et al (36) believe that quantititative but not qualitative changes occur in the membrane polypeptides, and further, that membranes are synthesized by a reorganization of the etioplast polypeptides rather than by de novo synthesis of proteins. Cobb & Wellburn (35) showed that only some chloroplast polypeptides are present in etioplasts, whereas some others appear or disappear during greening. Unfortunately, no function has yet been ascribed to any of the polypeptides in their studies. This last mentioned report accentuates the care which must be taken to ensure that one protein zone in a SDS gel electrophoretic pattern of one sample corresponds to a particular zone on electrophoresis of a slightly different sample. The membrane polypeptide studies have already given some insight into the biogenesis of the chloroplast membrane. The ultimate aim of such an approach is to provide a complete description of the composition and biosynthesis of this lamellar structure. But, since there is little agreement between the several groups working in this area on the number of polypeptides present in the membrane, it is essential that an early objective of this research should be an unequivocal determination of how many polypeptides are truly chloroplast membrane constituents, and what their functions are.

CONCLUDING REMARKS AND SUMMARY

An early notion that chlorophyll is conjugated with protein in a photosynthetic organism has been substantiated in the last decade. Much of the success in the area has come from the use of ionic detergents to solubilize the photosynthetic membranes and to isolate homogeneous chlorophyll-proteins. The isolated components are essentially composed only of chlorophyll and protein. Other methods for dissociating and fractionating the photosynthetic membranes (e.g. non-ionic detergents) yield a heterogeneous preparation of chlorophyll-protein in which any one such complex is enriched, but which also contains other pigment-protein complexes and/or colorless proteins in addition to lipids.

The earlier skepticism that chlorophyll-protein complexes were possibly formed by a nonspecific detergent-induced co-solubilization of chlorophyll and chloroplast protein(s) has now been mainly eliminated. It has been demonstrated that the absorption, fluorescence, and circular dichroism spectra (55; however, see 121) and the spectral forms of chlorophyll a in the two detergent-soluble chlorophyll-proteins precisely reflect the spectral characteristics of intact chloroplast lamellar membranes. This argues strongly that no change has occurred within these pigment-proteins during their isolation. In addition, the quantum efficiency of P700 photooxidation in the isolated chlorophyll-protein is comparable to that observed in vivo rather than to that of an artificially made chlorophyll-electron acceptor complex. Most compelling of all is the fact that not only have mutants been found which lack one but not the other chlorophyll-protein, but in each case the effect of the mutation on the photosynthetic activity of the organism is exactly what would be predicted for the absence of either the P700-chlorophyll *a*-protein or the light-harvesting chlorophyll *a*/*b*-protein.

To date, two detergent-soluble and two water-soluble chlorophyll-proteins have been isolated in a homogeneous form from plants and have been studied in some detail. The photochemically active P700-chlorophyll *a*-protein is almost certainly ubiquitous in the photosynthetic plant kingdom, whereas the other detergent-soluble complex is found only in chlorophyll *b*-containing organisms. The organization of that chlorophyll which is not located in either of these complexes in all chlorophyll *a*-containing organisms is largely unknown. However, initial reports of the existence of additional detergent-soluble chlorophyll-proteins have recently appeared (49, 56). If it is confirmed that a third chlorophyll *a*-protein exists, there is every reason to expect that such a component will represent a large proportion of the chlorophyll associated with photosystem I, and that it will occur in all those organisms in which the P700-chlorophyll *a*-protein is found. It is also to be anticipated that each class of photosynthetic plants will contain those pigments (e.g. chlorophyll *b* and *c*, fucoxanthin, and phycocyanobilin) whose distribution is limited within the plant kingdom, in a pigment-protein complex which will be the major component of the antenna of photosystem II. The chlorophyll *a*/*b*-protein, the peridinin-chlorophyll *a*-protein, and the biliproteins are examples of such components that have already been isolated and that function in this manner. It is expected that further detailed studies of the protein chemistry of those light-harvesting components will reveal whether they are phylogenetically related.

One particular urgent need for future research on chlorophyll-proteins is the development of a new technique for dissociating the chlorophyll-protein complexes from the membrane and for maintaining them in a stable form during their isolation as homogeneous entities. Anionic detergents have proved useful for obtaining the two major complexes, but such detergents apparently rapidly dissociate the chlorophylls from all other chlorophyll-protein complexes that may occur in the plant kingdom, although it must be borne in mind that some photosynthetic pigments in an organism may not necessarily occur in conjugation with protein. Such an organization for some of the photosynthetic pigments is unlikely, however, since it would make it difficult to envisage how their biosynthesis and deposition would be controlled. The biogenesis of chloroplast membranes has been briefly considered in this review. Studies on the biosynthesis of the polypeptides, particularly on that of the major membrane polypeptide, have revealed that their synthesis is precisely controlled. Two groups have published schemes depicting this control (45, 69), but since both studied the greening process in a mutant of *Chlamydomonas reinhardi* (y-1), their schemes await a confirmation that the same mechanism occurs in a normal chlorophyll *b*-containing plant. Further information on the control of biosynthesis has come from studies on chlorophyll-protein complexes. Since continuous illumination is required for the synthesis of chlorophyll *b* (14, 64, 154) and for synthesis of the major light-harvesting chlorophyll *a*/*b*-protein (14, 64), then synthesis of chlorophyll *b* and of the apoprotein of the major pigment-protein complex are intimately associated; apparently one component cannot exist without the other. A similar situation exists in the chlorophyll b-less mutant of barley (10, 150). This sheds new light on the lesion in the barley mutant, which has long been thought to be an enzyme in the biosynthetic pathway of chlorophyll b. It now appears equally possible that this lesion is the apoprotein of the light-harvesting chlorophyll a/b-protein.

Why are some chlorophyll-proteins not aqueously soluble? The initial thought (118, 148) that this was caused by the hydrophobic nature of the protein may not be correct, since the polypeptide chains of the detergent-soluble complexes after solvent extraction are apparently soluble in water. An alternate explanation is that other hydrophobic entities (carotenoids, lipids, or the phytyl tail of chlorophylls) are located on the exterior of the folded protein. In this respect the proposal (120, 130) that galactolipids localize the chlorophyll molecules in the membrane for photoreception may be relevant. This postulation was mainly based on the observation that the acyl groups of the galactolipids fit the phytyl group of chlorophyll like a key and lock. Thus, if the water-insolubility of the chlorophyll-proteins is due to the protrusion of phytyl chains, then interaction of the phytyl groups with galactolipids may well be used to position the pigment-proteins in the membrane. The porphyrin rings of the chlorophylls are almost certainly located in the interior of the folded protein. If this were not so, it would be hard to rationalize the existence of different spectral forms of chlorophyll a as well as the absence of pheophytin in the detergent-treated complexes.

Knowledge of the location of the chlorophyll-proteins in the thylakoid membrane has progressed little since it was last reviewed by Kirk (83). The recent data of Klein & Vernon (86) on the location of certain membrane polypeptides on the exterior of the thylakoid discs will be more useful once the function of these polypeptides has been unequivocally determined.

Obviously much remains to be discovered about the characteristics of chlorophyll-proteins. Certain data will be relatively easy to obtain. The basic protein chemistry, particularly determination of their subunit structure, is a pressing need, and the relationship, if any, of the known chlorophyll-proteins to protochlorophyllide holochrome(s), the peridinin-chlorophyll a-protein, and to other higher molecular weight chlorophyll-containing subfractions of the thylakoid membrane [e.g. the fractions of Koenig et al (87)] should soon be elucidated. On the other hand, certain necessary studies will be more difficult, such as the development of an improved technique for isolating the existing and new chlorophyll-proteins in a stable form, as well as the determination of the three-dimensional organization of the chlorophylls within the protein framework.

Finally, limited space has not permitted the inclusion of any data on the chlorophyll-proteins isolated from photosynthetic bacteria. Data on some bacteriochlorophyll-protein complexes is far in advance of that on plant proteins. For example, X-ray crystallographic studies are in progress on a water-soluble complex from green bacteria, and as mentioned previously, the photochemical reaction center of some bacteria has been isolated and characterized free of light-harvesting chlorophyll and contaminating proteins. The reviews of Clayton (34) and Sauer (133) will introduce the reader to chlorophyll-proteins in photosynthetic bacteria.

ACKNOWLEDGMENTS

The author is grateful for the help of Dr. Randall S. Alberte in the preparation of this manuscript. The author's research since 1971 referred to in this article and the preparation of this review was largely supported by funds from the National Science Foundation, grant number GB 31207.

Literature Cited

1. Aghion, J. 1963. *Biochim. Biophys. Acta* 66:212–17
2. Alberte, R. S. et al 1974. *Proc. Nat. Acad. Sci. USA* 71:2414–18
3. Alberte, R. S., Thornber, J. P. 1974. *Plant Physiol. Suppl.* 53:47
4. Alberte, R. S., Thornber, J. P., Naylor, A. W. 1972. *J. Exp. Biol.* 23:1060–69
5. Alberte, R. S., Thornber, J. P., Naylor, A. W. 1973. *Proc. Nat. Acad. Sci. USA* 70:134–37
6. Allen, M. B., Murchio, J. C., Jeffrey, S. W., Bendix, S. A. 1963. In *Studies on Microalgae and Photosynthetic Bacteria*, ed. Jap. Soc. Plant Physiol., 407–12. Univ. Tokyo Press
7. Anderson, D. R., Spikes, J. S., Lumry, R. 1954. *Biochim. Biophys. Acta* 15:298
8. Anderson, J. M., Goodchild, D. J., Boardman, N. K. 1973. *Biochim. Biophys. Acta* 325:573–85
9. Anderson, J. M., Levine, R. P. 1974. *Biochim. Biophys. Acta* 333:378–87
10. Ibid 1974. 357:118–26
11. Apel, K., Schweiger, H. G. 1972. *Eur. J. Biochem.* 25:229–38
12. Ibid 1973. 38:373–83
13. Argyroudi-Akoyunoglou, J. H., Akoyunoglou, G. 1973. *Photochem. Photobiol.* 18:219–28
14. Argyroudi-Akoyunoglou, J. H., Feleki, Z., Akoyunoglou, G. 1971. *Biochem. Biophys. Res. Commun.* 45:606–13
15. Bailey, J. L., Downton, W. J. S., Maisiar, E. 1971. In *Photosynthesis and Photorespiration*, ed. M. D. Hatch, C. B. Osmond, R. O. Slatyer, 382–86. New York: Wiley
16. Bailey, J. L., Kreutz, W. 1969. *Progr. Photosyn. Res.* 1:149–58
17. Bailey, J. L., Thornber, J. P., Whyborn, A. G. 1966. In *Biochem. Chloroplasts* 1:243–55
18. Baszynski, T., Brand, J., Barr, R., Krogmann, D. W., Crane, F. L. 1972. *Plant Physiol.* 50:410–11
19. Biggins, J., Park, R. B. 1965. *Plant Physiol.* 40:1109–15
20. Boardman, N. K. 1970. *Ann. Rev. Plant Physiol.* 21:115–40
21. Boardman, N. K., Anderson, J. M. 1964. *Nature* 293:166–67
22. Boardman, N. K., Highkin, H. R. 1966. *Biochim. Biophys. Acta* 126:189–99
23. Boquet, M., Guignery, G., Duranton, J. 1968. *Bull. Soc. Chim. Biol.* 50:531–46
24. Borg, D. C., Fajer, J., Dolphin, D., Felton, R. H. 1970. *Proc. Nat. Acad. Sci. USA* 67:813–20
25. Brangeon, J. 1973. *Photosynthetica* 7:365–72
26. Braunitzer, G., Bauer, G. 1967. *Naturwissenschaften* 54:70–71
27. Brown, J. S. 1973. *Photophysiology* 8:97–112
28. Brown, J. S., Alberte, R. S., Thornber, J. P., French, C. S. 1974. *Carnegie Inst. Wash. Yearb.* 73:694–706
29. Brown, J. S., Alberte, R. S., Thornber, J. P. 1974. *Proc. 3rd Int. Photosyn. Congr. Israel.* In press
30. Brown, J. S., Duranton, J. G. 1964. *Biochim. Biophys. Acta* 79:209–11
31. Butler, W. L. 1961. *Arch. Biochem. Biophys.* 73:413–22
32. Chiba, Y. 1955. *Arch. Biochem. Biophys.* 54:83–93
33. Ibid 1960. 90:294–303
34. Clayton, R. K. 1973. *Ann. Rev. Biophys. Bioeng.* 2:131–56
35. Cobb, A. H., Wellburn, A. R. 1973. *Planta* 114:131–42
36. Collet, D., Guignery, G., Duranton, J. 1970. *Bull. Soc. Chim. Biol.* 52:241–51
37. Criddle, R. S. 1966. *Biochem. Chloroplasts* 1:203–31
38. Criddle, R. S., Park, L. 1964. *Biochem. Biophys. Res. Commun.* 17:74–79
39. Davis, D. J., Gross, E. L. 1974. *Fed. Proc.* 33:1255
40. Dietrich, W. E. Jr., Thornber, J. P. 1971. *Biochim. Biophys. Acta* 245:482–93
41. Duysens, L. N. M. 1952. *Transfer of excitation energy in photosynthesis.* PhD thesis. State Univ., Utrecht, The Netherlands
42. Eaglesham, A. R. J., Ellis, R. J. 1974. *Biochim. Biophys. Acta* 335:396–407

43. Eytan, G., Ohad, I. 1970. *J. Biol. Chem.* 245:4297–4307
44. Ibid 1972. 247:112–21
45. Ibid, 122–29
46. French, C. S. 1968–69. *Carnegie Inst. Wash. Yearb.* 68:578–87
47. French, C. S., Brown, J. S., Lawrence, M. C. 1972. *Plant Physiol.* 49:421–29
48. French, C. S., Takamiya, A., Murata, T. 1972–73. *Carnegie Inst. Wash. Yearb.* 72:336–51
49. Genge, S., Pilger, D., Hiller, R. G. 1974. *Biochim. Biophys. Acta* 347:22–30
50. Givan, A. L., Levine, R. P. 1969. *Biochim. Biophys. Acta* 189:404–10
51. Goedheer, J. C. 1966. *The Chlorophylls,* ed. L. P. Vernon, G. Seeley, 399–411. New York: Academic
52. Goodchild, D. J., Bjorkman, O., Pyliotis, N. A. 1971–72. *Carnegie Inst. Wash. Yearb.* 71:102–7
53. Grahl, H., Wild, A. 1972. *Z. Pflanzenphysiol.* 67:443–53
54. Gregory, R. P. F., Raps, S., Bertsch, W. 1971. *Biochim. Biophys. Acta* 234:330–34
55. Gregory, R. P. F., Raps, S., Thornber, J. P., Bertsch, W. F. 1971. *Proc. 2nd Int. Congr. Photosyn. Res.* 2:1503–8
56. Guignery, G., Luzzati, A., Duranton, J. 1974. *Planta* 115:227–43
57. Hager, W. G., Briggs, W. 1974. *Plant Physiol.* 53:4 (Abstr.)
58. Haidak, D. J., Mathews, C. K., Sweeney, B. M. 1966. *Science* 152:212–13
59. Haxo, F. T., Kycia, J. H., Siegelman, H. W., Somers, G. F. 1972. *Third Int. Symp. Carotenoids,* 73
60. Herrmann, F. 1971. *FEBS Lett.* 19:267–69
61. Herrmann, F. 1972. *Exp. Cell Res.* 70:452–53
62. Herrmann, F., Meister, A. 1972. *Photosynthetica* 6:177–82
63. Highkin, H. R., Boardman, N. K., Goodchild, D. J. 1969. *Plant Physiol.* 44:1310–20
64. Hiller, R. G., Pilger, D., Genge, S. 1973. *Plant Sci. Lett.* 1:81–88
65. Hiller, R. G., Genge, S., Pilger, D. 1974. *Plant Sci. Lett.* 2:239–42
66. Hiyama, T., Ke, B. 1972. *Biochim. Biophys. Acta* 267:160–71
67. Hoober, J. K. 1970. *J. Biol. Chem.* 245:4327–34
68. Hoober, J. K. 1972. *J. Cell Biol.* 52:84–96
69. Hoober, J. K., Stegeman, W. J. 1973. *J. Cell Biol.* 56:1–12
70. Howell, S. H., Moudrianakis, E. N. 1967. *Proc. Nat. Acad. Sci. USA* 58:1261–68
71. Inoue, Y., Ogawa, T., Shibata, K. 1973. *Biochim. Biophys. Acta* 305:483–87
72. Itoh, M., Izawa, S., Shibata, K. 1963. *Biochim. Biophys. Acta* 69:130–42
73. Jacobi, G., Lehmann, H. 1968. *Z. Pflanzenphysiol.* 59:457–76
74. Jennings, R. C., Eytan, G. 1973. *Arch. Biochem. Biophys.* 159:832–36
75. Jhamb, S., Zalik, S. 1973. *Can. J. Bot.* 51:2147–54
76. Ji, T. H., Hess, J. L., Benson, A. A. 1968. *Biochim. Biophys. Acta* 150:676–85
77. Kahn, J. S. 1964. *Biochim. Biophys. Acta* 79:234–40
78. Kahn, J. S., Bannister, T. T. 1965. *Photochem. Photobiol.* 4:27–32
79. Kahn, J. S., Chang, I. C. 1965. *Photochem. Photobiol.* 4:733–38
80. Ke, B. 1973. *Biochim. Biophys. Acta* 301:1–33
81. Ke, B., Clendenning, K. A. 1956. *Biochim. Biophys. Acta* 19:74–83
82. Ke, B., Sahu, S., Shaw, E., Beinert, H. 1974. *Biochim. Biophys. Acta* 347:36–48
83. Kirk, J. T. O. 1971. *Ann. Rev. Biochem.* 40:161–96
84. Klein, S. M., Vernon, L. P. 1974. *Photochem. Photobiol.* 19:43–49
85. Klein, S. M., Vernon, L. P. 1974. *Plant Physiol.* 53:777–78
86. Klein, S. M., Vernon, L. P. 1974. *Ann. NY Acad. Sci.* 227:568–79
87. Koenig, F., Menke, W., Craubner, H., Schmid, G. H., Radunz, A. 1972. *Z. Naturforsch.* 27b:1225–88
88. Kok, B. 1961. *Biochim. Biophys. Acta* 48:527–33
89. Krasnovskii, A. A., Brin, G. P. 1954. *Dokl. Akad. Nauk SSR* 95:611–15
90. Kung, S. D., Thornber, J. P. 1971. *Biochim. Biophys. Acta* 253:285–89
91. Kung, S. D., Thornber, J. P., Wildman, S. G. 1972. *FEBS Lett.* 24:185–88
92. Kupke, D. W., French, C. S. 1960. *Encycl. Plant Physiol.* 1:298–322
93. Lagoutte, B., Duranton, J. 1971. *Biochim. Biophys. Acta* 253:232–39
94. Lagoutte, B., Duranton, J. 1972. *FEBS Lett.* 28:333–36
95. Levine, R. P., Burton, W. G., Duram, H. A. 1972. *Nature* 237:176–77
96. Levine, R. P., Duram, H. A. 1973. *Biochim. Biophys. Acta* 325:565–72
97. Lockshin, A., Burris, R. H. 1966. *Proc. Nat. Acad. Sci. USA* 56:1564–70
98. Machold, O. 1971. *Exp. Cell Res.* 65:466–67

99. Machold, O. 1971. *Biochim. Biophys. Acta* 238:324–30
100. Machold, O. 1972. *Biochim. Physiol. Pflanzen* 163:30–41
101. Machold, O., Aurich, O. 1972. *Biochim. Biophys. Acta* 281:103–12
102. Malkin, R., Aparicio, P. J., Arnon, D. I. 1974. *Proc. Nat. Acad. Sci. USA* 71:2362–66
103. Mani, R. S., Zalik, S. 1970. *Biochim. Biophys. Acta* 200:132–37
104. McEvoy, F. A., Lynn, W. S. 1973. *J. Biol. Chem.* 248:4568–73
105. Menke, W., Jordan, E. 1959. *Z. Naturforsch.* 14b:234–40
106. Menke, W., Ruppel, H. -G. 1971. *Z. Naturforsch.* 26b:825–31
107. Menke, W., Scholzel, E. 1971. *Z. Naturforsch.* 26b:378–79
108. Mohanty, P., Braun, B. Z., Govindjee, Thornber, J. P. 1972. *Plant Cell Physiol.* 13:81–91
109. Molchanov, M. I., Besinger, E. N. 1968. *Dokl. Akad. Nauk SSR* 178:475–78
110. Murata, T., Murata, N. 1971–72. *Carnegie Inst. Wash. Yearb.* 70:504–7
111. Murata, T., Odaka, Y., Uchino, K., Yakushiji, E. 1968. In *Comparative Biochemistry and Biophysics of Photosynthesis,* ed. K. Shibata, A. Takamiya, A. T. Jagendorf, R. C. Fuller, 222–28. Univ. Tokyo Press
112. Myers, J. 1971. *Ann. Rev. Plant Physiol.* 22:289–312
113. Noguchi, I., Takashima, S. 1958. *Arch. Biochem. Biophys.* 74:478–86
114. Ogawa, T., Obata, F., Shibata, K. 1966. *Biochim. Biophys. Acta* 112:223–34
115. Oku, T., Yoshida, M., Tomita, G. 1972. *Plant Cell Physiol.* 13:183–86
116. Ibid, 773–82
117. Oku, T., Yoshida, M., Tomita, G. 1974. *Photochem. Photobiol.* 19:177–79
118. Olson, J. M. et al. 1969. *Progr. Photosyn. Res. Proc. Int. Congr.* 1:217–25
119. Park, R. B., Biggins, J. 1964. *Science* 144:1009–11
120. Patton, S. 1968. *Science* 159:221
121. Philipson, K. D., Sauer, K. 1973. *Biochemistry* 12:3454–58
122. Prezelin, B. L., Haxo, F. T. 1974. *J. Phycol. Suppl.* 10:14
123. Reger, B. J., Krauss, R. W. 1970. *Plant Physiol.* 46:568–75
124. Remy, R. 1971. *FEBS Lett.* 13:313–17
125. Ibid 1973. 31:308–12
126. Remy, R. 1973. *Photochem. Photobiol.* 18:409–16
127. Remy, R., Phung-Nhu-Hung-Suong, Moyse, A. 1972. *Physiol. Veg.* 10:269–90

128. Ridley, S. M., Thornber, J. P., Bailey, J. L. 1967. *Biochim. Biophys. Acta* 140:62–79
129. Rogers, L. J., Kersley, J., Lees, D. N. 1973. *Physiol. Veg.* 11:327–60
130. Rosenberg, A. 1967. *Science* 157:1191–96
131. Sane, P. V., Goodchild, D. J., Park, R. B. 1970. *Biochim. Biophys. Acta* 216:162–78
132. Sane, P. V., Park, R. B. 1970. *Biochem. Biophys. Res. Commun.* 41:206–10
133. Sauer, K. 1974. In *Bioenergetics of Photosynthesis,* ed. Govindjee, 115–81. New York: Academic
134. Seeley, G. R. 1973. *J. Theor. Biol.* 40:189–99
135. Shibata, K. 1971. *Methods Enzymol.* 23:296–302
136. Shiozawa, J. A., Alberte, R. S., Thornber, J. P. 1974. *Arch. Biochem. Biophys.* 165:388–97
137. Sironval, C., Clijsters, H., Michel, J.-M., Bronchart, R., Michel-Wolwertz, M.-R. 1967. In *Le Chloroplaste,* ed. C. Sironval, 99–123. Masson et Cie
138. Smith, E. L. 1941. *J. Gen. Physiol.* 24:565–82
139. Smith, E. L., Pickels, E. G. 1941. *J. Gen. Physiol.* 24:753–64
140. Strasser, R. J., Sironval, C. 1972. *FEBS Lett.* 28:56–60
141. Takamiya, A. 1971. *Methods Enzymol.* 23:603–13
142. Takamiya, A. 1972–73. *Carnegie Inst. Wash. Yearb.* 72:330–36
143. Takamiya, A., Obata, H., Yakushiji, E. 1963. In *Photosynthetic Mechanisms in Green Plants,* ed. B. Kok, A. Jagendorf, 479–85. Washington, D.C.: Nat. Acad. Sci. Nat. Res. Counc.
144. Takashima, S. 1952. *Nature* 169:182–83
145. Terpstra, W. 1966. *Biochim. Biophys. Acta* 120:317–25
146. Ibid 1967. 143:221–28
147. Thirkell, D. 1964. *Phytochemistry* 3:301–5
148. Thornber, J. P. 1969. *Biochim. Biophys. Acta* 172:230–41
149. Thornber, J. P., Gregory, R. P. F., Smith, C. A., Bailey, J. L. 1967. *Biochemistry* 6:391–96
150. Thornber, J. P., Highkin, H. R. 1974. *Eur. J. Biochem.* 41:109–16
151. Thornber, J. P., Olson, J. M. 1971. *Photochem. Photobiol.* 14:329–41
152. Thornber, J. P., Smith, C. A., Bailey, J. L. 1966. *Biochem. J.* 100:14
153. Thornber, J. P., Stewart, J. C., Hatton,

M. W. C., Bailey, J. L. 1967. *Biochemistry* 6:2006–14
154. Thorne, S. W., Boardman, N. K. 1971. *Plant Physiol.* 47:252–61
155. Van Gorkom, H. J., Donze, M. 1971. *Nature* 234:231–32
156. Van Wyk, D. 1966. *Z. Naturforsch.* 21b:700–3
157. Vernon, L. P., Shaw, E. R., Ogawa, T., Raveed, D. 1971. *Photochem. Photobiol.* 14:343–57
158. Vernon, L. P., Yamamoto, H. Y., Ogawa, T. 1969. *Proc. Nat. Acad. Sci. USA* 63:911–17

159. Weber, P. 1962. *Z. Naturforsch.* 17b:683–88
160. Ibid 1963. 18b:1105–11
161. Wessels, J. S. C., von Alphen-van Waveren, O., Voorn, G. 1973. *Biochim. Biophys. Acta* 292:741–52
162. Wolken, J. J. 1958. *Brookhaven Symp. Biol.* 11:87–100
163. Wolken, J. J., Schwartz, F. A. 1956. *Nature* 177:136–38
164. Yakushiji, E., Uchino, K., Sugimura, Y., Shiratori, I., Takamiya, F. 1963. *Biochim. Biophys. Acta* 75:293–98
165. Yakushiji, Y. 1970. *Physiol. Plant.* 8:73–78

Ann. Rev. Plant Physiol. 1975. 26:159–86

THE REGULATION OF ❖7587
CARBOHYDRATE METABOLISM

John F. Turner and Donella H. Turner
Department of Agricultural Chemistry, University of Sydney, N.S.W. 2006, Australia

CONTENTS

159

INTRODUCTION

There is little doubt that most major pathways of carbohydrate metabolism are now known. These sequences of enzyme catalyzed reactions have been established by a variety of techniques and experimental approaches involving both in vitro and in vivo investigations. For some time increasing attention has been given to those aspects of this work which are relevant to the regulation of metabolic rates: it is axiomatic that the rates of both synthetic and degradative processes do not remain constant but vary according to the physiological requirements of the cell. This review is concerned with the regulation of enzyme activity which in turn controls the flux through metabolic pathways. The article will concentrate on the metabolic significance of observations rather than on the interpretation of the phenomena in terms of the theoretical predictions of particular models. The possible molecular mechanisms involved in the regulation of enzyme activity have been discussed elsewhere (87, 88, 115). Emphasis will be given to work on metabolic regulation with higher plants and the enzymes obtained therefrom, but consideration will be given, where appropriate, to investigations on mammalian tissues and microorganisms and their enzymes. In addition to results with probable regulatory enzymes, mention will be made of the effects of metabolites on other enzymes, the regulatory significance of which is doubtful or remains to be elucidated. Most major pathways of carbohydrate breakdown and synthesis are encompassed, but photosynthesis will not be considered as regulatory aspects of this process have been the subject of recent reviews (17, 131). The effects of gibberellins, other hormones, and ethylene will not be dealt with as these have been discussed in this series (74, 128, 140).

Although a detailed consideration of the mechanisms of enzyme regulation is outside the scope of this article, it may be useful to mention briefly a few aspects of this topic pertinent to this review. For more information the reader is referred to the review by Stadtman (148). Metabolic control can be achieved in a number of ways, but it is believed that regulation of rates is brought about mainly by the control of enzyme activity by factors such as substrate and product concentration, cofactor concentration (including metallic ions) and inhibition or activation of enzymes by other metabolic intermediates, and by change in the concentration or amount of an enzyme ("coarse control"). A valuable lead in questions of regulation is often provided by the identification of nonequilibrium reactions in metabolic pathways. The term "nonequilibrium" refers to those reactions which are far from equilibrium in tissues as judged by estimations of substrates and products. Such calculations may be criticized because usually it has to be assumed that the substances being determined are evenly distributed throughout the cells or tissues and factors such as compartmentation and adsorption are ignored. Also if there is a flux through a metabolic pathway, the reactions cannot be at equilibrium. It has been suggested that a twentyfold difference between the equilibrium constant and the observed ratio of the concentrations of the products and substrates of a reaction is necessary before a reaction can be considered as nonequilibrium (115, 138), but nonequilibrium reactions usually have much greater differentials. The failure to attain approximate equilibrium is due to the low activity of the enzyme involved, and therefore the enzyme is likely to be regulatory. Although it is commonly held that nonequilibrium reactions control metabolic rates, there may be situations where equilibrium steps make important contributions to metabolic regulation (90).

The early work of Adelberg & Umbarger (5) on amino acid metabolism established that the end product of a process may regulate its own biosynthesis: this effect became known as feedback control and is of fundamental importance. The most effective point of control is the first committed step in a pathway as this will determine the flux through the sequence. Monod, Changeux & Jacob (104) observed that the biological activity of many enzymes is controlled by specific metabolites which do not interact directly with the substrates or products of the reactions. These metabolites, having no structural resemblance to the normal substrates for the enzymes, were termed allosteric effectors. Allosteric enzymes exhibit cooperativity, i.e. a plot of initial velocity against substrate (or effector) concentration yields a sigmoid or S-shaped curve rather than the hyperbolic curve given by an enzyme obeying the usual Michaelis-Menten kinetics. Cooperativity indicates that the binding of one molecule to the enzyme in some manner facilitates the binding of the next. Koshland (86) has pointed out that an enzyme having hyperbolic kinetics will require a large change (81-fold) in ligand concentration to alter the rate from 10% to 90% of full activity, whereas for an enzyme having sigmoid kinetics the same degree of rate variation may be achieved by a three to sixfold change in ligand concentration (149). The implications of sigmoid response curves are therefore most important for metabolic regulation as a relatively small change in ligand concentration will produce a large change in enzyme activity. It should be noted that some regulatory enzymes show hyperbolic (Michaelis-Menten) kinetics and also that allosteric properties of an enzyme may be lost during purification or as the result of other procedures.

REGULATION OF GLYCOLYSIS

Carbohydrates are the principal energy source for most living organisms, and the main pathway of carbohydrate degradation is through glycolysis (Embden-Meyerhof-Parnas pathway). This sequence was originally proposed as a mechanism for the conversion of sugars to ethanol and CO_2 by yeast and for the breakdown of glycogen to lactate by muscle. There is firm evidence for the occurrence and operation of glycolysis in higher plant tissues (14, 16, 59, 68).

Enzymes of Glycolysis

HEXOKINASE   The reaction catalyzed by hexokinase is the principal method for incorporating hexoses into metabolism and is generally considered to be a nonequilibrium reaction. The hexokinases from yeast and mammalian tissues have been studied extensively (26, 136). There are two naturally occurring isozymes of yeast hexokinase with different catalytic and physical properties. Glucose-6-P is an inhibitor of yeast hexokinase but since the inhibitor constant (K_i) is relatively high (8.0 mM), it is unlikely that inhibition by glucose-6-P plays an important regulatory role in yeast (134, 136). One isozyme of the yeast enzyme is activated at pH 6.6 by low concentrations of a number of metabolites including citrate, malate, P_i, 2-P-glycerate, 3-P-glycerate, CTP, and GTP (89). Strong inhibition is also obtained at low pH by relatively low concentrations (0.2–0.5 mM) of ADP or GDP (26). Four isozymes of mammalian hexokinase are present in varying proportions in different tissues such as liver, brain, kidney, and skeletal muscle, and these have pronounced differences in K_m values for glucose (26). Mammalian hexokinases are strongly inhibited by glucose-6-P, and it is accepted that in mammalian tissues metabolic control of phosphofructokinase regulates the glucose-6-P concentration which in turn regulates hexokinase activity (26, 136, 139). Glucose-6-P inhibition is decreased by P_i which usually also affects phosphofructokinase. *Euglena gracilis* contains a soluble fructokinase and a particulate glucokinase with a high K_m (8 mM) for glucose (98). The glucokinase is activated allosterically by P_i. Less work has been carried out on hexokinases from higher plants. Wheat germ hexokinases have been purified (102) and hexokinase preparations have been obtained from the endosperm and scutellum of developing and germinating maize seeds (27). These maize hexokinase preparations had K_m values for glucose varying from 0.1–0.2 mM with an additional K_m of 3.4 mM for the enzyme from scutellum of germinating seeds. The hexokinases of pea seeds have been separated into three main fractions: (i) a glucokinase which has little activity with fructose; this fraction also phosphorylates mannose but not galactose; (ii) a fraction which phosphorylates glucose, mannose, and fructose; and (iii) a fructokinase which has very slight activity with glucose (Q. J. Chensee, D. D. Harrison & J. F. Turner, in preparation).

PHOSPHOFRUCTOKINASE   Hexose monophosphates may undergo glycolysis, enter the pentose phosphate pathway, be utilized for oligosaccharide and polysaccharide formation, or be hydrolyzed to free hexoses. The reaction between fructose-6-P and ATP to give fructose-1,6-P_2 is the first unique step in glycolysis and is a nonequilibrium reaction. Regulation of phosphofructokinase could determine the rate of glycolysis and is appropriate as a major control, and indeed the enzyme has attributes which enable it to fulfil this function. The properties of phosphofructokinase from a variety of sources have been reviewed recently (18, 99). Phosphofructokinase from animal tissues is affected by a number of intracellular constituents, e.g. the enzyme from sheep brain and sheep liver is inhibited by ATP, citrate, and Mg^{2+}, and is stimulated by fructose-1,6-P_2, ADP, AMP, 3',5'-AMP, NH_4^+, K^+, and P_i (126). P-Enolpyruvate (1 mM) inhibits rabbit muscle phosphofructokinase by 41% (167). There is now a considerable amount of information on phosphofructokinase from higher plants, and it is clear that although the plant enzyme has some properties similar to those of the enzyme from mammalian sources, there are also pronounced differences. The enzyme from parsley (97), carrots (42), Brussels sprouts (43), corn (56), and pea seeds (78, 79) is inhibited by ATP, ADP, and to a lesser extent by AMP, and is stimulated by P_i. The plant enzyme is inhibited by citrate (42, 43, 56, 78) and this inhibition is highly cooperative (79). A distinctive feature of the plant enzyme is the strong cooperative inhibition given by low concentrations of P-enolpyruvate; with the pea seed enzyme 50% inhibition may be obtained with 1.3 μM P-enolpyruvate, and with 4 μM P-enolpyruvate complete inhibition is approached (79). The presence of P-enolpyruvate induces a number of other changes including transformation of the fructose-6-P saturation curve from hyperbolic to sigmoid (79, 81). By this means the sensitivity of the enzyme to variation in substrate concentration can be greatly increased. Increasing the concentration of fructose-6-P or of Mg^{2+} decreases the inhibition given by P-enolpyruvate whereas low and noninhibitory concentrations of both ATP and citrate enhance the inhibition given by P-enolpyruvate. On the other hand, stimulators such as P_i, K^+, and NH_4^+ decrease the P-enolpyruvate inhibition. P-Enolpyruvate also increases the sensitivity of the enzyme to change in pH (79). In addition to P-enolpyruvate, pea seed phosphofructokinase is strongly inhibited by the glycolytic intermediates 2-P-glycerate, 3-P-glycerate, and 2,3-P_2-glycerate with concentrations of 9 μM, 23 μM, and 40 μM respectively needed for 50% inhibition (80). The enzyme is also inhibited by 6-P-gluconate (50% by 1.2 mM) (80). 2-P-Glycerate, 3-P-glycerate, and 6-P-gluconate all give sigmoid inhibition curves and increasing the concentration of ATP enhances each inhibition. In the presence of P_i and at low ATP concentrations, P-enolpyruvate, 2-P-glycerate, 3-P-glycerate, and 6-P-gluconate stimulate phosphofructokinase activity (79, 80).

It is apparent that the properties and kinetics of phosphofructokinase are complex and that a considerable variety of controls is possible. The activity of plant phosphofructokinase is affected by at least ten metabolites, the concentrations of which may vary in vivo. These may be grouped conveniently into the phosphorylated intermediates of glycolysis, the adenine nucleotides and P_i, and the initial metabolites of the pentose phosphate pathway (6-P-gluconate) and the tricarboxylic acid cycle (citrate). These effects, coupled with the strategic position of the enzyme, point to an important and involved role for phosphofructokinase in the regulation of glycolysis.

TRIOSE PHOSPHATE ISOMERASE   This enzyme is involved in photosynthetic CO_2 fixation in addition to glycolysis and gluconeogenesis. Triose phosphate isomerase from pea seeds is inhibited by low concentrations of sulfate, chloride, P_i, and arsenate (158). Pea leaf chloroplast and cytoplasmic triose phosphate isomerases are competitively inhibited (with glyceraldehyde-3-P as substrate) by ATP, ADP, AMP, fructose-6-P, fructose-1,6-P_2, P-enolpyruvate, ribulose-1,5-P_2, and 3-P-glyc-erate (6). Strong inhibition by 2-P-glycolate was found for both the chloroplast and cytoplasmic triose phosphate isomerases, the K_i values being 15.2 μM and 4.1 μM respectively. Triose phosphate isomerase of pea seeds is subject to strong competitive inhibition by low concentrations of P-enolpyruvate and the phospho-glycerates (J.D. Tomlinson & J.F. Turner, in preparation). The concentrations required for 50% inhibition are: P-enolpyruvate, 0.2 μM; 2-P-glycerate, 10 μM; 2,3-P_2-glycerate, 40 μM; and 3-P-glycerate, 100 μM. The pea seed enzyme ($K_i = 0.1 \mu M$) is more sensitive to inhibition by P-enolpyruvate than the pea leaf chloro-plast ($K_i = 1.3$ mM) or cytoplasmic ($K_i = 0.66$ mM) enzymes.

GLYCERALDEHYDE 3-PHOSPHATE DEHYDROGENASE   Schulman & Gibbs (143) found that the reversible NAD- and NADP-linked D-glyceraldehyde 3-phos-phate dehydrogenases of pea seeds and pea shoots are competitively inhibited by sedoheptulose-7-P and sedoheptulose-1,7-P_2. The heptose phosphates may inhibit glyceraldehyde-3-P formation causing accumulation of 1,3-P_2-glycerate and 3-P-glycerate which may in turn inhibit ribulose 5-phosphate kinase and ribulose 1,5-diphosphate carboxylase and thus provide a mechanism for control of the Calvin cycle (143). Cysteine and other amino acids such as glycine and alanine are reported to activate the NAD- and NADP-glyceraldehyde 3-phosphate dehydrogenases from *Chlorella pyrenoidosa, Scenedesmus acutus*, and spinach leaves (154). Duggleby & Dennis (49) purified the NAD-glyceraldehyde 3-phosphate dehydrogenase of pea seeds. Glyceraldehyde-3-P and, to a lesser extent, P_i were inhibitory at high concen-trations when the concentrations of the other substrates were low. In addition to the two types (NAD- and NADP-linked) of reversible P_i-dependent D-glyceralde-hyde 3-phosphate dehydrogenases there is an NADP-specific D-glyceraldehyde 3-phosphate dehydrogenase which does not require P_i and which is not reversible. Kelly & Gibbs (76) have shown that commercial preparations of glyceraldehyde-3-P are unsuitable for determining the activity of this enzyme because of the presence of L-glyceraldehyde-3-P which is inhibitory. The nonreversible D-glyceraldehyde 3-phosphate dehydrogenase from pea shoots is strongly inhibited by erythrose-4-P, a component of the pentose phosphate pathway (77).

3-PHOSPHOGLYCERATE KINASE   AMP and ADP inhibit rabbit muscle and yeast 3-phosphoglycerate kinases with ATP and 3-P-glycerate as substrates (91, 92). Pea leaf cytoplasmic and chloroplast 3-phosphoglycerate kinases are inhibited by AMP and ATP in the ATP-generating direction and by AMP and ADP in the ATP-utilizing direction (124).

PYRUVATE KINASE   The reaction catalyzed by this enzyme is difficult to reverse and is usually considered to be a nonequilibrium step. It is bypassed in gluconeogenic tissues and microorganisms by pyruvate carboxylase and phosphoenolpyruvate carboxykinase. As the product pyruvate can proceed into several metabolic pathways, pyruvate kinase is at a major metabolic intersection and is a logical control point. The properties of pyruvate kinases from mammalian tissues and yeast have been the subject of reviews (75, 139, 144). In mammalian tissues there are two immunologically distinct forms of pyruvate kinase designated as the muscle and liver types. Muscle pyruvate kinase is inhibited by ATP and is typical of the enzyme from most mammalian nongluconeogenic tissues. The liver enzyme has more regulatory properties and shows sigmoid kinetics with P-enolpyruvate, strong allosteric inhibition by ATP and alanine, and allosteric activation by fructose-1,6-P_2, K^+, NH_4^+, and H^+. Free fatty acids and possibly acetyl CoA inhibit the liver enzyme whereas citrate has no effect (144). Yeast pyruvate kinase shows properties similar to those of the enzyme from liver and is allosterically activated by P-enolpyruvate, ADP, fructose-1,6-P_2, and monovalent cations, and is subject to allosteric inhibition by ATP, citrate, $NADP^+$, and Ca^{2+}. Pyruvate kinase from *Euglena gracilis* shows sigmoid kinetics with P-enolpyruvate, allosteric activation with fructose-1,6-P_2, but no inhibition by ATP (120). It has been known for some years that pyruvate kinase from higher plants requires K^+, NH_4^+, or Rb^+ for activity (103), and there are recent reports on other properties of pyruvate kinases from cotton seeds (47, 48) and from pea seeds, carrots, and a number of other plant tissues (153). The enzyme from higher plants does not show cooperativity with the substrates P-enolpyruvate and ADP and requires Mg^{2+} or Mn^{2+} in addition to a monovalent cation. Ca^{2+} inhibits the pea seed and carrot enzymes and these have K_m values for P-enolpyruvate and ADP ten to a hundredfold lower than the pyruvate kinases from mammalian tissues or yeast (153). Citrate inhibits plant pyruvate kinase and 50% inhibition is given by 1.3 mM, 2 mM, and 9 mM citrate with the enzyme from pea seeds, cotton seeds, and carrots respectively (47, 153). ATP inhibits pyruvate kinase from higher plants; AMP activates the cotton seed enzyme (47) but shows no effect with the enzyme from pea seeds or carrots (153). No plant pyruvate kinase was activated by fructose-1,6-P_2; the tissues examined included germinating castor beans (153) in which there is an active phase of fat breakdown and gluconeogenesis (23, 85). The lack of any effect of fructose-1,6-P_2 in the germinating castor bean is perhaps surprising but may be compatible with the existence of separate metabolic compartments (gluconeogenic and glycolytic) proposed by Kobr & Beevers (83). The two processes could then be regulated independently, and this may eliminate a need for fructose-1,6-P_2 activation of pyruvate kinase. Although work so far has not shown allosteric kinetics for plant pyruvate kinases, the enzyme is inhibited by ATP and relatively low concentrations of citrate, a component of the tricarboxylic acid cycle and the glyoxylate cycle. These phenomena are of likely metabolic significance, especially as variations in ATP and citrate will be reinforced by changes in the opposite direction of the substrate ADP. The ADP content of some plant tissues and the K_m values for plant pyruvate kinase (153) suggest that the ADP concentration may be an important determinant of pyruvate kinase activity in vivo. Since pyruvate kinase has a requirement for monovalent cations and Mg^{2+} and is inhibited by Ca^{2+} (153), changes in the concentrations of metallic ions, perhaps through mitochondrial action, may also play a part in the regulation of this enzyme. Meli & Bygrave (100) have shown that by altering the ratio of $[Mg^{2+}]/[Ca^{2+}]$ in the immediate environment of pyruvate kinase, rat liver mitochondria can dictate the final rate of the enzyme reaction. The mitochondria can accumulate Ca^{2+} and the addition of uncouplers of oxidative phosphorylation such as carbonyl cyanide m-chlorophenylhydrazone causes a rapid release of Ca^{2+}. By this and similar means, including addition of Ca^{2+} and EDTA, the pyruvate kinase can be switched on and off.

LACTATE DEHYDROGENASE   Potato tubers produce lactic acid under anaerobic conditions (11). Davies & Davies (32) have purified lactate dehydrogenase from potatoes and with pyruvate as substrate found slight inhibition by ATP and NAD^+ at alkaline pH values. At slightly acid pH (6.1–6.5) there is strong cooperative inhibition of the enzyme by ATP and the product NAD^+. These inhibitions may well be of metabolic significance in controlling the diversion of pyruvate into lactate. The lowered NADH level under aerobic conditions would also militate against lactate formation from pyruvate.

Regulation of Glycolysis in vivo

A number of studies with intact plant tissues have been reported which are relevant to the regulation of glycolysis. Barker, Khan & Solomos (13) followed changes in a series of glycolytic intermediates in pea seeds transferred from air to nitrogen and from nitrogen to air. A substantial fall in P-enolpyruvate, 2-P-glycerate, and 3-P-glycerate and an increase in fructose-1,6-P_2 occurred when the peas were transferred from air to nitrogen. There was also an increase in ADP and a fall in ATP. The changes in P-enolpyruvate, 2-P-glycerate, and 3-P-glycerate were strikingly similar to each other [as they were in potatoes and apples (12)] and tended to precede the change in fructose-1,6-P_2. It was concluded that pyruvate kinase is of primary importance in controlling the rate of glycolysis, and the probable significance of the change in the level of ADP on the pyruvate kinase reaction was emphasized (13). It is clear from this work that phosphofructokinase is also a control point.

Adams & Rowan (4) followed changes in the respiration of aging carrot slices and identified regulatory reactions by applying the crossover theorem of Chance (25). The first stage of induced wound respiration involved an increase in ADP and an activation of the pyruvate kinase reaction. A response to an increase in ADP concentration is kinetically feasible as the endogenous concentration of ADP in carrot slices is approximately 25 μM (3) and the K_m for ADP of carrot pyruvate kinase is 55 μM (153). The respiration of carrot slices was stimulated 20 to 85% by ADP, and Adams (2, 3) proposed that this was due to the relief of ADP limitation of the pyruvate kinase reaction. This was supported by the observation that added pyruvate produced a similar increase in the respiration rate (2). Dinitrophenol also increased the respiration rate of carrot slices. Chalmers & Rowan (24) studied changes in a number of metabolites in ripening tomato fruits during the climacteric and used Chance's crossover theorem to identify phosphofructokinase as the regulatory site. It was suggested that an increased concentration of P_i in the cytoplasm stimulated phosphofructokinase which contributed to the climacteric rise in respiration.

Kobr & Beevers (83), in an elegant investigation of the glycolytic pathway in germinating castor bean endosperms, used two approaches to determine control points. In the first approach a number of glycolytic intermediates were measured and results indicated that the phosphofructokinase and pyruvate kinase reactions were nonequilibrium. In the second approach, crossover points in glycolytic intermediates were determined in endosperms following transfer to anaerobic conditions, when gluconeogenesis is prevented and a glycolytic flux in the direction of pyruvate is induced. There were dramatic changes soon after the endosperms were placed in nitrogen. ATP dropped to one-third of its aerobic value within 15 min. Glucose-6-P and fructose-6-P fell immediately while dihydroxyacetone-P and fructose-1,6-P_2 rose considerably and then returned to new levels approximately twice the aerobic control levels. Marked changes occurred in the amounts of P-enolpyruvate and pyruvate. The P-enolpyruvate level dropped sharply for 30 min and remained at approximately one-quarter of the aerobic value. The pyruvate content doubled in 15 min, and this coincided with the time required to establish linear rates of ethanol and lactate formation. Thus the anaerobic evidence also pointed to the fructose-6-P —fructose-1,6-P_2 and P-enolpyruvate—pyruvate reactions as being the regulatory steps. Davies (31) followed changes in glycolytic intermediates in carrot discs shortly after transfer from air to nitrogen. With 3 min of anaerobiosis there was a 350% increase in fructose-1,6-P_2 and a 60% increase in pyruvate. After 6 min the fructose-1,6-P_2 was 500% above the control, pyruvate fell below the control level, and there were decreases in P-enolpyruvate and the phosphoglycerates. This work points to phosphofructokinase as a controlling enzyme and suggests that pyruvate kinase is also involved.

The control of aerobic glycolysis by ATP and P_i in cell-free extracts of germinating pea seeds was studied by Givan (60). At high Mg^{2+} concentrations ATP accelerated glycolysis, whereas at lower Mg^{2+} concentrations ATP severely inhibited the process. P_i stimulated glycolysis under all conditions studied and the inhibitory effect of ATP was relieved by P_i. Analysis of glycolytic intermediates showed that low ratios of fructose-6-P to fructose-1,6-P_2 were associated with a high rate of glycolysis, suggesting a regulatory role for phosphofructokinase.

Mechanism of the Regulation of Glycolysis

The in vivo evidence that phosphofructokinase and pyruvate kinase are the enzyme steps regulating the rate of glycolysis is compatible with the established properties of these enzymes. Both enzymes catalyze nonequilibrium reactions, have strategic locations in metabolism which are appropriate as control points, and have attributes which are consistent with the control function required. Phosphofructokinase has allosteric properties and the activity is affected in a complex manner by a large number of metabolites. Plant pyruvate kinase, although allosteric properties have not been revealed, is inhibited by ATP and citrate, will be affected in vivo by change in the concentration of ADP, and has the potential for regulation by the divalent cations Ca^{2+} and Mg^{2+} and by monovalent cations, especially K^+.

From the data available a scheme (Figure 1) is proposed showing the main regulatory points in glycolysis and the possible effectors involved. The relative importance of the various controlling factors (metabolites and cations) will depend on the concentrations at the site of enzyme action. These may well vary from tissue to tissue and could determine the inhibition sequence. Consideration of this scheme will be from the aspect of inhibiting glycolysis. Inhibition of the pathway will occur, for example, in the transfer of a tissue from nitrogen to air; this transfer will result in an increase in ATP and citrate and a decrease in ADP and P_i. It should be pointed

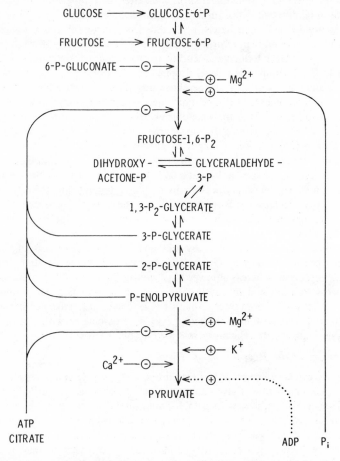

Figure 1   A scheme for the regulation of glycolysis. ⊕, Positive effector; ⊖, negative effector. The participation of ADP as substrate is indicated by the dotted line.

out that in aerobic tissues the enzymes will generally be in an inhibited state and that these inhibitions will be released, to a greater or lesser extent, in appropriate circumstances. In this scheme ATP and citrate inhibit pyruvate kinase. Inhibition of phosphofructokinase may be exerted by ATP and citrate and there may also be effective inhibition of phosphofructokinase through a decrease in the stimulator P_i. At physiological concentrations the decrease in rate of pyruvate kinase produced by the fall in the substrate ADP (and perhaps by cation changes) may, on occasions, be more significant than the inhibition by citrate and ATP. Lowered pyruvate kinase activity results in an accumulation of P-enolpyruvate and, as the reactions catalyzed by enolase and phosphoglyceromutase are freely reversible, this produces an increase in the levels of 2-P-glycerate and 3-P-glycerate. It has been shown that these three compounds respond similarly in several tissues. Control of pyruvate kinase in itself cannot regulate the earlier stages of glycolysis leading to fructose-1,6-P_2. Regulation of phosphofructokinase is, however, achieved through inhibition by P-enolpyruvate, 2-P-glycerate, and 3-P-glycerate. This can be considered as a sequential effect (as opposed to primary effects by compounds such as ATP, citrate, and P_i) and may be viewed as a mechanism for "inhibition-transfer," viz the inhibition of phosphofructokinase through the inhibition of pyruvate kinase. With phosphofructokinase, a major role at physiological concentrations for ATP, citrate, and P_i may be to affect the inhibition produced by P-enolpyruvate. This would involve the synergistic enhancement of P-enolpyruvate inhibition of the enzyme by increased concentrations of ATP and citrate and an increase in the inhibition due to a lowered P_i concentration. In a number of the investigations using intact tissues described in the previous section, the transfer was from aerobic to anaerobic conditions, i.e. in the opposite direction from the change discussed here. In such a transition from inhibition to de-inhibition it may well be that in some tissues the initial effect is a decrease in the synergistic inhibition of phosphofructokinase by ATP and P-enolpyruvate and by citrate and P-enolpyruvate and an increase in P_i relief of P-enolpyruvate inhibition. It will be recalled also that in the presence of P_i and at low ATP concentrations, phosphofructokinase is stimulated by P-enolpyruvate. The metabolic significance of observations with other enzymes of glycolysis is speculative. The very strong inhibition of triose phosphate isomerase by P-enolpyruvate may, for example, affect the carbon flow through the aldolase-triose phosphate isomerase system.

The Pasteur Effect

The Pasteur effect, which may be defined as the aerobic inhibition of glycolysis (or fermentation), is of wide occurrence in plant tissues (166). With the more recent knowledge of the enzymes of glycolysis and the results of tissue analyses, the Pasteur effect is no longer a mystery and may be explained in terms of Figure 1 as discussed in the preceding section. Phosphofructokinase is the key enzyme. Davies (31) has proposed that a fall in pH in anaerobiosis leading to an increase in phosphofructokinase activity (79) may be a factor.

REGULATION OF GLUCONEOGENESIS

Hexoses and hexose phosphates can be formed from pyruvate by gluconeogenesis, a pathway which is largely a reversal of glycolysis. In glycolysis there are three reactions which are difficult to reverse:

$$\text{hexose} + \text{ATP} \xrightarrow{\text{hexokinase}} \text{hexose-6-P} + \text{ADP} \qquad 1.$$

$$\text{fructose-6-P} + \text{ATP} \xrightarrow{\text{phosphofructokinase}} \text{fructose-1,6-P}_2 + \text{ADP} \qquad 2.$$

$$\text{P-enolpyruvate} + \text{ADP} \xrightarrow{\text{pyruvate kinase}} \text{pyruvate} + \text{ATP} \qquad 3.$$

In gluconeogenesis the reversal of reaction 1 is brought about by phosphatase and the reversal of reaction 2 by fructose 1,6-diphosphatase which occurs in both photosynthetic and nonphotosynthetic tissues. Apart from photosynthesis (which involves a reversal of part of the glycolytic sequence) the best known gluconeogenic process in plant tissues is associated with the conversion of fatty acids to carbohydrate in the germinating castor bean; in this tissue oxaloacetate from the glyoxylate cycle is converted to P-enolpyruvate by phosphoenolpyruvate carboxykinase. The metabolism of the germinating castor bean has been the subject of a comprehensive series of investigations by Beevers and his co-workers (14, 15, 84, 85).

FRUCTOSE 1,6-DIPHOSPHATASE   Two fructose 1,6-diphosphatases have been found in the castor bean endosperm (141). One of these enzymes is present in the ungerminated bean endosperm, has an optimum pH of 7.5, and is competitively inhibited by AMP. A second fructose 1,6-diphosphatase is synthesized in the endosperm during 3-day germination, has an optimum pH of 6.7, and is not inhibited by AMP. A third fructose 1,6-diphosphatase is present in the leaves and cotyledons of castor beans, has an optimum pH of 8.8, and is not inhibited by AMP. Since the fructose 1,6-diphosphatase formed during germination is not sensitive to AMP, it is unlikely that AMP regulation of this enzyme is an important factor in regulating gluconeogenesis in the castor bean.

Regulation of Gluconeogenesis in vivo

Castor bean and other seedlings high in fat contain proplastids possessing a second set of enzymes of the glycolysis sequence (82). Kobr & Beevers (83) suggested that gluconeogenesis occurs in this inclusion and that glycolysis occurs elsewhere in the cell. This separation of the two processes could account for the observation that the rate of gluconeogenesis is ten times the rate of glycolysis even under conditions which favor maximum glycolysis. Gluconeogenesis and glycolysis could occur simultaneously in the same cell and be subject to independent regulation (83). Thomas & ap Rees (150) studied gluconeogenesis in the cotyledons of germinating marrow (*Cucurbita pepo*) and also favor separate compartments in the cell for the two processes (151). Sucrose and stachyose were the main products of gluconeogenesis, and the formation of these sugars was accompanied by an increase in fructose

1,6-diphosphatase activity. It was suggested that a change in the amount of this enzyme may contribute to the control of gluconeogenesis (150).

REGULATION OF THE PENTOSE PHOSPHATE PATHWAY

The pentose phosphate pathway, often referred to as the hexose monophosphate shunt, can effect the complete breakdown of glucose-6-P to CO_2. It is considered that the primary function of this sequence is to supply NADPH in the cytoplasm for use in biosynthetic processes. Another important function in many tissues is the provision of pentose phosphates (particularly ribose-5-P) for nucleotide synthesis. As the requirements of the cell for NADPH and pentose phosphates will vary, it is probable that mechanisms exist to regulate flux through the pathway. Glucose-6-P, the initial substrate of the pentose phosphate pathway, is at a branch point in metabolism. It may also enter the glycolytic sequence, be a substrate for oligo- or polysaccharide formation (through transformation to glucose-1-P or fructose-6-P), or be hydrolyzed to glucose. The logical point of control of the pentose phosphate pathway would be at the initial step, i.e. the reaction in which glucose-6-P is utilized, and there is considerable enzymic evidence consistent with this.

GLUCOSE 6-PHOSPHATE DEHYDROGENASE   This enzyme catalyzes reaction 4 leading to the production of 6-P-glucono-γ-lactone. The lactone is then hydrolyzed by 6-phosphoglucono-lactonase (reaction 5) although the nonenzymic hydrolysis of 6-P-glucono-γ-lactone is very rapid.

$$\text{glucose-6-P} + \text{NADP}^+ \rightleftharpoons \text{6-P-glucono-}\gamma\text{-lactone} + \text{NADPH} \qquad 4.$$

$$\text{6-P-glucono-}\gamma\text{-lactone} + H_2O \longrightarrow \text{6-P-gluconate} \qquad 5.$$

The overall reaction is irreversible. Determination of the overall mass-action ratio of the two enzymes has shown that it is a nonequilibrium reaction in Krebs ascites cells (64) and in rat liver cells (61).

It has been known since 1935, many years before the pentose phosphate pathway was formulated, that yeast glucose 6-phosphate dehydrogenase is inhibited by NADPH (113). The regulatory properties of glucose 6-phosphate dehydrogenases from microorganisms and mammalian cells have been the subject of a recent review (19). The enzyme from brewers' yeast (*Saccharomyces carlsbergensis*) is competitively inhibited by NADPH (with respect to NADP$^+$) and by ATP (with respect to glucose-6-P). P_i is also competitive with glucose-6-P. The rat liver enzyme is competitively inhibited by NADPH. Coarse control is achieved through variation in the amount of glucose 6-phosphate dehydrogenase. Starvation leads to a decrease in the enzyme in rat liver, but on refeeding the starved rats there is an "overshoot" of glucose 6-phosphate dehydrogenase activity. With a balanced diet a threefold increase compared with unstarved rats is observed, and with a high carbohydrate diet the increase may be as high as ten to twelvefold. A marked increase in lipid synthesis coincides with the increase in enzyme activity. Osmond & ap Rees (121) grew *Candida utilis* on various media and observed changes in amounts of enzymes involved in the pentose phosphate pathway. It was suggested that the capacity of the pathway in this yeast may be controlled by variation in the amount of glucose 6-phosphate dehydrogenase.

There are distinct isozymes of glucose 6-phosphate dehydrogenase in the chloroplasts and cytosol of spinach leaves, and this finding led to the suggestion that plants contain a pentose phosphate pathway in the chloroplasts as well as in the cytosol (142). Muto & Uritani (109) purified the glucose 6-phosphate dehydrogenase from sweet potato 290-fold and found that the enzyme exists in at least two forms with different sedimentation coefficients. In the presence of excess $NADP^+$, and depending on the glucose-6-P concentration, the aggregate could dissociate to a low molecular weight species (110). There were indications of negative cooperativity between glucose-6-P and the enzyme (109, 111). Glucose 6-phosphate dehydrogenase from sweet potato is competitively inhibited by NADPH and ATP (112). With saturating levels of glucose-6-P, NADPH was a competitive inhibitor but ATP did not inhibit. The K_i values for NADPH were 10.0 μM at pH 7.0 (the physiological pH) and 11.0 μM at pH 8.0 (the optimum pH). When $NADP^+$ was saturating and glucose-6-P was varied at pH 7.0, NADPH was a noncompetitive inhibitor and ATP a competitive inhibitor. At pH 8.0 the kinetics were more complex. A preliminary report indicates that oleic and caprylic acids and β-carotene may competitively inhibit glucose 6-phosphate dehydrogenase of the ripening mango (57).

Eggleston & Krebs (50) have investigated mechanisms for the fine control of glucose 6-phosphate dehydrogenase in rat liver. The inhibition by NADPH was competitively reversed by $NADP^+$ and was governed by the ratio [NADPH]/[NADP$^+$]: the inhibition approached 100% when the value of this ratio was 9. The ratio of [NADPH] to [NADP$^+$] in rat liver cytoplasm is approximately 100 so that glucose 6-phosphate dehydrogenase would be almost completely inhibited. This suggested that regulation of glucose 6-phosphate dehydrogenase is achieved by de-inhibition. Of more than 100 cellular constituents tested in an attempt to relieve the NADPH inhibition only oxidized glutathione was effective at physiological concentrations, and it was established that this effect was not due to the removal of NADPH by the glutathione reductase reaction. The activity of glucose 6-phosphate dehydrogenase, and hence regulation of the pentose phosphate pathway, may therefore be controlled by the concentration of oxidized glutathione (50).

REGULATION OF THE TRICARBOXYLIC ACID CYCLE

In aerobic tissues most of the pyruvate produced in glycolysis enters the tricarboxylic acid cycle. This sequence is closely related to the glyoxylate cycle and is a source of intermediates for the synthesis of other cellular components such as amino acids. A number of the enzymes participating in and associated with the tricarboxylic acid cycle are affected by metabolites and may play a part in regulation.

Enzymes of the Tricarboxylic Acid Cycle

PYRUVATE DEHYDROGENASE COMPLEX This is a complex of several enzymes, one of which is pyruvate dehydrogenase (EC 1.2.4.1). The overall reaction catalyzed by the pyruvate dehydrogenase complex is

$$\text{pyruvate} + \text{CoA} + \text{NAD}^+ \longrightarrow \text{acetyl-CoA} + \text{NADH} + \text{H}^+ + \text{CO}_2 \qquad 6.$$

This reaction determines whether pyruvate produced from carbohydrate is converted to acetyl-CoA or is diverted to lactate or ethanol formation. It is therefore a possible control point. Crompton & Laties (30) found that NADH is a strong competitive inhibitor (with NAD^+) for the pyruvate dehydrogenase complex from potato tubers and suggested that the mole fraction $[\text{NAD}^+]/[\text{NAD}^+]+[\text{NADH}]$ may be a fine control mechanism.

CITRATE SYNTHASE This enzyme catalyzes the entry of acetyl-CoA into the tricarboxylic acid cycle and the glyoxylate cycle. It is found in mitochondria and in glyoxysomes (15, 20). As citrate synthase is the first step in the tricarboxylic acid cycle, it is a logical point for control by ATP, the end product. Citrate synthases from mammalian tissues, yeast, *Escherichia coli* and lemon fruit are inhibited by ATP (125, 147), and the inhibition is competitive with respect to CoA. Axelrod & Beevers (10) found that in germinating castor bean seedlings citrate synthase from the mitochondria is inhibited by ATP while citrate synthase from glyoxysomes is not inhibited. The function of the glyoxysomes is to convert acetyl-CoA to succinate while that of the mitochondria is, inter alia, to produce ATP. These observations are consistent with the hypothesis that inhibition by ATP of citrate synthase of mitochondria is of regulatory significance in the control of the tricarboxylic acid cycle (10). Citrate synthase from the mitochondria of wheat shoots, cauliflower buds, and bean hypocotyls is also inhibited by ATP (62).

ISOCITRATE DEHYDROGENASE The NAD-specific isocitrate dehydrogenase of higher plants shows a sigmoid response with increasing isocitrate concentration and is inhibited by NADH (131). Cox & Davies (28) studied activation by citrate and the effect of pH on the enzyme from pea mitochondria (29). With increase in pH (especially above pH 8.0) the affinity of the enzyme for ligands is diminished and isocitrate dehydrogenase could then be of regulatory significance. The authors conclude that assessment of the controlling effects of this enzyme must await more information on intramitochondrial pH. Duggleby & Dennis (46) found that the activity of the isocitrate dehydrogenase of pea stems is controlled in a complex manner by NADH, citrate, and Mg^{2+} and they consider that the enzyme may have a regulatory role.

α-KETOGLUTARATE DEHYDROGENASE The α-ketoglutarate dehydrogenase-lipoyl transsuccinylase complex from cauliflower florets is activated by AMP (169). With 1 mM AMP the rate is increased sixfold.

SUCCINATE DEHYDROGENASE   A considerable part of the succinate dehydrogenase in mung bean and cauliflower mitochondria as isolated is in a deactivated form which can be activated by incubation of the mitochondria with succinate (145). ADP, ATP, IDP, ITP, reduced coenzyme Q_{10}, lowered pH (6.1–6.5), and anions also activate plant mitochondrial succinate dehydrogenases (119). NADH may also activate but the extent of activation varies with the type of preparation.

MALATE DEHYDROGENASE   The possible role of malate dehydrogenases in peroxisomes and glyoxysomes has been discussed recently (152). In more recent studies, Yang & Scandalios (170) found a number of isozymes of malate dehydrogenase in maize; in etiolated seedlings there were two soluble forms, five mitochondrial forms, and two glyoxysomal forms. These were subject to varying degrees of inhibition by oxaloacetate and NAD^+, and it was suggested that these inhibitions may play a role in regulation of the tricarboxylic acid cycle. The malate dehydrogenase of orange juice vesicles is inhibited by NADH (21).

Regulation of the Tricarboxylic Acid Cycle in vivo

There is little real evidence on the in vivo regulation of the tricarboxylic acid cycle in any tissue, mammalian or plant. Control points are difficult to establish. On the basis of the enzymic evidence available it seems that NADH will affect the pyruvate dehydrogenase complex and so determine acetyl-CoA production. The entry of acetyl-CoA into the tricarboxylic acid cycle, and hence activity of the cycle itself, will be governed by the degree of ATP inhibition of citrate synthase. Douce & Bonner (45) reported that oxaloacetate penetrates the mitochondria of potato tubers and mung bean hypocotyls and inhibits all tricarboxylic acid cycle oxidations. It was postulated that this inhibition of the NAD-linked dehydrogenases could be a control mechanism in plant cells. Eloff (51) obtained evidence for the operation of the tricarboxylic acid cycle in the South African plant Gifblaar (*Dichapetalum cymosum*). This plant is remarkable in that it contains fluoroacetate which is extremely toxic to animals due to in vivo conversion to fluorocitrate, a powerful competitive inhibitor of aconitase. The aconitase of plants is also inhibited by fluorocitrate.

REGULATION OF THE BIOSYNTHESIS AND DEGRADATION OF OLIGOSACCHARIDES AND POLYSACCHARIDES

This section will deal mainly with the oligosaccharide sucrose and the polysaccharide starch. Sucrose is quantitatively the most important sugar present in plants and is significant both in transport and in storage. Starch is a reserve polysaccharide which can be readily metabolized.

Sucrose Metabolism

The initial reaction in sucrose biosynthesis is the formation of UDP-glucose by UDP-glucose pyrophosphorylase (reaction 7). Sucrose may be synthesized directly from UDP-glucose and fructose by sucrose synthase (reaction 8) or indirectly through sucrose phosphate synthase catalyzing a reaction between UDP-glucose and fructose-6-P to yield sucrose-P (reaction 9). The sucrose-P is then hydrolyzed by phosphatase to give free sucrose (reaction 10).

$$\text{glucose-1-P} + \text{UTP} \rightleftharpoons \text{UDP-glucose} + \text{PP}_i \qquad\qquad 7.$$

$$\text{UDP-glucose} + \text{fructose} \rightleftharpoons \text{sucrose} + \text{UDP} \qquad\qquad 8.$$

$$\text{UDP-glucose} + \text{fructose-6-P} \rightleftharpoons \text{sucrose-P} + \text{UDP} \qquad\qquad 9.$$

$$\text{sucrose-P} + \text{H}_2\text{O} \longrightarrow \text{sucrose} + \text{P}_i \qquad\qquad 10.$$

As reaction 10 is essentially irreversible, degradation of sucrose will be by reversal of sucrose synthase. Sucrose may also be hydrolyzed directly by invertase.

UDP-GLUCOSE PYROPHOSPHORYLASE This enzyme is widely distributed in plant tissues (160) and has been purified from the shoots of etiolated *Sorghum vulgare* seedlings (65). There are indications that in some tissues coarse control of UDP-glucose pyrophosphorylase activity is achieved by variation in the amount of enzyme present (164, 165). In developing pea seeds a rapid increase in UDP-glucose pyrophosphorylase coincides with a period of intense starch synthesis (164). Starch formation in the developing pea seed is an example of a sucrose-starch conversion, and the increased activity of UDP-glucose pyrophosphorylase may be necessary for UDP-glucose breakdown rather than synthesis. An analogous coarse control may operate in the developing wheat grain where an increase in UDP-glucose pyrophosphorylase appears to be closely related to a decrease in sucrose and an increase in starch (165). Human erythrocyte and dog cardiac muscle UDP-glucose pyrophosphorylases are subject to strong product inhibition by UDP-glucose, with K_i values of 15 μM and 23 μM respectively (157). Mung bean seedling UDP-glucose pyrophosphorylase is less sensitive to UDP-glucose inhibition ($K_i = 0.16$ mM) (157). The enzymes from *Lilium longiflorum* pollen ($K_i = 0.13$ mM) and from the shoots of *Sorghum vulgare* seedlings ($K_i = 0.050$ mM) are also less sensitive to UDP-glucose inhibition than the mammalian enzymes (65, 71). A number of other UDP-sugars and sugar acids inhibit *Lilium* UDP-glucose pyrophosphorylase and these inhibitions are additive when the concentrations of inhibitors and substrates are low (71). It was suggested by Hopper & Dickinson (71) that UDP-sugars and sugar acids, the last soluble precursors of cell wall polysaccharide synthesis, may regulate UDP-glucose pyrophosphorylase activity and so control the flow of hexose monophosphates into pathways leading to cell wall polysaccharide formation. UTP inhibits both mammalian and plant UDP-glucose pyrophosphorylases (65, 157).

SUCROSE SYNTHASE Since evidence for the existence of this enzyme was first obtained (94, 162, 163), it has been found in a large number of plant tissues. Both UDP-glucose and ADP-glucose are effective glucose donors (37, 146), but UDP-glucose inhibits the synthesis of sucrose from other nucleoside diphosphate sugars (63). The ready reversibility of sucrose synthase led to suggestions that it may have an important role in sucrose breakdown (37, 58, 159), and there is now considerable evidence that UDP-glucose rather than ADP-glucose is the physiological product of the reverse reaction. De Fekete & Cardini (37) found that sucrose synthase from sweet corn had a much lower affinity for ADP than UDP and also that the reaction of sucrose with ADP was strongly inhibited by UDP. These authors concluded that the production of ADP-glucose in vivo by a reaction of ADP with sucrose was improbable. Kinetic parameters of mung bean sucrose synthase provided further evidence that UDP-glucose would be the major nucleoside diphosphate-glucose produced in vivo by the reverse reaction (40). Distribution studies have supported the significance of sucrose synthase as a degradative enzyme. In mung bean seedlings sucrose synthase activity was high only in nonphotosynthetic tissues of the plant (41). This suggested that sucrose synthase may function mainly in the metabolism of sucrose transported from the leaves to the nonphotosynthetic tissues. Changes in enzyme activity in maturing potatoes (133), developing maize endosperm (156), and developing sugar beet roots (127) were also consistent with the degradative role of sucrose synthase. Murata (106) found hyperbolic saturation curves for the substrates fructose and UDP-glucose with the sucrose synthase from sweet potatoes. On the other hand, in the sucrose cleaving reaction the substrate saturation curves for sucrose and UDP were sigmoid, which would render this reaction more sensitive to variation in substrate concentration.

SUCROSE PHOSPHATE SYNTHASE   The reaction of UDP-glucose with fructose-6-P (95) followed by phosphatase action is probably the main pathway of sucrose synthesis, particularly in leaves (131). Formation of sucrose by sucrose phosphate synthase involves an irreversible phosphatase step and explains the accumulation of large amounts of sucrose in the presence of relatively low concentrations of the logical precursors. Sigmoid saturation curves for the substrates UDP-glucose and fructose-6-P were reported for wheat germ sucrose phosphate synthase by Preiss & Greenberg (130) and for fructose-6-P with the enzyme from several sources by Murata (107). On the other hand, hyperbolic substrate curves have been obtained by some other workers (101, 116, 146). Nomura & Akazawa (116) have noted that enzyme assays yielding Michaelis-Menten kinetics were performed at pH 7.4–7.5, whereas pH 6.0–6.5 was used when deviations from these kinetics were observed. De Fekete (34) purified the sucrose phosphate synthase of broad bean cotyledons; after freezing and thawing an activator was separated from the enzyme by centrifuging. In the absence of activator the substrate saturation curves for UDP-glucose and fructose-6-P were sigmoid but the addition of activator produced hyperbolic curves. The labile activator could be replaced by citrate, and a plot of citrate concentration against enzyme activity gave a pronounced sigmoid curve indicating cooperative interaction. The activator-bound sucrose phosphate synthase (the enzyme which was active in the absence of added citrate) was inhibited by UTP, UDP, ATP, ADP, and P_i; all the inhibition curves (except that with ADP) were sigmoid. P-Enolpyruvate and 3-P-glycerate also inhibited the activator-bound enzyme. The activator-devoid enzyme was activated about threefold by ATP and 3-P-glycerate although the maximum extent of activation was less than with citrate. De Fekete (34–36) has made the interesting suggestion that citrate may control the utilization of fructose-6-

P by activating sucrose phosphate synthase and inhibiting phosphofructokinase (79).

Starch Metabolism

For some years after the discovery of starch phosphorylase by Hanes (66, 67) it was accepted that this enzyme was responsible for the synthesis of starch in plants. Starch phosphorylase utilizes glucose-1-P as substrate and transfers a glucosyl residue to a primer molecule (an α-1,4-glucan), according to reaction 11.

$$\text{glucose-1-P} + \alpha\text{-1,4-glucan} \rightleftharpoons \alpha\text{-1,4-glucosylglucan} + P_i \qquad 11.$$

In a number of plant tissues starch is formed when the ratio of P_i to glucose-1-P is so high that it would be kinetically impossible for starch phosphorylase to bring about any net production of starch (52). Although compartmentation may mean that the results of gross analyses of tissues bear little relation to the actual concentration of reactants at the site of enzyme action, observations such as this did stimulate the search for another starch-forming system. Subsequently the biosynthesis of glycogen from UDP-glucose by liver preparations was demonstrated by Leloir's group, and later the synthesis of starch by a similar reaction (starch synthase - reaction 12) was reported with starch granule preparations from plant tissues (38, 96).

$$\text{UDP-glucose} + \alpha\text{-1,4-glucan} \rightleftharpoons \text{UDP} + \alpha\text{-1,4-glucosylglucan} \qquad 12.$$

Further work showed that glucose was transferred at a faster rate from ADP-glucose (reaction 13) than from UDP-glucose (137).

$$\text{ADP-glucose} + \alpha\text{-1,4-glucan} \rightleftharpoons \text{ADP} + \alpha\text{-1,4-glucosylglucan} \qquad 13.$$

The formation of ADP-glucose (reaction 14) is catalyzed by ADP-glucose pyrophosphorylase, and the presence of this enzyme has been demonstrated in a considerable number of plant tissues (129).

$$\text{glucose-1-P} + \text{ATP} \rightleftharpoons \text{ADP-glucose} + PP_i \qquad 14.$$

There are thus three systems capable of forming starch viz starch phosphorylase, starch synthase using UDP-glucose and starch synthase using ADP-glucose. Evidence concerning the relative importance of each of these pathways of starch biosynthesis is to a certain extent circumstantial, but there is general support for the major contribution being through ADP-glucose-starch synthase. The role of starch phosphorylase in starch biosynthesis is not resolved, and the participation of this enzyme has been suggested when the level of ADP-glucose (or UDP-glucose)-starch synthase extracted from tissues is insufficient to account for the observed rate of starch formation, or in the initial steps in the formation of the starch granule (35, 39, 55). In some tissues there is enough ADP-glucose pyrophosphorylase and ADP-glucose-starch synthase to account for the observed starch formation (122). Soluble ADP-glucose-starch synthase has been separated into four forms, and one of these is able to effect the unprimed synthesis of starch (54, 123).

Phosphorylase is believed to play a major role in starch breakdown in vivo. In germinating seeds the main route of starch breakdown is by amylases (108, 117).

ADP-GLUCOSE PYROPHOSPHORYLASE   The work of Preiss' laboratory on this enzyme was discussed in an earlier volume of this series (131) and has been the subject of a recent review (129). ADP-glucose pyrophosphorylases from plants are activated by 3-P-glycerate and by other glycolytic intermediates such as P-enolpyruvate, fructose-1,6-P_2, and fructose-6-P. The spinach leaf enzyme may be stimulated up to eightyfold by 3-P-glycerate, the stimulation is dependent on pH, and 3-P-glycerate decreases the K_m values for all substrates of the ADP-glucose pyrophosphorylase reaction. P_i is a strong inhibitor of spinach leaf ADP-glucose pyrophosphorylase; 3-P-glycerate reverses this effect and increases the cooperativity of the P_i inhibition. In general the ADP-glucose pyrophosphorylases from non-photosynthetic tissues are not activated to the same extent as those from leaves, e.g. the enzyme from maize endosperm is activated 1.5 to 4-fold by 3-P-glycerate and is not activated by P-enolpyruvate (44). It is thought that a decrease in P_i and an increase in the levels of glycolytic intermediates and ATP in the chloroplast in the light bring about increased ADP-glucose pyrophosphorylase activity and so lead to an increased rate of starch synthesis (129).

ADP-glucose pyrophosphorylase activity is much lower than the activity of UDP-glucose pyrophosphorylase in bean and rice leaves (118), maize endosperm (168), developing pea seeds (164), developing wheat grains (165), and developing maize kernels (122). The relatively low enzyme activity increases the likelihood of ADP-glucose pyrophosphorylase functioning as a regulatory enzyme. The ADP-glucose pyrophosphorylase content of tissues also appears to be associated with active starch synthesis. Mature wheat grains show only very slight ADP-glucose pyrophosphorylase activity, whereas highly active preparations can be obtained from immature wheat in which starch is being synthesized (105). In the developing pea seed the ADP-glucose pyrophosphorylase level is initially very low but increases very rapidly to a maximum which coincides with the maximum rate of starch synthesis (164). In the developing wheat grain there is an almost sixfold increase in ADP-glucose pyrophosphorylase activity over a 5-day period, and this also coincides with the phase of rapid starch synthesis (165). In both these tissues ADP-glucose pyrophosphorylase falls markedly as the rate of starch synthesis declines. With ADP-glucose pyrophosphorylase in starch-forming tissues there is thus the possibility of coarse control through variation in the amount of enzyme as well as fine control through activators such as 3-P-glycerate and inhibitors such as P_i. These factors indicate an important regulatory role for this enzyme.

STARCH PHOSPHORYLASE   Tsai & Nelson (155) have separated the phosphorylases from maize endosperm into three forms. One form, which appears only at the stage of rapid starch biosynthesis and which is not present during germination, is inhibited by ATP, ADP, GTP, and GDP and to a lesser extent by UTP, UDP, CTP, and CDP. The inhibitions are pH-dependent. ADP-glucose ($K_i = 0.15$ mM) and, to a lesser extent, UDP-glucose inhibit maize endosperm phosphorylases and are competitive with glucose-1-P and P_i (22). Sweet corn phosphorylase is inhibited by ADP-glucose and 2,3-P_2-glycerate (30% by 5mM) but 3-P-glycerate has no effect (93).

Sucrose-Starch Interconversions

Sucrose-starch interconversions are widespread in plants. In developing seeds there is a period of starch formation which involves a conversion of sugars (principally sucrose) to starch. In ripening fruits such as banana or apple, there is a conversion of starch to sugars. Other polysaccharides as well as starch are also subject to change during these processes. In potato tubers transferred from 10–18° to 0–2° there is a decrease in starch and an increase in sugars. When tubers are taken from the lower to the higher temperature there is an increase in starch while sugars decline. Isherwood (72) has recently shown that the sum of starch + sugar + CO_2 (respiration) does not change in potato tubers throughout the various changes in temperature.

More information, particularly at the enzyme level, is available for the sucrose-starch conversion occurring in developing seeds. A mechanism for this transformation is shown in Figure 2; this is based on changes in the developing pea seed (164) and wheat grain (165) and is an extension of pathways proposed earlier for sweet corn (37) and peas (159). Sucrose synthase and UDP-glucose pyrophosphorylase may be considered as the means for the mobilization of sucrose. The UDP-glucose formed by sucrose synthase represents only half of the sucrose molecule; the fructose formed in this reaction is converted to fructose-6-P and glucose-6-P which may be substrates for glycolysis, the pentose phosphate pathway, oligosaccharide synthesis, etc. This scheme envisages a coupling of PP_i utilization and production in the UDP-glucose pyrophosphorylase and ADP-glucose pyrophosphorylase reactions. Substantial increases in UDP-glucose pyrophosphorylase and ADP-glucose pyrophosphorylase activities coincide with the onset of rapid starch synthesis in the developing pea seed and wheat grain, indicating that the process may be under coarse control, particularly with respect to ADP-glucose pyrophosphorylase (164, 165). Related to these enzyme changes are the observations of Jenner (73) that the amounts of UDP-glucose and ADP-glucose increase in wheat grains at the time when active starch synthesis commences. There are a number of possibilities for fine control, e.g. activation of ADP-glucose pyrophosphorylase by an increase in the level of fructose-6-P (44). This mechanism is also in accord with the changes in the developing endosperm of maize observed by Tsai, Salamini & Nelson (156). In addition to the pyrophosphorylases, these workers found substantial increases in sucrose synthase, hexokinase, and ADP-glucose-starch synthase which coincided with the period of intense starch formation. Preiss et al (132) also observed a marked increase in UDP-glucose pyrophosphorylase, ADP-glucose pyrophosphorylase, and ADP-glucose-starch synthase during the period of starch synthesis in developing maize kernels. It is of interest that in both the developing maize endosperm (156)

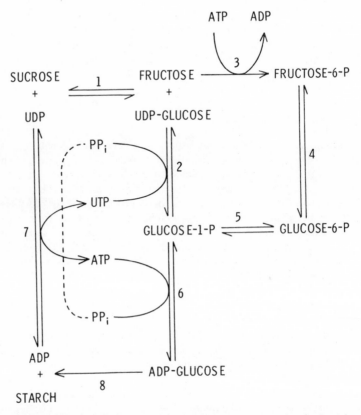

Figure 2 Mechanism of starch formation from sucrose. The enzymes involved are: 1. sucrose synthase; 2. UDP-glucose pyrophosphorylase; 3. fructokinase; 4. glucose phosphate isomerase; 5. phosphoglucomutase; 6. ADP-glucose pyrophosphorylase; 7. nucleoside diphosphate kinase; 8. ADP-glucose-starch synthase.

and the developing pea seed (159) there is an increase in phosphorylase activity at the time of starch synthesis.

Regulation of Related Enzymes

UDP-GLUCOSE 4-EPIMERASE   This enzyme, which catalyzes the interconversion of UDP-glucose and UDP-galactose, requires NAD^+ as a cofactor. NADH is a competitive inhibitor ($K_i = 2 \ \mu M$) of UDP-glucose 4-epimerase from wheat germ (53). The $NADH/NAD^+$ ratio may participate in controlling the formation of UDP-galactose from UDP-glucose in the plant.

UDP-GLUCOSE DEHYDROGENASE   The oxidation of UDP-glucose to UDP-glucuronate is the initial step of pathways leading to the production of cell-wall polysaccharides containing glucuronic and galacturonic acid residues and the formation of pentoses and pentose polysaccharides. UDP-glucose dehydrogenase from pea cotyledons (114), pea stems (1), and lily pollen (33) is inhibited by UDP-xylose and to a lesser extent by UDP-glucuronate and UDP-galacturonate. The UDP-glucose saturation curve, normally hyperbolic, is sigmoid in the presence of UDP-xylose. As UDP-xylose is formed by the decarboxylation of UDP-glucuronate, this suggests feedback control of UDP-glucose dehydrogenase.

MANNAN SYNTHESIS   *Phaseolus aureus* hypocotyls contain a GDP-mannose mannosyltransferase which can form a β-1,4-mannan, and GDP-glucose is a strong competitive inhibitor (69). Heller & Villemez (70) have suggested that this inhibition, in association with GDP-glucose glucosyltransferase which requires a mannose-containing acceptor, will promote the synthesis of a glucomannan rather than of a mannan plus a glucan when both GDP-mannose and GDP-glucose are present. Plant cell wall polysaccharides are, in general, heteropolysaccharides and could be formed by the controlled and cooperative action of individual glycosyltransferases.

OTHER ASPECTS OF THE REGULATION OF CARBOHYDRATE METABOLISM

Regulation by Energy Charge

The concept of energy charge ([ATP]+½[ADP]/[ATP]+[ADP]+[AMP]) developed by Atkinson (7) was discussed in an earlier volume in this series (131) and has been recently reviewed (8, 9). The activation of phosphofructokinases from mammalian tissues and yeast by AMP was one of the factors leading to this hypothesis. Plant phosphofructokinases are not activated by AMP, are stimulated by P_i, and are inhibited by P-enolpyruvate (which, in vivo, is probably related to the ATP concentration). The concept may therefore require some modification before it is applicable to some plant enzyme systems and regulatory processes. Purich & Fromm (135) have studied the energy charge responses of yeast hexokinase, rabbit muscle pyruvate kinase, soluble rat brain hexokinase, and rabbit muscle phosphofructokinase. Several additional factors including pH, total adenylate concentration, levels of nonadenylate reaction products, and nonadenine nucleotides were shown to affect markedly the responses to adenylate energy charge. The authors point out that no study has established the central regulatory role of energy charge in vivo, and they suggest that the hypothesis requires substantial amendment to account for practical and theoretical deficiencies.

Regulation by pH

Davies (31) has provided a stimulating discussion on aspects of control of and by pH. Several enzymes have the potential to act as pH-stats. The pH in the cell may also vary and so affect the relative rates of a number of processes, e.g. the initial production of lactate in anaerobiosis may result in a fall in pH to a value more favorable for pyruvate decarboxylase and for ethanol formation.

Phosphatase Action

The rate of hydrolysis of glucose-1-P and glucose-6-P by pea seed phosphatase falls markedly as the reaction proceeds, and this decline in rate is not predominantly due to substrate depletion (161). P_i is a strong competitive inhibitor of the phosphatase, and 0.15 mM P_i inhibits the hydrolysis of glucose-1-P and glucose-6-P by 52% and 29% respectively. If the concentration of P_i in the cell fell to a low level, there would be a tendency for the phosphate esters to be hydrolyzed and so increase the P_i concentration. Conversely, if the P_i concentration were increased (as in anaerobiosis) there would be a sparing effect on the phosphate esters. Thus the cellular P_i concentration could determine the extent of hydrolysis of phosphate esters, and the inhibition of phosphatases by P_i could be a contributing control factor in a number of aspects of carbohydrate metabolism.

Integration of Regulation

Many instances of control phenomena involve the regulation of an enzyme in a metabolic pathway by a metabolite in that pathway. Just as there must be integration of the various metabolic pathways in the cell, so there must be integrated regulation of these pathways. This may be seen where metabolites from one pathway have an effect on an enzyme of another pathway, as in the inhibition of pyruvate kinase and phosphofructokinase by citrate and ATP and the inhibition of the latter enzyme by 6-P-gluconate. Information is also accumulating which may indicate the existence of antagonistic interactions when a metabolite may have opposite effects on key enzymes competing for a common substrate (or its equivalent) and so amplify the effect of the metabolite. Two examples will suffice. In plants several enzyme systems may compete for the available hexose monophosphates (glucose-1-P, glucose-6-P, and fructose-6-P). As phosphoglucomutase and glucose phosphate isomerase probably catalyze equilibrium reactions, the three hexose monophosphates may be regarded as forming a common pool. Two of the enzymes competing for this pool are ADP-glucose pyrophosphorylase (for glucose-1-P) and phosphofructokinase (for fructose-6-P), and these enzymes may be affected in opposing ways by a number of effectors. 3-P-Glycerate, P-enolpyruvate, and 2,3-P_2-glycerate activate ADP-glucose pyrophosphorylase and inhibit phosphofructokinase. P_i inhibits ADP-glucose pyrophosphorylase and activators (such as 3-P-glycerate) can reverse this inhibition: P_i stimulates phosphofructokinase and relieves inhibition by 3-P-glycerate, P-enolpyruvate, etc. The response to effectors varies between tissues, and some metabolites do not have opposing effects with these two enzymes. The second example concerns the enzyme pair sucrose phosphate synthase and phosphofructokinase, both of which have fructose-6-P as substrate. Citrate activates sucrose phosphate synthase and inhibits phosphofructokinase. This inhibition may be reinforced by the inhibition of pyruvate kinase by citrate which in turn may amplify the citrate effect on phosphofructokinase through sequential inhibition by P-enolpyruvate, 2-P-glycerate, and 3-P-glycerate. Thus sucrose formation as well as glycolysis may be regulated by citrate, a component of the tricarboxylic acid cycle.

CONCLUSION

Studies with individual enzymes and enzyme systems and analyses of tissues during development and anaerobiosis have yielded much valuable information and will continue to provide the guide to in vivo regulation. Information about the environment at the site of enzyme action—including factors such as concentration of substrates and products, effector and cofactor concentrations, and pH—and questions of enzyme association and integration would provide very useful complementary evidence.

ACKNOWLEDGMENT

The authors wish to thank Miss D. D. Harrison for her help in the preparation of this review.

Literature Cited

1. Abdul-Baki, A. A., Ray, P. M. 1971. *Plant Physiol.* 47:537–44
2. Adams, P. B. 1970. *Plant Physiol.* 45:495–99
3. Ibid, 500–3
4. Adams, P. B., Rowan, K. S. 1970. *Plant Physiol.* 45:490–94
5. Adelberg, E. A., Umbarger, H. E. 1953. *J. Biol. Chem.* 205:475–82
6. Anderson, L. E. 1971. *Biochim. Biophys. Acta* 235:237–44
7. Atkinson, D. E. 1968. *Biochemistry* 7:4030–34
8. Atkinson, D. E. 1970. *Enzymes* 1:461–89
9. Atkinson, D. E. 1971. *Metab. Pathways* 5:1–21
10. Axelrod, B., Beevers, H. 1972. *Biochim. Biophys. Acta* 256:175–78
11. Barker, J., el Saifi, A. F. 1952. *Proc. Roy. Soc. B* 140:362–403
12. Barker, J., Khan, M. A. A. 1968. *New Phytol.* 67:205–12
13. Barker, J., Khan, M. A. A., Solomos, T. 1967. *New Phytol.* 66:577–96
14. Beevers, H. 1961. *Respiratory Metabolism in Plants.* Evanston, Ill.: Row, Peterson. 232 pp.
15. Beevers, H. 1969. *Ann. NY Acad. Sci.* 168:313–24
16. Beevers, H., Gibbs, M. 1954. *Plant Physiol.* 29:318–21
17. Black, C. C. 1973. *Ann. Rev. Plant Physiol.* 24:253–86
18. Bloxham, D. P., Lardy, H. A. 1973. *Enzymes* 8:239–78
19. Bonsignore, A., De Flora, A. 1972. *Curr. Top. Cell. Regul.* 6:21–62
20. Breidenbach, R. W., Beevers, H. 1967. *Biochem. Biophys. Res. Commun.* 27:462–69
21. Bruemmer, J. H., Roe, B. 1971. *Phytochemistry* 10:255–59
22. Burr, B., Nelson, O. E. 1973. *Ann. NY Acad. Sci.* 210:129–38
23. Canvin, D. T., Beevers, H. 1961. *J. Biol. Chem.* 236:988–95
24. Chalmers, D. J., Rowan, K. S. 1971. *Plant Physiol.* 48:235–40
25. Chance, B., Holmes, W., Higgins, J., Connelly, C. M. 1958. *Nature* 182:1190–93
26. Colowick, S. P. 1973. *Enzymes* 9:1–48
27. Cox, E. L., Dickinson, D. B. 1973. *Plant Physiol.* 51:960–66
28. Cox, G. F., Davies, D. D. 1969. *Biochem. J.* 113:813–20
29. Ibid 1970. 116:819–24
30. Crompton, M., Laties, G. G. 1971. *Arch. Biochem. Biophys.* 143:143–50
31. Davies, D. D. 1973. *Symp. Soc. Exp. Biol.* 27:513–29
32. Davies, D. D., Davies, S. 1972. *Biochem. J.* 129:831–39
33. Davies, M. D., Dickinson, D. B. 1972. *Arch. Biochem. Biophys.* 152:53–61
34. de Fekete, M. A. R. 1971. *Eur. J. Biochem.* 19:73–80
35. de Fekete, M. A. R. 1972. *Biochemistry of the Glycosidic Linkage,* ed. R. Piras, H. G. Pontis, 291–95. New York: Academic. 787 pp.
36. de Fekete, M. A. R. 1972. *Metabolic Interconversion of Enzymes,* ed. O. Wieland, E. Helmreich, H. Holzer, 347–51. Berlin: Springer. 448 pp.
37. de Fekete, M. A. R., Cardini, C. E. 1964. *Arch. Biochem. Biophys.* 104:173–84
38. de Fekete, M. A. R., Leloir, L. F., Cardini, C. E. 1960. *Nature* 187:918–19

39. de Fekete, M. A. R., Vieweg, G. H. 1973. *Ann. NY Acad. Sci.* 210:170–80
40. Delmer, D. P. 1972. *J. Biol. Chem.* 247:3822–28
41. Delmer, D. P., Albersheim, P. 1970. *Plant Physiol.* 45:782–86
42. Dennis, D. T., Coultate, T. P. 1966. *Biochem. Biophys. Res. Commun.* 25:187–91
43. Dennis, D. T., Coultate, T. P. 1967. *Biochim. Biophys. Acta* 146:129–37
44. Dickinson, D. B., Preiss, J. 1969. *Arch. Biochem. Biophys.* 130:119–28
45. Douce, R., Bonner, W. D. 1972. *Biochem. Biophys. Res. Commun.* 47:619–24
46. Duggleby, R. G., Dennis, D. T. 1970. *J. Biol. Chem.* 245:3751–54
47. Duggleby, R. G., Dennis, D. T. 1973. *Arch. Biochem. Biophys.* 155:270–77
48. Duggleby, R. G., Dennis, D. T. 1973. *Plant Physiol.* 52:312–17
49. Duggleby, R. G., Dennis, D. T. 1974. *J. Biol. Chem.* 249:162–66
50. Eggleston, L. V., Krebs, H. A. 1974. *Biochem. J.* 138:425–35
51. Eloff, J. N. 1972. *Z. Pflanzenphysiol.* 67:207–11
52. Ewart, M. H., Siminovitch, D., Briggs, D. R. 1954. *Plant Physiol.* 29:407–13
53. Fan, D-F., Feingold, D. S. 1969. *Plant Physiol.* 44:599–604
54. Fox, J. et al 1973. *Ann. NY Acad. Sci.* 210:90–103
55. Frydman, R. B., Slabnik, E. 1973. *Ann. NY Acad. Sci.* 210:153–69
56. Garrard, L. A., Humphreys, T. E. 1968. *Phytochemistry* 7:1949–61
57. Ghai, G., Modi, V. V. 1970. *Biochem. Biophys. Res. Commun.* 41:1088–95
58. Gibbs, M. 1966. *Plant Physiology,* ed. F. C. Steward, 4B:3–115. New York: Academic. 599 pp.
59. Gibbs, M., Turner, J. F. 1964. *Modern Methods of Plant Analysis,* ed. H. F. Linskens, B. D. Sanwal, M. V. Tracey, 7:520–45. Berlin: Springer. 735 pp.
60. Givan, C. V. 1972. *Planta* 108:29–38
61. Greenbaum, A. L., Gumaa, K. A., McLean, P. 1971. *Arch. Biochem. Biophys.* 143:617–63
62. Greenblatt, G. A., Sarkissian, I. V. 1973. *Physiol. Plant.* 29:361–64
63. Grimes, W. J., Jones, B. L., Albersheim, P. 1970. *J. Biol. Chem.* 245:188–97
64. Gumaa, K. A., McLean, P. 1969. *Biochem. J.* 115:1009–29
65. Gustafson, G. L., Gander, J. E. 1972. *J. Biol. Chem.* 247:1387–97
66. Hanes, C. S. 1940. *Proc. Roy. Soc. B* 128:421–50
67. Ibid, 129:174–208
68. Hatch, M. D., Turner, J. F. 1958. *Biochem. J.* 69:495–501
69. Heller, J. S., Villemez, C. L. 1972. *Biochem. J.* 128:243–52
70. Ibid, 129:645–55
71. Hopper, J. E., Dickinson, D. B. 1972. *Arch. Biochem. Biophys.* 148:523–35
72. Isherwood, F. A. 1973. *Phytochemistry* 12:2579–91
73. Jenner, C. F. 1968. *Plant Physiol.* 43:41–49
74. Jones, R. L. 1973. *Ann. Rev. Plant Physiol.* 24:571–98
75. Kayne, F. J. 1973. *Enzymes* 8:353–82
76. Kelly, G. J., Gibbs, M. 1973. *Plant Sci. Lett.* 1:253–57
77. Kelly, G. J., Gibbs, M. 1973. *Plant Physiol.* 52:111–18
78. Kelly, G. J., Turner, J. F. 1968. *Biochem. Biophys. Res. Commun.* 30:195–99
79. Kelly, G. J., Turner, J. F. 1969. *Biochem. J.* 115:481–87
80. Kelly, G. J., Turner, J. F. 1970. *Biochim. Biophys. Acta* 208:360–67
81. Ibid 1971. 242:559–65
82. Kobr, M. J., Beevers, H. 1968. *Plant Physiol.* 43:S–17
83. Ibid 1971. 47:48–52
84. Kornberg, H. L., Beevers, H. 1957. *Nature* 180:35–36
85. Kornberg, H. L., Beevers, H. 1957. *Biochim. Biophys. Acta* 26:531–37
86. Koshland, D. E. 1968. *Advan. Enzyme Regul.* 6:291–301
87. Koshland, D. E. 1969. *Curr. Top. Cell. Regul.* 1:1–27
88. Koshland, D. E. 1970. *Enzymes* 1:341–96
89. Kosow, D. P., Rose, I. A. 1971. *J. Biol. Chem.* 246:2618–25
90. Krebs, H. A. 1969. *Curr. Top. Cell. Regul.* 1:45–55
91. Krietsch, W. K. G., Bücher, Th. 1970. *Eur. J. Biochem.* 17:568–80
92. Larsson-Raznikiewicz, M., Arvidsson, L. 1971. *Eur. J. Biochem.* 22:506–12
93. Lee, E. Y. C., Braun, J. J. 1973. *Arch. Biochem. Biophys.* 156:276–86
94. Leloir, L. F., Cardini, C. E. 1953. *J. Am. Chem. Soc.* 75:6084
95. Leloir, L. F., Cardini, C. E. 1955. *J. Biol. Chem.* 214:157–65
96. Leloir, L. F., de Fekete, M. A. R., Cardini, C. E. 1961. *J. Biol. Chem.* 236:636–41
97. Lowry, O. H., Passonneau, J. V. 1964.

Arch. Exp. Pathol. Pharmakol.
248:185–94
98. Lucchini, G. 1971. *Biochim. Biophys. Acta* 242:365–70
99. Mansour, T. E. 1972. *Curr. Top. Cell. Regul.* 5:1–46
100. Meli, J., Bygrave, F. L. 1972. *Biochem. J.* 128:415–20
101. Mendicino, J. 1960. *J. Biol. Chem.* 235:3347–52
102. Meunier, J. C., Buc, J., Ricard, J. 1971. *FEBS Lett.* 14:25–28
103. Miller, G., Evans, H. J. 1957. *Plant Physiol.* 32:346–54
104. Monod, J., Changeux, J-P., Jacob, F. 1963. *J. Mol. Biol.* 6:306–29
105. Moore, C. J., Turner, J. F. 1969. *Nature* 223:303–4
106. Murata, T. 1971. *Agr. Biol. Chem.* 35:1441–48
107. Ibid 1972. 36:1877–84
108. Murata, T., Akazawa, T., Fukuchi, S. 1968. *Plant Physiol.* 43:1899–1905
109. Muto, S., Uritani, I. 1970. *Plant Cell Physiol.* 11:767–76
110. Ibid 1971. 12:803–6
111. Ibid 1972. 13:111–18
112. Ibid, 377–80
113. Negelein, E., Haas, E. 1935. *Biochem. Z.* 282:206–20
114. Neufeld, E. F., Hall, C. W. 1965. *Biochem. Biophys. Res. Commun.* 19:456–61
115. Newsholme, E. A., Start, C. 1973. *Regulation in Metabolism.* London: Wiley. 349 pp.
116. Nomura, T., Akazawa, T. 1974. *Plant Cell Physiol.* 15:477–83
117. Nomura, T., Kono, Y., Akazawa, T. 1969. *Plant Physiol.* 44:765–69
118. Nomura, T., Nakayama, N., Murata, T., Akazawa, T. 1967. *Plant Physiol.* 42:327–32
119. Oestreicher, G., Hogue, P., Singer, T. P. 1973. *Plant Physiol.* 52:622–26
120. Ohmann, E. 1969. *Arch. Mikrobiol.* 67:273–92
121. Osmond, C. B., ap Rees, T. 1969. *Biochim. Biophys. Acta* 184:35–42
122. Ozbun, J. L. et al 1973. *Plant Physiol.* 51:1–5
123. Ozbun, J. L., Hawker, J. S., Preiss, J. 1972. *Biochem. J.* 126:953–63
124. Pacold, I., Anderson, L. E. 1973. *Biochem. Biophys. Res. Commun.* 51:139–43
125. Parvin, R., Atkinson, D. E. 1968. *Arch. Biochem. Biophys.* 128:528–33
126. Passonneau, J. V., Lowry, O. H. 1964. *Advan. Enzyme Regul.* 2:265–74

127. Pavlinova, O. A., Prasolova, M. F. 1972. *Fiziol. Rast.* 19:920–25
128. Pratt, H. K., Goeschl, J. D. 1969. *Ann. Rev. Plant Physiol.* 20:541–84
129. Preiss, J. 1973. *Enzymes* 8:73–119
130. Preiss, J., Greenberg, E. 1969. *Biochem. Biophys. Res. Commun.* 36:289–95
131. Preiss, J., Kosuge, T. 1970. *Ann. Rev. Plant Physiol.* 21:433–66
132. Preiss, J., Ozbun, J. L., Hawker, J. S., Greenberg, E., Lammel, C. 1973. *Ann. NY Acad. Sci.* 210:265–78
133. Pressey, R. 1969. *Plant Physiol.* 44:759–64
134. Purich, D. L., Fromm, H. J. 1972. *Curr. Top. Cell. Regul.* 6:131–67
135. Purich, D. L., Fromm, H. J. 1973. *J. Biol. Chem.* 248:461–66
136. Purich, D. L., Fromm, H. J., Rudolph, F. B. 1973. *Advan. Enzymol.* 39:249–326
137. Recondo, E., Leloir, L. F. 1961. *Biochem. Biophys. Res. Commun.* 6:85–88
138. Rolleston, F. S. 1972. *Curr. Top. Cell. Regul.* 5:47–75
139. Rose, I. A., Rose, Z. B. 1969. *Compr. Biochem.* 17:93–161
140. Sacher, J. A. 1973. *Ann. Rev. Plant Physiol.* 24:197–224
141. Scala, J., Patrick, C., Macbeth, G. 1968. *Arch. Biochem. Biophys.* 127:576–84
142. Schnarrenberger, C., Oeser, A., Tolbert, N. E. 1973. *Arch. Biochem. Biophys.* 154:438–48
143. Schulman, M. D., Gibbs, M. 1968. *Plant Physiol.* 43:1805–12
144. Seubert, W., Schoner, W. 1971. *Curr. Top. Cell. Regul.* 3:237–67
145. Singer, T. P., Oestreicher, G., Hogue, P., Contreiras, J., Brandao, I. 1973. *Plant Physiol.* 52:616–21
146. Slabnik, E., Frydman, R. B., Cardini, C. E. 1968. *Plant Physiol.* 43:1063–68
147. Srere, P. A. 1972. *Curr. Top. Cell. Regul.* 5:229–83
148. Stadtman, E. R. 1970. *Enzymes* 1:397–459
149. Taketa, K., Pogell, B. M. 1965. *J. Biol. Chem.* 240:651–62
150. Thomas, S. M., ap Rees, T. 1972. *Phytochemistry* 11:2177–85
151. Ibid, 2187–94
152. Tolbert, N. E. 1971. *Ann. Rev. Plant Physiol.* 22:45–74
153. Tomlinson, J. D., Turner, J. F. 1973. *Biochim. Biophys. Acta* 329:128–39
154. Tomova, N., Setchenska, M., Krusteva, N., Christova, Y., Detchev, G. 1972. *Z. Pflanzenphysiol.* 67:113–16, 117–19
155. Tsai, C. Y., Nelson, O. E. 1968. *Plant Physiol.* 43:103–12

156. Tsai, C. Y., Salamini, F., Nelson, O. E. 1970. *Plant Physiol.* 46:299–306
157. Tsuboi, K. K., Fukunaga, K., Petricciani, J. C. 1969. *J. Biol. Chem.* 244:1008–15
158. Turner, D. H., Blanch, E. S., Gibbs, M., Turner, J. F. 1964. *Plant Physiol.* 40:1146–50
159. Turner, D. H., Turner, J. F. 1957. *Aust. J. Biol. Sci.* 10:302–9
160. Turner, D. H., Turner, J. F. 1958. *Biochem. J.* 69:448–52
161. Ibid 1960. 74:486–91
162. Turner, J. F. 1953. *Nature* 172:1149–50

163. Ibid 1954. 174:692–93
164. Turner, J. F. 1969. *Aust. J. Biol. Sci.* 22:1145–51
165. Ibid, 1321–27
166. Turner, J. S. 1960. *Encycl. Plant Physiol.* 12/2:42–87
167. Uyeda, K., Racker, E. 1965. *J. Biol. Chem.* 240:4682–88
168. Vidra, J. D., Loerch, J. D. 1968. *Biochim. Biophys. Acta* 159:551–53
169. Wedding, R. T., Black, M. K. 1971. *J. Biol. Chem.* 246:1638–43
170. Yang, N-S., Scandalios, J. G. 1974. *Arch. Biochem. Biophys.* 161:335–53

Ann. Rev. Plant Physiol. 1975. 26:187–208

PLANT NUCLEASES[1]

❖7588

Curtis M. Wilson

Agricultural Research Service, U.S. Department of Agriculture, and Department of Agronomy, University of Illinois, Urbana, Illinois 61801

CONTENTS

[1]Abbreviations used: ABA (abscisic acid); DNPP (dinitrophenyl phosphate); α-naphthyl-pU (α-naphthyl-5'-uridylic acid); nitrophenyl-pT (*p*-nitrophenyl-5'-thymidylic acid); and abbreviations and symbols for nucleic acids etc, as recommended by the Commission on Biochemical Nomenclature (66), including: A (adenosine); Cp (cytidine-3'-monophosphate); N (unspecified nucleoside); pG (guanosine-5'-monophosphate); R (unspecified purine nucleoside); U>p (uridine 2':3'-cyclic monophosphate); Y (unspecified pyrimidine nucleoside).

INTRODUCTION

Several reviews of the past decade have presented basic facts on various nuclease activities, but the plant nucleases have not been covered in detail (12, 78, 116, 134). Dove (38) reviewed the cellular distribution of plant nucleases and their changes during development, but concluded that we lack sufficient knowledge of the specific character of the enzymes to make definite conclusions. This review will concentrate on the biochemical nature of plant nucleases and on the problems that must be overcome before the role of nucleases in plant metabolism can be clarified. An annotated bibliography includes many papers omitted from this review for lack of space.[2] A classification of nucleases is presented along with a recommended standard enzyme unit to assist in comparisons of work from different laboratories. It is hoped that this review will provide a framework for future in-depth studies of the various plant nucleases and their functions.

NOMENCLATURE AND CLASSIFICATION

Many names have been used for enzymes that degrade nucleic acids, but the complex nature of the enzyme-substrate interactions and our increasing knowledge of the mode of action of individual enzymes make a simple and unambiguous nomenclature virtually impossible. The recommended names and the numbering system in the recently revised *Enzyme Nomenclature* (67) appear inadequate. I use a system taken in part from previous reviews (12, 78, 134).

Definitions

NUCLEASE   Any enzyme cleaving a polynucleotide when the details of the reaction are not known. In addition, nuclease is used as a specific name for enzymes that split both RNA and DNA.

RIBONUCLEASE (RNASE)   RNase should be used only for enzymes specific for RNA as a substrate. Unfortunately, in this review, as in many papers, the name RNase will be used when it is not known whether or not the activity measured is a mixture of nucleases and RNases, because only RNA was tested as a substrate.

[2]See NAPS document No. 02515 for 24 pages of supplementary material. Order from ASIS/NAPS c/o Microfiche Publications, 440 Park Avenue South, New York, NY 10016. Remit in advance for each NAPS accession number. Make checks payable to Microfiche Publications. Photocopies are $5.00. Microfiche are $1.50. Outside of the United States and Canada, postage is $2.00 for a photocopy or $.50 for a fiche.

DEOXYRIBONUCLEASE (DNASE)   A DNase is specific for DNA and its products. No plant DNase has been identified.

EXONUCLEASE   An exonuclease attacks the terminal diester bond of a nucleic acid (see phosphodiesterase).

PHOSPHODIESTERASE (PDASE)   A PDase, in the strict sense, is any enzyme which splits a bond involving diesterified phosphate (75), which includes the enzymes given above. Operationally, a PDase is usually an enzyme which hydrolyzes substrates such as DNPP and α-napthyl-pT, but not necessarily nucleic acids (118). When an enzyme known as a PDase does hydrolyze a nucleic acid, it acts as an exonuclease—the latter term will then be used in this review.

3'-NUCLEOTIDASE   A 3'-nucleotidase specifically hydrolyzes the 3'-phosphate of a nucleotide.

I have adopted a Procrustian system for placing the plant nucleases into four categories: Plant RNase I, Plant RNase II, Plant Nuclease I, and Plant Exonuclease I (Table 1). This simplification is justified by certain basic properties shared by most of the known plant enzymes. Minor differences in methods can give the appearance of different properties for enzymes that are similar.

Any system of classification is tentative and subject to change as more enzymes and species are studied. The enzymes considered here have been at least partially purified and, in my opinion, also separated from other enzymes that might interfere. The classification was first presented in 1968 (177). It is based on criteria proposed by Laskowski (78) and Reddi (124). More complete discussions of the following criteria may be found elsewhere (12, 75, 116, 163).

1. Substrate (sugar specificity). The substrate may be RNA, DNA, or both. The cyclizing enzymes are restricted to RNA as a substrate. Mung bean nuclease I is sugar-nonspecific to the extent that it also hydrolyzes arabinosyl oligonucleotides (172). Denatured DNA is usually more easily attacked than is native DNA. The relationship between secondary structure of the substrate and the activity of plant enzymes has received almost no study (see 78; see also the many bacterial nucleases reported in recent years).

2. Products. The difference between enzymes liberating products with a phosphate group on the 3' or on the 5' position of the terminal nucleotide or mononucleotide is quite clear-cut, with no reports of enzymes that produce both types. A third product is a 2':3'-cyclic nucleotide, formed by splitting the polynucleotide chain without hydrolysis and transferring the phosphate group from the 5' position of one nucleotide to the 2' position on the adjacent nucleotide. The cyclic products may be hydrolyzed later by the same enzyme to give a 3'-nucleotide as the final product. The hydrolysis reaction is slower than the transfer reaction and may not be detected with enzyme preparations of low activity. On the other hand, excess enzyme might give the appearance of direct hydrolysis.

3. Mode of action. Endonucleases attack polynucleotides at random, releasing oligonucleotides and, sooner or later, mononucleotides. Exonucleases hydrolyze the

Table 1 Classification of plant nucleases, with selected references keyed to enzyme properties (discussed in text) and to species

	RNase I	RNase II	Nuclease I	Exonuclease I
1. Substrate	RNA; R>p (68, 141, 158, 162, 174, 189)	RNA (125, 176); N>p (68)	RNA≥denatured DNA>native DNA (18, 19, 53, 70, 73, 105, 139, 155, 170, 174, 184, 188); Np (53, 69, 76, 84, 101, 105, 106, 156)	DNA>RNA (55, 58, 81, 167, 188); Nitrophenyl-pT (56, 167); Nitrophenyl-pN (81); DNPP (55, 167)
2. Products	N>p (43, 68, 77, 87, 100, 120, 141, 153, 160, 162, 170, 177, 189)	N>p (68, 177)	pN, 3'-OH terminals (18, 155, 171, 177, 184)	pN (55, 58, 81, 167)
Secondary products	Rp (68, 77, 87, 100, 120, 121, 153, 189)	Np (68, 125, 177)		
3. Mode of action	Endonuclease (68, 77, 141, 160, 189)	Endonuclease (68, 177)	Endonuclease (19, 70, 155, 156, 177)	Exonuclease (55, 58, 81, 167)
4. Base specificity	G>A=U>C (68, 77, 120, 153, 158, 160, 162, 174, 188, 189)	G>A=U>C (68, 177)	A>U(T), G, C (18, 70, 84, 101, 171, 184, 188)	None
5. Molecular weight	9,000 (160); 24,000 (77, 176); 19,700 (68)	17,000 (176); 21,000 (68)	31–35,000 (105, 106, 176, 184); 54,000 (104)	>100,000 (56, 58, 81)
6. pH optimum	5.0–6.0 (68, 77, 100, 152, 158, 160, 162, 166, 170, 176, 188)	6.0–7.0 (68, 125, 176)	5.0–6.5 (84, 105, 155, 176, 184, 188)	7.0–9.0 (55, 58, 81, 167, 188)
7. Intracellular location	Soluble (68, 144, 178)	Microsomes (68, 125, 176, 178)	Particles, membranes? (144, 176, 178, 184)	?

Table 1 (Continued)

	RNase I	RNase II	Nuclease I	Exonuclease I
8. EDTA sensitivity	Low (77, 145, 166, 170)	Low (145, 178)	High (17, 53, 84, 106, 145, 155, 178, 184) Low (105)	High (56, 58, 81, 167)
9. General	(68, 77, 141, 175, 189)	(68, 176–178)	(8, 69, 70, 84, 101, 176, 177, 184)	(56, 58, 167)
10. Enzyme nomenclature classification	2.7.7.17 (65) 3.1.4.23 (67)	2.7.7.17 (65) 3.1.4.23 (67)	3.1.4.9 (67)	3.1.4.1 (67)
11. Species[a] and enzyme names[b]				
Barley (Hordeum vulgare)	RNase I (77)		DNase (19)	
Carrot				Exophosphodiesterase (58) Phosphodiesterase (55, 56)
Corn (Zea mays)	RNase A (173, 175), RNase I (176–178)	RNase II (176–178)	RNase B (173, 174), Nuclease I (176, 177) Nuclease (104)	
Muskmelon (Cucumis melo)				
Mung bean (Phaseolus aureus)	RNase M$_1$ (170), RNase (151–153)		Nuclease I (69, 70, 101, 155), RNase M$_2$ (84, 170, 171)	
Oats (Avena sativa)	Relative purine specific endoribonuclease (166, 187, 189)		Sugar nonspecific endonuclease I (184, 187)	Alkaline phosphodiesterase (167, 187)
Peas	Ribonuclease (87)			

Table 1 (Continued)

	RNase I	RNase II	Nuclease I	Exonuclease I
Potato			Nuclease I (17, 18), 3'-nucleotidase-nuclease (105, 156)	
Ryegrass (*Lolium* spp.)	RNase (43, 139, 141)		RNase II (43)	
Soybean	Ribonuclease (100)			
Spinach	Ribonuclease (162)			
Sugar beet (*Beta vulgaris*)				Phosphodiesterase (81)
Sugar cane (*Saccharum officinarum*)	Ribonuclease (158)			
Tobacco (*Nicotiana tabacum*)	RNase 1 (68), TL-RNAase-I (44, 120, 125), Relatively guanine-specific endoribonuclease (188)	RNase 2 (68), TL-RNAase-II (125),	Extracellular nuclease (106), Relatively adenine specific endonuclease (188)	
Wheat (*Triticum* spp.)	Acid ribonuclease (160), WL-RNase I (144)		Nuclease (53), WL-RNase II (144)	Alkaline phosphodiesterase (188)

a Latin names as designated by authors.
b Enzyme names are those used in the references cited.

bond adjacent to the terminal nucleotide to release mononucleotides directly.

4. Base specificity. There are four common bases and 16 combinations for the two bases on either side of the bond being split. One enzyme may have a wide range of activity toward the various base pair combinations (77, 101, 160). Base specificity is usually determined by the relative rates of release of the four mononucleotides, but the bases at either end of the oligonucleotides released during early stages of hydrolysis may also be measured (77, 155, 160). Pancreatic RNase is specific for a pyrimidine on the 3' side of the bond being split, RNase T1 is specific for guanosine on the 3' side, and plant RNases I and II appear to have a preference for guanosine and a low activity on cytosine. Base specificity may be expressed by differential rates of attack on synthetic homo- and heteropolymers, but differences in secondary structure of these substrates may produce spurious results. Plant RNase I will hydrolyze only R>p, whereas plant RNase II will hydrolyze both R>p and Y>p.

5. Molecular weight. The reported molecular weights for plant nucleases vary from very small to more than 100,000. The smaller enzymes appear to have the greater specificity.

6. pH optimum. The pH optimum is often affected by the substrate, buffer, ionic strength, and cations (12, 43, 84, 173).

7. Intracellular location. This factor is only tentatively known for plant nucleases. In no case have both the enzyme and the cell particle on which it is located been adequately identified.

8. EDTA sensitivity. It is usually assumed that EDTA inhibits an enzyme by removing a metallic ion required for activity. The converse experiment, requirement for cations, is less clear-cut. Reports differ widely as to whether cations accelerate, inhibit, protect, or inactivate plant nucleases. Impurities, pH, ionic strength, and substrate may all interact with EDTA or added cations.

9. General. Several key references to purified preparations of each enzyme type are given.

10. *Enzyme Nomenclature* classification.

11. Species and enzyme names.

Plant RNase I

Plant RNase I may be briefly described as a soluble endoribonuclease that releases 2':3'-cyclic nucleotides, with a preference for G>p as the major early product. Only purine cyclic nucleotides are hydrolyzed. The pH optimum is near 5 and is lower than that for the other plant nucleases. The molecular weight is 20,000–25,000.

RNase I is usually the first specific RNase to be isolated, because it is soluble and more stable than the other nucleases (177). It has been obtained in pure form from corn endosperm (175) and barley leaves (77) (the second reference includes a good description of the kinetic properties). The ability of a semipurified preparation of an RNase to hydrolyze Y>p (124, 125) does not by itself imply that the enzyme is an RNase II. Other plant enzymes are known that can hydrolyze Y>p (82, 150), and they may act after RNase I to produce RNase II-like activity.

Plant RNase II

Plant RNase II is a microsomal endoribonuclease that releases 2':3'-cyclic nucleotides, with a preference for G>p as the major early product. Both R>p and Y>p are hydrolyzed, with a preference for R>p. The pH optimum is near 6. The molecular weight is 17,000–20,000.

RNase II is very much like RNase I in its action. It has been distinguished from RNase I in corn (176–178) and tobacco (68, 125). Corn RNase II was reported as slightly smaller than RNase I (176), but tobacco RNase II appeared slightly larger than the RNase I (68). Crude microsomal preparations may contain Nuclease I as well as RNase II (see *Subcellular Localization*).

Plant Nuclease I

Plant Nuclease I is a particle-bound sugar-nonspecific endonuclease that releases 5'-mononucleotides with a preference for pA as the major early product. It is about equally active on RNA and denatured DNA, but is much less active on native DNA. The pH optimum is near 6 and is higher than that of RNase I. The molecular weight is about 33,000. The enzyme is usually inhibited by EDTA and often requires a divalent cation for maximum activity or stability. The nuclease is accompanied by a 3'-nucleotidase activity.

Some doubts will be raised as to whether the nucleases listed as nuclease I in Table 1 should be lumped together, but they do share a number of properties. Usually all three activities (RNase, DNase, 3'-nucleotidase) are isolated together, are inhibited or activated in parallel, and show the same base specificities. One mung bean nuclease lost DNase activity in storage (170), but this has not been repeated.

The best characterized nuclease is that isolated and purified from frozen mung bean sprouts by Laskowski and co-workers (8, 69, 101, 155). The enzyme has a preference for the bond pA-pX (69). When X is H, the enzyme is called a 3'-nucleotidase, but because the bond attacked may be at the 3' end of a nucleotide chain, the name should be (3')-ω-monophosphatase (101). Coenzyme A (84) and various synthetic esters at the 3' position (76) are substrates. The nuclease has a tenfold sugar preference for ribodinucleotides over deoxyribodinucleotides, and this preference increases to 50- to 100-fold for hydrolysis of the 3'-mononucleotides (101). The mung bean nuclease has an extremely high specificity for denatured regions in DNA that are rich in A and T (70).

Variable reports of inhibition by EDTA are probably caused by differences in experimental conditions. At least 30 min were required for EDTA to inhibit a wheat nuclease fully (145). Zn^{++} specifically protected mung bean (84) and wheat (145) nucleases. Zn^{++} inhibited the RNase activity but not the DNase activity of an oat nuclease (184). Mg^{++} restored activity to a potato nuclease inhibited by EDTA (17), but another potato nuclease was not affected by EDTA (105). Laskowski's group focused on nuclease activity that did not require a divalent cation (69), but they reported some inhibition by EDTA and Mg^{++} (155). Studies on a wheat nuclease suggest some reasons for these conflicts (53). A buffer containing 10^{-4} M Zn^{++} was required for stability during purification, and 10^{-3} M Zn^{++} stimulated the less pure preparations. Yet the purest fractions were inhibited by even 10^{-5} M Zn^{++}. Ca^{++} and Mg^{++} stimulated the RNase activity only. Sulfhydryl reagents and Zn^{++} were needed for stability at pH 4.5, but sulfhydryl reagents inactivated the nuclease at pH 8 (53). Perhaps Zn^{++} is a part of the enzyme, but its binding is affected only by the more drastic isolation methods. Divalent cations may influence the configuration of the substrate and thus differentially affect the three types of activity.

3' Nucleotidase

Three types of enzyme may hydrolyze 3'-nucleotides: (a) nuclease I, (b) a specific 3'-mononucleotidase without nuclease activity, and (c) a nonspecific phosphomonoesterase. The nuclease I 3'-nucleotidase pH optimum is usually near 8 (69, 104, 105), but it varies with the substrate (84) and may be as low as 5.4 (53). The well-known 3'-nucleotidase from ryegrass (140) may be a nuclease or a mixture of enzymes. Mung beans contain a 3'-nucleotidase that is not inhibited by NaF (89, 154), but the nuclease activity was not examined. Two enzymes active against Ap were obtained from peas (82). One appeared to be a nuclease I, whereas the other, active at pH 5.4, may not have been specific. A specific 3'-nucleotidase that did not hydrolyze RNA was isolated from peas (42). Relative changes in 3'-nucleotidase and RNase activities in germinating ryegrass (139) and corn (64) and in kinetin-treated potato stolons (107) suggest that not all of the 3'-nucleotidase activity was accounted for by a nuclease I.

Plant Exonuclease

Plant exonucleases hydrolyze polynucleotide chains from the 3' end, releasing 5'-mononucleotides. The pH optimum is 7–9. The molecular weight is 100,000 or above. Each of the plant exonucleases was reported to resemble snake venom phosphodiesterase in many properties.

The plant exonucleases were all isolated with the use of simple diesters such as nitrophenyl-pT, α-naphthyl-pU, or DNPP as substrates. The rates of release of mononucleotides from a polynucleotide were not compared with the rates of hydrolysis of the simple diesters or dinucleotides. Lerch & Wolf (81) concluded that highly polymerized nucleic acids would not be the normal substrate in the living cell. The exonuclease activity of oat PDase appeared to be too low to be detected in the assay used for the other nucleases (167). The removal of a few key terminal nucleotides from a nucleic acid might function in the regulation of nucleic acid metabolism, but no reports suggest that the exonucleases make any but a minor contribution to the total nuclease activity in plants.

Miscellaneous Enzymes

An RNase with a molecular weight near 12,500 and a specificity for the release of Up (and some G>p) was purified from cucumber seedlings (72). This is the only report of a plant enzyme that released a pyrimidine nucleotide as an early major product.

Barley (malt) had a PDase that hydrolyzed DNPP but not nucleic acids (45), and potato had a similar enzyme active on nitrophenyl-pN (119). A mung bean PDase hydrolyzed DNPP and N>p, but not RNA (150). Cyclic PDases active against 2':3'- and 3':5'-cyclic nucleotides were isolated from peas (82) and barley (168). Pea roots contain a factor that specifically degraded a plant leucyl tRNA (9), but whether this is a highly specific nuclease or a nonspecific nuclease acting on a particularly sensitive substrate were not determined. Exonucleases with pH optima at 5.5 were found in oats (167) and tobacco (188), but were not isolated and characterized. An enzyme from oats with properties similar to those of nuclease I gave a separate peak on a Sephadex column, but little is known about it.

I know of no isolated and partially purified plant DNase. Some reports are suggestive, but RNA was not specifically excluded as a substrate (23, 60).

Intraspecific Comparisons

Only tobacco has been reported to have all four nucleases. RNase I was identified at least three times, RNase II twice, nuclease I twice, and exonuclease once (Table 1).

That RNase I, nuclease I, and exonuclease of oats have been studied in one laboratory (167, 184, 189) lends confidence to the hypothesis that these different types do exist. Studies on the increase in RNase activity in plant tissues after treatment might lead to premature conclusions if the nature of the various nucleases involved was not determined (167). RNase I is the major nuclease of oat leaves, but nuclease I may increase to a major constituent under some conditions (184).

Mung bean RNase I was separated from nuclease I on a DEAE column (170). A mung bean extract yielded three fractions from a Sephadex G-100 column: a nuclease I, an RNase (RNase I?), and a small nuclease equally active on native and denatured DNA (69).

Corn RNase I, RNase II, and nuclease I were obtained from seedling parts and cell fractions by use of ion exchange columns, gel electrophoresis, and gel exclusion columns (175–178).

METHODS

Assays

The assay used most often for nucleases, the determination of the acid-soluble nucleotides liberated, does not distinguish among the different nucleases. Further, the many modifications used differ in volume (sometimes not completely specified), temperature, time, pH, buffer, cations, and other details, so that it is difficult or impossible to compare activities reported by different authors.

The activity of most nucleases cannot be defined by the recommendations in *Enzyme Nomenclature* (67), because there is not a one-to-one relationship between the number of bonds split and the number of nucleotides that become soluble. I recommend that a standard unit of nuclease be that amount of enzyme that, under standard conditions (substrate fully identified and temperature, pH, buffer, ionic strength, and time specified), releases one A_{260} unit of acid-soluble nucleotide/min (77). One A_{260} unit is that amount of (soluble) nucleotide that has an A_{260} of 1.0

in a volume of 1.0 ml. If activity is low, milliunits may be used. Spectrophotometer readings may be converted to standard units as follows:

$$\text{Units/ml} = \frac{\Delta \ A_{260} \ \times \ (\text{ml assay solution} + \text{ml precipitating agent}) \ \times \ (\text{dilution factor})}{(\text{ml enzyme solution assayed}) \ \times \ \text{min}}$$

Conversion factors for other units may be used. For example, the conversion factor for the nuclease unit that I have used [where a Δ A_{260} reading of 0.100 = 1 unit (173, 176)] would be calculated as follows:

$$1 \text{ old "Wilson" unit} = \frac{0.100 \ \times \ (2.5 \text{ ml} + 0.5 \text{ ml}) \ \times \ 15}{30 \text{ min}} = 0.15 \text{ standard unit}$$

The four common nucleotides have ϵ_{260} values of 6.8–14.2 $\times$ 10^3, with an average of 10.6 $\times$ 10^3. One A_{260} unit of a mixture of equal amounts of the four is about 0.1 μmole, allowing a rough comparison with other enzyme units. However, if the acid-soluble fraction is rich in purines, as it would be for plant nucleases, one A_{260} unit would contain less than 0.1 μmole. The acid-solubles will also include variable amounts of oligonucleotides, depending upon the precipitating agent, the ionic strength, and the RNA concentration at the end of the assay (36, 39, 59, 149). About three times more product was soluble in the commonly used uranyl acetate-perchloric acid reagent (6) than in a $Ba(ClO_4)_2$–2-ethoxyethanol precipitating agent (39). The former reagent makes a more sensitive assay, but the actual number of bonds split cannot be determined. Particulate fractions may produce artifacts when acid-soluble products are measured (11).

A rapid spectrophotometric assay, based on the increase in absorbance at 260 nm after hydrolysis of nucleic acids, appears to be quite useful for studies of kinetic properties (30). However, it may measure a slightly different aspect of the enzymatic action. Both the RNase I (166) and the nuclease (184) from oats appeared far more active on poly(A) than on poly(U) by this method, but the differences were much less when measured by an acid-soluble nucleotide assay (184, 189). Differential hyperchromicity of the substrates may influence the relative reaction rates.

Acetate buffer may give lower rates than other buffers with some nucleases (4; C. M. Wilson, unpublished). Ionic strength and pH affect the rate of reaction (12, 43). A mixture of enzymes in a partially purified ryegrass extract liberated N>p at pH 5.5, but both N>p and pN at pH 7.5 (43).

Inhibition and Stability

Phosphate and EDTA are often used in extraction buffers, though EDTA is known to inhibit nucleases under certain conditions. Phosphate (0.01 M) also inhibits some nucleases (41, 53, 103, 148) but not others, even from the same plant (54). EDTA in the extraction medium slightly inhibited the total activity of a crude corn extract, had little effect on a soybean extract, but tripled the activity of a cucumber extract

(182). *p*-Chloromercuribenzoate may activate nucleases or release them from an enzyme-inhibitor complex (46, 52, 74), but the optimum concentration for a wheat extract was only half that of the inhibitory concentration (52).

Pancreatic RNase is noted for its heat stability. Plant nucleases also tend to be heat stable, but to a lesser extent. The RNases from barley (40) and mung bean (152) showed variable stability toward heat, depending upon the relative purity of the preparation and the concentration of cations. Wheat nuclease was more stable at 70° than at 65° (53), suggesting that instability was caused by a thermolabile factor (152). RNase activity of flax cotyledon homogenates increased 50% in activity after 10 min at 60°, but the RNase activity of rust-infected cotyledons was temperature sensitive (137).

The relative stabilities of different enzymes in a mixture can be important in purification and separation procedures. Methods for quantitative assay of the individual nucleases of oats and tobacco after Sephadex gel filtration have been reported (184, 185, 188). However, no data were presented to confirm that recovery was quantitative, and in some studies (167, 184) more than half of some enzymes appears to have been lost. Even if the results are only semiquantitative, the techniques of Udvardy, Wyen, and co-workers are an important advance in illustrating the changes in the different nucleases after treatment or aging of plants (164, 165, 184, 185, 188, 189).

Gel Electrophoresis and Gel Filtration

Polyacrylamide gel electrophoresis separated the nucleases of wheat leaves into 11 bands, three of which corresponded to bands of PDases acting on α-naphthyl-pU (183). Bean root, kidney bean root, and sugar beet leaf nucleases separated into four, two, and two bands, respectively (183). Hybrid corn roots contain one RNase I band, two RNase II bands, and three nuclease bands, the last distinguished by EDTA sensitivity (178). Some of the electrophoretically separated enzymes from corn and tobacco were quite sensitive to changes in buffers and salt concentration during the selective enzymatic staining test, so that quantitation was difficult or impossible (178).

Molecular weights of enzymes in crude homogenates and semipurified preparations are conveniently estimated by gel filtration on Sephadex columns. Sephadex may retard basic groups and accelerate acid groups unless the eluting agent has an ionic strength of at least 0.05 (5). Wheat leaf RNase I was separated from a nuclease on a Sephadex column, eluted with 0.1 M sodium acetate, as if it were a CM-cellulose column (145). Proteases may reduce the size of a protein during purification steps, sometimes without affecting the activity of the protein (20). A purified wheat RNase was reported to have a molecular weight of 9000 after gel filtration on Sephadex G-75, yet the RNase in the crude extract had been eluted from a Sephadex G-100 column at a position corresponding to a higher molecular weight (160).

Sephadex columns are used to distinguish between the products of exo- and endonucleases (16), but artifacts are possible here also. Corn nuclease I is an endonuclease (177), but apparent exonuclease products were detected when the columns were mistakenly eluted with a low ionic strength buffer (C. M. Wilson, unpublished).

Subcellular Localization

The work in this area is confusing and, without exception, incomplete. Recommended reading is de Duve's essay on tissue fractionation (34). He emphasizes the importance of being quantitative, analyzing all fractions, and keeping an accurate balance sheet. Also, he believes that results should not be reported exclusively in terms of specific activity, which gives incomplete information and often leaves the reader in doubt as to how much of a particular enzyme is present in a fraction (34). The four major cell fractions are the nuclear fraction (plus chloroplasts, if present), the mitochondrial fraction (including lysosomes and peroxisomes), the microsomal fraction (the particles isolated from the postmitochondrial supernatant), and the supernatant fraction (34).

The first fraction, obtained with a low-speed centrifugation, is often discarded, though it may contain one-third (182) or most (129) of the total plant nuclease activity. A nuclease was associated with chromatin from barley leaves (148). In contrast, so little nuclease was found in pea nuclei that it appeared to be random contamination from the soluble phase (85). Nucleases with a high specific activity and a high pH optimum were found in wheat chloroplasts (52).

The mitochondrial fraction was an enriched source for the isolation of purified nuclease I from corn (176, 178). Subfractions called lysosomes (91, 93), phytolysosomes (28), spherosomes (93), vacuoles (90), or aleurone vacuoles (92) contain nucleases with higher specific activities than some other fractions, but the results were not quantitative, the particles were heterogeneous, and the enzymes were not identified. Little of the total RNase activity of wheat aleurone (47) or of potato shoots (115) was recovered from a lysosome fraction.

The microsomal fraction contains RNase II (68, 125, 176, 178) and other nucleases in a mixture of ribosomes with various membrane fragments and small pieces. Purified ribosomes may be associated with a mixture of RNases and nucleases, apparently adsorbed from the cells from which they were isolated (38a, 61). Ribosomes with different levels of RNase activity can be isolated from various tissues, but ribosomes low in activity may (38a) or may not (99) adsorb RNase. Oat ribosomes were associated with an RNase (186) that could be either RNase I or RNase II. Pea microsomes were fractionated into an endoplasmic reticulum fraction and a ribonucleoprotein particle fraction (98). The former contained an EDTA-sensitive enzyme (nuclease I?) and the latter an EDTA-insensitive enzyme with a pH optimum at 5.6 (RNase II?).

That the soluble or supernatant fraction is the usual source of purified enzyme preparations suggests that all plant nucleases may become soluble under appropriate conditions and that the supernatant nucleases may be quite a mixture. The proportion of the total nuclease recovered in the supernatant fraction may range from 25% to 95%, with most of the activity soluble in high salt (0.3–0.5 M) buffers (182).

The distribution of nucleases among the various fractions of a homogenate can be drastically affected by changes in buffers and ionic concentration (182). The RNase activity recovered from pea root mitochondrial and microsomal fractions varied from 13 to 30% and from 9 to 20%, respectively, depending upon the buffer used (57). Selection of buffers made it possible to separate three types of enzymes from corn seedlings (176). RNase I is the predominant nuclease in many soluble preparations, but suggestions that it may have been released by rupture of lysosome-like bodies have not been accompanied by clear distinctions between RNase I and the other nucleases. When the soluble fraction is compared to particulate fractions containing the other nucleases, the former almost invariably has the lower pH optimum, and the latter are often EDTA sensitive (52, 74, 96, 176). When the different nucleases from a species have been separated, their relative activities in different pH buffers and on RNA and DNA can be compared with the same activities in a mixture of nucleases to give some idea about the proportions of the different enzymes (52, 64, 176).

METABOLIC FUNCTIONS OF PLANT NUCLEASES

The major metabolic function for the nucleases is presumably the degradation of nucleic acids, but little is known about the in vivo rates of reaction or the control mechanisms that prevent the nucleases from rapidly degrading all cellular nucleic acid. Given the variety of enzymes known, we may speculate that each has a certain function, is active on specific substrates at the appropriate time in the life cycle, and is kept inactive at other times by physical separation, natural inhibitors, lack of suitable intracellular conditions, binding of substrates to proteins, or substrates in resistant conformational states. Without knowledge of these things, we cannot with certainty place the nucleases in an overall picture of cellular metabolism.

The potential degradation reactions of the known nucleases and associated enzymes are summarized by seven reactions:

$$RNA\ (DNA) \xrightarrow{\text{nuclease I, exonuclease}} pN\ (pdN) \qquad 1.$$

$$RNA \xrightarrow{\text{RNase I, RNase II}} N{>}p\ (R{>}p\ +\ Y{>}p) \qquad 2.$$

$$R{>}p \xrightarrow{\text{RNase I, RNase II, cyclic PDase}} Rp \qquad 3.$$

$$Y{>}p \xrightarrow{\text{RNase II, cyclic PDase}} Yp \qquad 4.$$

$$Np \xrightarrow{\text{nucleoside phosphotransferase}} pN\ (21,\ 128) \qquad 5.$$

$$Np \xrightarrow{\text{3'-nucleotidase, nuclease I}} N\ +\ P_i \qquad 6.$$

$$N\ +\ ATP \xrightarrow{\text{nucleoside kinase}} pN\ (125) \qquad 7.$$

Once a nucleic acid has lost its biological value, its constituent parts may be recycled into a new nucleic acid if the breakdown products can be converted into 5'-nucleotides. Nuclease I and exonuclease yield the desired products in one step (reaction

1), but several enzymes and two pathways (reactions 2, 3, and 4, followed by 5 or 6 and 7) may be used when the RNases degrade RNA. No data exist that allow us to decide which pathway predominates in vivo.

Nucleases in the Life Cycle of the Plant

The resting seed contains some nuclease (usually only RNase activity is examined), and it increases upon germination (13, 64, 95). The early increase in activity in the storage organ might be caused by activation of pre-existing enzyme (10, 10a) or merely by a nonspecific hydration that allows the enzyme to be extracted (62). As germination continues, the RNA content of the major storage organ declines, presumably the result of the increasing RNase activity (50, 80, 108, 157). In corn endosperm, the initial low nucleic acid content is only slightly affected by germination (64). Barley aleurone released RNase in response to gibberellic acid (later than it secreted amylase), but the specific enzyme was not identified (27). This enzyme would accumulate in the endosperm. The release of RNase from the aleurone was associated with proliferation of dictyosome vesicles (71). The cotyledons of peas (11) and the scutella of corn (64) and of barley (80) had increasing amounts of RNase activity throughout germination, with the later increase in peas caused by new synthesis (10, 10a). RNase I is secreted by corn scutella (173). The developing axis greatly increases in size and in RNase activity during the first 2 weeks (13, 79), and it is often noted that the RNase activity increases in parallel with RNA synthesis and accumulation (50, 64, 79, 80). The RNase bound to microsomes isolated from rye seed embryos had a higher specific activity than the RNase bound to microsomes from germinating embryos (143), but ribosomes from rye (38a) or wheat (169) embryos had little or no detectable RNase.

Nuclease activity is usually low in the root tip and increases in the regions of cell elongation and maturation (88, 111, 127, 132). However, when the extreme tip (0.2 mm or root cap) of *Lens culinaris* root was assayed, it was found to have a high RNase activity, whereas the meristem and quiescent center (0.2–0.5 mm) had much lower RNase activity and higher RNA and auxin contents (110). Corn root tips may have a nuclease or DNase that is different from that in the more mature portions of the root (176). Cells in corn mesocotyl tissue increased in RNase content as they expanded and matured (138). DNase activity was highest in synchronous *Chlorella* cultures at the time of DNA synthesis (135).

When etiolated lupin hypocotyls were illuminated, growth was inhibited, polysomes decreased, and there was a large increase in an RNase associated with ribosomes (3). This may be a phytochrome controlled system (2). After illumination of etiolated oat leaves, RNase I increased whereas nuclease I decreased (184). As oat leaves aged naturally, attached to the plant, the relative level of nuclease I increased whereas RNase I decreased (184). In bean leaves RNase (assayed at pH 6.8) tended to parallel protein, chlorophyll, and RNA content during senescence (109). RNase and DNase activities dropped in the first leaf of barley from 7 days after germination (83, 146), whereas a nuclease associated with chromatin increased dramatically (146). The chromatin-associated nuclease also increased in excised leaves, where most of the increase was prevented by kinetin (147). Little is known about the properties of this enzyme or whether it makes up more than a minor part of the total nuclease activity.

Only a few papers relate to the flowering process. A single inductive long night caused changes in RNase activity opposite to changes in the rate of nucleic acid synthesis in *Pharbitis nil* (51) and *Chenopodium rubrum* (159). During the synchronous meiotic cycle of *Lilium,* a specific endonuclease appeared that produced a repairable single-strand break in double-stranded DNA (60). It was suggested that this enzyme, together with a polynucleotide kinase and a polynucleotide ligase, might be responsible for genetic recombination activities in meiosis. The RNase activity in the style varied greatly among species (136). The corolla of *Ipomoea* (morning glory) wilted rapidly at a time when RNase, DNase, and β-glucosidase activities rose rapidly (94). Cycloheximide inhibited the increase in discs from the corolla, suggesting that the enzymes were newly synthesized. The RNA content began dropping before the time of the increase in RNase activity, whereas the loss of DNA coincided with the period of high DNase activity (94).

During rapid seed development, RNase increased in the seeds of wheat (97), rice (29), and corn (31, 63) at the same time that protein, starch, and RNA were increasing. In corn endosperm the RNA content dropped at maturity whereas the RNase level remained relatively high (63); the reverse occurred in wheat (97) and rice (29). The RNase content appears to be far out of proportion to the rather low RNA levels found in seeds, and its function during development is unknown. The maximum RNase level in corn endosperm was genetically controlled (179). RNase was not uniformly distributed in corn endosperm in the early stages of development (180), suggesting that the rate equations for RNase synthesis (31) may not be relevant. Maturing corn (178) and wheat (96) endosperms contain mostly RNase I.

Corn endosperms homozygous for the opaque-2 mutation, which disrupts zein synthesis, contained two to five times as much RNase as their normal counterparts (22, 31, 32, 179). Although it was suggested that the high RNase content might inhibit zein synthesis by acting on a specific messenger RNA (33, 181), no evidence supports the idea of any direct link between the high RNase and the changes in protein synthesis (32). Floury-2, another mutant that inhibits zein synthesis, does not affect the RNase level (31, 179).

RNase activity, as well as lipase and an acid phosphatase, first increased and then decreased with respiration during the climacteric period in apple peel tissue (126).

Two isoenzymes of RNase II were separated from hybrid corn roots, one derived from each inbred parent (178). No isoenzymes of RNase I are known; the second band in a purified preparation is probably an artifact (175, 178). The nature of the multiple electrophoretic bands of corn nuclease I (178) and nucleases of several other species (106, 183) is not known. No subunits of any nuclease have been reported, except for the exonucleases of carrot (56) and sugar beet (81).

A few comparisons may be made on relative enzyme levels within a plant or between plant parts. Nuclease I may make up over half of the total RNase activity of corn roots, along with 30% RNase I and 10–20% RNase II (C. M. Wilson, unpublished). Young oat leaves have 6.7 times as much RNase I as nuclease I, but this ratio drops with age (184). In tobacco the ratio is 5:1 (188). However, the ratio of RNase I to RNase II in tobacco leaves was reported to be 1:1 in young leaves and up to 5:1 in older leaves (68). Frozen mung bean sprouts were reported by two laboratories to have more of nuclease I than of RNase I (69, 170). RNase II and nuclease I tend to be unstable and are more easily lost during extraction; thus they may be underestimated (176).

Corn roots contain up to 30 standard units/g fresh weight (total nuclease activity), young leaves about 7.5 units/g, the scutella of 3-day seedlings about 60 units/g, and the developing endosperms of various inbreds from 43 to 195 units/g (176, 179; C. M. Wilson, unpublished). The endosperms of opaque-2 mutants may contain 750 units/g (179). The total RNase in the different organs of barley (80) and corn (64) increased throughout germination.

Interactions of Stress and Hormones

Plant parts that are stressed, as by rubbing or excision, usually have an increased level of nuclease activity after a few hours, and hormones are usually tested on plant parts that have been stressed. A somewhat complicated interaction of influences must be considered, and only a few of the many possible combinations can be covered. Other references are found in a recent review (38).

Three hours after excision, the RNase I content of excised and illuminated oat leaves doubled (185). The increase was prevented by kinetin, but was accelerated by ABA or benzimidazole. Nuclease I was unchanged during this period, but increased 96 hr after excision (184). When the ABA effect on excised oat leaves was examined over 24 hr, the early increase in nuclease activity was an RNase I, and the later increase was a nuclease I (164). Interestingly, 2 mM ATP produced nuclease changes in excised oat leaves similar to those produced by ABA (165). Nuclease I increased in excised wheat leaves kept in the dark for several days, but RNase I increased if the leaves were illuminated (144). Kinetin, tested only in the dark, prevented most of the increase in the nuclease (144). The most rapid rise in RNase in *Rhoeo discolor* leaves occurred between 3 and 6 hr after excision, and a later increase was stimulated by ABA and reversed by auxin (35). The ABA effect was seen only if the leaves were taken from slowly growing plants. The RNase level in the abscission zone of *Phaseolus* leaf explants increased during the first few hours, then increased to a still higher level at 24 hr; only the latter increase was prevented by auxin (1).

A very large increase in RNase activity followed damage to tubers and leaves of potato (114). Attempts were made to determine whether the new activity represented activation of preformed enzyme (perhaps in lysosomes) or de novo synthesis of new molecules. Immunochemical assays of potato RNase after tuber damage (112) and assays for incorporation of ^{14}C-leucine into newly formed RNase in leaves (113) suggested that new RNase molecules were not synthesized. Unfortunately, the purification procedures appeared to be inadequate and the results are not conclusive.

When growth was inhibited by irradiation (26, 48) or fluoride (25), the RNase activity increased. Auxin reversed the effects of irradiation on RNA loss and rooting without preventing the increase in RNase (49). Water deficiency also produced high RNase activity, but only after severe drought symptoms had appeared (86) or changes in polysomes were detected (102). Although ABA increased and kinetin decreased the RNase levels in intact leaves or in humid detached leaves, the two hormones had the opposite effects if the leaves were allowed to dry (7).

Wheat coleoptile sections or pea stem sections treated with auxin had higher growth rates and lower RNase levels than the control sections (161). In contrast, levels of 2,4-dichlorophenoxyacetic acid that stimulated growth of corn mesocotyl sections also increased the RNase level, and a concentration that inhibited growth also produced lower RNase levels (138). RNase activity increased in corn coleoptiles treated with ABA, but only after other changes had taken place in RNA metabolism, suggesting that the increase in RNase was a secondary effect (14).

Auxin stimulated the growth of decapitated pea stems accompanied by a dramatic increase in the RNase recovered in the microsomal fractions (15). Auxin plus cytokinin, cytokinin alone, or gibberellic acid alone also stimulated growth, but with no increase in the microsomal RNase. The microsomal enzyme was not identified. When polysomes were isolated from the microsomes, the RNase was found in the soluble and membrane portions of the density gradient (15).

Pathology

It has often been noted that plants infected with pathogens have elevated RNase levels, though care must be taken to avoid confusing a specific change caused by the host-pathogen interaction with that caused by mechanical damage during inoculation. Both RNase (24, 129–131) and DNase (130, 131) activities increased in wheat seedlings infected with compatible strains of *Puccinia* rust, whereas the change was later (131), smaller (129), or only very slight (24) with incompatible strains. The increased RNase activity in infected wheat appeared in only three of ten fractions separated on a hydroxyapatite column (131) or had a lower heat stability and an increased ability to hydrolyze poly(C) (24). Heat stability and poly(C) hydrolysis changed also in rust-infected flax (137). In both wheat and flax, a small increase in RNase activity occurred 3 or 4 days after either resistant or susceptible seedlings were inoculated, but a second and larger increase occurred only in the susceptible seedlings (24, 137). The levels of DNase, PDase, phosphatase, and RNase increased in mechanically damaged flax cotyledons, whereas only RNase increased in infected cotyledons (137).

RNase levels are also elevated in virus-infected tissues. It has been suggested that the high RNase, by breaking down host RNA into products that can be incorporated into newly synthesized virus RNA, may favor the establishment of the virus (122, 133). Others reported little difference between the RNase changes induced by mechanical injury and those induced by a virus, because virus lesions are also damaged tissues (37, 188). Increased RNase did not appear to be associated with increased virus infection (37, 188). The virus-induced enzyme appeared to be a soluble RNase (117) or RNase I (188). The longer lasting effects of the virus on RNase levels, compared to those caused by the inoculation damage, may be caused by the continued stress applied to the infected plant. There are no reports that RNases degrade the viral RNA in vivo as part of a protective reaction to infection.

The complications of the wounding-virus-host-RNase interactions were discussed by Diener (37).

Transformed tissue cultures derived from *Vinca rosea* inoculated with *Agrobacterium tumefaciens* (crown gall bacterium) had a greatly increased RNase activity, with a lower pH optimum than that in normal and transformed tissues (123). In contrast, rapidly dividing cells of crown gall tumors on tobacco plants contained little RNase, as revealed by a histochemical test, though differentiating tissues were RNase positive (142).

CONCLUDING REMARKS

The major known endonucleases of higher plants can be grouped into three classes —plant RNase I, plant RNase II, and nuclease I—each with distinctive properties. Differences reported for nucleases from different species are as likely to be caused by variations in analytical techniques as by true differences between enzymes. The relative proportions of these three nucleases vary in different organs, cell fractions, and after periods of growth, senescence, and stress, as shown by the work of Udvardy, Wyen, and co-workers on oats and of Wilson on corn. Cytokinins reduce total nuclease levels in stressed plant parts, ABA increases nuclease levels, and auxin may either increase or decrease nuclease levels, depending upon conditions. No direct correlation between RNase levels and RNA metabolism has been established. In the storage organs of germinating seeds and in excised plant parts, RNase activity increases and RNA decreases, but in the developing axis of a germinating seedling, RNase activity and RNA content increase together. Further understanding will come only as the different nucleases are assayed separately and their individual functions in RNA metabolism are established.

Literature Cited

1. Abeles, F. B., Holm, R. E., Gahagan, H. E. 1968. In *Biochemistry and Physiology of Plant Growth Substances*, ed. F. Wightman, G. Setterfield, 1515–23. Ottawa: Runge
2. Acton, G. J. 1972. *Nature New Biol.* 236:255–56
3. Acton, G. J., Lewington, R. J., Myers, A. 1970. *Biochim. Biophys. Acta* 204:144–55
4. Ambellan, E. 1972. *Anal. Biochem.* 50:453–59
5. Andrews, P. 1970. *Methods Biochem. Anal.* 18:1–53
6. Anfinsen, C. B., Redfield, R. R., Choate, W. L., Page, J., Carroll, W. R. 1954. *J. Biol. Chem.* 207:201–10
7. Arad, S. M., Mizrahi, Y., Richmond, A. E. 1973. *Plant Physiol.* 52:510–12
8. Ardelt, W., Laskowski, M. Sr. 1971. *Biochem. Biophys. Res. Commun.* 44:1205–11
9. Babcock, D. F., Morris, R. O. 1973. *Plant Physiol.* 52:292–97
10. Barker, G. R., Bray, C. M. 1972. *Nucleic Acids and Proteins in Higher Plants*, ed. G. L. Farkas, 61–68. Budapest: Akademai Kiado. 372 pp.
10a. Barker, G. R., Bray, C. M., Walter, T. J. 1974. *Biochem. J.* 142:211–19
11. Barker, G. R., Hollinshead, J. A. 1967. *Biochem. J.* 103:230–37
12. Barnard, E. A. 1969. *Ann. Rev. Biochem.* 38:677–732
13. Beevers, L., Splittstoesser, W. E. 1968. *J. Exp. Bot.* 19:698–711
14. Bex, J. H. M. 1972. *Planta* 103:1–10
15. Birmingham, B. C., Maclachlan, G. A. 1972. *Plant Physiol.* 49:371–75
16. Birnboim, H. 1966. *Biochim. Biophys. Acta* 119:198–200
17. Bjork, W. 1965. *Biochim. Biophys. Acta* 95:652–62
18. Bjork, W. 1967. *Ark. Kemi* 27:539–54

19. Brawerman, G., Chargaff, E. 1954. *J. Biol. Chem.* 210:445–54
20. Briggs, W. R., Rice, H. V. 1972. *Ann. Rev. Plant Physiol.* 23:293–334
21. Brunngraber, E. F., Chargaff, E. 1967. *J. Biol. Chem.* 242:4834–40
22. Cagampang, G. B., Dalby, A. 1972. *Can. J. Plant Sci.* 52:901–5
23. Carell, E. F., Egan, J. M., Pratt, E. A. 1970. *Arch. Biochem. Biophys.* 138:26–31
24. Chakravorty, A. K., Shaw, M., Scrubb, L. A. 1974. *Nature* 247:577–80
25. Chang, C. W. 1970. *Can. J. Biochem.* 48:450–54
26. Cherry, J. H. 1962. *Biochim. Biophys. Acta* 55:487–94
27. Chrispeels, M. J., Varner, J. E. 1967. *Plant Physiol.* 42:398–406
28. Coulomb, P. 1971. *J. Microsc. (Paris)* 11:299–318
29. Cruz, L. J., Cagampang, G. B., Juliano, B. O. 1970. *Plant Physiol.* 46:743–47
30. Cuatrecasas, P., Fuchs, S., Anfinsen, C. B. 1967. *J. Biol. Chem.* 242:1541–47
31. Dalby, A., Cagampang, G. B. 1970. *Plant Physiol.* 46:142–44
32. Dalby, A., Cagampang, G. B., Davies, I. ab I., Murphy, J. J. 1972. *Symposium: Seed Proteins,* ed. G. E. Inglett, 39–51. Westport, Conn.: Avi. 320 pp.
33. Dalby, A., Davies, I. Ab I. 1967. *Science* 155:1573–75
34. de Duve, C. 1971. *J. Cell Biol.* 50:20D–55D
35. De Leo, P., Sacher, J. A. 1970. *Plant Physiol.* 46:806–11
36. Dickman, S. R., Ring, B. 1958. *J. Biol. Chem.* 231:741–50
37. Diener, T. O. 1961. *Virology* 14:177–89
38. Dove, L. D. 1973. *Phytochemistry* 12:2561–70
38a. Dyer, T. A., Payne, P. I. 1974. *Planta* 117:259–68
39. Fiers, W. 1961. *Anal. Biochem.* 2:126–39
40. Fiers, W., Lepoutre, L. 1961. *Arch. Int. Physiol. Biochim.* 69:582–84
41. Fiers, W., Vandendriessche, L. 1961. *Arch. Int. Physiol. Biochim.* 69:339–63
42. Forti, G., Tognoli, C., Parisi, B. 1962. *Biochim. Biophys. Acta* 62:251–60
43. Freeman, K. B. 1964. *Can. J. Biochem.* 42:1099–1109
44. Frisch-Niggemeyer, W., Reddi, K. K. 1957. *Biochim. Biophys. Acta* 26:40–46
45. Georgatsos, J. G. 1963. *Arch. Int. Physiol. Biochim.* 71:674–79
46. Geri, C., Parenti, R., Durante, M. 1973. *Ital. J. Biochem.* 22:23–35

47. Gibson, R. A., Paleg, L. G. 1972. *Biochem. J.* 128:367–75
48. Gordon, S. A., Buess, E. M. 1972. *Radiat. Bot.* 12:361–64
49. Ibid 1973. 13:283–86
50. Grellet, F., Julien, R., Milhomme, H. 1968. *Physiol. Veg.* 6:11–17
51. Gressel, J., Zilberstein, A., Arzee, T. 1970. *Develop. Biol.* 22:31–42
52. Hadziyev, D., Mehta, S. L., Zalik, S. 1969. *Can. J. Biochem.* 47:273–82
53. Hanson, D. M., Fairley, J. L. 1969. *J. Biol. Chem.* 244:2440–49
54. Hara, A., Yoshihara, K., Watanabe, T. 1969. *Bull. Fac. Agr. Kagoshima Univ.* No. 19:81–88
55. Harvey, C., Malsman, L., Nussbaum, A. L. 1967. *Biochemistry* 6:3689–94
56. Harvey, C. L., Olson, K. C., Wright, R. 1970. *Biochemistry* 9:921–25
57. Hirai, M., Asahi, T. 1973. *Plant Cell Physiol.* 14:1019–29
58. Holbrook, J. J., Ortanderl, F., Pfleiderer, G. 1966. *Biochem. Z.* 345:427–39
59. Holden, M., Pirie, N. W. 1955. *Biochem. J.* 60:53–62
60. Howell, S. H., Stern, H. 1971. *J. Mol. Biol.* 55:357–78
61. Hsiao, T. C. 1968. *Plant Physiol.* 43:1355–61
62. Ingle, J. 1963. *Biochim. Biophys. Acta* 73:331–34
63. Ingle, J., Beitz, D., Hageman, R. H. 1965. *Plant Physiol.* 40:835–39
64. Ingle, J., Hageman, R. H. 1965. *Plant Physiol.* 40:48–53
65. International Union of Biochemistry 1965. *Enzyme Nomenclature.* New York: Elsevier. 219 pp.
66. IUPAC-IUB Commission on Biochemical Nomenclature 1970. *J. Biol. Chem.* 245:5171–76
67. IUPAC-IUB Commission on Biochemical Nomenclature 1973. *Enzyme Nomenclature.* Amsterdam: Elsevier. 443 pp.
68. Jervis, L. 1974. *Phytochemistry* 13:709–14
69. Johnson, P. H., Laskowski, M. Sr. 1968. *J. Biol. Chem.* 243:3421–24
70. Ibid 1970. 245:891–98
71. Jones, R. L., Price, J. M. 1970. *Planta* 94:191–202
72. Kado, C. I. 1968. *Arch. Biochem. Biophys.* 125:86–93
73. Kedzierski, W., Laskowski, M. Sr., Mandel, M. 1973. *J. Biol. Chem.* 248:1277–80
74. Kessler, B., Engelberg, N. 1962. *Biochim. Biophys. Acta* 55:70–82

75. Khorana, H. G. 1961. *Enzymes* 5B:79–94
76. Kole, R., Sierakowska, H., Szemplinska, H., Shugar, D. 1974. *Nucleic Acids Res.* 1:699–706
77. Lantero, O. J., Klosterman, H. J. 1973. *Phytochemistry* 12:775–84
78. Laskowski, M. Sr. 1967. *Advan. Enzymol.* 29:165–220
79. Ledoux, L., Galand, P., Huart, R. 1962. *Biochim. Biophys. Acta* 55:97–104
80. Ledoux, L., Galand, P., Huart, R. 1962. *Exp. Cell Res.* 27:132–36
81. Lerch, B., Wolf, G. 1972. *Biochim. Biophys. Acta* 258:206–18
82. Lin, P. P-C., Varner, J. E. 1972. *Biochim. Biophys. Acta* 276:454–74
83. Lontai, I., van Loon, L. C., Bruinsma, J. 1972. *Z. Pflanzenphysiol.* 67:146–54
84. Loring, H. S., McLennan, J. E., Walters, T. L. 1966. *J. Biol. Chem.* 241:2876–80
85. Lyndon, R. F. 1966. *Biochim. Biophys. Acta* 113:110–19
86. Maranville, J. W., Paulsen, G. M. 1972. *Crop Sci.* 12:660–63
87. Markham, R., Strominger, J. L. 1956. *Biochem. J.* 64:46–47P
88. Maroti, M. 1969. *Acta Biol. Acad. Sci. Hung.* 20:263–68
89. Masuda, T., Hagiwara, K. 1972. *J. Agr. Chem. Soc. Jap.* 46:437–41
90. Matile, P. 1966. *Z. Naturforsch. B* 21:871–78
91. Matile, P. 1968. *Planta* 79:181–96
92. Matile, P. 1968. *Z. Pflanzenphysiol.* 58:365–68
93. Matile, P., Balz, J. P., Semadeni, E., Jost, M. 1965. *Z. Naturforsch. B* 20:693–98
94. Matile, P., Winkenbach, F. 1971. *J. Exp. Bot.* 23:759–71
95. Matsushita, S. 1959. *Mem. Res. Inst. Food Sci. Kyoto* 17:23–28
96. Ibid, 29–41
97. Ibid. 19:1–4
98. Matsushita, S., Ibuki, F. 1960. *Biochim. Biophys. Acta* 40:358–59
99. Matsushita, S., Mori, T., Hata, T. 1966. *Plant Cell Physiol.* 7:533–45
100. Merola, A. J., Davis, F. F. 1962. *Biochim. Biophys. Acta* 55:431–39
101. Mikulski, A. J., Laskowski, M. Sr. 1970. *J. Biol. Chem.* 245:5026–31
102. Morilla, C. A., Boyer, J. S., Hageman, R. H. 1973. *Plant Physiol.* 51:817–24
103. Mukai, J-I. 1965. *J. Fac. Agr. Kyushu Univ.* 13:361–68
104. Muschek, L. D. 1970. *The purification and characterization of a nuclease from the seeds of muskmelon.* PhD thesis.

Michigan State Univ., East Lansing, Mich. 164 pp.
105. Nomura, A., Suno, M., Mizuno, Y. 1971. *J. Biochem. (Tokyo)* 70:993–1001
106. Oleson, A. E., Janski, A. M., Clark, E. T. 1974. *Biochim. Biophys. Acta* 366:89–100
107. Palmer, C. E., Barker, W. G. 1972. *Plant Cell Physiol.* 13:681–88
108. Palmiano, E. P., Juliano, B. O. 1972. *Plant Physiol.* 49:751–56
109. Phillips, D. R., Fletcher, R. A. 1969. *Physiol. Plant.* 22:764–67
110. Pilet, P. E., Braun, R. 1970. *Physiol. Plant.* 23:245–50
111. Pilet, P. E., Pratt, R., Roland, J. C. 1972. *Plant Cell Physiol.* 13:297–309
112. Pitt, D. 1971. *Planta* 101:333–51
113. Ibid 1974. 117:43–55
114. Pitt, D., Galpin, M. 1971. *Planta* 101:317–32
115. Ibid 1973. 109:233–58
116. Privat de Garilhe, M. 1967. *Enzymes in Nucleic Acid Research.* San Francisco: Holden-Day. 393 pp.
117. Randles, J. W. 1968. *Virology* 36:556–63
118. Razzell, W. E. 1966. *Biochim. Biophys. Res. Commun.* 22:243–47
119. Razzell, W. E. 1968. *Can. J. Biochem.* 46:1–7
120. Reddi, K. K. 1958. *Biochim. Biophys. Acta* 28:386–91
121. Ibid. 30:638
122. Reddi, K. K. 1963. *Proc. Nat. Acad. Sci. USA* 50:75–81
123. Ibid 1966. 56:1207–14
124. Reddi, K. K. 1966. *Procedures in Nucleic Acid Research,* ed. G. L. Cantoni, D. R. Davies, 71–78. New York: Harper & Row. 667 pp.
125. Reddi, K. K., Mauser, L. J. 1965. *Proc. Nat. Acad. Sci. USA* 53:607–13
126. Rhodes, M. J. C., Wooltorton, L. S. C. 1967. *Phytochemistry* 6:1–12
127. Robinson, E., Cartwright, P. M. 1958. *J. Exp. Bot.* 9:430–35
128. Rodgers, R., Chargaff, E. 1972. *J. Biol. Chem.* 247:5448–55
129. Rohringer, R., Samborski, D. J., Person, C. O. 1961. *Can. J. Bot.* 39:775–84
130. Sachse, B., Wolf, G. 1970. *Phytopathol. Z.* 68:276–79
131. Sachse, B., Wolf, G., Fuchs, W. H. 1971. *Acta Phytopathol. Acad. Sci. Hung.* 6:39–49
132. Sahulka, J. 1971. *Biol. Plant.* 13:243–46
133. Santilli, V., Nepokroeff, C. M., Gagliardi, N. C. 1962. *Nature* 193:656–58
134. Schmidt, G., Laskowski, M. Sr. 1961. *Enzymes* 5B:3–35

135. Schonherr, O. T., Wanka, F., Kuyper, C. M. A. 1970. *Biochim. Biophys. Acta* 224:74–79
136. Schrauwen, J., Linskens, H. F. 1972. *Planta* 102:277–85
137. Scrubb, L. A., Chakravorty, A. K., Shaw, M. 1972. *Plant Physiol.* 50:73–79
138. Shannon, J. C., Hanson, J. B., Wilson, C. M. 1964. *Plant Physiol.* 39:804–9
139. Shuster, L. 1957. *J. Biol. Chem.* 229:289–303
140. Shuster, L., Kaplan, N. O. 1955. *Methods Enzymol.* 2:551–55
141. Shuster, L., Khorana, H. G., Heppel, L. A. 1959. *Biochim. Biophys. Acta* 33:452–61
142. Simard, A. 1973. *Rev. Can. Biol.* 32:261–66
143. Siwecka, M. A., Golaszewski, T., Szarkowski, J. W. 1973. *Bull. Acad. Pol. Sci. Ser. Sci. Biol.* 21:171–75
144. Sodek, L., Wright, S. T. C. 1969. *Phytochemistry* 8:1629–40
145. Sodek, L., Wright, S. T. C., Wilson, C. M. 1970. *Plant Cell Physiol.* 11:167–71
146. Srivastava, B. I. S. 1968. *Biochem. J.* 110:683–86
147. Srivastava, B. I. S. 1968. *Biochem. Biophys. Res. Commun.* 32:533–38
148. Srivastava, B. I. S., Matsumoto, H., Chadha, K. C. 1971. *Plant Cell Physiol.* 12:609–18
149. Stockx, J., Dierick, W., Vandendriessche, L. 1964. *Arch. Int. Physiol. Biochim.* 72:647–60
150. Ibid 1965. 73:421–31
151. Stockx, J., Vandendriessche, L. 1961. *Arch. Int. Physiol. Biochim.* 69:493–504
152. Ibid, 521–44
153. Ibid, 545–62
154. Stockx, J., Van Parijs, R. 1961. *Arch. Int. Physiol. Biochim.* 69:194–202
155. Sung, S–C., Laskowski, M. Sr. 1962. *J. Biol. Chem.* 237:506–11
156. Suno, M., Nomura, A., Mizuno, Y. 1973. *J. Biochem. (Tokyo)* 73:1291–97
157. Sutcliffe, J. F., Baset, Q. A. 1973. *Plant Sci. Lett.* 1:15–20
158. Tang, W. J., Maretzki, A. 1970. *Biochim. Biophys. Acta* 212:300–7
159. Teltscherová, L., Pleskotová, D. 1973. *Biol. Plant.* 16:136–39
160. Torti, G., Mapelli, S., Soave, C. 1973. *Biochim. Biophys. Acta* 324:254–66
161. Truelsen, T. A. 1967. *Physiol. Plant.* 20:1112–19

162. Tuve, T. W., Anfinsen, C. B. 1960. *J. Biol. Chem.* 235:3437–41
163. Uchida, T., Egami, F. 1971. *Enzymes* 4:205–50
164. Udvardy, J., Farkas, G. L. 1972. *J. Exp. Bot.* 23:914–20
165. Udvardy, J., Farkas, G. L. 1973. *Z. Pflanzenphysiol.* 69:394–401
166. Udvardy, J., Farkas, G. L., Marre, E. 1969. *Plant Cell Physiol.* 10:375–86
167. Udvardy, J., Marre, E., Farkas, G. L. 1970. *Biochim. Biophys. Acta* 206:392–403
168. Vandepeute, J., Huffaker, R. C., Alvarez, R. 1973. *Plant Physiol.* 52:278–82
169. Vold, B. S., Sypherd, P. S. 1968. *Plant Physiol.* 43:1221–26
170. Walters, T. L., Loring, H. S. 1966. *J. Biol. Chem.* 241:2870–75
171. Ibid, 2881–85
172. Wechter, W. J., Mikulski, A. J., Laskowski, M. Sr. 1968. *Biochem. Biophys. Res. Commun.* 30:318–22
173. Wilson, C. M. 1963. *Biochim. Biophys. Acta* 68:177–84
174. Ibid. 76:324–26
175. Wilson, C. M. 1967. *J. Biol. Chem.* 242:2260–63
176. Wilson, C. M. 1968. *Plant Physiol.* 43:1332–38
177. Ibid, 1339–46
178. Ibid 1971. 48:64–68
179. Wilson, C. M. 1973. *Biochem. Genet.* 9:53–62
180. Wilson, C. M. 1974. *Plant Physiol.* Ann. Suppl. 9 (Abstr.)
181. Wilson, C. M., Alexander, D. E. 1967. *Science* 155:1575–76
182. Wilson, C. M., Shannon, J. C. 1963. *Biochim. Biophys. Acta* 68:311–13
183. Wolf, G. 1968. *Experientia* 24:890–91
184. Wyen, N. V., Erdei, S., Farkas, G. L. 1971. *Biochim. Biophys. Acta* 232:472–83
185. Wyen, N. V., Erdei, S., Udvary, J., Bagi, G., Farkas, G. L. 1972. *J. Exp. Bot.* 23:37–44
186. Wyen, N. V., Farkas, G. L. 1971. *Biochem. Physiol. Pflanz.* 162:220–24
187. Wyen, N. V., Udvardy, J., Erdei, S., Farkas, G. L. See Ref. 10, 293–97
188. Wyen, N. V., Udvardy, J., Erdei, S., Farkas, G. L. 1972. *Virology* 48:337–41
189. Wyen, N. V., Udvardy, J., Solymosy, F., Marre, E., Farkas, G. L. 1969. *Biochim. Biophys. Acta* 191:588–97

Ann. Rev. Plant Physiol. 1975. 26:209–36

PLANT STEROLS

❖7589

C. Grunwald[1]

Botany and Plant Pathology, Illinois Natural History Survey, Urbana, Illinois 61801

CONTENTS

INTRODUCTION

The topic of sterols in plants was last discussed in this series by Heftmann (137) 12 years ago. During this period interest in the biochemistry and physiology of phytosterols has greatly increased and much progress has been made in unraveling the sterol biosynthetic pathway. Chromatography and mass spectrometry are the techniques mostly responsible for this progress; but much of the information could not

[1]Written while in the Department of Agronomy, University of Kentucky, Lexington, Kentucky 40506. Part of the literature review was carried out while on sabbatical leave at the Université Louis Pasteur, Strasbourg, France.

have been obtained without stereospecifically labeled sterol precursors. Sitosterol is the most often reported higher plant sterol, but campesterol and stigmasterol are also very common. Additionally a considerable number of diverse sterols have been found in the plant kingdom. A number of reviews have dealt with the distribution of sterols (22, 36, 112, 117, 140, 253) and, therefore, this area will not be discussed in this paper. The biosynthesis of sterols has been reviewed recently by several authors (113, 117, 163).

Over the last few years there has been increased interest in the physiological role of sterols. Several physiological roles for sterols have been proposed and will be discussed, but only two hypotheses find much support in the literature. They are (a) that sterols might act as hormones, or more likely be precursors to steroids which act as hormones; and (b) that sterols might be involved in the structural arrangement of membrane. The free 4-demethyl sterols have been proposed for both roles, but that does not answer the question as to the function of steryl esters and steryl glycosides. In this respect a few working hypotheses will be presented later.

In this review no attempt is made to give a detailed discussion of any particular phase of sterol physiology or biochemistry. For this the reader is referred to the original works. The objective of this review is to touch on as many areas of sterol physiology as possible and to give a general understanding as to its present state. The physiology (152) and biosynthesis (270) of sterols in fungi, as well as sterol synthesis in algae (117), have been reviewed recently and, therefore, whenever possible work with higher plants will be cited in this paper.

CHEMISTRY

In 1959, Fieser & Fieser (104) defined sterols as a group of C-27 to C-29 secondary alcohols of plant or animal origin which differed from common alcohols in being crystalline solids at room temperature. Sterols have a 1,2-cyclopentanophenanthrene system arrangement consisting of rings A, B, C, and D (Figure 1). The major sterols of animal and plant origin have a secondary 3β-hydroxy group, angular methyl groups at C-10 and C-13, and an eight to ten carbon side chain at C-17. The two methyl groups and the side chain are also β-oriented. With few exceptions, naturally occurring sterols have at least one double bond, and this is usually in the B-ring. The C- and D-ring may have a double bond; however, no sterols have been found with the double bond in the A-ring. The structures for the major higher plant sterols—sitosterol, campesterol, and stigmasterol—are given in Figure 1. The structure of cholesterol, which is found in small quantities in many plants (22, 42, 112, 117, 164, 253), and which may be of great physiological importance, is also shown. The major sterol in yeast and fungi is ergosterol (36, 117). This sterol is similar to campesterol except that it has two additional double bonds, one in the nucleus $\Delta^{7(8)}$ and the other in the side chain $\Delta^{22(23)}$. Many other sterols, varying in number and position of double bonds and in modification in the side chain, have been isolated. For more discussion see Goad & Goodwin (117), Goad (112, 113), Stoll & Jucker (253), Bergmann (36), Fieser & Fieser (104), or Heftmann (137).

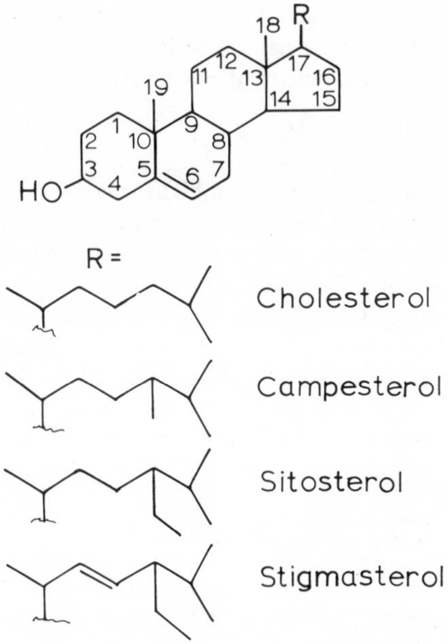

Figure 1   Structure of phytosterols.

Because of the ubiquity of phytosterols, the classification of sterols into groups or forms is of utmost necessity for any discussion of their physiological importance. Sterols can be grouped two ways: first according to their substituted moiety at the C-3 position, or second according to the number of methyl groups at the C–4 position. The latter classification is more convenient if the discussion is based on sterol biosynthesis. In this scheme, the sterols are divided into three groups: the 4-demethyl sterols or simply sterols, the 4α-methyl sterols, and the 4,4-dimethyl sterols (Figure 2). As will be seen later, demethylation at the C-4 position is an important step in the metabolism of sterols since the latter two groups are generally considered intermediates in the formation of 4-demethyl sterols (121, 192, 208, 279).

In the above classification, the moiety at the C–3 position is not considered. From a functional point of view, however, substitutions at the C-3 position are very important (128, 129, 198). A classification based on the modification at the C–3 position is convenient in considering physiological functions. At present, four groups are recognized: free sterol, or simply sterol which has a β-C-3-hydroxyl group; steryl ester, in which case a fatty acid forms an ester linkage with the sterol moiety at the C-3 position; steryl glycoside which contains a carbohydrate moiety at C-3; and the acylated steryl glycosides, which are linked to a fatty acid via ester linkage at the C-6 position of the sugar moiety. The latter classification of sterols can be used in conjunction with the former, and both free and esterified 4-demethyl, 4α-methyl, and 4,4-dimethyl sterols are found in nature (172, 177, 214). At present, only the 4-demethyl steryl glycosides and the 4-demethyl acylated steryl glycosides have been isolated from plants or natural products. However, this does not mean that the mono- and dimethyl acylated and nonacylated steryl glycosides do not exist.

BIOGENESIS

Most of the work on sterol biosynthesis in higher plants has been concerned with the formation of sitosterol and stigmasterol, while in yeast and fungi biosynthesis has been concerned with the formation of ergosterol. Weete (270) recently reviewed the biosynthesis of ergosterol, and the present review deals mainly with sterol biosynthesis in higher plants. Mevalonic acid is the substrate most frequently used in the study of sterol metabolism, but CO_2 or acetate can also be employed (4, 21, 101, 180). In any case, it has been established that all the carbon atoms of plant sterols, except carbon 28 and 29, are derived from mevalonic acid (21, 112, 117).

In 1959 Agrigoni (4) reported that he obtained labeled sterols from *Glycine max* treated with acetate-^{14}C and mevalonate-[14]C. Using mevalonate-2-^{14}C to study the

Figure 2	Biosynthetic pathway of sterols.	
Structure No.	Sterol	Compound Name
I	–	squalene
II	–	squalene 2,3-oxide
III	4,4-dimethyl	cycloartenol
IV	4,4-dimethyl	24-methylene cycloartanol
V	4-methyl	cycloeucalenol
VI	4-methyl	31-norcycloartenol
VII	4-methyl	31-norlanosterol
VIII	4-methyl	obtusifoliol
IX	4-methyl	24-methylene lophenol
X	4-methyl	24-ethylidene lophenol
XI	4-demethyl	stigmasta-7,24(28)-dien-3β-ol
XII	4-demethyl	stigmasta-5,7,24(28)-trien-3β-ol
XIII	4-demethyl	stigmasta-7-en-3β-ol
XIV	4-demethyl	avenasterol
XV	4-demethyl	stigmasta-5,7-dien-3β-ol
XVI	4-demethyl	spinasterol
XVII	4-demethyl	stigmasta-5,7,22-trien-3β-ol
XVIII	4-demethyl	sitosterol
XIX	4-demethyl	stigmasterol
XXI	4-methyl	24,25-dehydrolophenol
XXII	4-methyl	lophenol
XXIII	4-demethyl	desmosterol
XXIV	4-demethyl	24-methylene cholesterol
XXV	4-demethyl	cholesterol
XXVI	4-demethyl	campesterol

XXV CHOLESTEROL

XXVI CAMPESTEROL

XVIII SITOSTEROL

XIX STIGMASTEROL

biosynthesis of sitosterol, and upon degradation of the sterol molecule, it was found that the labeling pattern was the same as that observed in cholesterol (21). This led to the conclusion that sitosterol, and probably other sterols, arise in plants by a similar biosynthetic pathway established for cholesterol synthesis in animal systems (38). This general conclusion still holds true today, but a specific major difference is that the first cyclic intermediate, which is lanosterol in animals, is cycloartenol in plants.

Free Sterol Biosynthesis

CONVERSION OF ACETATE TO SQUALENE   In the overall process of sterol biosynthesis, the first phase is the conversion of acetate to mevalonic acid. This reaction is well established in animals and plants (38, 63). The second stage is the formation of squalene from mevalonic acid. The initial work, which is well documented, was carried out with liver and yeast preparations (58). Cornforth et al (59, 60, 63) used mevalonic acid which was stereospecifically labeled with deuterium and tritium to describe the mechanism of squalene biosynthesis. The incorporation of mevalonic acid into squalene with labeled precursors has since been demonstrated in a large number of in vivo and in vitro plant systems (26, 30, 32, 34, 47, 54, 122, 124, 175, 205, 227, 230, 275).

Nicholas (209–212), Bennett et al (24, 27), Baisted et al (17), and Capstack et al (50) have shown that in plants mevalonic acid is converted to squalene and the latter to sterols. Capstack et al (51) demonstrated that the squalene formed in plants had the same labeling pattern as squalene synthesized in animal tissue. Mevalonic kinase and phosphomevalonic kinase are the first two enzymes involved in the conversion of mevalonic acid to mevalonic-5-pyrophosphate through mevalonate-5-phosphate. Both of the reactions require ATP as a phosphate donor, and Mg^{2+} or Mn^{2+} promote greater enzyme activity (258). In the activation of the enzymes, low concentrations of Mn^{2+} are more effective than Mg^{2+} (258, 276). The pH optimum of this reaction depends upon the tissue, but generally it is in the range of 5.5 to 7.5 (106, 155, 199, 223). It has been reported that two isoenzymes exist, one a chloroplastidic kinase with an optimum pH at 7.5, and another kinase, not associated with the chloroplasts, with an optimum pH of 5.5 (155, 233). The observation is noteworthy since it has been suggested that membranes of chloroplasts are impermeable to mevalonic acid (232), giving rise to biosynthetic compartmentation and a possible way to regulate terpenoid biosynthesis at the subcellular level in germinating seedlings (261, 264).

Not all studies agree, however, with the two isoenzyme kinase proposition. Gray & Kekwick (125) in *Phaseolus vulgaris,* which is the same species studied by Rogers et al (232, 233), could not demonstrate the mevalonic kinase isoenzyme with the acid pH optimum. Furthermore, after disrupting chloroplasts, prepared both by aqueous and nonaqueous methods, they observed only very low kinase activity. The question of compartmentation of chloroplasts in the biosynthesis of terpenoids needs further study since Charlton et al (53) found evidence for the incorporation of mevalonic acid into phytoene by disrupted chloroplasts.

One of the most interesting reactions in the biosynthesis of squalene is the conversion of mevalonate-5-pyrophosphate to isopentenyl pyrophosphate. This reaction has been eloquently demonstrated by Cornforth et al (63), using stereochemically labeled mevalonic acid. The concerted transformation of mevalonate-5-pyrophosphate to isopentenyl pyrophosphate is a transelimination of the carboxyl and hydroxyl group in a highly stereospecific ATP-requiring process. A 3-phospho-5-pyrophosphomevalonate intermediate is postulated since 5-pyrophosphomevalonate labeled with ^{18}O in the tertiary hydroxyl group gave ^{18}O-labeled orthophosphate after decarboxylation (197).

In the presence of isopentenyl pyrophosphate isomerase, isopentenyl pyrophosphate is isomerized to 3,3-dimethylallyl pyrophosphate (3); and both isomers condense to form geranyl pyrophosphate (51, 115). A third isopentenyl pyrophosphate unit condenses with geranyl pyrophosphate to form farnesyl pyrophosphate (1, 119). In essence, three molecules of isopentenyl pyrophosphate are used in the biosynthesis of one farnesyl pyrophosphate molecule. During the overall reaction, one hydrogen atom is added and three are eliminated in a stereospecific manner (58). The formation of squalene is the result of a complex dimerization of two molecules of farnesyl pyrophosphate (38, 59). In addition to the carbon-carbon reaction, there is the cleavage of two C-O bonds with the release of pyrophosphate. Also there is the cleavage of a C-H bond with the ejection of one proton, and the formation of one C-H bond with a hydrogen donated by NADPH. The multiplicity of events suggests an intermediate and, by omitting NADPH, the intermediate prequalene pyrophosphate has recently been isolated from higher plants (148).

The major difference between requirements for the biosynthesis of sterols and carotenoids is NADPH. In soluble enzyme systems, squalene and phytoene were identified as products from mevalonate, and squalene was obtained only when the incubation mixture contained NADPH, ATP, and a metallic divalent cation (28). For carotene formation, no NADPH is required since the reaction is nonreductive. The requirements for squalene synthesis have been demonstrated in pea (124), tomato, carrot (23), and tobacco cell tissue culture (28, 34). Knapp et al (175) utilized an in vivo "pulse labeling" method, similar to that used successfully in studying cholesterol synthesis in mammalian tissue, and demonstrated that squalene biosynthesis occurs in the supernatant and microsomal fraction. Subsequently, Benveniste et al (34) reported that the soluble fraction is involved in the biosynthesis of farnesyl pyrophosphate and the microsomes promote the condensation reaction to form squalene. The conversion of acetate to squalene is anaerobic (32).

CYCLIZATION OF SQUALENE   The third phase in the biosynthesis of sterols is the cyclization of squalene (I) to form the cyclopentanophenanthrene ring system (Figure 2). A mechanism was proposed in which cyclization was initiated by a formal cation OH^+ attacking squalene at the position which becomes C-3 of the sterol ring system (237). A series of interacting electrophilic centers result in the cyclization, and once the tetracyclic system is established through molecular rearrangement, a migration of two hydrogen atoms and two methyl groups occurs (38, 67, 100, 277). Two research groups exploited the technique that molecular masses which differ in one or more isotopic atoms can be distinguished by mass spectroscopy. By this means they were able to conclude that two 1,2-methyl shifts occurred from C-8 to C-14 and from C-14 to C-13 (62, 202). In 1966 it was reported that in rat liver homogenates squalene was oxidized to the intermediate squalene 2,3-oxide before further conversion to lanosterol (57, 256). Tchen & Bloch (259) demonstrated that atmospheric oxygen, and not water, provides the 3β-hydroxyl group of sterol. This observation has been confirmed (55, 71, 257). The conversion of squalene to squalene 2,3-oxide (II) is aerobic and requires NADPH, while cyclization of squalene oxide to lanosterol is anaerobic (71). Cyclization to the sterol skeleton involves no loss or gain of elements and is a proton initiated reaction (to yield the sterol intermediate squalene 2,3-oxide) (55, 257), and the enzyme involved in this reaction, 2,3-oxide sterol cyclase, is microsomal in nature (71, 280).

The 2,3-oxidosqualene sterol intermediate has also been isolated from higher plants (33, 231). When inhibitors of sterol biosynthesis were added to the medium in the presence of ^{14}C-squalene or ^{14}C-mevalonate labeled squalene 2,3-oxide accumulated (80, 99, 231). In general, the conditions in plants under which the oxidative step occurs (33, 99, 231) are the same as those described for the liver system. Cyclization of squalene 2,3-oxide has been demonstrated with homogenates from higher plants (228), and the enzyme is localized in the microsomes (144). Furthermore, the step is anaerobic and does not require ATP, NADPH, or Mg^{2+} (144). Structural modifications in the epoxy group of the 2,3-oxide affected cyclization (56), and molecular rearrangements in the terminal group resulted in compounds with a low substrate specificity for plant cyclase (149).

The first cyclic product in animals (38) and fungi (270) is lanosterol, while in higher plants (10, 32, 115) it is cycloartenol (III). At one time it was postulated that lanosterol also played an important role in the biosynthesis of higher plant sterols; the main evidence for this was the reported isolation of lanosterol from coffee bean (167), Paul's Scarlet Rose tissue (274), *Nicotiana tabacum* (28), and the latex of *Euphorbia* species (221). In more recent studies it has been demonstrated that the compound which was believed to be lanosterol in *N. tabacum* is 24-methylene cycloartanol (30). The original identification of lanosterol in coffee bean is also in error (5, 208). The finding of lanosterol in Paul's Scarlet Rose was based on chromatographic techniques only and may need reexamination. The latex of *Euphorbia* contains lanosterol and cycloartenol, but apparently lanosterol is not in the actual plant tissue (221). The biosynthesis of triterpenes in the latex seems to be independent of that in the tissue (220, 222) because in the latex labeled cycloartenol is converted to lanosterol, but not vice versa (222). Apparently lanosterol in latex is the "end product" and not a sterol intermediate. It appears that the lanosterol pathway in latex of *Euphorbia* is different from that in animals and fungi and may represent a specialization of higher plant sterol biosynthesis.

At present it is generally accepted that cycloartenol rather than lanosterol is the first cyclic intermediate in the biosynthesis of higher plant sterols, and some of the most convincing results are those reported by Rees et al (227), Armarego et al (11) and Knapp & Nicholas (178, 180). These workers used $2\text{-}^{14}C$, $(4R)\text{-}4\text{-}^3H$,-mevalonic acid and found in *Solanum tuberosum* (227), *Musa sapientum* (178, 180), *Spinacea oleracea,* and *Medicago sativa* (11) a $^3H/^{14}C$ atomic ratio of 6:6 in squalene and cycloartenol. Lanosterol obtained with the rat liver system using the same precursor gave a $^3H/^{14}C$ ratio of 5:6 (61). Furthermore, several laboratories have studied a number of in vivo and in vitro plant systems using ^{14}C labeled precursors and have been unable to detect labeled lanosterol (32, 91, 92, 99, 227–229) even when unlabeled lanosterol was used as a trap (153). The reports that plants can convert lanosterol (18, 110, 153, 225) and 24-methylene lanosterol (6, 75) to 4-demethyl sterols may only demonstrate the low substrate specificity of the various enzymes involved.

TRANSMETHYLATION OF CYCLOARTENOL   The fourth and final phase in the biosynthesis of sterols is the conversion of the first cyclic intermediate, cycloartenol, to the respective sterols (Figure 2). A large number of phytosterols have been isolated (22, 112, 117), and there is just not enough space to describe all possible pathways. However, the pathway can be greatly simplified if one considers only the major phytosterols (sitosterol, campesterol, and stigmasterol) and cholesterol. Much is known about the metabolism of these higher plant sterols and, therefore, the discussion can be quite complete. Cholesterol is not a major plant sterol, but interest in its biosynthesis exists because of its possible importance as a precursor for steroid metabolism. The conversion of cycloartenol to 4-demethyl sterols requires introduction of an alkyl group at C-24, demethylations at the C-4 and C-14 positions, opening of the 9β, 19-cyclopropane ring and introduction of the Δ^5 bond (Figure 2). Introduction of an alkyl group at C-24 does not occur in the biosynthesis of cholesterol.

The formation of campesterol, sitosterol, and stigmasterol requires the alkylation of C-24 by means of a transmethylation reaction involving methionine (183, 213). The importance of methionine as the methyl donor in sterol metabolism was first shown in the biosynthesis of ergosterol in yeast (7, 8, 217). A number of researchers have elaborated on the alkylation mechanism in higher plants and the involvement of S-adenosyl methionine as the methyl donor (192, 279). The C-24 ethyl unit of sitosterol and stigmasterol is derived from two C_1-units (52). Before introduction of the first methyl unit can occur a $\Delta^{24(25)}$ bond is required (236). As transmethylation proceeds a cationic site is created at C-24 which is stabilized by hydrogen migration (118, 227) from C-24 to C-25 (225, 230). During this overall reaction one hydrogen atom is lost from the transferred methyl group (112, 113, 178, 236, 273, 279) and the product thus formed is 24-methylene cycloartanol (IV). This 24-methylene sterol intermediate has been isolated from a number of plants (5, 10, 31, 145, 208, 220).

DEMETHYLATION AT C-4   Demethylation at C-4 is required of all sterol precursors including the cholesterol precursor (Figure 2). The first demethylation in the campesterol-sitosterol-stigmasterol pathway is the removal of the 4α-methyl group forming cycloeucalenol (V) from 24-methylene cycloartanol (178). In the cholesterol pathway 31-norcycloartenol (VI) is formed from cycloartenol. During the 4 α-demethylation process the 4β-methyl group is epimerized to the 4α-position and the 3α-hydrogen is lost (178) by a process which possibly involves 3-ketone interme-diates (196, 227, 255). The overall reaction would be the same as reported for cholesterol synthesis in the mammalian system in which lanosterol is the cyclic intermediate (226). However, an alternate mechanism has been proposed (180) to account for the fact that the 3-ketones of 4,4-dimethyl and 4α-methyl triterpenes are not demethylated under nonreducing conditions (255). As with the former mechanism, sequential oxidation occurs through the 4α-primary alcohol to the carboxylic acid. But during decarboxylation the expulsion of the C-3 hydride ion occurs, resulting in the $\Delta^{3(4)}$-intermediate which is the enol of the 3-keto tautomer (180). Formation of the ketone would make it a product rather than the early intermediate in the demethylation process.

OPENING OF CYCLOPROPANE RING   Opening of the 9β-19-cyclopropane ring is the next step in the biosynthesis of plant sterols (Figure 2). The enzyme mediated reaction was observed in a cell-free system (147). In a comparative study using cycloartenol, 24-methylene cycloartanol, and cycloeucalenol, the most efficient sub-strate was found to be cycloeucalenol (143, 146). Obtusifoliol (VIII) is the product in the campesterol-sitosterol-stigmasterol pathway, and in the cholesterol pathway (77) it is probably 31-norlanosterol (VII). However, not all results agree with the proposed pathway for major phytosterol biosynthesis (29). It was observed that the specific radioactivities of 24-methylene cycloartanol and cycloeucalenol were lower than those of cycloartenol and obtusifoliol. The suggestion is made that a branch point exists at cycloartenol. Therefore, obtusifoliol can be synthesized via the path-way discussed above or through the 31-norcycloartenol to 31-norlanosterol pathway (143, 146).

Cholesterol biosynthesis in higher plants has not been studied extensively and, therefore, uncertainties exist in the sequence of intermediates. The cholesterol path-way in Figure 2 is not supported by all of the evidence, and it has been suggested that a second demethylation at C-4 can occur before the cyclopropane ring is opened (76). The sequence in this scheme would be cycloartenol⟶31-norcyloartenol⟶ pollinastanol⟶cholesterol. The conversion of pollinastanol to cholesterol has been demonstrated in leaves of *Nicotiana tabacum* (74). However, in the same system, the use of radioactive cycloartenol did not result in labeled pollinastanol even though the label was recovered from cholesterol (73). It is quite possible that cholesterol synthesis via this pathway proceeds only in pollen since pollinastanol has been isolated mainly from pollen (75–77, 159).

TRANSFORMATION OF 4α-METHYL TO 4,4-DEMETHYL STEROL   Results indi-cate that demethylation of C-14 is hindered by the 9β,19-cyclopropane ring. Since obtusifoliol and 31-norlanosterol no longer have the cyclopropane ring, demethyla-tion at C-14 can proceed. After rearrangement of $\Delta^{8(9)}$ to $\Delta^{7(8)}$, 24-methylene lo-phenol (IX) is formed in the pathway for the major plant sterols (115, 191, 273) and 24,25-dehydrolophenol (XXI) as the cholesterol intermediate (Figure 2). At this point the campesterol-sitosterol-stigmasterol pathway branches, because the biosyn-thesis of sitosterol and stigmasterol requires a second alkylation at C-28. A number of transmethylation processes have been considered (31, 118, 192), but probably the most widely accepted mechanism in higher plants is the mechanism proposed by Nes and his co-workers (52, 236). As in the first alkylation reaction, methionine is the methyl donor creating a cationic center at C-24. Stabilization of the cation occurs through the loss of a hydrogen atom from C-28 (2, 10, 29, 31, 75, 115, 116, 118, 193, 235, 271), forming 24-ethylidene lophenol (X). The presence of the 24-ethylidene sterol intermediate (52, 236) has been established with dual-labeled (11, 118) and deuterium labeled methionine (246). With both experimental approaches, only four of the tritium or deuterium atoms from methionine were recovered from the C-24 ethyl group of sterols. These results implicate 24-ethylidene lophenol as the sterol intermediate.

Next is the oxidative removal of the remaining C-4 methyl group which occurs only if the methyl is in the α-configuration (226, 242, 243, 255). This mechanism in the sitosterol-stigmasterol pathway gives rise to stigmasta-7,24(28)-dien-3β-ol (XI) which has been isolated (176, 181, 183, 254). From this point the overall pathway for sterol biosynthesis is ill defined in higher plants. But if the data obtained with algae is also considered (114, 133, 193, 234, 247, 248) a $\Delta^{5,7(8)} \longrightarrow \Delta^{5,7} \longrightarrow \Delta^{5(6)}$ sequence probably occurs first, and eventually gives rise to avenasterol (XIV), which has been found in small quantities in higher plants (20, 161, 181) usually in conjunction with sitosterol (39, 42, 162). Labeling experiments with mevalonic acid suggest a close biosynthetic relationship (15) which has been confirmed by the conversion of labeled avenasterol to sitosterol in *Pinus pinea* (9).

At present it is not clear whether sitosterol (XVIII) is the only precursor for stigmasterol (XIX), or if both sterols arise from a common intermediate. It has been postulated that 22,23-dehydrogenase converts sitosterol to stigmasterol (26, 235). Direct conversion of labeled sitosterol to stigmasterol generally has been very low and unconvincing (25, 269); however, recently in our laboratory a conversion of 10% was observed in seedlings of *Nicotiana tabacum* (unpublished data). In the same biological system (47) and in *Musa sapientum* (179) using ^{14}C-mevalonic acid as a precursor, a higher specific radioactivity was found in sitosterol than stigmasterol, indicating a precursor-product relationship.

A second pathway for stigmasterol and sitosterol biosynthesis has been proposed (Figure 2). After stigmasta-7,24(28)-dien-3β-ol is formed, through the removal of the C-4 methyl group from 24-ethylidene lophenol, a reduction of $\Delta^{24(28)}$ takes place (73, 78, 79) giving rise to stigmast-7-en-3β-ol (XIII). At this point the biosynthetic pathway for sitosterol and stigmasterol probably branches. To form sitosterol a $\Delta^{7(8)} \longrightarrow \Delta^{5,7} \longrightarrow \Delta^{5(6)}$ migration occurs, and to form stigmasterol the introduction of a double bond at C-22,23 is required followed by the $\Delta^{7(8)} \longrightarrow \Delta^{5,7} \longrightarrow \Delta^{5(6)}$ sequence. The order in which these reactions take place is not known but all postulated intermediates have been isolated (117).

The last few steps in the biosynthesis of cholesterol (XXV) and campesterol (XXVI) are quite similar to those described for stigmasterol and sitosterol. In the cholesterol pathway a second demethylation at C-4 of 24,25-dehydrolophenol (XXI) occurs, followed by rearrangement in $\Delta^{7(8)} \longrightarrow \Delta^{5,7} \longrightarrow \Delta^{5(6)}$ bonds, and a reduction in the C-17 side chain. The sequence of events has not been established, but demosterol (XXIII) (109, 159) and lophenol (XXII) (79, 239) have been isolated and are probably intermediates. The formation of campesterol very likely goes through 24-methylene cholesterol (183, 249) because if this sterol is labeled it is converted to campesterol (6). But other pathways may be involved, depending upon tissue or species. The interconversion of cholesterol to campesterol and stigmasterol also has been reported (265).

Steryl Glycoside and Acylated Steryl Glycoside Biosynthesis

The existence of steryl glycosides has been known for some time (224), and a little over 10 years ago the acylated steryl glycosides were first reported (194, 204). Since then, steryl glycosides both in the free and acylated forms have been isolated from a variety of plants (96, 173), and steryl glycosides probably exist in all plant species. The incorporation of mevalonate into steryl glycosides was greater in the roots than shoots when the labeled precursor was supplied to the petiole of intact *Digitalis purpurea* (102). At the cellular level, chloroplasts contained the lowest quantity of steryl glycosides (96, 97), while the bulk of the glycosides were found in the microsome fraction (88, 96, 127). The acylated steryl glycosides were quantitatively highest in the mitochondria (88). The sterol moiety of both glycoside forms can be any 4-demethyl sterol such as campesterol, sitosterol, stigmasterol, cholesterol, etc (48, 174, 194). Glycosides of the 4,4-dimethyl and 4α-methyl sterols have not been reported. The sugar moiety of free and acylated steryl glycosides is generally glucose (173, 174, 194); however, mannose has also been found (97) and the sterol is attached to C-1 of the sugar moiety. The acylated steryl glycosides have been identified as steryl-6-acyl D-glycoside and the main acyl moieties are palmitic, stearic, oleic, linoleic, and linolenic acid (174, 194).

The biosynthesis of steryl glycosides was first demonstrated with a particulate enzyme preparation from immature soybeans (157, 158), and has been confirmed with organelles prepared from *Phaseolus aureus* (168), *Pisum sativum, Brassica cauliflora, Spinacia oleracea, Triticum aestivum* (219), and *Nicotiana tabacum* (48) and inner mesocarp of *Persea americana* (215). In cell-free systems the glycoside bond is mediated through nucleotide sugars, and uridine diphosphate-glucose is the most active glycosyl donor. Steryl glycoside formation is stimulated by ATP (93, 158), but the exact reason for its requirement has not been established (48). Overall, similar results have been obtained whether the radioactive label was in the sugar (157, 158, 168, 215, 219) or sterol (48) moiety. All 4-demethyl sterols are incorporated at about the same rate. The pH of the incubation medium is quite critical and optimum conditions are near the neutral point (48, 158, 168, 215). The exact pH requirement depends upon the tissue (94, 95), but if the pH is too low, acyl steryl glycoside formation is favored.

Theoretically, acylated steryl glycosides could be formed via two pathways (Figure 3). The generally accepted theory for acyl steryl glycoside formation involves steryl glycoside as the precursor (94, 95, 215) and cell-free experiments support this hypothesis (13, 158). Galactolipids were required for acylation to occur and digalactosyl diglyceride was more active than monogalactosyl diglyceride (94, 95). Phos-

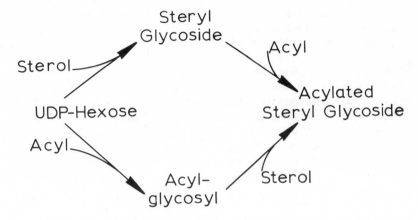

Figure 3   Theoretical biosynthetic pathway for acylated steryl glycosides.

phatidyl ethanolamine was needed when acetone powder was used as an enzyme preparation (13).

In vivo experiments, however, with labeled uridine diphosphate-glucose (98) or labeled sterol (48), do not support the above hypothesis. In intact systems incorporation of the label was higher into the acylated steryl glucosides than into the steryl glycosides. If synthesized cholesterol-^{14}C glucose was fed to *Nicotiana tabacum* seedlings, only little of the label was recovered as the acylated steryl glycoside, while more than twice as much was found as free sterol (48). In the in vivo system at least acylated steryl glycosides may be formed through the transfer of an acyl-glycosyl unit to the sterol moiety. This pathway of acylated steryl glycoside biosynthesis does not exclude the former pathway.

Steryl Ester Biosynthesis

Higher plants not only contain free and glycosidated sterols but also terpene esters. The 4,4-dimethyl, 4α-methyl, and 4-demethyl steryl esters have been identified (35, 112, 170–172, 177, 273). The esters of sitosterol, campesterol, stigmasterol, etc have been isolated from *Zea mays* (170, 171), *Pisum sativum* (112), *Nicotiana tabacum* (46), flowers of *Calendula officinalis* (278), peel of *Musa sapientum* (177), and many other species. The esters have been isolated from a variety of tissues such as roots, shoots, scutellum, and endosperm (169, 171), as well as from intracellular organelles—nuclei, chloroplasts, mitochondria, and microsomes (172). But of even greater interest is the isolation of 4,4-dimethyl and 4-methyl steryl esters. The ester of cycloartenol, the first cyclic intermediate, as well as 24-methylene cycloartanol, the second 4,4-dimethyl sterol intermediate, have been isolated (35, 170–172). The esters of the 4α-methyl intermediates, cycloeucalenol, obtusifoliol, 24-methylene lophenol, and 24-ethylidene lophenol have also been found in plants (171, 172, 177). The most important fatty acid moieties of the steryl esters are palmitic, oleic, linoleic, and linolenic (46, 214); however, the exact composition depends upon the species or tissue in question.

Information about the biosynthesis of steryl esters is meager. In the peel of *Musa sapientum* (179), and in leaves of *Zea mays* (261) and *Nicotiana tabacum* (47), labeled mevalonic acid was rapidly incorporated into the steryl ester fraction. Incubation periods of 3 to 6 hours gave a higher specific radioactivity in the free sterols (47, 179), but at longer periods the specific radioactivity was higher in the esters (47, 179, 261). Of the steryl esters, the 4,4-dimethyl sterols had the highest radioactivity, followed in decreasing order by the 4-methyl and 4-demethyl sterols (170, 179). It is interesting to note that the 4,4-dimethyl and 4-methyl sterols had the highest radioactivity, but it is very unlikely that these esters are intermediates during phytosterol biosynthesis, since the proposed C-4 demethylation mechanism has a 3-ketone intermediate.

The 4,4-dimethyl and 4-methyl steryl esters, however, may play a role in the regulation of sterol biosynthesis. Competition between steryl ester formation and demethylation of C_{28}-sterol intermediates has been demonstrated (43). Aerobic conditions favor demethylation while anaerobic conditions favor esterification. Esterification renders the sterol intermediates inert for the demethylation enzymes, thereby possibly controlling the formation of 4-demethyl sterols.

Although the presence of steryl esters in plants has been demonstrated, virtually nothing is known about their biosynthetic pathway. Very low esterification of sterols, about 1 to 3%, has been observed in 6-day-old *Nicotiana tabacum* seedlings incubated in labeled cholesterol or sitosterol (48). All intracellular organelles had about the same percentage of radioactive label in their steryl ester fraction and no localization as to esterification could be established. However, when mevalonic acid was used as the substrate, higher radioactivity was recovered in the steryl esters (47, 179, 261). Similar results have been obtained with cell-free systems (C. Grunwald, unpublished results). While no difference in ester formation was observed whether labeled cholesterol or sitosterol was supplied (48), when mevalonic acid was used the cholesteryl ester fraction had a much higher specific radioactivity than the sitosteryl esters (47). Since the specific radioactivity of the various free and esterified 4-demethyl sterols were not the same, compartmentation in the biosynthesis of sterols may be involved.

The conversion of about 50 to 60% of cholesteryl-^{14}C palmitate to free cholesterol-^{14}C has been reported with *N. tabacum* seedlings (45). Available information suggests that steryl esterase probably functions mainly as a hydrolase. This interpretation would be in agreement with results obtained for cholesterol esterase in mammalian systems (251, 252). One possible reason for the low level of sterol esterification (48) may be the requirement for anaerobic conditions (43).

PHYSIOLOGICAL FUNCTION OF STEROLS

The physiological importance of sterols in higher plants is only rarely mentioned in the literature (22, 137, 139, 200). It has not been too many years ago that phytosterols were considered secondary metabolites and waste products. The general biosynthetic sequence for selected sterols in higher plants is quite well defined; however, studies on the physiological role of these compounds has not received much attention. It is difficult to understand why plant physiologists have largely neglected this area, because steroids are very important in mammalian and microbial physiology. It seems only reasonable that botanists would look for sterol functions similar to those postulated for animals and fungi (152), that is, in the growth and reproduction of organisms. With this in mind, it should be asked: Are phytosterols hormones or hormone precursors? And are they components of plant membranes?

Sterols as Hormones and Hormone Precursors

Experimental data relating to the question of sterols acting directly as hormones is scarce. The higher plant sterols, sitosterol and stigmasterol, had no affect on the growth of excised *Pisum sativum* embryo (150) and did not influence stem elongation in *Chrysanthemum* (37). But sitosterol stimulated growth in 6-day-old dwarf *P. sativum,* whereas cholesterol was inactive (184). However, in excised *P. sativum* embryo cholesterol and the fungal sterol ergosterol inhibited root growth (150). In the initiation of flower buds in *Chrysanthemum,* sitosterol was active while cholesterol was inactive (37). Based on these few experiments it is difficult to say whether or not sterols act directly as hormones or are an intermediate metabolite for steroid hormones, since sterols are precursors to steroids.

Steroid biosynthesis in higher plants in many ways resembles that in animals (138, 140). Sterols through oxidation and aromatization are converted to a number of steroids which have been recognized as hormones in animal systems. The more recent isolations of insect molting hormones (ectysterols) from plants is further proof that plants are capable of steroid synthesis (142, 154). There are several well-established systems in which estrogenic compounds influenced vegetative growth (40, 111, 150, 184, 272) and flowering (66, 160, 187, 188, 200), including sex expression. In many plants flowering is controlled by day length, and inhibitors of sterol biosynthesis suppressed floral induction in *Pharbitis nil* (41, 238), *Lolium temulentum* (103), and *Xanthium pennsylvanicum* (41). Steroid inhibitors also inhibited the development of tubers in *Solanum andigena* grown under short-day condition (14). The age of the plant at the time of treatment with steroid inhibitors appears to be very important. *Pharbitis nil* treated 7 days after germination with tris-(2-diethylaminoethyl)-phosphate hydrochloride showed a high rate of inhibition in flower initiation, but if treated 20 days after germination, no inhibition was observed (238). *Ecballium elaterium* treated with estradiol-17β, estriol, and estrone showed a large increase in the number of flowers. These steroids also influenced the female to male sex ratio in favor of female flowers (188). Plant treatment with adrenosterone, androstenedione, and testosterone had only a very slight increase in the number of flowers but the sex ratio changed in favor of maleness. Cortisone increased the number of flowers but not the sex ratio. The economic importance of these observations could be great, but a much greater research effort is needed in this direction to elucidate the mechanism involved in controlling flower number and sex expression.

Cold-requiring long-day *Cichorium intybus* plants were grown under noninductive condition and treated with gibberellic acid, estradiol-17β, and estrone (187). Gibberellic acid was twice as effective in inducing flower formation as estrone and slightly more active than estradiol-17β. Estrone has been isolated from seeds of *Phoenix dactylifera* (141). In *Phaseolus vulgaris* estrogen-like substances appeared at time of flower bud formation and reached a maximum at flower bud development and pod formation (189). It has been shown that estrone increased the endogenous level of gibberellins in dwarf *Pisum sativum* (185).

In addition to the effects which estrogens exert on flowering, there are additional responses on plant growth. Estrone increased, while testosterone decreased stem growth of excised *P. sativum* embryo (150). Intact dwarf *P. sativum* plants also showed an increase in growth with estrone and estradiol-17β treatment (184). Testosterone, however, was inactive in this system. Growth of *Nicotiana tabacum* which was inhibited with growth retardants could be reversed with gibberellic acid and sterols (81). It is quite possible that the steroids do not in themselves act as hormones but may influence the biosynthesis of gibberellins, thereby acting on flowering and plant growth. An alternate explanation is possible which would also involve auxin metabolism. Steroid derivatives have been isolated from *Coleus* which had auxin-like activity in the *Avena* curvature test (267). Furthermore, *P. sativum* and *Pinus silvestris* seedlings treated with estrone or estradiol-17β had an increased amount of extractable auxin (186). Steroid-stimulated growth may be due to a raised auxin activity or a lowered auxin oxidase activity, which might be caused by the initial estrogen-induced increase in gibberellins.

Sterol Role in Membranes

Another important function of sterols in the mammalian system is their involvement in biological membranes (201, 216). For example, removal of cholesterol from erythrocytes increases their permeability and fragility. There is also evidence suggesting that sterols in higher plants play an important role in giving structure to membranes. The polyene antibiotic filipin acts by interacting with cellular sterols (123), and use of this antibiotic has made it possible to increase the permeability of tuber discs of *Solanum tuberosum* and root sections of *Beta vulgaris* (207). A low concentration of alcohol increases the permeability of cells, and it is suggested that alcohol acts at the membrane level, influencing the sterol-phospholipid interaction. The classic membrane stabilizer $CaCl_2$ prevented the leakage of betacyanin from alcohol-treated *B. vulgaris* discs, but cholesterol at the same concentration was much more effective (126). Cholesterol also restored the K^+ and NO_3^- uptake capability in etiolated *Pisum sativum* stem sections treated with filipin (151). Other sterols such as sitosterol, campesterol, ergosterol, and stigmasterol were less effective than cholesterol, but the degree of activity not only depends upon the sterol in question but also upon the tissue under study (126, 130).

Plant tissues contain not only free sterols but also steryl glycosides and steryl esters; however, only the free sterols are effective plant membrane stabilizers (128). Similar conclusions were reached with cholesteryl esters in pure and mixed lipid monolayer (190). From all evidence available it appears that the hydroxyl group at

C-3 of the sterol molecule must be free to allow for interaction with membrane phospholipids, and this interaction is probably by means of ion-dipole and hydrogen bonding (105, 241, 266). The C-3 hydroxyl requirement is quite specific since the C-3 sulfhydryl analog of cholesterol was unable to influence permeability in *Hordeum vulgare* roots, indicating that the bonding requirements of sterol molecules must be precise as to bond energy and bond angle (130).

Another structural requirement for sterol molecules to be active is a flat molecular configuration. Calculations predict only molecules in configuration similar to cholesterol fit the phospholipid cavity (105, 266). Increasing the bulkiness of the C-17 side chain of the sterol molecule should decrease its activity at the membrane level. Campesterol was less effective in decreasing plant membrane permeability than cholesterol, but more effective than sitosterol and stigmasterol (128, 130). Similar results have been obtained through sterol incorporation studies with model membranes (90). To influence membrane structure, sterols must have at least one double bond in the perhydro-cyclopentanophenanthrene ring system (130). Ergosterol which has a $\Delta^{5,7}$ conjugated bond system was as potent as similar sterols with only one double bond, while cholestanol which does not have a double bond was inactive (130). For sterols to be active at the membrane level, exact molecular configurations are required. Furthermore, the fatty acid composition as well as the pairing of fatty acids in the phospholipid molecule are important in membrane physiology (72).

Kemp & Mercer (172) separated the intracellular organelles of *Zea mays* and found that microsomes and mitochondria contained the bulk of the sterols. This observation has been confirmed by other laboratories using *Nicotiana tabacum* (127), *Phaseolus vulgaris* (44, 136), and *Solanum tuberosum* (88). All cellular fractions isolated from plants contained free, esterified, glycosidated, and acyl glycosidated sterols (88, 127), as well as free and esterified 4α-methyl and 4,4-dimethyl sterols (44, 172). The relative sterol composition of the various intracellular organelles is not the same, indicating that the sterol composition of membranes is affected by a number of membrane functions. If sterols in plants play a role in the structure and function of membranes, then the sterols found in any particular membrane will be determined by all the functions performed by the membrane, rather than by any single function. Generally nuclei contain most of the cholesterol, while the other cellular fractions are rich in sitosterol, stigmasterol, or campesterol (44, 88, 172). The localization of sterols in membranes is further complicated by the fact that some sterols from chloroplasts are easily extractable while others are more difficult to extract (64, 182). It might be that these qualitatively distinct forms have different functions at the membrane level.

Seeds, Germination, and Plant Growth

The physiological role of sterols has interested plant physiologists and biochemists for a number of years, and probably one of the best systems to study is the developing and germinating seed. If, as already discussed, free sterols are an integral part of cell membranes, then as a consequence of increased organelle and membrane production, an increase in sterols should be observed during germination and seedling development. As postulated, an increase in free sterols was observed with germination of *Pisum sativum* (87) and *Nicotiana tabacum* seeds (46) and in growing shoots and roots of *Zea mays* seedlings (169). Likewise, in etiolated *Phaseolus aureus* plants the young growing tissue contained more sterol than the older tissue (107). During germination of *Phaseolus vulgaris* (84, 85) and *Raphanus sativus* (86) the sterol content increased, and the increase was higher in seedlings grown with nutrient solution than where grown with distilled water (82, 83). However, nitrogen fertilization did not influence the sterol level in older plants (132). The sterol increase in seedlings was proportional to the increase in protein and dry matter. For germination, *Corylus avellana* seeds require prechilling or gibberellic acid treatment. Sterol biosynthesis occurred only in those seeds that were induced to germinate, and hydration of seeds alone did not stimulate sterol synthesis (245). The gibberellin effect apparently is at the mevalonic kinase level (244).

Sterol accumulation does not occur during the first 2 to 4 days of germination (46, 83, 169), and it has been shown that biosynthesis of sterols during these early stages is insignificant (16). However, in *C. avellana* seeds sterol synthesis was detected after only 12 hr (245). Chemical examination of dried seeds has revealed that sterols are present in the seed (87, 162) and one must conclude that sufficient quantities of sterols are on hand to satisfy the initial demands made during germination. The overall process of germination is very complex; many factors control it (203) and virtually nothing is known about the composition of membranes in seeds. But an important early event in germination is the formation of new membranes and transformation of existing membranes to allow for changes in permeability to water and gases.

Since sterols are present in mature seeds, they had to be translocated to or synthesized in the developing seeds. From time of flowering to seed maturity, an accumulation of sterols was observed in *Zea mays* (70) and *Gycine max* (166). If the sterol content is expressed on the basis of dry matter or total lipids, accumulation of sterols preceded accumulation of dry matter and lipids. The capacity for sterol biosynthesis is highest in the least mature seeds and decreases with seed maturity (16). The decrease in sterol biosynthesis is not due to a decrease in mevalonic acid incorporation, but rather is due to an increased synthesis of α-amyrin. Squalene 2,3-oxide is the common metabolic precursor for α-amyrin and sterols (117), and experimental evidence indicates that final regulation of sterol synthesis is at the cyclization step of squalene 2,3-oxide (16). Could it be that the steryl esters, which continue to accumulate after the rapid free sterol accumulation stops (166), play some role in controlling sterol biosynthesis? As already pointed out, esterification of sterol intermediates prevents the demethylation reaction from proceeding (43). In developing *Pisum sativum* seeds the enzyme responsible for the transformation of 24-methylene cycloartanol into sterols becomes less active first, followed by the deactivation of squalene oxide cyclase (16). In this respect it should be noted that in the tubers of *Solanum tuberosum* the steryl ester concentration is much higher than that of free sterols (89) and the biosynthetic activity for sterols is insignificant (134). The sterol synthesizing capacity increases with aging of the tuber tissue, and it appears that the action is at the C-24 methyltransferase level.

The steryl ester content in germinating seeds does not change (46, 83, 245), except for the scutellum of *Zea mays* in which a rapid increase in esters was observed (169).

During germination the dry matter and protein content of cotyledons of *Phaseolus vulgaris* and *Raphanus sativus* decreased rapidly, but during the same period the total amount of sterol and steryl ester did not change (83). In essence the sterol and steryl ester content on a dry matter basis increased. These results do not support the suggestion that steryl esters are involved in the transport of sterols (169). In more recent experiments it has been found that when labeled sitosterol and sitosteryl palmitate was applied to leaves of *Pelargonium hortorum* and *Helianthus annuus,* mostly the label of the free sterol was recovered from other plant parts (12). Isolation of the various sterol fractions revealed that the steryl glycoside fraction contained the bulk of the radioactivity, and only a small amount of the label was recovered from the ester fraction. Similar results have been obtained with *Hordeum vulgare* plants which were fed sitosterol to their roots (Grunwald, unpublished results). The translocation of sterols, especially to the root, has also been found with *Nicotiana tabacum* plants fed by the wick method through the stem (265). With the available data no conclusion can be made on whether sterols are translocated as their glycosides or whether the glycosides are formed in situ in the various plant parts after translocation of the free sterol.

Very little is known about steryl glycoside changes (free and acylated) during germination, partly because, quantitatively speaking, they occur in small amounts. During germination of *N. tabacum* the steryl glycosides decreased, but this decrease could not account for the total increase in free sterols (46). Since steryl glycosides accumulated during seed maturation (166), and occur in large quantities in the tubers of *Solanum tuberosum* (89), it has been proposed that they are a storage form of sterols (158). However, this increase in steryl glycosides with seed maturation is not observed with all species (70).

Changes of individual sterols during germination have been studied, but no consistent picture emerges. In *Brassica napus* and *Sinapis alba* sitosterol accounted for most of the sterol increase during germination (162), while in *Nicotiana tabacum* stigmasterol showed the largest increase (46). However, other species of *Brassica* did not show a change in sterol composition with germination (162). Various tissues of *Zea mays* have been analyzed for sterol changes during germination, and the ratio of free and esterified sitosterol:stigmasterol:campesterol was constant in shoot, endosperm, and scutellum (169). However, in the root there was an increase in the percentage of stigmasterol and a decrease in sitosterol in both the free and the ester form. In 6-day-old seedlings of *N. tabacum* the incorporation of labeled mevalonic acid into the various sterols was quite different (47). Over a 12-hr incubation period the specific activity of campesterol and sitosterol was about the same, but that of stigmasterol was much lower and did not increase with incubation time, indicating a precursor product relationship. But these results do not reveal anything about the large increase in stigmasterol during germination. Compartmentalization is one possible explanation.

The youngest growing tissue in immature *Phaseolus aureus* plants contained the largest amount of sterol (107), and naphthalene acetic acid treatment increased the sterol content to the largest extent in those hypocotyl sections which showed maximum amount of elongation (108). This picture changes as plants become older. For instance, in *Solanum andigena* (14) and *N. tabacum* (132) the total sterol content increased from time of planting to maturity. Comparing *N. tabacum* leaves from lower stalk position (older) to leaves from upper stalk position (younger), a higher sterol content was found in the lower leaves (68, 131). In tubers of *Solanum tuberosum* the sterol content increased with storage time (89), and it has been suggested that the increase in sterols is due to senescence and disorganization of cellular organelles (83, 86). In tubers (89) and cotyledons (83) the steryl esters increased the most during aging, while in leaves the free sterols accounted for the largest increase (131).

Young growing tissue of *Phaseolus aureus* had a high ratio of sitosterol to stigmasterol, but this ratio gradually decreased with the age of the tissue (107). Similar decreases in the ratio of sitosterol to stigmasterol are found if one compares young and old *N. tabacum* leaves (68). However, the change in the sterol ratio is primarily an increase in stigmasterol and not a decrease in sitosterol (68), and most of the increase is due to increased free stigmasterol (131). A possible explanation to account for the increase in free sterols during aging is that in mature leaves the free sterols not only are structural components of membranes but are also part of a nonmetabolic sterol pool.

Experiments with abscisic acid are of interest in connection with senescence. Abscisic acid can induce dormancy in seedlings of a number of species and accelerate leaf abcission. Treating *Zea mays* seedlings with abscisic acid did not change the capacity to synthesize sterols, but it did depress the synthesis of chlorophyll and carotenoids (206).

The steryl ester, glycoside, and acylated glycoside contents do not change significantly during plant aging (14, 131); however, the ratios of sitosterol to stigmasterol in these fractions decreased. The sterol profiles indicate that the different sterol forms are not in simple equilibrium. Even though the sterol changes are small, it is quite possible that these slight changes influence the overall physiology of the plant.

Environmental Influence on Sterols

The environment has a very pronounced influence on the growth and development of plants. Therefore, it is only natural to ask if and what environmental factors influence the quantity and quality of sterols. If sterols are physiologically as important as suggested, one should expect that these compounds are not static in nature. For example, within 56 hr after feeding cholesterol to intact *Nicotiana tabacum*, less than 10% of the ^{14}C activity was recovered as the sterol (265). These and other data indicate a rapid transformation of sterols.

TEMPERATURE   While temperature is a very influential factor in determining the normal development of plants, essentially nothing is known about its effect on sterol biosynthesis. *Triticum aestivum* plants grown at 1°C had different sterol profiles from plants grown at 10°C (69). With lowering of the temperature the shoot tissue showed a decrease in sterols. Roots at the lower temperature had an initial lower sterol level, but this tissue had the ability to acclimate to the lower temperature, and the sterol level returned more to a "normal" level. The influence of temperature on physiochemical processes is well documented, and evidence has been presented which shows changes in membrane lipids with changes in temperature (195). It is to be expected that if sterols are membrane components they could be influenced by temperature, allowing for changes in cell permeability. In yeast the temperature for maximum rate of sterol synthesis coincided with the optimal temperature for growth (250); however, similar results have not yet been reported for higher plants.

LIGHT   The influence of light on certain physiological and morphological characteristics of plants is well documented. Plants grown in the dark are etiolated and do not have chloroplasts nor chlorophyll. Development of chloroplasts is a light-initiated reaction and with illumination, etiolated plants rapidly become green. Etiolated seedlings have sterols (49) and if grown with nutrients the sterol content is even higher (82, 156). Upon illumination the sterol level drops in greening shoot tissue (49), but the sterol content in the cotyledons is generally not affected. It is evident that green as well as etiolated plants have the ability to synthesize sterols.

Etiolated plants do not have chloroplastic terpenoids such as carotenoids and other closely related compounds (chlorophyll, tocopherol, etc). With greening of *Zea mays* and *Avena sativa*, $^{14}CO_2$ incorporation into sterols was low, while carotenoids and related compounds were highly labeled (264). The stage of chloroplast development controlled the degree of CO_2 incorporated into sterols. More developed chloroplasts had less CO_2 incorporation and increased light intensity gave a larger incorporation into sterols of *Calendula officinalis* (165). Using labeled mevalonic acid as the precursor, the results were reversed, more radioactivity was in the sterols and the chloroplasts (264), whereas light intensity (165) had no influence on the rate of ^{14}C incorporation. Incorporation of acetate into sterols was lower than incorporation of mevalonate but higher than CO_2, and light intensity had no influence on total incorporation (165). Labeled CO_2 and acetate under high illumination gave maximum labeling of sterols earlier than under low light intensity. It is postulated that CO_2, since supplied as a gas, can penetrate the chloroplasts by the natural photosynthetic pathway; whereas acetate and mevalonate which were supplied through the stem must be transported to the chloroplastic area. As discussed earlier, two mevalonic kinase enzyme systems exist: apparently one extrachloroplastic which gives rise to squalene and eventually to sterols, and another which is chloroplastic and forms chloroplastic terpenoids (38, 155). Chloroplasts are essentially impermeable to mevalonic acid (232); therefore, the terpenoid precursor remains in the cytoplasm where it is utilized for sterol biosynthesis. Acetate, however, is able to pass through the chloroplastic membrane and, as in the case of CO_2, can be involved in the biosynthesis of chloroplastic terpenoids. As chloroplasts develop their membranes become less permeable to mevalonic acid, creating biosynthetic compartmentalization (232, 264). The direct relationship between the intensity of photosynthesis and rate of sterol synthesis is unexplained, but triterpene precursors might be translocated from chloroplasts to the cytoplasm (165). However, the spatial separation of the two mevalonic kinase enzyme systems is not supported by all studies (53, 125).

The influence of light on the biosynthesis of sterols is even more complex than just described. The specific radioactivity of sterols in light grown *Nicotiana tabacum* seedlings was quite different whether mevalonic acid was supplied in the light or in the dark (47). Sitosterol and campesterol had the highest specific radioactivity in the light, while in the dark cholesterol had the highest specific radioactivity. The specific radioactivity of stigmasterol was not affected by light. On the other hand, the specific radioactivity of campesteryl and sitosteryl ester were not affected by illumination, while that of cholesteryl ester was stimulated by light. Only a complex sterol pool system could explain such results. Far-red light, as compared to dark treatment, stimulated the incorporation of labeled acetate into sterols of tuber discs of Jerusalem artichoke (135). Some of these effects which are caused by light have been linked to phytochrome-mediated systems (120, 135); but at present the evidence does not warrant such a conclusion.

POLLUTANTS   Environmental pollutants influence the biochemistry as well as the morphology of plants. It is known that smog and photooxidants, such as ozone, cause much plant damage and can become critical factors in plant growth. Ozone induces premature senescence and degradation of chloroplasts (260); however, certain plants can be protected against this injury with antisenescence compounds which inhibit tissue yellowing and loss of chlorophyll (263). These antisenescent compounds seem to prevent ozone damage by inhibiting changes in membrane permeability. With ozonation the sterol content in *Phaseolus vulgaris* leaves decreased and the steryl glycosides and acylated steryl glycosides increased (262). The free sterol content did not decrease in intact plants treated with antisenescence compounds benzimidazole, N-6-benzyladenine, and kinetin, and ozone injury was also prevented (263). It appears only reasonable to postulate that ozone increased membrane permeability by the conversion of free sterols to steryl glycosides and acylated steryl glycosides, and compounds which inhibit senescence inhibit the formation of steryl glycosides.

Metabolic Role

Mitochondria, like all other intracellular plant organelles, contain sterols (44, 127, 172). The polyene antibiotic filipin which binds sterols did not affect respiration in stem sections of *Pisum sativum* even if they were treated for long periods (151). A reduction in respiration with filipin, however, was found in yeast and rat liver mitochondria (19). Filipin can exist in a sterol binding and sterol nonbinding form, and the discrepancy in results could be due to the involvement of the two forms (240). If sterols are required for mitochondrial function, one might expect that changes in sterols would influence the respiratory capacity of the tissue. In roots of *Hordeum vulgare* ethanol decreased respiration and low concentrations of cholesterol had the ability to slightly suppress the ethanol effect (129). High cholesterol concentrations had the opposite influence on oxygen consumption, while steryl esters and steryl glycosides were inactive at any concentration. At present not much is known about the involvement of sterols in respiration, but in sterol-deficient yeast clones which are also low in respiration the respiratory competency can be restored with ergosterol (218). It would appear that the effect is at the membrane level, but the possibility that enzymes or coenzymes are directly influenced cannot be completely rejected (268). In this respect the increased influx of K^+ in *Pisum sativum* stem sections treated with cholesterol is of interest (151).

CONCLUDING REMARKS

In higher plants the biosynthesis of free sterols and steryl glycosides has been investigated in some detail, but the pathway for steryl ester formation is rarely studied. Most plant physiologists interested in the physiological role of sterols consider mainly the free sterols and often treat the steryl esters and steryl glycosides as "waste products." A number of working hypotheses have been suggested to explain the physiological function of steryl esters and steryl glycosides, but no experiments have been designed to test these suggestions. While it might not be easy to prove or disprove any of the working hypotheses, research in this area certainly would give more insight into the physiological role of these compounds. From what is known at the present time it is difficult to postulate why plants require such a diverse number of sterols, and why sterols occur as fatty acid esters and as glycosides. It is hoped that during the next 10 years some of these questions can be answered.

ACKNOWLEDGMENTS

The author expresses his appreciation to his associates who have assisted in preparing this review, with special thanks to Mrs. Louise Browinski for helping prepare the manuscript.

Literature Cited

1. Adams, S. R., Heinstein, P. F. 1973. *Phytochemistry* 12:2167–72
2. Aexel, R., Evans, R. A., Kelley, M., Nicholas, H. J. 1967. *Phytochemistry* 6:511–24
3. Agranoff, B. W., Eggerer, H., Henning, U., Lynen, F. 1960. *J. Biol. Chem.* 235:326–32
4. Arigoni, D. 1959. In *Ciba Foundation Symposium on Biosynthesis of Terpenes and Sterols*, ed. G. E. W. Wolstenholme, C. M. O'Conner, 235–36. Boston: Little, Brown
5. Alcaide, A., Devys, M., Barbier, M., Kaufmann, H. P., Sen Gupta, A. K. 1971. *Phytochemistry* 10:209–10
6. Alcaide, A., Devys, M., Bottin, M., Barbier, M., Lederer, E. 1968. *Phytochemistry* 7:1773–77
7. Alexander, G. J., Schwenk, E. 1957. *J. Am. Chem. Soc.* 79:4554–55
8. Alexander, G. J., Schwenk, E. 1958. *J. Biol. Chem.* 232:611–16
9. Aller, R. T. van, Chikamatsu, H., de Souza, N. J., John, J. P., Nes, W. R. 1968. *Biochem. Biophys. Res. Commun.* 31:842–44
10. Ardenne, M. von, Steinfelder, K., Trümmler, R. 1965. *Kulturpflanze* 13:115–29
11. Armarego, W. L. F., Goad, L. J., Goodwin, T. W. 1973. *Phytochemistry* 12: 2181–87
12. Atallah, A., Aexel, R. T., Ramsey, R. B., Threlkeld, S., Nicholas, H. J. 1975. *Phytochemistry*. In press
13. Axelos, M., Péaud-Lenoel, M. C. 1971. *C. R. Acad. Sci. Paris* 273D:1434–37
14. Bae, M., Mercer, E. I. 1970. *Phytochemistry* 9:63–68
15. Baisted, D. J. 1969. *Phytochemistry* 8:1697–1703
16. Baisted, D. J. 1971. *Biochem. J.* 124:375–83
17. Baisted, D. J., Capstack, E., Nes, W. R. 1962. *Biochemistry* 1:537–41
18. Baisted, D. J., Gardner, R. L., McReynolds, L. A. 1968. *Phytochemistry* 7:945–49

19. Balcavage, W. X., Beale, M., Chasen, B., Mattoon, J. R. 1968. *Biochim. Biophys. Acta* 162:525–32
20. Bates, R. B., Brewer, A. D., Knights, B. R., Rowe, J. W. 1968. *Tetrahedron Lett.* 59:6163–67
21. Battersby, A. R., Parry, G. V. 1964. *Tetrahedron Lett.* 14:787–90
22. Bean, G. A. 1973. *Advan. Lipid Res.* 11:193–218
23. Beeler, D. A., Anderson, D. G., Porter, J. W. 1963. *Arch. Biochem. Biophys.* 102:26–32
24. Bennett, R. D., Heftmann, E. 1965. *Phytochemistry* 4:475–79
25. Bennett, R. D., Heftmann, E. 1969. *Steroid* 14:403–7
26. Bennett, R. D., Heftmann, E., Preston, W. H., Haun, J. R. 1963. *Arch. Biochem. Biophys.* 103:74–83
27. Bennett, R. D., Heftmann, E., Purcell, A. E., Bonner, J. 1961. *Science* 134:671–73
28. Benveniste, P., Durr, A., Hirth, L., Ourisson, G. 1964. *C.R. Acad. Sci. Paris* 259:2005–8
29. Benveniste, P., Hewlins, M. J. E., Fritig, B. 1969. *Eur. J. Biochem.* 9:526–33
30. Benveniste, P., Hirth, L., Ourisson, G. 1964. *C. R. Acad. Sci. Paris* 259:2284–87
31. Benveniste, P., Hirth, L., Ourisson, G. 1966. *Phytochemistry* 5:31–44
32. Ibid, 45–58
33. Benveniste, P., Massy-Westrop, R. A. 1967. *Tetrahedron Lett.* 37:3553–56
34. Benveniste, P., Ourisson, G., Hirth, L. 1970. *Phytochemistry* 9:1073–82
35. Bergman, J., Lindgren, B. O., Svahn, C. M. 1965. *Acta Chem. Scand.* 19:1661–66
36. Bergmann, W. 1953. *Ann. Rev. Plant Physiol.* 4:383–426
37. Biswas, P. K., Paul, K. B., Henderson, J. H. M. 1967. *Nature* 213:917–18
38. Bloch, K. 1965. *Science* 150:19–28
39. Bolger, L. M., Rees, H. H., Goad, L. J., Goodwin, T. W. 1969. *Biochem. J.* 114:892–93
40. Bonner, J., Axtman, G. 1937. *Proc. Nat. Acad. Sci.* 23:453–57
41. Bonner, J., Heftmann, E., Zeevaart, J. A. D. 1963. *Plant Physiol.* 38:81–88
42. Bowden, B. N. 1971. *Phytochemistry* 10:3135–37
43. Brady, D. R., Gaylor, J. L. 1971. *J. Lipid Res.* 12:270–76
44. Brandt, R. D., Benveniste, P. 1972. *Biochim. Biophys. Acta* 282:85–92
45. Bush, P. B. 1971. *Physiology of sterols in germinating seeds and developing seed-lings of Nicotiana tabacum L.* PhD thesis. Univ. Kentucky, Lexington. 99 pp.
46. Bush, P. B., Grunwald, C. 1972. *Plant Physiol.* 50:69–72
47. Ibid 1973. 51:110–14
48. Ibid 1974. 53:131–35
49. Bush, P. B., Grunwald, C., Davis, D. L. 1971. *Plant Physiol.* 47:745–49
50. Capstack, E. et al 1962. *Biochemistry* 1:1178–83
51. Capstack, E., Rosin, N. L., Blondin, G. A., Nes, W. R. 1965. *J. Biol. Chem.* 240:3258–63
52. Castle, M., Blondin, G. A., Nes, W. R. 1963. *J. Am. Chem. Soc.* 85:3306–8
53. Charlton, J. M., Treharne, K. J., Goodwin, T. W. 1967. *Biochem. J.* 105:205–12
54. Coolbaugh, R. C., Moore, T. C., Barlow, S. A., Ecklund, P. R. 1973. *Phytochemistry* 12:1613–18
55. Corey, E. J., de Montellano, P. R., Yamamoto, H. 1968. *J. Am. Chem. Soc.* 90:6254–55
56. Corey, E. J., Lin, K., Jautelat, M. 1968. *J. Am. Chem. Soc.* 90:2724–26
57. Corey, E. J., Russey, W. E., de Montellano, P. R. 1966. *J. Am. Chem. Soc.* 88:4750–51
58. Cornforth, J. W. 1969. *Quart. Rev.* 23:125–40
59. Cornforth, J. W., Cornforth, R. H., Donninger, C., Popják, G. 1966. *Proc. Roy. Soc. B* 163:492–514
60. Cornforth, J. W. et al 1966. *Proc. Roy. Soc. B* 163:436–64
61. Cornforth, J. W. et al 1965. *J. Am. Chem. Soc.* 87:3224–28
62. Cornforth, J. W., Cornforth, R. H., Pelter, A., Horning, M. G., Popják, G. 1959. *Tetrahedron* 5:311–39
63. Cornforth, J. W., Cornforth, R. H., Popják, G., Yengoyan, L. 1966. *J. Biol. Chem.* 241:3970–87
64. Costes, C., Bazier, R., Lechevallier, D. 1972. *Physiol. Vég.* 10:291–317
65. Croteau, R., Loomis, W. D. 1973. *Phytochemistry* 12:1957–65
66. Czygan, F. 1962. *Naturwissenschaften* 49:287–88
67. Dauben, W. G. et al 1953. *J. Am. Chem. Soc.* 75:3038
68. Davis, D. L. 1972. *Phytochemistry* 11:489–94
69. Davis, D. L., Finkner, V. C. 1972. *Plant Physiol.* 52:324–26
70. Davis, D. L., Poneleit, C. G. 1975. *Plant Physiol.* In press
71. Dean, P. D. G., de Montellano, P. R., Bloch, K., Corey, E. J. 1967. *J. Biol. Chem.* 242:3014–19

72. Deenen, L. L. M. van 1972. *Naturwissenschaften* 59:485–91
73. Devys, M., Alcaide, A., Barbier, M. 1968. *Bull. Soc. Chim. Biol.* 50:1751–57
74. Devys, M., Alcaide, A., Barbier, M. 1969. *Phytochemistry* 8:1441–44
75. Devys, M., Alcaide, A., Barbier, M., Lederer, E. 1968. *Phytochemistry* 7: 613–17
76. Devys, M., Alcaide, A., Pinte, F., Barbier, M. 1969. *C. R. Acad. Sci. Paris* 269D:2033–35
77. Devys, M., Andre, D., Barbier, M. 1969. *C. R. Acad. Sci. Paris* 269D:798–801
78. Devys, M., Barbier, M. 1967. *Bull. Soc. Chim. Biol.* 49:865–71
79. Djerassi, C., Krakower, G. W., Lemin, A. J., Lin, L. H., Mills, J. S. 1958. *J. Am. Chem. Soc.* 80:6284–92
80. Douglas, T. J., Paleg, L. G. 1972. *Plant Physiol.* 49:417–20
81. Ibid 1974. 54:238–45
82. Dupéron, P. 1968. *C. R. Acad. Sci. Paris* 266D:1658–61
83. Dupéron, P. 1971. *Physiol. Vég.* 9:373–99
84. Dupéron, P., Dupéron, R. 1965. *C. R. Acad. Sci. Paris* 261:3459–62
85. Ibid 1969. 269D:157–60
86. Ibid 1970. 270D:1108–11
87. Dupéron, P., Renaud, M. 1966. *C. R. Soc. Biol.* 160:1710–14
88. Dupéron, R., Billard, M., Dupéron, P. 1972. *C. R. Acad. Sci. Paris* 274D:2321–24
89. Dupéron, R., Dupéron, P., Thiersault, M. 1971. *C. R. Acad. Sci. Paris* 273D:580–83
90. Edwards, P. A., Green, C. 1972. *FEBS Lett.* 20:97–99
91. Ehrhardt, J. D., Hirth, L., Ourisson, G. 1965. *C. R. Acad. Sci. Paris* 260:5931–34
92. Ehrhardt, J. D., Hirth, L., Ourisson, G. 1967. *Phytochemistry* 6:815–21
93. Eichenberger, W., Grob, E. C. 1968. *Chimia* 22:46–48
94. Ibid 1969. 23:368–69
95. Ibid 1970. 24:394–96
96. Eichenberger, W., Grob, E. C. 1970. *FEBS Lett.* 11:177–80
97. Eichenberger, W., Menke, W. 1966. *Z. Naturforsch.* 216:859–67
98. Eichenberger, W., Newman, D. W. 1968. *Biochem. Biophys. Res. Commun.* 32:366–74
99. Eppenberger, U., Hirth, L., Ourisson, G. 1969. *Eur. J. Biochem.* 8:180–83
100. Eschenmoser, A., Ruzicka, L., Jeger, D., Arigoni, D. 1955. *Helv. Chem. Acta* 38:1850–56
101. Evans, F. J. 1973. *Planta* 111:33–40
102. Ibid 1974. 116:99–104
103. Evans, L. T. 1964. *Aust. J. Biol. Sci.* 17:24–35
104. Fieser, L. F., Fieser, M. 1959. *Steroids.* New York: Reinhold. 945 pp.
105. Finean, J. B. 1953. *Experientia* 9:17–19
106. Garcia-Peregrin, E., Suárez, M. D., Aragón, M. C., Mayor, F. 1972. *Phytochemistry* 11:2495–98
107. Genus, J. M. C. 1973. *Phytochemistry* 12:103–6
108. Genus, J. M. C., Vendrig, J. C. 1974. *Phytochemistry* 13:913–22
109. Gibbons, G. F., Goad, L. J., Goodwin, T. W. 1967. *Phytochemistry* 6:677–83
110. Gibbons, G. F., Goad, L. J., Goodwin, T. W., Nes, W. R. 1971. *J. Biol. Chem.* 246:3967–76
111. Gioelli, F. 1942. *Arch. Sci. Biol. (Italy)* 28:311–16
112. Goad, L. J. 1967. In *Terpenoids in Plants,* ed. J. B. Pridham, 159–90. London:Academic
113. Goad, L. J. 1970. In *Natural Substances Formed Biologically from Mevalonic Acid,* ed. T. W. Goodwin, 45–77. London: Academic
114. Goad, L. J., Gibbons, G. F., Bolger, L. M., Rees, H. H., Goodwin, T. W. 1969. *Biochem. J.* 114:885–93
115. Goad, L. J., Goodwin, T. W. 1966. *Biochem. J.* 99:735–46
116. Goad, L. J., Goodwin, T. W. 1967. *Eur. J. Biochem.* 1:357–62
117. Goad, L. J., Goodwin, T. W. 1972. In *Progress in Phytochemistry,* ed. L. Reinhold, Y. Liwschitz, 3:113–98. London: Interscience
118. Goad, L. J., Hammam, A. S. A., Dennis, A., Goodwin, T. W. 1966. *Nature* 210:1322–24
119. Goodman, W. S. de, Popják, G. 1960. *J. Lipid Res.* 1:286–300
120. Goodwin, T. W. 1967. In *Terpenoids in Plants,* ed. J. B. Pridham, 1–23. London: Academic
121. Goodwin, T. W. 1971. *Biochem. J.* 123:293–329
122. Goodwin, T. W., Williams, R. J. H. 1966. *Proc. Roy. Soc. B* 163:515–23
123. Gottlieb, D., Shaw, P. D. 1970. *Ann. Rev. Phytopathol.* 8:371–402
124. Graebe, J. E. 1968. *Phytochemistry* 7:2003–20
125. Gray, J. C., Kekwick, R. G. O. 1973. *Biochem. J.* 133:335–47
126. Grunwald, C. 1968. *Plant Physiol.* 43:484–88

127. Ibid 1970. 45:663–66
128. Ibid 1971. 48:653–55
129. Grunwald, C. 1974. *Plant Physiol. Suppl.* 53:48 (Abstr.)
130. Grunwald, C. 1974. *Plant Physiol.* 54:624–28
131. Grunwald, C. 1975. *Phytochemistry* 14: 79–82
132. Grunwald, C., Bush, L. P., Keller, C. J. 1971. *J. Agr. Food Chem.* 19:216–21
133. Hall, J., Smith, A. R. H., Goad, L. J., Goodwin, T. W. 1969. *Biochem. J.* 112:129–30
134. Hartmann, M. A., Benveniste, P. 1973. *C. R. Acad. Sci. Paris* 276D:3143–46
135. Hartmann, M. A., Benveniste, P., Durst, F. 1972. *Phytochemistry* 11: 3003–5
136. Hartmann, M. A., Ferne, M., Gigot, C., Brandt, R., Benveniste, P. 1973. *Physiol. Vég.* 11:209–30
137. Heftmann, E. 1963. *Ann. Rev. Plant Physiol.* 14:225–48
138. Heftmann, E. 1968. *Lloydia* 31:293–317
139. Heftmann, E. 1971. *Lipids* 6:128–33
140. Heftmann, E. 1973. In *Phytochemistry-Organic Metabolites*, ed. L. P. Miller, 2:171–226. New York: Van Nostrand-Reinhold
141. Heftmann, E., Ko, S., Bennett, R. D. 1965. *Naturwissenschaften* 52:431–32
142. Heftmann, E., Sauer, H. H., Bennett, R. D. 1968. *Naturwissenschaften* 55: 37–38
143. Heintz, R. 1973. *Utilization de fractions subcellulaires pour l'etude de la biosynthesis des sterols des vegetaux superieurs.* PhD thesis. Univ. Louis Pasteur, Strasbourg, France. 118 pp.
144. Heintz, R., Benveniste, P. 1970. *Phytochemistry* 9:1499–1503
145. Heintz, R., Benveniste, P. 1972. *C. R. Acad. Sci. Paris* 274D:947–50
146. Heintz, R., Benveniste, P. 1974. *J. Biol. Chem.* 249:4267–74
147. Heintz, R., Benveniste, P., Bimpson, T. 1972. *Biochem. Biophys. Res. Commun.* 46:766–72
148. Heintz, R., Benveniste, P., Robinson, W. H., Coates, R. M. 1972. *Biochem. Biophys. Res. Commun.* 42:1547–53
149. Heintz, R., Schaeffer, P. C., Benveniste, P. 1970. *J. Chem. Soc. Chem. Commun.* 15:946–47
150. Helmkamp, G., Bonner, J. 1953. *Plant Physiol.* 28:428–36
151. Hendrix, D. L., Higinbotham, N. 1973. *Plant Physiol.* 52:93–97
152. Hendrix, J. W. 1970. *Ann. Rev. Phytopathol.* 8:111–30

153. Hewlins, M. J. E., Ehrhardt, J. D., Hirth, L., Ourisson, G. 1969. *Eur. J. Biochem.* 8:184–88
154. Hikimo, H., Takemoto, T. 1972. *Naturwissenschaften* 59:91–98
155. Hill, H. M., Rogers, L. J. 1974. *Phytochemistry* 13:763–77
156. Hirayma, O., Suzuki, T. 1968. *Agr. Biol. Chem.* 32:549–54
157. Hou, C. T., Umemura, Y., Nakamura, M., Funahashi, S. 1967. *J. Biochem.* 62:389–91
158. Ibid 1968. 63:351–60
159. Hügel, M. F., Vetter, W., Audier, H., Barbier, M., Lederer, E. 1965. *Phytochemistry* 3:7–16
160. Hylmö, B. 1940. *Bot. Notis.* 389–94
161. Idler, D. R. et al 1953. *J. Am. Chem. Soc.* 75:1712–15
162. Ingram, D. S., Knights, B. A., McEvoy, I. J., McKay, P. 1968. *Phytochemistry* 7:1241–45
163. Jacobsohn, G. M. 1970. In *Recent Advances in Phytochemistry*, ed. C. Steelink, V. C. Runeckles, 3:229–47. New York:Appleton-Century-Crofts
164. Johnson, D. F., Bennett, R. D., Heftmann, E. 1963. *Science* 140:198–99
165. Kasprzyk, Z., Wojciechowski, Z., Jerzmanowski, A. 1971. *Phytochemistry* 10:797–805
166. Katayama, M., Kotoh, M. 1973. *Plant Cell Physiol.* 14:681–88
167. Kaufmann, H. P., Sen Gupta, A. K. 1964. *Fette Seifen Anstrichm* 66:461–66
168. Kauss, H. 1968. *Z. Naturforsch.* 23B:1522–26
169. Kemp, R. J., Goad, L. J., Mercer, E. I. 1967. *Phytochemistry* 6:1609–15
170. Kemp, R. J., Hammam, A. S. A., Goad, L. J., Goodwin, T. W. 1968. *Phytochemistry* 7:447–50
171. Kemp, R. J., Mercer, E. I. 1968. *Biochem. J.* 110:111–18
172. Ibid, 119–25
173. Kiribuchi, T., Chen, C. S., Funahashi, S. 1965. *Agr. Biol. Chem.* 29:265–67
174. Kiribuchi, T., Mizunaga, T., Funahashi, S. 1966. *Agr. Biol. Chem.* 30:770–78
175. Knapp, F. F., Aexel, R. T., Nicholas, H. J. 1969. *Plant Physiol.* 44:442–46
176. Knapp, F. F., Greig, J. B., Goad, L. J., Goodwin, T. W. 1971. *J. Chem. Soc. D* 13:707–9
177. Knapp, F. F., Nicholas, H. J. 1969. *Phytochemistry* 8:2091–93
178. Knapp, F. F., Nicholas, H. J. 1970. *J. Chem. Soc. Chem. Commun.* 7:399–400
179. Knapp, F. F., Nicholas, H. J. 1971. *Phytochemistry* 10:85–95

180. Ibid, 97–102
181. Knights, B. A. 1965. *Phytochemistry* 4:857–62
182. Knights, B. A. 1971. *Lipids* 6:215–18
183. Knights, B. A., Laurie, W. 1967. *Phytochemistry* 6:407–16
184. Kopcewicz, J. 1969. *Naturwissenschaften* 56:287
185. Ibid, 334
186. Ibid 1970. 57:48
187. Ibid, 136
188. Ibid 1971. 65:92–94
189. Kopcewicz, J. 1971. *Phytochemistry* 10:1423–27
190. Kwong, C. N., Heikkila, R. E., Cornwell, D. G. 1971. *J. Lipid Res.* 12:31–35
191. Laseter, J. L., Evans, R., Walkinshaw, C. H., Weete, J. D. 1973. *Phytochemistry* 12:2255–58
192. Lederer, E. 1969. *Quart. Rev. Chem. Soc.* 23:453–81
193. Lenton, J. R. et al 1971. *Arch. Biochem. Biophys.* 143:664–74
194. Lepage, M. 1964. *J. Lipid Res.* 5:587–92
195. Levitt, J. 1972. *Responses of Plants to Environmental Stresses.* New York: Academic. 697 pp.
196. Lindberg, M., Gautschi, F., Bloch, K. 1963. *J. Biol. Chem.* 238:1661–64
197. Lindberg, M., Yuan, C., DeWaard, A., Bloch, K. 1962. *Biochemistry* 1:182–88
198. Long, R. A., Hruska, F., Gesser, H. D. 1970. *Biochem. Biophys. Res. Commun.* 41:321–27
199. Loomis, W. D., Battaile, J. 1963. *Biochim. Biophys. Acta* 67:54–63
200. Löve, A., Löve, D. 1945. *Ark. Bot.* 32A:1–60
201. Masiak, S., LeFevre, P. G. 1974. *Arch. Biochem. Biophys.* 162:442–47
202. Maudgal, R. K., Tchen, T. T., Bloch, K. 1958. *J. Am. Chem. Soc.* 80:2589–90
203. Mayer, A. M., Shain, Y. 1974. *Ann. Rev. Plant Physiol.* 25:167–93
204. McKillican, M. E. 1964. *J. Am. Oil Chem. Soc.* 41:554–57
205. Mercer, E. I., Davies, B. H., Goodwin, T. W. 1963. *Biochem. J.* 87:317–25
206. Mercer, E. I., Pughe, J. E. 1969. *Phytochemistry* 8:115–22
207. Mudd, J. B., Kleinschmidt, M. G. 1970. *Plant Physiol.* 45:517–18
208. Nagasampagi, B. A., Rowe, J. W., Simpson, R., Goad, L. J. 1971. *Phytochemistry* 10:1101–7
209. Nicholas, H. J. 1961. *Nature* 189:143–44
210. Nicholas, H. J. 1962. *J. Biol. Chem.* 237:1476–80
211. Ibid, 1481–84
212. Ibid, 1485–88
213. Nicholas, H. J., Moriarty, S. 1963. *Fed. Proc.* 22:529
214. Nordby, H. E., Nogy, S. 1974. *Phytochemistry* 13:443–52
215. Ongun, A., Mudd, J. B. 1970. *Plant Physiol.* 45:255–62
216. Papahadjopoulos, D. 1974. *J. Theor. Biol.* 43:329–37
217. Parks, L. W. 1958. *J. Am. Chem. Soc.* 80:2023–24
218. Parks, L. W., Starr, P. R. 1963. *J. Cell. Comp. Physiol.* 61:61–65
219. Péaud-Lenoel, M. C., Axelos, M. 1971. *C. R. Acad. Sci. Paris* 273D:1057–60
220. Ponsinet, G., Ourisson, G. 1967. *Phytochemistry* 6:1235–43
221. Ibid 1968. 7:89–98
222. Ibid, 757–64
223. Potty, V. H., Bruemmer, J. H. 1970. *Phytochemistry* 9:99–105
224. Power, F. B., Salway, A. H. 1913. *J. Chem. Soc.* 103:399–406
225. Raab, K. H., de Souza, N. J., Nes, W. R. 1968. *Biochim. Biophys. Acta* 152:742–48
226. Rahman, R., Sharpless, K. B., Spencer, T. A., Clayton, R. B. 1970. *J. Biol. Chem.* 245:2667–71
227. Rees, H. H., Goad, L. J., Goodwin, T. W. 1968. *Biochem. J.* 107:417–26
228. Rees, H. H., Goad, L. J., Goodwin, T. W. 1968. *Tetrahedron Lett.* 6:723–25
229. Rees, H. H., Goad, L. J., Goodwin, T. W. 1969. *Biochim. Biophys. Acta* 176:892–94
230. Rees, H. H., Mercer, E. I., Goodwin, T. W. 1966. *Biochem. J.* 99:726–34
231. Reid, W. W. 1968. *Phytochemistry* 7:451–52
232. Rogers, L. J., Shah, S. P. J., Goodwin, T. W. 1966. *Biochem. J.* 99:381–88
233. Ibid. 100:14c–17c
234. Rohmer, M., Brandt, R. D. 1973. *Eur. J. Biochem.* 36:446–54
235. Rowe, J. W. 1965. *Phytochemistry* 4:1–10
236. Russell, P. T., van Aller, R. T., Nes, W. R. 1967. *J. Biol. Chem.* 242:5802–6
237. Ruzicka, L. 1953. *Experientia* 9:357–67
238. Sachs, R. M. 1966. *Plant Physiol.* 41:1392–94
239. Schreiber, K., Osske, G. 1964. *Tetrahedron* 20:2575–84
240. Schroeder, F., Holland, J. F., Bieber, L. L. 1973. *Biochemistry* 12:4785–89
241. Shah, D. O., Schulman, J. H. 1968. *Advan. Chem. Ser.* 84:189–209
242. Sharpless, K. B. et al 1968. *J. Am. Chem. Soc.* 90:6874–75

243. Sharpless, K. B. et al 1969. *J. Am. Chem. Soc.* 91:3394–96
244. Shewry, P. R., Pinfield, N. J., Stobart, A. K. 1974. *Phytochemistry* 13:341–46
245. Shewry, P. R., Stobart, A. K. 1974. *Phytochemistry* 13:347–55
246. Smith, A. R. H., Goad, L. J., Goodwin, T. W. 1967. *Biochem. J.* 104:56c–58c
247. Smith, A. R. H., Goad, L. J., Goodwin, T. W. 1968. *Chem. Commun.* 15:926–27
248. Smith, A. R. H., Goad, L. J., Goodwin, T. W. 1972. *Phytochemistry* 11:2775–81
249. Standifer, L. N., Devys, M., Barbier, M. 1968. *Phytochemistry* 7:1361–65
250. Starr, P. R., Parks, L. W. 1962. *J. Cell. Comp. Physiol.* 59:107–10
251. Stokke, K. T. 1972. *Biochim. Biophys. Acta* 270:156–66
252. Ibid. 280:329–35
253. Stoll, A., Jucker, E. 1955. In *Modern Methods of Plant Analysis*, ed. K. Paech, M. V. Tracy, 3:141–271. Berlin: Springer Verlag
254. Sucrow, W. 1968. *Tetrahedron Lett.* 20:2443–45
255. Swindell, A. C., Gaylor, J. L. 1968. *J. Biol. Chem.* 243:5546–55
256. Tamelen, E. E. van, Willett, J. D., Clayton, R. B., 1966. *J. Am. Chem. Soc.* 88:4752–54
257. Ibid 1967. 89:3371–73
258. Tchen, T. T. 1958. *J. Biol. Chem.* 233:1100–3
259. Tchen, T. T., Bloch, K. 1957. *J. Biol. Chem.* 226:931–39
260. Thomson, W. W., Dugger, W. M. Jr., Palmer, R. L. 1966. *Can. J. Bot.* 44:1677–82
261. Threlfall, D. R., Griffiths, W. T., Good-win, T. W. 1967. *Biochem. J.* 103:831–51
262. Tomlinson, H., Rich, S. 1971. *Phytopathology* 61:1404–5
263. Ibid 1973. 63:903–6
264. Treharne, K. J., Mercer, E. I., Goodwin, T. W. 1966. *Biochem. J.* 99:239–45
265. Tso, T. C., Cheng, A. L. S. 1971. *Phytochemistry* 10:2133–37
266. Vandenheuvel, F. A. 1963. *J. Am. Oil Chem. Soc.* 40:455–72
267. Vendrig, J. C. 1967. *Ann. NY Acad. Sci.* 144:81–93
268. Wade, R., Jones, H. W. Jr. 1956. *J. Biol. Chem.* 220:553–62
269. Waters, J. A., Johnson, D. F. 1965. *Arch. Biochem. Biophys.* 112:387–91
270. Weete, J. D. 1973. *Phytochemistry* 12:1843–64
271. Weizmann, A., Mazur, Y. 1958. *J. Org. Chem.* 23:832–34
272. Weyland, H. Z. 1948. *Krebsforschung* 56:148–64
273. Williams, B. L., Goad, L. J., Goodwin, T. W. 1967. *Phytochemistry* 6:1137–45
274. Williams, B. L., Goodwin, T. W. 1965. *Phytochemistry* 4:81–88
275. Williams, R. J. H., Britton, G., Goodwin, T. W. 1967. *Biochem. J.* 105:99–105
276. Williamson, I. P., Kekwick, R. G. O. 1965. *Biochem. J.* 96:862–71
277. Woodward, R. B., Bloch, K. 1953. *J. Am. Chem. Soc.* 75:2023–24
278. Wojciechowski, Z., Bochénska-Hryniewicz, M., Kucharczak, B. 1972. *Phytochemistry* 11:1165–68
279. Wojciechowski, Z., Goad, L. J., Goodwin, T. W. 1973. *Biochem. J.* 136:405–12
280. Yamamoto, S., Lin, K., Bloch, K. 1969. *Nat. Acad. Sci.* 63:110–17

Ann. Rev. Plant Physiol. 1975. 26:237–58

THE "WASHING" OR "AGING"   ❖7590
PHENOMENON IN PLANT TISSUES[1]

Reinhard F. M. Van Steveninck

Department of Botany, University of Queensland, Brisbane, Australia

CONTENTS

INTRODUCTION

Since the first plant tissue slice was launched into an aqueous environment, probably at the turn of the century (120), in a climate dominated by Sachs and by "whole plant" physiology, an extensive literature has evolved around the use of this material. It was soon perceived that prolonged washing of tissue slices caused profound changes in metabolism and physiological competence of the component cells (161, 162), and the characterization of this washing phenomenon and research aimed at

---

[1]Abbreviations used: ABA (abscisic acid); ADP (adenosine diphosphate); ATP (adenosine triphosphate); CCCP (carbonyl cyanide- m-chlorophenylhydrazone); ER (endoplasmic reticulum); GA_3 (gibberellic acid); mRNA (messenger ribonucleic acid); NAA (α-napthalene acetic acid); NADH (nicotinamide adenine dinucleotide); poly U (poly uridylic acid); TCAC (tricarboxylic acid cycle); tris (tris-hydroxymethyl-aminomethane).

        237

finding an explanation for the changes observed became a primary challenge for many years (84, 86, 87, 165, 166).

It is significant that an increasing number of plant physiologists may question the scientific usefulness of information based on what may appear to be a highly artificial set of conditions. Such a view was brought to our attention in this series by Kramer (81), who expressed his concern about the present lack of understanding regarding the overall regulatory aspects of morphogenesis and physiology of the whole plant. While tissue slice physiologists are not likely to be alarmed by this expression of concern, the time seems right to place more emphasis on interpretations of induced changes in excised tissues, which relate to cellular differentiation in situ. An admirable example of this approach was provided by Laties (91), who, in his provocative application of the concept of dual mechanisms, provided much additional insight into the problem of long distance salt transport in whole plants. Another fine example of the integrative approach is provided by Steward & Mott (160) in their comprehensive review on salt accumulation in cultured cells and in tissue explants.

The latest reviews on the subject of aging in tissue slices have dealt mainly with genetic and metabolic regulation (derepression) in differentiating cells (76) and with the development of ion transport capacities (184). In the latter it was proposed to define the stimulated synthesis and increased physiological competence resulting from derepression during washing of tissue slices as "adaptive aging" as opposed to "senescent aging" where a pathway of reduced synthesis leads to an irreversible diminution of physiological competence and death.

Slicing and washing of storage tissue provides one of the most convenient, easily repeatable methods of inducing a certain measure of cell differentiation under controlled conditions. Therefore it seems appropriate to focus attention on changes in cellular compartmentation and membrane structures during this induced cell differentiation, in the hope that further clues may be obtained in the future regarding the relative significance of the endomembrane complex and the various organelles in determining the physiological competence of cells suddenly exposed to surface conditions. Until recently it was not possible to obtain unambiguous biochemical separation of specific membrane fractions, but with the increasing ability to characterize various membrane fractions, techniques for their accurate separation are certain to improve rapidly (51, 64, 150).

As yet, information on the biogenesis and turnover of organelles in plant cells is almost completely lacking, although there are already some indications that the properties and functional capacities of organelles may change during the process of differentiation. While the study of differentiation at the organelle level should be a central issue of this discussion, one may reflect that at their first appearance, organelles of plants and animals may be so much alike that it is sometimes difficult to distinguish portions of a meristematic plant cell from an animal cell. Moreover, biochemically speaking, isolated plant mitochondria perform nearly all the same tricks as animal mitochondria, and hence one might argue that cytoplasmic organelles are basically similar for all organisms and therefore only of secondary importance in matters of cell differentiation and morphogenesis. At present the endomembrane system remains ill-defined, but gradually a picture is emerging that, especially in meristematic cells, a membranous continuity exists between nucleus, ER, Golgi apparatus and provacuoles (29, 52, 107, 118), and also between ER and mitochondria (22, 117). Already a certain polarity is implied by this continuity from the nucleus via ER and Golgi apparatus to its distal saccules which, especially in plant cells, are endowed with a capacity to give rise to provacuoles and/or to produce cell wall materials. For a further discussion on the polarity of the Golgi apparatus see Northcote (121). It seems particularly relevant that the membrane system has proved to be continuous from cell to cell through the presence of plasmodesmata (25, 139), thus allowing a measure of control between groups of cells and providing the outermost or surface cells with a mechanism for sensing their structural position at the outside of the tissue. It seems premature to suggest that the totality of the endomembrane system may provide the "skeleton" of the morphogenetic system, but in using this analogy, one could envisage the nucleus and ER continuity as the brain and nervous system which, via the endomembrane complex as a closely integrated part of the skeleton, would be capable of controlling organelle function. It is time to bring this little fantasy to an end and consider the facts available. However, one need not apologize for the fact that throughout the universe structural features tend to repeat themselves at different levels of organization and dimension.

MEMBRANES, ORGANELLES, AND LIPID METABOLISM

Endomembrane System

Direct observations of changes in the membrane system of sliced storage tissue are rare, mainly because adequate fixation of large vacuolated plant parenchyma cells is notoriously difficult to achieve, and also because the cytoplasmic phase usually consists of a relatively featureless layer of less than one micrometer thick. Organelles and membrane structure in such cells are relatively sparse, and not randomly distributed, and hence do not invite quantitative assessment unless the changes are dramatic and easily recognized. Such a marked change, however, was observed by Jackman & Van Steveninck (68), who compared cytoplasmic structures in parenchyma cells of beetroot tissue in situ with parenchyma cells in slices which had been aged for various periods of time. The extent of the ER was drastically reduced to occasional small cytoplasmic vesicles immediately after cutting and rinsing the slices, while after about 24 hr this was followed by a reorganization into a lamellar system which became quite extensive from 48 hr onwards. Fowke & Setterfield (47) observed a similar response in slices of Jerusalem artichoke tuber. Cells fixed directly from intact tubers exhibited only a rudimentary ER and dictyosomes were practically absent; slices aged 24 hr, however, showed a distinct increase in amount of ER, especially of the rough type. The latter is of interest since it has been shown that the microsomal fraction isolated from aged slices has a considerably increased capacity for protein synthesis compared with fractions from fresh slices of beetroot (41) and Jerusalem artichoke tissue (28), which were practically inactive. It is not known to what extent the increased efficiency is due to molecular changes in the microsome membranes during adaptive aging, but it is now generally accepted that mRNA is associated with membranes of the rough ER (100), and it seems likely that progressively more mRNA is incorporated into the ER structures which yield the microsome fraction. Protein synthesis by ribosomal preparations from intact roots and tubers can be greatly stimulated by addition of artificial mRNA (poly U) [carrot (93), potato (75)], while the synthesis of new mRNA during aging is also reflected by the increased number of polyribosome profiles in aged artichoke slices (47) or potato slices (11) and by the increased percentage of extractable polyribosomes compared to total ribosomes [e.g. from 10% to 65% in carrot slices (93), sugar beet slices (30), potato slices (74)].

From earlier studies (93) it seemed that ribosomes did not alter during aging because ribosomes extracted from fresh and aged slices of carrot were equally efficient in the incorporation of ^{14}C-phenylalanine when they were supplied with poly U. However, recent work has shown that differences between the in vitro amino acid incorporation by ribosomes and by the supernatant from fresh and aged carrot slices were mainly confined to the ribosome fraction (98). The ribosomes from the aged slices were more effective in binding ^{3}H-poly U than ribosomes from fresh slices, and they also contained two components which were not normally associated with ribosomes from fresh slices (98). It now appears increasingly important to establish whether the composition and character of the ER system changes during the period of adaptive aging. In vivo observation on the formation of crystalloid protein inclusions in the ER of aging beetroot slices indicate that new mRNA is not an essential requirement for the continued functioning of this protein synthesizing system (172). Whereas actinomycin D added to fresh slices may strongly inhibit or prevent the synthesis of several important enzymes, e.g. invertase in carrot, beet, potato, sugar cane and Jerusalem artichoke tissue (50, 149, 189, 190) and the development of an ion transport capacity (188), it was found to have no effect on the assembly of the ER during the aging of beetroot slices, and furthermore it actually promoted the formation of crystalloid protein bodies inside the ER system. It was therefore suggested that the mRNA associated with the lamellar system of the ER provided a complex which would prevent its rapid degradation. The observation by Blobel & van Potter (20) that membrane bound polyribosomes are more resistant to exogenous ribonuclease than free polyribosomes in rat liver tissue lent support to this hypothesis. Although new mRNA does not seem to be an essential requirement for the assembly of the ER system, it was shown unequivocally that the process of assembly could not proceed without active protein synthesis. Both cycloheximide ($3.5 \times 10^{-6}M$) and puromycin ($10^{-4}M$) prevented the formation of ER lamellae as well as the production of crystalloid protein bodies (188).

The characterization of electron transport in turnip microsomes has led to some interesting observations which indicate that drastic changes may occur in the properties of microsomal fractions during the aging of slices. On slicing of turnip, swede, and beetroot tissue 20–100% of the microsomal NADH-dehydrogenase activity was lost within 10 min, but this activity recovered during the washing of slices to reach a maximum at 24 hr and then again declined (147). However, a salt-stimulated Mg^{++}-activated alkaline ATPase showed little change in activity during the aging of the slices (148). Leonard & Hanson (96, 97) ascribed the development of an enhanced ion absorption induced by the washing of excised corn root tissue to an increased activity of $(Mg^{++} + K^+)$-stimulated ATPase. It was claimed that the increased activity was confined to the microsomal fraction, but characterization of the membrane fraction from oat roots indicated that the enrichment occurred in the plasma membranes rather than the endomembrane system (64).

A few negative results should not deter future workers from making a determined effort to characterize the molecular and biochemical changes of specific membrane fractions during the adaptive aging of tissue slices, since it is bound to be an essential prerequisite for understanding the role of the endomembrane system in genetic and metabolic regulation during cell differentiation.

Plasma Membrane and Tonoplast

It is just over 10 years ago that Laties (87) opened his outstanding critical discussion on physiological aspects of membrane function with the sentence: "In any consideration of membrane characteristics of plant tissues there is, at once, the compelling need to distinguish between the two main bounding membranes which characterize plant cells." Leakage of large quantities of ions from fresh tissue slices or isolated steles was then thought to originate from the free space and from damaged cells, but continuation of this leakage for many hours and its subsequent reduction after a period of at least 24 hr showed that, especially for cations, significant changes in passive permeability characteristics occurred during the aging of slices (174, 175). It was unfortunate that the results of this early work did not allow a clear-cut distinction between tonoplast and plasmalemma fluxes. It was tacitly assumed that the large quantities of K^+ and Cl^- lost from fresh slices originated from the vacuole, implying leakiness of both tonoplast and plasmalemma, and evidence of a "Cl^- absorption shoulder" in aged slices indicated that the plasmalemma was a lesser permeability barrier than the tonoplast (175). With the introduction by Pitman (127) of a more elaborate flux analysis based on a series model of vacuolar and cytoplasmic compartments and external solution, the tonoplast became firmly established as the main permeability barrier in aged slices. However, this method of efflux analysis has not always yielded straightforward results (34, 123). Apart from a necessity to consider greater complexity of cytoplasmic compartmentation, the possibility that the permeability characteristics of the plasmalemma may be modified by the experimental procedure should not be dismissed lightly. Possibly because of a greater membrane permeability, fresh tissue slices especially have proved to be sensitive to external influences [e.g. the frequent repetitive rinses with deionized water or single salt solutions may cause removal of Ca^{++} during rinses (43, 176), or may have an effect on leakage through removal of hormones (177)].

An even greater uncertainty exists with respect to the location of ion transport capacities which develop during the washing of tissue slices or the isolated steles of main roots. This topic was recently reviewed by Anderson (3) in this series, and hence only some salient features will be mentioned here. The proposal by Laties (89, 91) of a series model based on a dual ion transport system in roots, with a high affinity for ions in the plasmalemma and a low affinity in the tonoplast, is still strongly disputed by Epstein (43), who maintains that both systems are situated at the plasmalemma. According to Lüttge & Laties (101), the capacity for ion transport acquired by isolated steles during aging is due to the development of a high affinity transport system located in the plasmalemma. Osmond & Laties (122) also proposed that the ion uptake capacity in aging beetroot slices is located in the plasmalemma. Hiatt (61), on the other hand, in considering organic acid compartmentation, argued that a high affinity system should also operate at the tonoplast. It is self-evident that the above arguments will be more easily settled when the ion activities for each compartment and the appropriate transmembrane potentials can be determined in fresh and in aged slices (cf review in this series by Higinbotham 63), e.g. Pitman et al (128) measured a pronounced hyperpolarization in the cortex of barley root segments washed in 0.5 mM $CaSO_4$ for 6–8 hr.

A valuable and equally direct approach would be to establish changes of ion stimulated ATPase activities during aging and to combine such studies with the cytochemical localization of these activities at the ultrastructural level (55). Leonard & Hanson (96, 97), who reported that washing of excised or intact maize roots doubled the rate of accumulation of $^{86}Rb^+$, $^{36}Cl^-$, and $^{32}P_i$, correlated this development with a similar increase in (Mg^{++} + K^+)-stimulated ATPase activity. These authors (56) also observed a change in the electron density of the tonoplast in fixed and stained tissue and hence speculated that the reported increases in solute absorption rate may be due to changes in the properties of the tonoplast induced by the washing treatment. Earlier it was claimed that in oat roots the enriched ATPase activity occurred in the plasmamembrane fraction which was identified by glucan synthetase activity, a high sterol to phospholipid ratio, and by a staining technique which was specific for plant plasma membranes (64). Clearly, recent advances in the reliability of the localization of specific ATPase activities are of great interest. Apparently it is now possible to separate membrane fractions on the basis of differential increases in their densities caused by nucleotidase-induced attachment of Pb-phosphate (150). This method combined with techniques of positive identification of plasma membranes in situ and as isolated fractions by the periodate-phosphotungstate-chromate staining technique (142), or of the Golgi apparatus by zinc iodide-osmium (36), offers excellent scope for future research.

Organelles

Few in vivo changes of the larger organelles have been observed during the aging of tissue slices. Verleur (192) reported that the number of mitochondria per cell had greatly increased in the layer of cells with cambial activity just below the surface of potato slices. Fowke & Setterfield (47), on the other hand, reported a marked increase in the granularity and size of the nucleoli, but could not detect any change in the appearance and number of dictyosomes or mitochondria in slices of Jerusalem artichoke. The latter seemed surprising since the respiratory rise in tissue slices was shown to involve activation of the tricarboxylic acid cycle (88). Van Steveninck & Jackman (186), however, observed a marked difference in the activity of mitochondrial suspensions isolated from whole beetroot tissue and from fresh and aged slices. Mitochondria from fresh slices disintegrated during the isolation procedure, and lacked activity and respiratory control, whereas the same procedures yielded active mitochondria from aged slices. Castelfranco, Tang & Bolar (27) observed a spread of succinoxidase activity when an homogenate obtained from fresh potato slices was centrifuged on a sucrose density gradient. An homogenate obtained from aging slices showed a sharpening of the peak, a suitable indicator for an increasing mitochondrial intactness. Successful mitochondrial isolation very much depends on technique and use of suitable additives to the isolation medium (65), but the fact remains that the observed differential fragility between mitochondria from fresh and aged slices could reflect a partial replacement of existing mitochondria after slicing.

Information on turnover in plant mitochondria is still very scarce, although Ben Abdelkader & Mazliak (14) established a half life for potato mitochondria of 10.4 days, which is similar to values found for animal cells (143). That the newly synthesized mitochondria in slices may differ from those in whole tissue was indicated by the heterogeneity of mitochondrial populations isolated from cut potato tissue (192, 193). Similarly, a new fraction of mitochondrial particles with increased cytochrome oxidase and dehydrogenase activities arose during the aging of sweet potato slices (152, 153). Nakano & Asahi (119) reported that the newly formed mitochondria, which were heavier than the pre-existing ones, were insensitive to CN^-, probably due to an acquired bypass of electron transport at the level of cytochrome b.

Lipid Metabolism and Membrane Turnover

An excellent review on lipid metabolism by Mazliak (109) appeared in this series, hence only a brief outline will be presented here.

Provided suitable corrections are applied for recycling of CO_2 and for the dissolved CO_2 and HCO_3^- already present in the tuber, $^{13}C/^{12}C$ carbon isotope ratios can indicate the nature of the carbon reserves which are being utilized in maintaining the respiratory rise during the aging of potato slices. Jacobson et al (69), in using this elegant method, made a particularly significant finding when they provided a measure of the relative contribution of α-oxidation of endogenous fatty acids to the basal respiration of potato slices. It is to be expected that immediately after slicing, as a result of extensive damage to the membrane system of neighboring cells [severed plasmodesmata, rapid disappearance of cisternae of the ER (cf 68)], unusually large amounts of lipids will be released in the cytoplasmic phase, and one might speculate that α-oxidation constitutes a mechanism for the rapid removal of excess fatty acids which could be important for at least two reasons: (a) to remove an inhibiting influence on the activity of mitochondria, which are known to be inactivated by the presence of excessive amounts of free fatty acids (8, 35, 40, 164); and (b) to provide an immediate source of energy and suitable lipid molecules for the synthesis of membranes according to specifications provided by the derepression procedure.

Earnshaw & Truelove (39) showed by means of phospholipase A added to mitochondrial suspensions that the release of lysophosphatides and free fatty acids had a more damaging effect on mitochondrial membranes than the loss of phospholipids per se. It seems pertinent that a phospholipase A-like enzyme system is greatly stimulated by Ca^{++} in rat liver tissue (18), and one might be tempted to suggest that in plant cells Ca^{++} may also play an important role in speeding up the process of adaptation in fresh slices by allowing a rapid conversion of superfluous endogenous phospholipids to fatty acids for subsequent α-oxidation. The significance of the presence of Ca^{++} in the medium during washing procedures was discussed extensively by Epstein (43) and by Van Steveninck (184).

A marked stimulation and change of pattern of lipid synthesis has been well documented for aging potato slices (10, 12, 168, 195). Willemot & Stumpf (196) showed by means of cycloheximide and actinomycin D that a severalfold increase in fatty acid synthetase activity depended on protein synthesis involving derepression. A strong stimulation of the capacity for synthesis of free sterols and their precursors, resulting in a fivefold increase of free sterol content after 32 hr of aging, was possibly related to the induction or activation of C_{24}-methyl transferase which is not functioning in fresh tissue (58). This, however, may be characteristic only of potato slices, as indicated by the unusually high sterol glycoside content of their mitochondria (155).

It appears that there is a shift in the incorporation of phospholipids from the microsomal fraction to the mitochondrial fraction at a later stage [24 hr of aging (27)]. This agrees with the earlier finding that mitochondria in vivo as well as in vitro become increasingly capable of $[1-^{14}C]$-acetate incorporation into lipids and fatty acids during the aging of potato slices (15).

The work of Mazliak's group especially has provided much additional insight into the dynamic aspects of membranes and their components in plant cells. Their experiments have shown that lipids can be rapidly transferred from microsomes to mitochondria and vice versa in an in vitro system. Because the kinetics of exchange were roughly similar with three different precursors ($[1-^{14}C]$-acetate, ^{32}P-phosphate, and $[1-^{14}C]$-glycerol), they believe that entire phospholipid molecules (mainly phosphatidylcholine and phosphatidylethanolamine) are exchanged between cellular fractions (13), although all types of fatty acids may be exchanged without apparent breakdown or resynthesis (110). It is of particular interest that isolated mitochondria show an active synthesis of fatty acids and phospholipids (mainly saturated and monounsaturated fatty acids and phosphatidylethanolamine) which differs from that of mitochondria in vivo (saturated, mono, di, tri-unsaturated fatty acids and phosphatidylcholine). These discrepancies indicate that mitochondrial synthesis in vivo partly depends on cooperation from other organelles of the endomembrane system (111, 112). Data for turnover of membrane lipids in aging potato slices show that the "half lives" of microsomes, mitochondria, and supernatant fraction are 32, 10.4, and 5.2 days respectively (14). However, the earliest lipid changes occur immediately after slicing predominantly in the microsomal fraction (27), and may coincide with significant changes in the distribution of sterols (38). This information suggests that the endomembrane system may assume an important role in controlling cellular differentiation during an early stage of adaptive aging of tissue slices. The hypothesis is further supported by the fact that at least in potato slices the endomembrane system is relatively stable once developmental changes have taken place (14).

CONTROL OF RESPIRATORY DEVELOPMENT, PROTEIN SYNTHESIS, AND ION TRANSPORT

Respiratory Development

The widespread occurrence of induced respiration in slices of plant tissue (4) and the availability of such detailed information on the process (for reviews see 21, 84, 86, 89, 141) raised hopes many times that an unequivocal characterization of its controlling principles would be forthcoming. In 1967 Laties (89) wrote: "thus we face a dismayingly complex picture regarding the events which lead to the respiratory rise and the burgeoning of physiological competence following slicing." Yet a review on this topic 6 years later by Kahl (76) shows that the intervening years have not added much to provide a stringent overall description of the controls. The salient features are that large quantities of stored products (starch, sugars) are rapidly degraded via glycolysis, the pentose phosphate pathway, and tricarboxylic acid cycle. In this connection it is interesting to note that kinetin, ABA, and GA_3 seem to have marked effects on the storage and hydrolysis of starch (54, 72, 115, 169), while the same substances have also been shown to control the development of ion transport capacities resulting from the washing of slices (see next section and Van Steveninck 185).

Although it was shown that the activity of starch phosphorylase as initial catalyst in starch breakdown increases after slicing, the activity of the glycolytic enzymes showed a bewildering array of increases, decreases, or no change (77), and yet all changes were shown by inhibitor studies to be dependent on protein synthesis. A comparison of ratios of concentrations of glycolytic intermediates with their apparent mass action constants, however, did show that only the ATP-transferases (hexokinase, phosphofructokinase, and pyruvate kinase) provided rate-limiting steps (76). Phosphofructokinase activity especially may constitute a major control in glycolysis with ATP exerting a highly sensitive negative allosteric control on the enzyme in intact Jerusalem artichoke tissue, while 18 hr after slicing the enzyme does not respond to ATP (19). Adams & Rowan (1) believe that the glycolytic control of respiration in aging carrot slices involves two stages: (a) active synthesis resulting in preponderance of ADP which would activate pyruvate kinase; and (b) decline in synthesis resulting in accumulation of ATP which would activate phosphofructokinase. These authors also suggested that the well-known activation of the pentose phosphate pathway (5) is involved in the first stage of the induced respiration only. The washing and aging of red beet slices also causes a dramatic increase in 6-phosphogluconate dehydrogenase activity (80, 136).

In connection with the observed stimulation of α-oxidation of fatty acids immediately after slicing (69, 92), it seems pertinent that in liver tissue, Na-octanoate most severely inhibits the key glycolytic enzymes [hexokinase, phosphofructokinase, and pyruvate kinase (194)].

It is considered almost axiomatic that freshly sliced tissue is incapable of oxidation of TCAC components while aged slices are highly active. It appears that the TCAC is blocked rather than absent in fresh tissue since mitochondria isolated from both fresh and aged tissue are readily capable of oxidizing TCAC-acids (88). It is believed that the blockage results from the action of γ-hydroxy-α-ketoglutarate which is a powerful inhibitor of several TCAC enzymes [aconitase, isocitrate, and α-ketoglutarate dehydrogenase (90, 125)] and is present in fresh tissue probably as a condensation product of glyoxylate and pyruvate (80). Isocitrate dehydrogenase seems to lack activity in bulk tissue until its allosteric effector, citrate, activates the enzyme, possibly because of its release after breakdown of compartmentation by slicing (137). ap Rees & Royston (7), however, maintain that no qualitative differences exist in the operation of the TCAC in fresh and aged carrot slices, and hence it appears that other factors determine the respiratory rise resulting from the washing of tissue slices.

Intriguing changes occur in the respiratory chain assembly, which in fresh potato tissue is typified by severe CN^- and light-reversible carbon monoxide inhibition (53). During adaptive aging the sensitivity to these inhibitors gradually disappears and may be related to the formation of new mitochondria (see previous section). The development of this CN^- resistant pathway has been reviewed recently by Ikuma (66), and apparently depends on the incorporation of a second nonheme iron oxidase which mediates electron transfer between reduced flavoproteins and oxygen (16, 154). It has been suggested that the TCAC-independent α-oxidation of long chain fatty acids may represent the respiratory pathway after CN^- application (76). But this seems unlikely since this component of respiration typically occurs in fresh slices, and moreover, mitochondria isolated from fresh tissue and aged slices show all the characteristic changes allied to CN^- sensitivity and CN^- resistance respectively (53, 119). The development of CN^- resistant respiration requires energy from oxidative phosphorylation, and is linked to derepression events and the aerobic biogenesis of mitochondria (32, 119). Although both the phosphorylative CN^- sensitive pathway and the nonphosphorylative CN^- resistant pathway may exist side by side, it appears that the CN^- resistant pathway is virtually inoperative in vivo until either CN^- or antimycin A are added. Since adenylates can act as allosteric effectors for a number of enzymes (146), it has been suggested that the CN^- resistant pathway in newly formed mitochondria may play a role in controlling relative availabilities of adenylates. At an early stage after slicing, when the requirement for ATP is high, ADP and/or AMP may inhibit the CN^- resistant pathway in order to allow oxidative phosphorylation to proceed maximally, while at a later stage when ATP concentrations increase, mitochondrial electron transfer would proceed through the resistant path. This would maintain a high rate of metabolic flux while the activity of the regular CN^- sensitive path would continue at a much reduced (state 4) rate (for further details see 66). This hypothesis is of major interest in connection with the energy requirements for the ion pumps which develop during adaptive aging, and could explain phenomena such as the lack of coincidence of respiratory peaks (104) with the maximum capacity for ion transport, or the observation that added ATP inhibits cation uptake in red beet tissue slices (70).

Apparently, development of the CN^- resistant pathway requires O_2 (53) and on the basis of measured diffusion coefficients, O_2 supplies were calculated to be more than adequate to maintain high rates of respiration even in pieces of tuber up to 10 cm diameter (85, 197). But later work by MacDonald (102) employing high partial pressures of O_2 indicated that the development of respiratory capacity in aging slices varied with thickness from 0.75 to 3.0 mm. It appears that the O_2 requirements for the development of the CN^- resistant pathway may exceed those required for the saturation of cytochrome oxidase.

Protein Synthesis and Ion Transport

In the early 1930s, Steward (162) established that the onset of salt absorption depended on a revival of metabolic activity, and since the extent and rapidity of the changes varied with the surface to volume ratio of the slices, it was concluded that they were mainly confined to the cut surface of the tissue. Because potato tissue responds by cell division at the surface, Steward (157) became convinced that cell division and the ensuing protein synthesis provided an indispensable link with salt uptake. But slices of several other tissues (e.g. beetroot) show a pronounced rise in salt uptake without concomitant cell division (165). The ensuing issue of whether or not a direct link exists between ion absorption and protein turnover has been kept alive by conflicting results obtained with a variety of inhibitors of protein synthesis (6, 42, 71, 103, 105, 167, 188). The ability to delay the development of anion uptake capacity for an almost indefinite period by means of actinomycin D has firmly indicated that a derepression of the genome is required for the formation or activation of carrier molecules or ion pumps (188).

Various storage tissues exhibit appreciable differences in the duration of the lag phase required for the development of a salt uptake capacity, e.g. on a particular occasion parsnip phloem tissue required a lag of more than 40 hr while artichoke required less than 10 hr to establish a net K^+ uptake capacity (173). The timing very likely depends also on varietal characteristics and on the growing conditions, time of harvest, and storage of roots (144, 157, 181), and proper consideration should be given to reduce experimental variables of this nature to a minimum. It seems worthwhile to emphasize that the duration of the lag may be different for each particular ion even when the slices are derived from a single root. For instance, it was shown that in slices obtained from individual beetroots, net uptake of K^+, Na^+, and Cl^- may develop sequentially with a time difference of 10 hr or more, and with net Na^+ uptake commencing before net K^+, and net Cl^- uptake commencing last (188). This contrasts with Poole's (132) observation that in slices of beetroot a Na^+ transport mechanism develops several days after the establishment of a K^+ transport mechanism which only requires one day of washing. It has been shown, however, that the sequence can be altered by hormone treatment (181), the addition of abscisic acid to the washing medium inducing a reversal in the order of onset of net K^+ and Na^+ uptake. Poole's work is especially valuable since it includes a compartmental analysis based on short (15 min) influx experiments, which indicate an immediate inhibiting effect of Na^+ on K^+ influx (131). Apparently a high level of K^+ in the cytoplasm was maintained during Na^+ influx through the inhibition of K^+ transfer to the vacuole. A similar inhibitory effect of Na^+ on K^+ influx was observed by Van Steveninck (183), especially when ABA was added to the external medium. Poole's (131) final conclusion was that although cation selectivity was primarily determined at the plasma membrane, there was no evidence that Na^+ and K^+ interacted by competition for a single carrier at the plasma membrane, and hence uptake would be by distinct transport mechanisms. Further support for this contention came from inhibitor studies (188). Puromycin and cycloheximide prevented the development of an ion uptake capacity when added to fresh slices of beetroot, but when added after the establishment of the ion transport capacities, the lag before inhibition varied in duration for each particular ion (e.g. 3 hr for Cl^-, 9 hr for Na^+, and 24 hr for K^+), and also progressively lengthened when the inhibitor was added during aging. The lag was not due to lack of penetration of the inhibitors since inhibition of protein synthesis by cycloheximide in aged beetroot slices was shown to be almost immediate (129). These results were taken as evidence that synthesis and decay characteristics of specific proteins required for the ion uptake mechanisms were different for each ion species (188).

Many examples show that ion selectivity will change when cells or tissues are exposed to different environmental conditions. The observations of Rains on bean stem tissue during the aging of slices in 0.5 mM $CaSO_4$ are particularly relevant (48, 133). Fresh stem slices had a propensity for Na^+ uptake, especially from low external concentrations, while in aged stem slices the situation was reversed, i.e. the Na^+-absorbing capacity was lost and replaced by a K^+-absorbing capacity. This change in selectivity was not due to a significant change in the rate of efflux of either ion, and it was demonstrated by use of cycloheximide that protein synthesis was involved in the development of the K^+-absorbing capacity by the aging tissue. The treatment of slices with benzyladenine reversed the trend by suppressing the development of the K^+-absorbing capacity in aged tissue without preventing loss of the Na^+-absorbing capacity (133).

It is evident from the above examples that an intricate system of controls may govern the selection of ions through cellular metabolism. Especially when further cellular differentiation is triggered by changes of the cellular environment, e.g. by slicing of tissue and washing the slices in an aqueous medium, one might expect the ensuing derepression of the genome combined with a stimulated protein synthesis to result in adjustments and changes of ion selectivity. At present it appears that plant hormones may play an important role at both the transcriptional and translational level in controlling the synthesis and turnover of proteinaceous components which form part of specific ion transport mechanisms, and thus determine characteristic changes in ion selectivity of tissue slices.

HORMONAL REGULATION AND THE EFFECT OF TRIS BUFFER

Extensive reviews have already appeared on the topics of hormonal regulation and ion transport (185) and on the effect of tris buffer on ion transport (187), and hence only some salient points regarding the washing phenomenon will be discussed here.

Hormonal Regulation

The detailed work of Gayler & Glasziou (49) offered considerable insight into the regulatory aspects of hormones on enzyme synthesis in slices of sugar cane internodal tissue. Both NAA and GA_3 increased invertase, possibly by causing a stabilization of mRNA for invertase. ABA also increased the rate of synthesis of invertase, possibly by stimulating a step subsequent to invertase-mRNA formation and prior to invertase destruction. Kinetin, on the other hand, seemed to have little effect on invertase-mRNA synthesis. Palmer (124) observed that incubation of freshly cut slices of beetroot tissue in solutions of IAA and kinetin completely prevented the development of the respiratory rise, invertase activity, and the capacity for inorganic phosphate uptake. On the other hand, GA_3 had no inhibitory effect. These contrasting results are most likely due to the fact that Gayler & Glasziou's experiments were of a relatively short duration (less than 5 hr) and on tissue taken from rapidly expanding immature internodes of sugar cane stalks, while Palmer's experiments were carried out over a period of at least 8 hr on dormant storage tissue. Both results, however, do illustrate that hormones may have a profound effect on the synthesis of specific proteins in differentiating cells. These effects, of course, are distinctly different from the rapid responses elicited by plant hormones, e.g. auxin induced proton pumps and ABA effects on stomatal action which may occur within minutes (for recent reviews see 45, 134).

It is generally agreed that cytokinins retard senescence by stimulating protein synthesis and by promoting the uptake and retention of low molecular weight metabolites (156). Curiously enough, the long-term effects of cytokinins on washing tissue slices is to maintain a status quo. For instance, the washing of bean stem slices in 0.5 mM $CaSO_4$ for 20 hr normally results in the development of a K^+-absorbing capacity which is suppressed by the presence of 5 μM benzyladenine in the external medium (133). Similarly, Van Steveninck (182) found that kinetin and benzyladenine (both 38 μM) completely prevented the development of ion transport capacities for K^+ and Na^+ when added to the washing medium of fresh slices of beetroot, and in this respect the effects very much resembled those obtained with cycloheximide or actinomycin D (188). In aged slices which had already developed an ion uptake capacity, benzyladenine and kinetin had no immediate inhibitory effect on net uptake, i.e. the inhibition became apparent only after several hours and this period became more extended the longer the ion transport mechanism had been in operation. In this respect the action of kinetin and benzyladenine followed a pattern identical to cycloheximide inhibition (188). In both cases kinetin and benzyladenine appear to induce a condition of flux-equilibrium (influx = efflux) at relatively high ion concentrations of the external medium (183). The flux equilibrium was probably achieved by an increased efflux, indicating that kinetin may play a role in facilitating passage of ions across the outer membrane (183). Further flux analyses are required to confirm this impression, and it is hoped that this may ultimately contribute to an understanding of the regulation of different "salt saturation" levels in various cell types within a tissue.

ABA and kinetin are often shown to have contrasting effects on cell differentiation and metabolism, and it is interesting therefore to note that ABA strongly stimulates net uptake of Na^+, K^+, and Cl^- in slices of beetroot tissue, but only when the cells have already acquired a capacity for net uptake of these ions. Thus ABA did not shorten the lag phase (181). The stimulations of net uptake were shown to be due to dramatic increases in influx, especially in the case of Cl^-, while stimulation of net cation uptake was due primarily to a reduced efflux of K^+ and Na^+. ABA also induced a selectivity for Na^+ over K^+, and the uptake of Na^+ and Cl^- stimulated by ABA was most likely directed into the vacuole (183). The effect of IAA and GA_3 on slices of beetroot tissue were not nearly so marked (117, 185), but clearly one should not generalize because hormone specificity is greatly dependent on species differences [e.g. benzyladenine has no effect on ion transport in swede slices (182)], on origin of tissue (e.g. root vs stem), and on the hormone status of the experimental material.

Notwithstanding these characteristic specificities, it seems reasonable to predict that plant hormones will prove to play an important role in the regulatory aspects of ion transport which control salt saturation, selectivity, and distribution of ions between various part of a plant (185).

Effect of Tris Buffer

Tris buffer on the alkaline side of pH 7 has the remarkable property of inducing an immediate net uptake of cations in fresh slices of beetroot, parsnip, or carrot tissue but not in swede, artichoke, potato, or turnip slices. The induction of net K^+ or Na^+ uptake is due to a spectacular increase of influx and an equally spectacular decrease in efflux of these ions while Cl^- fluxes remain relatively unaffected (179).

The timing of the response and the rate of induced net uptake depend on the ratio of tissue to volume of medium, an increase in this ratio causing the tris treatment to be more effective (Table 1). This dependence on quantity of tissue in a certain volume of medium is still unexplained, but suggests that an interaction may occur between tris and a possible component leaching from the tissue into the medium which may depend for its activity on a threshold concentration.

In tissues which respond to tris the induced excess cation uptake was mediated by a proton-cation exchange mechanism and balanced by an accumulation of organic acids in the slices (179). Because the nonprotonated form of tris was the effective molecular species, it was implied that tris base (RNH_2) would act as a proton acceptor and thereby facilitate the exchange process. Considering that nonprotonated amines including tris rapidly equilibrate across membranes (116, 145), one might expect that tris would act in the cytoplasmic phase, thus facilitating the transfer of cations and organic acids across the tonoplast. This induced excess cation and organic acid accumulation may amount to 30 meq. kg^{-1} fresh weight in storage tissue and is unlikely to be located in the cytoplasm because of the relatively small volume of this compartment in storage tissue cells (23).

It has also been shown that the OH-groups of tris are important for their effectiveness (173), implying that tris molecules may affect membrane structure causing

Table 1 The dependence of the tris response on the amount of red beet tissue slices in solution ($10^{-2}M$ tris-Cl, pH 7.80–7.90)

Fresh weight of tissue g/liter	Rates of net K^+ uptake (me. kg^{-1} hr^{-1})		
	4–8 hr	8–18 hr	18–24 hr
13.6	0.04	0.04	0.33
27.2	0.04	0.16	0.24[a]
54.4	0.12	0.21	0.10[a]
16.4	0.00	0.10	0.28
32.8	0.08	0.17	0.21[a]
65.5	0.12	0.17	0.06[a]

[a] Rate of uptake limited by depletion of K^+ in the external solution.

conformational changes in the protein components through interactions of the OH-groups with H-bonds and side chains containing –COO^- and $\equiv N^+H$ groups (79, 135), or by its action on the water structure associated with membranes (chaotropic effect; see 57). Ultrastructural evidence (187) supported this contention and indicated that in storage slices tris interacts with plasmalemma membranes preventing normal fixation with OsO_4. This may be due largely to competition between tris and OsO_4 in the process of H-bonding which is necessary for an adequate structural preservation of membrane proteins (99). The tonoplast, however, was unaffected.

Many instances have been reported of tris affecting enzymatic reactions including those of the glycolytic pathway (26). The cationic hydroxamine, D-glucosamine, completely prevents tris-induced K^+ uptake in beetroot slices (180), while this inhibition is completely reversed on addition of an equimolar concentration of D-glucose. The reversal is glucose specific, and therefore it appears that tris may act through a stimulative effect on a specific hexokinase. It was implied (46) and later confirmed that the activity of hexokinase increased considerably during the aging of carrot slices (138). This increased activity could stimulate the availability of protons in various ways, e.g. either as a result of phosphorylation processes (62) or through the creation of proton gradients (114, 141).

Tuli et al (170) reported that benzyladenine competes with ATP and ADP in the activities of hexokinase and pyruvate kinase respectively, and that this effect would tend to delay the respiratory rise or senescence. It was found that a combination of benzyladenine and tris would indeed result in the complete inhibition of the tris effect in slices of beetroot, although this inhibition was delayed and preceded by a 3–4 hr period of accelerated cation uptake (182). It is still too early to draw any conclusions regarding the mechanism which leads to this complex interaction, but it seems significant that slices of swede tissue which normally do not exhibit a lag phase and which do not show a tris stimulated cation uptake, did not respond to kinetin or benzyladenine treatment with respect to ion transport (182). In fact, in slices of swede tissue, tris had the reverse effect and severely inhibited cation uptake (182), and in potato slices tris totally suppressed cell division. In both cases, the inhibition was fully reversible and it was suggested that the mechanism of action might involve an uncoupling of electron transfer from phosphorylation (78). In this respect it seems possible that the tissues which normally show a response to tris (parsnip, carrot, beetroot) may operate a stimulated cation transport directly linked to electron transport as was shown by Polya & Atkinson (130) for red beet slices, while those tissues which do not show a response to tris (potato, swede, artichoke) may be more directly dependent on ATP as an energy source to drive the ion pumps. Lange, Kahl & Rosenstock (83) showed that tris does not change any aspect of glucose catabolism in potato slices and has no effect on the relevant enzymes. Conversely, a rapid decrease in activity of phosphoglucomutase, aldolase, pyruvate kinase, and glutamic-pyruvate transaminase, which normally occurs during the washing of potato slices, was prevented by tris, and the authors concluded that nucleic acid metabolism was a more likely target for the tris ion.

Another apsect which should be further explored is the possibility of tris partic-ipating in metabolic pathways which may lead to its incorporation into mac-romolecular components affecting protein synthesis and/or the properties of membranes. Interaction of tris with aldehydes (59) pyridoxal-5-phosphate (73) and valyl-RNA (126), its phosphorylation by alkaline phosphatase (37), its stimulating effect on amino acid incorporation (17), etc all point towards an actual molecular involvement in cellular metabolism. This is certainly true in the case of glucosamine which is readily incorporated in plant cell walls and cytoplasmic components (108, 140, 191). Glucosamine-containing glycoproteins have been isolated not only from chloroplasts (67), but also from storage tissues such as carrot and potato (2, 44). The hydroxyproline-containing glycoproteins recently isolated from potato (2) and carrot tissue (151) are of particular interest since they have been shown to accumu-late up to 2½-fold in the cytoplasm, and up to tenfold in the cell wall of carrot slices over a period of 6 days (31). Although not fully characterized, the carrot glyco-protein appears similar to the potato-lectin isolated by Allen & Neuberger (2). The latter consists of approximately 50% carbohydrate and, among the sugar residues which mainly consist of arabinose, small amounts of glucose and glucosamine are present, while 16% of the amino acid residues consists of OH-proline. The newly isolated glycoproteins obtained from unicellular coenocytic water molds (*Achlya* spp. and *Blastocladiella emersonii*) prove to be extremely sensitive to cytokinin in respect to the allosteric regulation of their Ca^{++} binding sites (94). Thus cytokinins will stimulate the release of Ca^{++} bound to this glycoprotein which is localized on the membrane surface (95), and this results in the prevention of energy-linked import of amino acids, nucleosides, and sugar by *Achlya* cells. Apparently the effect of cytokinins may be neutralized by Mg^{++}, and hence the metabolite import by the fungal cells appears to depend on a delicate balance between the concentrations of cytokinins, Ca^{++}, and Mg^{++} (95).

It should be interesting to explore the effect of tris and glucosamine in the above system. Already it has been shown that tris interferes with Ca^{++} metabolism in *Haematococcus pluvialis*. The growth of fresh water algae is generally inhibited by tris, but this inhibition can be reduced by increasing the Ca^{++} concentration of the medium. Other cations either alone or in combination with Ca^{++} had no effect on the tris inhibition (106). The effect of tris on K^+ fluxes in fresh beetroot slices was also strongly modified by Ca^{++}, and it was suggested that Ca^{++} and tris compete for a common site, the bivalent Ca^{++} being much more efficient in this respect than the univalent tris ions (178).

It must be emphasized that Ca^{++} plays a key role in the washing phenomenon of excised roots and tissue slices (for a critical discussion on this aspect see 43), and it should be stressed that the slicing and washing treatment is bound to have dramatic effects on the hormone balance of the component cells. These two factors are likely to interact with specific cell surface components and thus may constitute a very significant part of the regulatory system controlling the transport and permeability characteristics of sliced storage tissue.

TISSUE SLICES VERSUS INTACT ROOTS

It still remains to be seen whether induced cell differentiation in tissue slices can be equated to the whole organ situation. Processes of cell differentiation strongly affect metabolic regulation in the whole plant and, particularly in roots, it is evident that cells which are exposed to the surface and in direct contact with the external environment possess a capacity for active accumulation of ions. This is borne out by recent observations that only the surface cells (epidermis and outer cortical cells) of a root are actively engaged in the loading of the cytoplasmic continuum from the cortex to the stele (9, 171). Correspondingly, cells of bulky roots or storage organs which are not normally exposed to the external environment may develop a capacity for active accumulation of ions when exposed by slicing the tissue, provided the ability to differentiate through a process of derepression of the genome has been retained.

The primary stimulus which triggers the derepression is a dark secret. An important criterion appears to be that the induced changes are mainly confined to the surface layers of cells (158). One might speculate on why the surface cells are affected. In root cells in situ, the cytoplasmic continuum or symplasm may behave as the primary unit, and hence may have an overriding effect on the individuality of the component cells. This continuum may be capable of perceiving a polarity through gradients created by the aqueous medium towards the outside and by plasma connections inwards. Although this polarity may depend on diffusional aspects involving, for example, components of a hormonal nature, or simply the availability of metabolites or substrates, it is also likely that control is exercised by an intricate structure of a relatively permanent and characteristic nature. Gradients in differentiation also occur from extreme apex (nonvacuolated cells) to more basal regions of roots (vacuolated cells), and there are large differences in solute regulation between these extremes of cellular differentiation (159, 160). It has been suggested that the endomembrane system provides an operational link in the formation of vacuoles (29, 107), and hence it is tempting to invoke the endomembrane concept to explain matters of control and continuity. The degree of individuality of cells might then depend largely on the character and frequency of plasmodesmata (60), which may be indicated by a marked ATPase activity in the plasmodesmata region (33).

The slicing and washing of tissue clearly has a drastic effect on this intricate communication system. Diffusional aspects come to the fore. The ratio of tissue quantity to volume of medium may have profound effects as was shown with the onset of the tris effect (Table 1), and marked differences in response may also occur, depending on whether the tissue is washed or merely kept in a moist atmosphere (24, 82, 157, 162)

CONCLUDING REMARKS

The slicing and washing of storage tissue continues to provide a reliable method of inducing cell differentiation under controlled conditions. From its innocent beginnings (113, 163) when salt absorption was measured with considerable difficulty through lack of rapid analytical methods, and using slices in tightly stoppered flasks, we have now reached a stage where a significant advance is possible in our understanding of the overall regulatory aspects of morphogenesis and cellular differentiation. The combined information resulting from studies involving derepression phenomena, metabolic control, enzyme localization, cellular compartmentation, membrane potentials, ionic gradients etc must ultimately provide the necessary pieces to complete the jigsaw puzzle. Most likely, a number of apparently unconnected pieces of information will fall into place by the further characterization of the various membrane structures. Through improved methods of separation and identification of membrane fractions it should be possible to define the role of the endomembrane system with respect to function and turnover of organelles.

The washing phenomenon implies that diffusional aspects are involved and key regulatory substances may escape from the tissue causing shifts in metabolic and hormonal control. It is known that hormones have a profound effect on derepression phenomena and protein synthesis in tissue slices, while other substances of a similar nature may still have escaped our attention. Further research on the washing phenomenon should help to provide an understanding of the interaction between the endomembrane complex and organelle function, and so lead to a knowledge of cellular morphogenesis.

Literature Cited

1. Adams, P. B., Rowan, K. S. 1970. *Plant Physiol.* 45:490–94
2. Allen, A. K., Neuberger, A. 1973. *Biochem. J.* 135:307–14
3. Anderson, W. P. 1972. *Ann. Rev. Plant Physiol.* 23:51–72
4. ap Rees, T. 1966. *Aust. J. Biol. Sci.* 19:981–90
5. ap Rees, T., Beevers, H. 1960. *Plant Physiol.* 35:839–47
6. ap Rees, T., Bryant, J. A. 1971. *Phytochemistry* 10:1183–90
7. ap Rees, T., Royston, B. J. 1971. *Phytochemistry* 10:1199–1206
8. Baddeley, M. S., Hanson, J. B. 1967. *Plant Physiol.* 42:1702–10
9. Bange, G. G. J. 1973. *Acta Bot. Neer.* 22:529–42
10. Ben Abdelkader, A. 1968. *Physiol. Vég.* 6:417–42
11. Ben Abdelkader, A., Anderset, G. 1972. *C. R. Acad. Sci. D* 274:1311–14
12. Ben Abdelkader, A., Mazliak, P. 1968. *C. R. Acad. Sci. D* 267:609–12
13. Ben Abdelkader, A., Mazliak, P. 1970. *Eur. J. Biochem.* 15:250–62
14. Ben Abdelkader, A., Mazliak, P. 1971. *Physiol. Vég.* 9:227–40
15. Ben Abdelkader, A., Mazliak, P., Catesson, A-M. 1961. *Phytochemistry* 8:1121–33

16. Bendall, D. S., Bonner, W. D. Jr. 1971. *Plant Physiol.* 47:236–45
17. Bewley, J. D., Marcus, A. 1970. *Phytochemistry* 9:1031–33
18. Björnstad, P. 1966. *J. Lipid Res.* 7:612–20
19. Black, M. K., Wedding, R. T. 1968. *Plant Physiol.* 43:2066–69
20. Blobel, G., van Potter, R. 1967. *J. Mol. Biol.* 26:279–92
21. Bonner, J., Varner, J. E., Eds. 1965. *Plant Biochemistry,* 213–30. New York: Academic. 1054 pp.
22. Bracker, C. E., Grove, S. N. 1971. *Protoplasma* 73:15–34
23. Briggs, G. E., Hope, A. B., Pitman, M. G. 1958. *J. Exp. Bot.* 9:128–41
24. Bryant, J. A., ap Rees, T. 1971. *Phytochemistry* 10:1191–97
25. Burgess, J. 1971. *Protoplasma* 73:83–95
26. Buse, M. G., Buse, J., McMaster, J., Krech, L. H. 1964. *Metabolism* 13:339–53
27. Castelfranco, P. A., Tang, W-J., Bolar, M. L. 1971. *Plant Physiol.* 48:795–800
28. Chapman, J. M., Edelman, J. 1967. *Plant Physiol.* 42:1140–46
29. Chardard, R. 1973. *C.R. Acad. Sci. D* 276:2155–58
30. Cherry, J. H. 1968. In *Biochemistry and Physiology of Plant Growth Substances,* ed. F. Wightman, G. Setterfield, 417–31. Ottawa:Runge. 1642 pp.
31. Chrispeels, M. J. 1969. *Plant Physiol.* 44:1187–93
32. Click, R. E., Hackett, D. P. 1963. *Proc. Nat. Acad. Sci. USA* 50:243–50
33. Coulomb, P., Coulomb, C. 1972. *C.R. Acad. Sci. D* 275:1035–38
34. Cram, W. J. 1968. *Biochim. Biophys. Acta* 163:339–53
35. Dalgarno, L., Birt, L. M. 1963. *Biochem. J.* 87:586–96
36. Dauwalder, M., Whaley, W. G. 1973. *J. Ultrastruct. Res.* 45:279–96
37. Dayan, J., Wilson, I. B. 1964. *Biochim. Biophys. Acta* 81:620–23
38. Dupéron, P., Dupéron, R. 1973. *Physiol. Vég.* 11:487–505
39. Earnshaw, M. J., Truelove, B. 1970. *Plant Physiol.* 45:322–26
40. Earnshaw, M. J., Truelove, B., Butter, R. D. 1970. *Plant Physiol.* 45:318–21
41. Ellis, R. J., MacDonald, I. R. 1967. *Plant Physiol.* 42:1297–1302
42. Ibid 1970. 46:227–32
43. Epstein, E. 1973. *Int. Rev. Cytol.* 34:123–68
44. Ericson, M. C., Chrispeels, M. J. 1973. *Plant Physiol. Suppl.* 51:55
45. Evans, M. L. 1974. *Ann. Rev. Plant Physiol.* 25:195–223
46. Everson, R. G., Rowan, K. S. 1965. *Plant Physiol.* 40:1247–50
47. Fowke, L., Setterfield, G. 1968. In *Biochemistry and Physiology of Plant Growth Substances,* ed. F. Wightman, G. Setterfield, 581–602. Ottawa:Runge. 1642 pp.
48. Floyd, R. A., Rains, D. W. 1971. *Plant Physiol.* 47:663–67
49. Gayler, K. R., Glasziou, K. T. 1969. *Planta* 84:185–94
50. Glasziou, K. T. 1969. *Ann. Rev. Plant Physiol.* 20:63–88
51. Glaumann, H. 1973. In *Techniques in Protein Biosynthesis,* ed. P. N. Campbell, J. R. Sargent, 3:191–248. New York: Academic. 265 pp.
52. Grove, S. N., Bracker, C. E., Morré, D. J. 1970. *Am. J. Bot.* 57:245–66
53. Hackett, D. P., Haas, D. W., Griffiths, S. K., Niederpruem, D. J. 1960. *Plant Physiol.* 35:8–19
54. Hadačová, V., Luštinec, J., Kaminek, M. 1973. *Biol. Plant.* 15:427–29
55. Hall, J. L. 1973. In *Ion Transport in Plants,* ed. W. P. Anderson, 11–24. New York: Academic. 630 pp.
56. Hanson, J. B., Leonard, R. T., Mollenhauer, H. H. 1973. *Plant Physiol.* 52:298–300
57. Hanstein, W. G., Davis, K. A., Hatefi, Y. 1971. *Arch. Biochem. Biophys.* 147:534–44
58. Hartmann, M-A., Benveniste, P. 1973. *C.R. Acad. Sci. D* 276:3143–46
59. Hauptmann, S., Gabler, W. 1968. *Z. Naturforsch.* B23:111–12
60. Helder, R. J., Boerma, J. 1969. *Acta Bot. Neer.* 18:99–107
61. Hiatt, A. J. 1967. *Plant Physiol.* 42:294–98
62. Hiatt, A. J. 1967. *Z. Pflanzenphysiol.* 56:233–45
63. Higinbotham, N. 1973. *Ann. Rev. Plant Physiol.* 24:25–46
64. Hodges, T. K., Leonard, R. T., Bracker, C. E., Keenan, T. W. 1972. *Proc. Nat. Acad. Sci. USA* 69:3307–11
65. Honda, S. I., Hongladarom, T., Laties, G. G. 1966. *J. Exp. Bot.* 17:460–72
66. Ikuma, H. 1972. *Ann. Rev. Plant Physiol.* 23:419–36
67. Izumi, K. 1971. *Phytochemistry* 10:1777–78
68. Jackman, M. E., Van Steveninck, R. F. M. 1967. *Aust. J. Biol. Sci.* 20:1063–68
69. Jacobson, B. S., Smith, B. N., Epstein, S., Laties, G. G. 1970. *J. Gen. Physiol.* 55:1–17

70. Jacoby, B. 1965. *J. Exp. Bot.* 16:243–48
71. Jacoby, B., Sutcliffe, J. F. 1962. *Nature* 195:1014
72. Jones, R. L. 1973. *Ann. Rev. Plant Physiol.* 24:571–98
73. Kabayashi, Y., Makino, K. 1970. *Biochim. Biophys. Acta* 208:137–40
74. Kahl, G. 1971. *Z. Naturforsch.* B26:1058–64
75. Ibid, 1064–67
76. Kahl, G. 1973. *Bot. Rev.* 39:274–99
77. Kahl, G., Lange, H., Rosenstock, G. 1969. *Z. Naturforsch.* B24:1544–49
78. Kahl, G., Rosenstock, G., Lange, H. 1969. *Planta* 87:365–71
79. Klotz, I. M., Ayers, J. 1953. *Discuss. Faraday Soc.* 13:189–96
80. Kolattukudy, P. E., Reed, D. J. 1966. *Plant Physiol.* 41:661–69
81. Kramer, P. J. 1973. *Ann. Rev. Plant Physiol.* 24:1–24
82. Lange, H., Kahl, G., Rosenstock, G. 1970. *Physiol. Plant.* 23:80–87
83. Ibid 1971. 24:1–4
84. Laties, G. G. 1957. *Surv. Biol. Progr.* 3:215–99
85. Laties, G. G. 1962. *Plant Physiol.* 37:679–90
86. Laties, G. G. 1963. In *Control Mechanisms in Respiration and Fermentation,* ed. B. Wright, 129–55. New York: Ronald
87. Laties, G. G. 1964. In *Cellular Membranes in Development,* ed. M. Locke, 229–320. New York: Academic. 382 pp.
88. Laties, G. G. 1964. *Plant Physiol.* 39:654–63
89. Laties, G. G. 1967. *Aust. J. Sci.* 30:193–203
90. Laties, G. G. 1967. *Phytochemistry* 6:181–85
91. Laties, G. G. 1969. *Ann. Rev. Plant Physiol.* 20:89–116
92. Laties, G. G., Hoelle, C., Jacobson, B. S. 1972. *Phytochemistry* 11:3403–11
93. Leaver, C. J., Key, J. L. 1967. *Proc. Nat. Acad. Sci. USA* 57:1338–44
94. Le John, H. B., Cameron, L. E. 1973. *Biochem. Biophys. Res. Commun.* 54:1053–60
95. Le John, H. B., Stevenson, R. M. 1973. *Biochem. Biophys. Res. Commun.* 54:1061–66
96. Leonard, R. T., Hanson, J. B. 1972. *Plant Physiol.* 49:430–35
97. Ibid, 436–40
98. Lin, C-Y., Travis, R. L., Chia, S. Y., Key, J. L. 1973. *Phytochemistry* 12:2801–7
99. Litman, R. B., Barrnett, R. J. 1972. *J. Ultrastruct. Res.* 38:63–86
100. Loening, U. E. 1968. *Ann. Rev. Plant Physiol.* 19:37–70
101. Lüttge, U., Laties, G. G. 1967. *Planta* 74:173–87
102. MacDonald, I. R. 1968. *Plant Physiol.* 43:274–80
103. MacDonald, I. R., Bacon, J. S. D., Vaughan, D., Ellis, R. J. 1966. *J. Exp. Bot.* 17:822–37
104. MacDonald, I. R., DeKock, P. C. 1958. *Ann. Bot. London* 22:429–48
105. MacDonald, I. R., Ellis, R. J. 1969. *Nature* 222:791–92
106. McLachlan, J. 1963. *Can. J. Bot.* 41:35–40
107. Marty, F. 1973. *C.R. Acad. Sci. D* 277:1749–52
108. Mayer, F. C., Bikel, I., Hassid, W. Z. 1968. *Plant Physiol.* 43:1097–1107
109. Mazliak, P. 1973. *Ann. Rev. Plant Physiol.* 24:287–310
110. Mazliak, P., Ben Abdelkader, A. 1971. *Phytochemistry* 10:2879–90
111. Mazliak, P., Oursel, A., Ben Abdelkader, A., Grosbois, M. 1972. *Eur. J. Biochem.* 28:399–411
112. Mazliak, P., Stoll, U., Ben Abdelkader, A. 1968. *Biochim. Biophys. Acta* 152:414–17
113. Meurer, R. 1909. *Jahrb. Wiss. Bot.* 46:503–67
114. Mitchell, P., Moyle, J. 1965. *Nature* 208:147–51
115. Mittelheuser, C. J., Van Steveninck, R. F. M. 1971. *Protoplasma* 73:253–62
116. Mittelheuser, C. J., Van Steveninck, R. F. M. 1972. *Aust. J. Biol. Sci.* 25:517–30
117. Morré, D. J., Merritt, W. D., Lembi, C. A. 1971. *Protoplasma* 73:43–49
118. Morré, D. J., Mollenhauer, H. H., Bracker, C. E. 1971. In *Results and Problems in Cell Differentiation II. Origin and Continuity of Cell Organelles,* ed. T. Reinert, H. Ursprung, 82–126. Berlin: Springer. 342 pp.
119. Nakano, M., Asahi, T. 1970. *Plant Cell Physiol.* 11:499–502
120. Nathanson, A. 1904. *Jahrb. Wiss. Bot.* 39:607–44
121. Northcote, D. H. 1971. *Endeavour* 30:26–33
122. Osmond, C. B., Laties, G. G. 1968. *Plant Physiol.* 43:747–55
123. Pallaghy, C. K., Lüttge, U., von Willert, K. 1970. *Z. Pflanzenphysiol.* 62:51–57
124. Palmer, J. M. 1966. *Plant Physiol.* 41:1173–78
125. Payes, B., Laties, G. G. 1963. *Biochem. Biophys. Res. Commun.* 10:460–66

126. Pinck, M., Schuber, F. 1971. *Biochimie* 53:887–91
127. Pitman, M. G. 1963. *Aust. J. Biol. Sci.* 16:647–68
128. Pitman, M. G., Mertz, S. M., Graves, J. S., Pierce, W. S., Higinbotham, N. 1970. *Plant Physiol.* 47:76–80
129. Polya, G. M. 1968. *Aust. J. Biol. Sci.* 21:1107–18
130. Polya, G. M., Atkinson, M. R. 1969. *Aust. J. Biol. Sci.* 22:573–84
131. Poole, R. J. 1971. *Plant Physiol.* 47: 731–34
132. Ibid, 735–39
133. Rains, D. W. 1969. *Plant Physiol.* 44: 547–54
134. Raschke, K. 1975 *Ann. Rev. Plant Physiol.* 26:309–40
135. Rawitch, A. B., Gleason, M. 1971. *Biochem. Biophys. Res. Commun.* 45: 590–97
136. Reed, D. J., Kolattukudy, P. E. 1966. *Plant Physiol.* 41:653–60
137. Ribéreau-Gayon, G., Laties, G. G. 1969. *C.R. Acad. Sci. D* 268:2612–15
138. Ricardo, C. P. P., ap Rees, T. 1972. *Phytochemistry* 11:623–26
139. Robards, A. W. 1971. *Protoplasma* 72:315–23
140. Roberts, R. M., Cetorelli, J. J., Kirby, E. G., Ericson, M. 1972. *Plant Physiol.* 50:531–35
141. Robertson, R. N. 1968. *Protons, Electrons, Phosphorylation and Active Transport.* Cambridge Univ. Press. 96 pp.
142. Roland, J-C., Lembi, C. A., Morré, D. J. 1972. *Stain Technol.* 47:195–200
143. Roodyn, D. B., Wilkie, D. 1968. *The Biogenesis of Mitochondria.* London: Methuen. 123 pp.
144. Rosenstock, G., Kahl, G., Lange, H. 1971. *Z. Pflanzenphysiol.* 64:130–38
145. Rottenberg, H., Grunwald, T., Avron, M. 1972. *Eur. J. Biochem.* 25:54–63
146. Rowan, K. S. 1966. *Int. Rev. Cytol.* 19:301–91
147. Rungie, J. M., Wiskich, J. T. 1972. *Aust. J. Biol. Sci.* 25:103–13
148. Rungie, J. M., Wiskich, J. T. 1973. *Plant Physiol.* 51:1064–68
149. Rutherford, P. P. 1971. *Phytochemistry* 10:1469–73
150. Ryan, J. W., Smith, U. 1972. *Biochim. Biophys. Acta* 249:177–80
151. Sadava, D., Chrispeels, M. J. 1973. In *Biogenesis of Plant Cell Wall Polysaccharides,* ed. F. Loewus, 165–74. New York: Academic. 379 pp.
152. Sakano, K., Asahi, T. 1969. *Agr. Biol. Chem.* 33:1433–39
153. Sakano, K., Asahi, T. 1971. *Plant Cell Physiol.* 12:417–26
154. Schonbaum, G. R., Bonner, W. D. Jr., Storey, B. T., Bahr, J. T. 1971. *Plant Physiol.* 47:124–28
155. Schwertner, H. A., Biale, J. B. 1973. *J. Lipid Res.* 14:235–42
156. Skoog, F., Armstrong, D. J. 1970. *Ann. Rev. Plant Physiol.* 21:359–84
157. Steward, F. C., Berry, W. E., Preston, C., Ramamurti, T. K. 1943. *Ann. Bot. London* 7:221–60
158. Steward, F. C., Harrison, J. A. 1939. *Ann. Bot. London* 3:427–53
159. Steward, F. C., Millar, F. K. 1954. *Symp. Soc. Exp. Biol.* 8:367–406
160. Steward, F. C., Mott, R. L. 1970. *Int. Rev. Cytol.* 28:275–370
161. Steward, F. C., Preston, C. 1940. *Plant Physiol.* 15:23–61
162. Steward, F. C., Wright, R., Berry, W. E. 1932. *Protoplasma* 16:576–611
163. Stiles, W. 1924. *Ann. Bot. London* 152:617–33
164. Strickland, R. G. 1961. *Biochem. J.* 81:286–91
165. Sutcliffe, J. F. 1954. *Symp. Soc. Exp. Biol.* 8:325–42
166. Sutcliffe, J. F. 1959. *Biol. Rev.* 34:159–220
167. Sutcliffe, J. F. 1960. *Nature* 188:294–97
168. Tang, W-J., Castelfranco, P. A. 1968. *Plant Physiol.* 43:1232–38
169. Tasseron-de Jong, J. G., Veldstra, H. 1971. *Physiol. Plant.* 24:235–38
170. Tuli, V., Dilley, D. R., Wittwer, S. H. 1964. *Science* 146:1477–79
171. Vakhmistrov, D. B. 1967. *Sov. Plant Physiol.* 14:103–7
172. Van Steveninck, M. E., Van Steveninck, R. F. M. 1971. *Protoplasma* 73:107–19
173. Van Steveninck, R. F. M. 1961. *Nature* 190:1072–75
174. Van Steveninck, R. F. M. 1962. *Physiol. Plant.* 15:211–15
175. Ibid 1964. 17:757–70
176. Ibid 1965. 18:54–69
177. Van Steveninck, R. F. M. 1965. *Nature* 205:83–84
178. Van Steveninck, R. F. M. 1965. *Aust. J. Biol. Sci.* 18:227–33
179. Ibid 1966. 19:271–81
180. Ibid, 283–90
181. Van Steveninck, R. F. M. 1972. *Z. Pflanzenphysiol.* 67:282–86
182. Van Steveninck, R. F. M. 1972. *Physiol. Plant.* 27:43–47
183. Van Steveninck, R. F. M. 1974. In *Membrane Transport in Plants and Plant Organelles,* ed. U. Zimmermann, J. Dainty, 450–56. Berlin: Springer. 473 pp.

184. Van Steveninck, R. F. M. 1975. In *Encyclopedia of Plant Physiology. New Series,* ed. U. Lüttge, M. G. Pitman. Berlin: Springer. In press
185. Ibid,
186. Van Steveninck, R. F. M., Jackman, M. E. 1967. *Aust. J. Biol. Sci.* 20: 749–60
187. Van Steveninck, R. F. M., Mittelheuser, C. J., Van Steveninck, M. E. 1973. In *Ion Transport in Plants,* ed. W. P. Anderson, 251–69. New York: Academic. 630 pp.
188. Van Steveninck, R. F. M., Van Steveninck, M. E. 1972. *Physiol. Plant.* 27:407–11
189. Vaughan, D., MacDonald, I. R. 1967. *Plant Physiol.* 42:456–58
190. Vaughan, D., Macdonald, I. R. 1967. *J. Exp. Bot.* 18:587–93
191. Veiga, L. A. 1968. *Plant Cell Physiol.* 9:1–12
192. Verleur, J. D. 1969. *Z. Pflanzenphysiol.* 61:299–309
193. Verleur, J. D., Van der Velde, H. H., Sminia, T. 1970. *Z. Pflanzenphysiol.* 62:352–61
194. Weber, G., Convery, H. J. H., Lea, M. A., Stamm, N. B. 1966. *Science* 154:1357–66
195. Willemot, C., Stumpf, P. K. 1967. *Can. J. Bot.* 45:579–84
196. Willemot, C., Stumpf, P. K. 1967. *Plant Physiol.* 42:391–97
197. Wolley, J. T. 1962. *Plant Physiol.* 37:793–98

Ann. Rev. Plant Physiol. 1975. 26:259–78

SEED FORMATION ❖7591

Leon ,S. Dure III

Department of Biochemistry, University of Georgia, Athens, Georgia 30602

CONTENTS

INTRODUCTION

This marks the first time that the subject of seed formation has been the object of review in this series. This can be attributed to the paucity of literature directly concerned with the physiology, biochemistry, and molecular biology of the ontogeny of seeds, which in turn can be attributed to the very slow realization of what a fruitful area seed development and embryo ontogeny affords the developmental biologist. The second half of this developmental story, seed germination, has on the other hand received considerably more attention. One aspect of the biochemistry of seed formation has not been neglected, and that concerns the characterization of the storage proteins and aspects of their synthesis. The attractiveness of the synthesis of storage proteins is readily apparent, not only from the point of view of human and animal nutrition, but also from that of the molecular biologist. This synthesis represents the one-time expression of but a few genes at a single point in the organism's life cycle, and further represents a massive translation of these few gene products over a short time period. In this sense this synthesis system may be the botanist's equivalent to the reticulocytes and silk gland cells of the animal molecular biologists. As a consequence, this area of seed development has warranted a review of its own in this volume in the most capable hands of Dr. Adele Millerd.

Scope, Limitations, and Other Considerations

Quite naturally, seed formation has received an enormous amount of attention from an anatomical and morphological standpoint. The fascinating sequence of cell positioning that brings about embryo sac formation, fertilization, and the early cleavages of the embryo and endosperm cells have been very well defined and collated for many species (for early reviews see 19,22,43,52,103; for more recent reviews see 14 and 86). Recently these early events have been examined ultrastructurally and cyto/histochemically (21). In many cases the anatomical and morphological descriptions have thoroughly covered the entire process of seed formation for individual species (for example, see 48). However, this review is put together from the point of view of the plant biochemist who wants the biological phenomena measured and expressed in molecular terms, in the rise and fall of metabolites, of enzyme activites, and finally viewed in terms of differential gene expression and its regulation. This viewpoint then brings together the principles of intermediary metabolism and molecular biology and attempts to use them to comprehend phenomena in developmental biology. Since the theme of this review is essentially molecular, ultrastructural and cyto/histochemical studies are not considered. Furthermore, because of the virtual nonexistence of molecular studies on the very lengthy embryogenesis of gymnosperms, this review finds itself unhappily restricted to angiosperm species. And even here the influence of human and animal nutrition on the choice of botanical species studied on the molecular level is found to be overwhelming. The commonly consumed legumes and the cereals are the species examined in over 90% of the studies reported on.

In order to narrow the focus of this review to the point where it is conceptually manageable and does not degenerate into a grab bag of loose-ended observations, it has been limited to a review of the normal development of the tissues that will constitute the seed, i.e. to the tissues of the ovule—the testa, integuments, nucellus, endosperm, and embryo. Development of the other tissues of the plant ovary—the pericarp, mesocarp, endocarp, placenta, and other elements of the fruit—are not considered. Furthermore, events prior to fertilization are not considered. At the other end, events after the formation of the desiccated mature seed such as dormancy, after ripening, vernalization, etc are not explored. To a small extent some aspects of seed germination must be considered, since germination is but the continuation of the overall process of establishing another generation which begins with zygote formation. And in many cases what has transpired in embryogenesis must be inferred from what subsequently has been found to take place in germination. The separation of embryogenesis from germination is clearly artificial in considering the development and function of such tissues as the endosperm of cereals or the cotyledons of legumes.

The abnormal or experimental generation of seedlings from single cells or cell masses in culture, although a most interesting subject fraught with potential, is not considered as part of this topic since it may not represent the regulated and defined interplay of tissues that brings about true embryogenesis. Embryo culture, on the other hand, since it may reveal nutritional requirements and enzymatic capabilities of developing embryos and may open up the study of the regulation of this development, is considered briefly.

Embryogenesis in angiosperms commences with the double fertilization that gives rise to the zygote and the triploid cell that will give rise to the endosperm tissue. These events take place within the embryo sac that is embedded in maternal tissue comprised of nucellus tissue and one or more integuments. This integumentary tissue in turn is connected to the remainder of the ovary through the stalk-like finiculus. At first seed development results from the rapid development of the integument tissue, sometimes the nucellar tissue, and the triploid endosperm tissue surrounding the embryo. This last tissue becomes very extensive. The embryo itself initially grows very slowly giving rise to a suspensor of unknown function, the development of which is arrested early, and the organogenetic portion of the embryo. The endosperm continues to expand in size and cell number, followed somewhat later by the embryo proper, the growth of which is at the expense of the endosperm. The outer integument cells differentiate into the testa (seed coat), and in many cases the inner integument and nucellar cells are lost to the proliferating endosperm. At some point the cells of the finiculus connecting the ovule to the ovary degenerate, making the ovule a closed system nutritionally. The extent to which the embryo consumes the endosperm during embryogenesis varies among angiosperm species. Finally, desiccation sets in as the ovule loses water to the surrounding environment and the seed coat tissue sclerifies and dies, encasing the endosperm/embryo nicely in protective armor.

Seed germination is perhaps the most precarious time in the life of a plant. Thus enbryogenesis can be viewed as a preparation for successful germination. Consequently, there is an obvious function for the massive assembly of nutrients upon which early germination is dependent. Yet in attempting to balance nutrient storage in embryogenesis with the nutritional requirements for germination, we should keep in mind that most of the species studied are agronomically important to man, and have been selected by man for several thousand years. Thus the nutritional capital of present day seeds may not reflect what is needed to establish the seedling in a state of nature, but may reflect man's selection pressure for seeds more nutritionally beneficial to man.

Two of the most fascinating aspects of seed development require that embryogenesis and germination be viewed together. One of these involves the flow of nutrients through the tissues of the ovule. There is no direct vascular connection between the vegetative plant and the embryo itself, rather the vascular tissue coming out of the ovarian tissue ends in the testa/integument tissues of the ovule. The proliferation of the nucellus and/or endosperm results from the nutrient flow from the vegetative plant. As mentioned before, there is a gradation in the extent to which the developing embryo absorbs the nucellus/endosperm during embryogenesis throughout the angiosperm world, ranging from total absorption (legumes, cotton) to almost no absorption as is found in the cereals. Thus we have the somewhat strange phenomenon in many seeds of nutritive tissues expanding in one zone of the ovule while being destroyed in the areas surrounding the embryo. Hence the nucellus/endosperm is a transient tissue in some cases, but a persistent one in others. It would seem that dicots that completely consume the nucellus/endosperm during embryogenesis expend a great deal more energy in forming seeds than do monocots and those dicots that enter dormancy with most of the seed nutrition still polymerized in the nucellus/endosperm. These species avoid the repackaging step in the cotyledons, since in germination the nucellus/endosperm hydrolysates move to the developing axis for the most part. It would appear that the latter process would have an evolutionary advantage. The detailed biochemistry of this nutrient flow, with its rise and fall of tissues, its integration and regulation, presents a unique and potentially rewarding system for developmental biologists.

Another fascinating aspect of seed development deals with the general metabolic reversal that takes place in the nutrient storage tissue, be it endosperm or cotyledons, when germination begins. Cells that have been synthesizing enormous amounts of protein and carbohydrate or lipid reserve material during embryogenesis completely reverse this process and commence a very rapid hydrolysis of these same materials in germination. This reversal is generally not accompanied by cell division in these tissues, and thus we have the very interesting phenomenon of a gross metabolic reversal taking place in an unchanging cell population. Obviously, a certain amount of gene activation/deactivation is involved in such a fundamental reversal, and when one considers that the gene activation/deactivation may not be buried in the continued synthesis of all necessary cell constituents that accompanies cell division, this phenomenon becomes even more attractive to those interested in the molecular biology of development.

By far most of the work performed to date has been done on dicots in which the cotyledons totally consume the endosperm, or upon monocots that leave most of the seed nutrition in the endosperm. Since these two situations represent the extremes in the endosperm/embryo relationship, it is convenient to employ the dicot/monocot dicotomy in examining the research performed to date.

LITERATURE REVIEW

Much of the early work consists of rather gross measurements of the levels in whole seeds or individual tissues of some of the major cellular components such as starch, total nitrogen, protein, soluble nitrogen, sugars, etc, along with measurements of fresh and dry weight. Considering the scientific time these reports were published, these gross measurements were necessary to provide the background for more sophisticated experiments and also were necessarily gross. Many of the early experiments measured parameters that today we recognize as obvious, i.e. that rapid cell division is accompanied by rapid RNA synthesis, high respiration rates, etc. But at the time they were not obvious relationships. Still other experiments wrestled with problems that were certainly poorly thought out. For example, to look for the existence of all amino acyl-tRNA synthetases in living tissues during stages of development in order to prove that they exist is surely foolish. If any of them did not exist, how could they ever get synthesized? The same goes for all the protein components of the protein synthesizing apparatus. It must exist complete in order to make more of its protein components. This literature has for the most part been ignored in this review.

Dicots

LEGUMES   As would be expected, most of the research on legumes has been confined to peas (*Pisum sativum, Pisum arvense*), beans (*Phaseolus vulgaris, Phaseolus coccineus, Vicia faba*), soybeans (*Glyine max*), and lupines (*Lupinus* species). The early workers with *P. sativum* measured the levels of starch, sugars, total nitrogen, and ash in whole seeds (16) and in individual seed tissues (23,24,61,94) against a background of fresh and dry weight changes. Raacke (74) followed the flow of nitrogenous compounds as they moved from the pod through the seed and accumulated in the cotyledons. In all cases the levels of constituents established the accumulation of starch and protein in the cotyledons and showed that soluble monomeric units which accumulate initially are depleted as the polymers accumulate. More importantly, these studies fixed the typical time scale for this sequence of events upon which further measurements could be based. Turner & Turner (93) attempted to follow the accumulation of starch in terms of the rise and fall of sucrose and of the enzymes starch phosphorylase and β amylase, since at the time the function of UDP, ADP sugars in 1–4 α glucan synthesis was not appreciated. Several years later, Turner (95) reexamined starch synthesis in terms of the levels of UDPG and ADPG pyrophosphorylase units per cotyledon during embryogenesis and noted that whereas UDPG pyrophosphorylase levels are consistently higher in all stages of embryo development, ADPG pyrophosphorylase activity shows a dramatic rise coincident with the time of major starch deposition, followed by an equally rapid decline. The roles of these two enzymes in starch synthesis remain unclear.

McKee et al (61) made another developmentally interesting observation that was subsequently found to hold for other legumes and other dicots as well. Namely, the final number of cells of the pea embryo (1.4×10^6) is reached less than half way through the period of seed formation, which is about 50–55 days, and long before the embryo has accumulated all its starch and protein (or its final dry weight). Bain & Mercer (3) repeated measurements of the levels of nitrogen and carbohydrate against the rise in fresh and dry weight but focused on the cotyledons exclusively. They divided the development of cotyledons into four phases that again emphasized the cessation of cell division in cotyledons early in embryogenesis and prior to the large build-up of the nutritional storage products. Ultrastructurally these authors noted a loss of fine structure in cotyledon mitochondria during the final phase of embryogenesis (maturation and desiccation). Kollöffel (50) looked further at the fate of these mitochondria as the seeds mature and dry up by measuring the activity of a number of mitochondrial enzymes during this period. He observed a decrease in enzyme units per cotyledon but no total loss of any activity measured. (The status of mitochondria during the phases of embryogenesis and through early germination has not been fully explored in depth for any species, unfortunately.)

More recently, Flinn & Pate (30) have established the levels of protein and amino acids for all the seed tissues and also for the pod during seed formation in *P. arvense*. They noted some differences in the pool sizes of several amino acids when compared with those of *P. sativum*, but more significantly they noted that the gain in nitrogenous compounds by the embryo during embryogenesis is much greater than that lost by the pod, testa, and endosperm, which brings attention to the fact that, although nutrients build up in the ovarian tissue and the maternal tissue of the seed itself, a continuous flow into the seed from the vegetative plant must take place until the finiculum desiccates and the seed becomes nutritionally a closed system. These authors found it convenient to divide embryogenesis in this species, which also takes about 55 days, into three phases.

As the rapid advances in molecular biology mounted, measurements of RNA and DNA levels were undertaken first in whole seeds (82, 107), showing that the most rapid accumulation of RNA and DNA occurred during the period of most rapid seed cell division, and subsequently on the individual seed tissues (89). When RNA levels of cotyledons were measured as ribosomes and polysomes, a marked drop in polysome levels was observed to occur during the desiccation phase of embryogenesis (4, 11). Out of the measurements of RNA and DNA levels came the novel and unanticipated finding that pea cotyledon cells (in common with other legumes, as will be seen) become highly polyploid during embryogenesis. Smith (88) noted that the cotyledon cells of *P. arvense* dry seeds contain up to a 16 C level of DNA, which decreases during germination to a 2–4 C level. He suggested that this polyploidy is a form of nucleotide storage for the incipient seedling. When DNA levels of developing *P. sativum* cotyledons were examined by the histophotometric measuring of Feulgen stained material by Sharpé & Van Parijs (85), these cells were found to contain DNA levels that averaged between 32 and 64 C. They further observed that these ploidy levels are reached shortly after the final number of cells is reached in the cotyledons and concomitant with the beginning of maximum storage protein synthesis in these cells. Smith (89) examined in detail the accumulation of DNA per cell in *P. arvense* in relation to the levels of RNA per cell and to the time of maximum starch and storage protein synthesis. He also observed that levels of DNA of 32–64 C are reached just after cell division ceases and coincident with the beginning of maximum starch and storage protein synthesis. When the RNA content of cotyledons was considered per cell, per unit DNA, or per unit protein, the only consistent ratio was found to be RNA per cell volume.

More recently, measurements of the dynamics of RNA and protein synthesis in developing pea cotyledons have appeared in the work of Beevers & Poulson (11, 73). They also have observed a drastic fall in the level of polysomes as the desiccation phase of seed maturation begins, and they further observed that the cell-free amino acid incorporating system extractable from desiccating cotyledons was much lower in rate than that extractable from cotyledons in their rapid growth phase or at the time of rapid storage protein accumulation. When RNA synthesis was measured by following the incorporation of radioisotopes into RNA, the maximum rate was found to occur just prior to the phase of storage protein synthesis, i.e. during the phase of rapid cell division.

The history of biochemical investigations of seed development in *Phaseolus vulgaris* and *Phaseolus coccineus* is similar to that of *Pisum* species, although some of the basic phenomena observed in peas were noticed first in *Phaseolus*. Loewenberg (51), who measured respiration rates, RQ values, and the levels of nitrogen and phosphate-containing compounds in *P. vulgaris,* noticed that the final number of cells of the embryo is reached when the dry weight of the embryo is only one-sixth of its final value. This number is reached only 21 days post anthesis whereas seed formation and maturation is not complete until about 44 days. As an obvious consequence, when all the parameters measured were put on a per cell basis, everything was seen to increase after 21 days (except for O_2 uptake which expectedly declines greatly during desiccation). He determined that the final cell number per cotyledon pair is 2.6×10^6, that the average mitotic cycle during the period of cell accumulation is 15.5 hours, and that 18.5 generations are needed to reach the final cell number. One would think that with such a basic foundation as presented by this paper, a great deal of work with *Phaseolus* embryos would follow shortly, but it did not.

Surprisingly, very high DNA levels were found to exist in the suspensor cells of both *P. vulgaris* and *P. coccineus* (17, 66). In fact, the chromosomes of these cells are now considered classically polytene. Öpik first suggested the possibility of polyploidy in *P. vulgaris* cotyledon cells (70) and also noted that cell division in these embryos ceases before embryogenesis reaches the halfway point in time and prior to the time of starch and protein deposition (71). She also noted the loss of polysome structures during the desiccation phase in this species. Attention was drawn by Makower (53) to the large build-up of phytic acid in *P. vulgaris* cotyledons during the time of starch and protein deposition. Since a similar build-up had been noted in peas (80), legume cotyledons could now be considered a storage sink for phosphate as well.

At this point the extensive work of Walbot on embryogenesis of both *P. vulgaris* and *P. coccineus* appeared, which presented a detailed examination of the dynamics of RNA synthesis during cotyledon and axis development and germination. She reported that the RNA content of cotyledons remains constant per cell once its maximum level is reached, that RNA does not turn over, and does not decrease during desiccation, and in fact showed that protein synthesis during the first hours of germination is carried out on ribosomes carried over undegraded from embryogenesis (97, 99). RNA ceases to increase in cotyledons before the cells have completed the synthesis of storage protein, and thus protein synthesis is seen to operate with stable RNA components in this late phase of embryogenesis. Furthermore, the ratio of rRNA to tRNA was shown to be constant throughout embryogenesis, and was calculated to provide 15 tRNA molecules per ribosome (99) [about the same value found for cotton cotyledons (62) and bacteria (105) and thus may represent some universally selected ratio.]

Since protein synthesis in the first hours of germination was found by Walbot (98) to occur prior to any resumption of RNA synthesis, a carryover of mRNA from embryogenesis into germination was also suggested. However, she also noted the loss of polysomes during desiccation. These two observations taken together hint at the possibility of unusual processes at work during the last days of embryogenesis. We shall explore the evidence for cotyledons as a source of specific mRNA for germination processes when examining the data obtained from cotton seed embryogenesis. These sorts of data illustrate the unnatural circumscription imposed from a developmental point of view when germination is separated from embryogenesis. Walbot also examined several parameters of precocious germination (101) which phenomenon had earlier been observed in *Phaseolus* by Klein & Pollack (49). (This phenomenon of immature embryos germinating when removed from their ovular environment has long been recognized by workers attempting to culture plant embryos, as we shall see further on.) She observed that cotyledon growth stopped immediately upon excision, which probably indicates a cessation of cell division as has been found in cotton (37), and also observed that the precocious germination of very small embryos was stimulated by incubating them in nutrient medium. This observation calls attention right away to the possible nonexistence of certain biosynthetic pathways in very small embryos that in vivo are living off endosperm hydrolysates!

The measurement of biochemical parameters of *Vicia faba* embryogenesis paralleled to some extent those of peas and *Phaseolus* species, and many of the same phenomena were found. Grzesink et al (33), who made measurements on both the endosperm and embryo, cataloged the accumulation in the cotyledons of starch at the expense of sucrose and of protein at the expense of free amino acids against the background of a 66-day period of seed formation. They noted the very high levels of alanine and glutamic acid relative to the other amino acids in these tissues and also observed that the suspensors cells are obviously polyploid.

A great deal of data emanated from the laboratory of Boulter and co-workers, who, although primarily interested in the synthesis and characterization of the storage protein, also established many of the basic macromolecular features of *V. faba* embryogenesis. These workers followed cell division, RNA, DNA, starch, total nitrogen, total phosphorus, and soluble sugar levels from anthesis to well into the dry seed period (130 days). They observed that cell division in the embryo lasts only for the first 35–40 days but that RNA and DNA levels increase extensively until almost 50 days (72, 106). This indicated the polyploidy of cotyledon cells as was found in peas. The rapid accumulation of storage protein and starch occurs after cell division is completed and continues after RNA and DNA levels have plateaued (18, 72). They also observed the conversion of membrane bound ribosomes to free ribosomes as desiccation sets in. The maximum amount of membrane bound ribosomes occurs at the time of maximum storage protein synthesis, suggesting that these proteins are made on rough endoplasmic reticula. The increase in free ribosomes may correspond to the loss of polysomes noted to occur in the other legumes when desiccation begins.

It remained for Millerd and co-workers to make a detailed analysis of this alleged polyploidy in *V. faba* cotyledon cells and relate it to cotyledon growth, cell number, and legumin and vicilin synthesis (63, 64). She showed that the broad bean cotyledon reaches a cell number of 2.8×10^6 and then increases its DNA level eight to tenfold. In a series of very elegant experiments, she established that the DNA increase is due to an increase in nuclear DNA (not mitochondrial or plastid), that it involves the increase in the total nuclear DNA, not of satellite DNA, and that it does not involve gene amplification. In other words, it is a clear case of endoreduplication. These facts were demonstrated by measurements of bouyant density and reassociation kinetics. Specifically, there are 4.5×10^{-11} g DNA per cotyledon cell prior to endoreduplication and an average of 33.2×10^{-11} g DNA per cell after endoreduplication has occurred. The latter represents a 16 C value. The DNA is 30% unique copy DNA and 70% DNA that is reiterated to various extents *both before and after endoreduplication.* Ribosomal RNA and tRNA per cell also increase eight to tenfold after cell division ceases, which seems to maintain the ratio of rRNA/DNA and rRNA/tRNA constant.

Soybean embryogenesis has received scant attention, with the laboratory of Howell and co-workers providing most of the data. They divided embryo development into phases that correspond to similar divisions in the other legumes, noting, for example, that cell division ceases about 2 weeks after flowering whereas desiccation does not begin until 50 days (15), that RNA accumulation is the most rapid during the period of rapid cell division (32), and that O_2 uptake was the most rapid during this same period of cell proliferation (69). The biochemistry of lipid deposition in these seeds has not received attention, unfortunately.

RNA, DNA, and protein levels during embryogenesis have also been established for several lupine species (96) where here again RNA and DNA values plateau prior to the deposition of most of the storage protein.

COMPOSITE VIEW OF LEGUME EMBRYOGENESIS   Out of these studies on legumes several consistent phenomena emerge. The initial development of the endosperm in mass and its function as a transient reservoir of sugars and amino acids becomes obvious. The growth of the embryo is found at first to be rather slow but then to accelerate. The final cell number of the embryo is found to be reached rather early in its ontogeny and its subsequent increase in mass to be the result of cell expansion and the concomitant deposition of starch and storage protein. The development of the cotyledons as a gigantic sump for protein, starch, and phosphate, devouring the endosperm and coming to occupy the entire intertestal cavity in the process, followed by its desiccation, completes the picture on this level. But the measurements of cell constituents that produced this picture actually only verify what can be predicted anatomically. However, a number of intriguing and puzzling features of this process in legumes have also emerged. The first of these is the polytenization of the chromosomes of the suspensor. When a phenomenon such as this is stumbled upon, one's reaction is to attempt to postulate "why?" and to concoct a reason. At this point none are available for this phenomenon. The polyploidization of the cotyledon cells that is accompanied by commensurate increases in RNA per cell is another unusual finding. Since these cotyledons at this point begin to synthesize large amounts of a very small number of gene products, the storage proteins, it might have been speculated that the increase in DNA was an example of gene amplification of the protein structural genes. This proved not to be the case (64). If it can be established that cotyledon DNA is hydrolyzed during germination back to its 2–4 C level, and that its substituent nucleotides/nucleosides are reused, as suggested by Smith (89), then perhaps we should consider legume cotyledons as a sump for nucleotide storage as well. The large increase in RNA per cell can be considered merely a means of keeping protein synthesis per volume of cytosol up to level, since cell size is increasing markedly during this period. Apparently it is not universally recognized that measurable increases in RNA per cell are obliged to be mostly increases in rRNA and tRNA rather than mRNA. This is true simply because it takes 15–17 times as much rRNA and 3–4 times as much tRNA (on a mass basis) to accomodate the mRNA in protein synthesis. Thus attempts to correlate RNA increases with storage protein mRNA were doomed from the start.

Another intriguing observation is that the ribosome population in cotyledons once established does not turn over during the remainder of embryogenesis nor during the first phase of germination. What is curious is the loss of polysomes noted by a number of workers to occur during desiccation. There is some evidence from legumes (and considerably more from cotton) that some mRNA for cotyledon germination is also provided for in embryonic cotyledons. If this mRNA is exclusively for proteins that function to bring about germination, then they should not be polysomal during embryogenesis, since a priori their protein products would not be wanted during embryogenesis.

Finally, the time period of synthesis of the major storage proteins has been shown to be limited and specific, occurring as the cotyledon cells reach their maximum RNA and DNA levels and continuing until desiccation sets in. It is tempting to speculate that the mRNA for these proteins is very stable indeed, being destroyed in some fashion only by desiccation.

Seen as a whole, the developmental pattern of legume seed formation is so similar among the species studied, so noncontradictory, that it is possible to diagram the sequence of many of the major events (Figure 1).

COTTON   Biochemical aspects of the embryogenesis of cotton insofar as they have been studied were arrived at in a backward fashion, i.e. observations made during germination forced the investigators to examine younger and younger embryos. Cotton embryogenesis is similar to that of legumes in that endosperm and nucellus develop first and are then totally consumed by the embryo, the bulk of which consists of the large cotyledons that accumulate lipid, storage protein, phytin, but no starch. Cotyledon morphology is quite different in that they do not develop into very thick structures but remain thin, become highly convoluted, and wrap themselves around the axis.

The initial observation that led into a study of cotton embryogenesis was the fact that protein synthesis in cotyledons during the first 3 days of germination is not inhibited by large doses of actinomycin D (28,104); in fact, it is slightly stimulated. This observation, while suggesting that mRNA is carried over into germination from embryogenesis, does not in itself imply anything of developmental significance. The mRNA carried over could be for constitutive enzymes of intermediary metabolism that are synthesized in both embryogenesis and germination and is stable enough to survive desiccation and dormancy. On the other hand, it could be mRNA

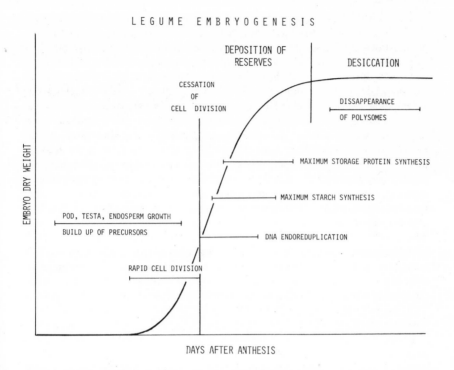

Figure 1   Diagrammatic representation of legume embryogenesis.

whose function is to bring about the synthesis of proteins unique to germination. To explore this, the synthesis of an enzyme unique to germination (carboxypeptidase C, probably involved in storage protein hydrolysis) was examined. It was found to be de novo synthesized in cotyledons during early germination (36), and its synthesis was found to be insensitive to actinomycin D (34). It was possible to show that actinomycin D inhibited rRNA and tRNA synthesis about 95% in these experiments and mRNA synthesis about 70% (102). Proof of its de novo synthesis required that the enzyme be purified to homogeneity from cotyledons germinated in the presence of radioactive precursors (36). Since so much of the phenotypic development in these cotyledons proceeds uninhibited for the first couple of days of germination in actinomycin D (in contrast to axis development which is strongly inhibited), and since total protein synthesis in cotyledons is not suppressed by the drug, it appears that the carboxypeptidase is representative of a large group of "germination" enzymes for which the mRNA is transcribed in embryogenesis.

More recently it has been shown that the drug cordycepin, the 3'd analog of adenosine, totally suppresses the appearance of the carboxypeptidase and decreases protein synthesis about 75% during early germination (100). This suggests that the putative mRNA for the germination enzymes is not polyadenylated in embryogenesis, when it is alledgedly transcribed, but polyadenylated during the first day of germination—a process that is sensitive to 3'd ATP formed from cordycepin. This polyadenylation of preexisting mRNA has now been demonstrated to occur in these cotyledons during the first 24 hr of germination (102). If indeed embryonic cotton cotyledons do synthesize and store mRNA for proteins to be used in bringing about germination, then still another facet of seed formation as a preparation for germination comes to light. These studies quite naturally led into an examination of cotton embryogenesis. By making use of the immature cotton embryo's proclivity for precocious germination when removed from its ovular encasement, it was possible to show when the germination mRNA first appears in embryo cotyledons. By precociously germinating successively younger embryos, it was shown that carboxypeptidase appearance and precocious germination of cotyledons in general becomes sensitive to actinomycin D when the cotyledons are about three-fifths their final size (34). Further it was shown that abscisic acid appears in the cotton ovule about this time and that precocious germination is totally sensitive to this growth regulator at 10^{-6} M concentration (35). In addition, DNA synthesis ceases in the embryo at this point (37), and the vascular connection between the ovule and the parent plant atrophies also at this point. The simultaneous occurrence of all these events is remarkable and may indicate that the vascular flow from the parent plants plays a fundamental role in regulating seed formation in addition to merely supplying nutrients for this process.

Younger cotton embryos will precociously germinate when removed from the ovule, but they do so slowly and this precocious germination is sensitive to actinomycin D. However, if the actinomycin D is not applied until the second day of precocious germination, much of the subsequent cotyledon development, including the advent of carboxypeptidase C, occurs (37). This is taken to mean that the germination mRNA, which presumably is not present in these young cotyledons, is made shortly after excision from the ovule. Ovule extracts from seeds containing these young embryos are not inhibitory to precocious germination, indicating that abscisic acid levels at this period are low. Furthermore, cell division in the cotyledons stops as soon as these embryos are excised, which gives rise to smaller cotyledons as successively younger embryos are precociously germinated (37).

Out of all this an undoubtedly oversimplified scheme can be erected that can be used as a basis for further experimentation (Figure 2). This scheme emphasizes several strictly conjectural points not heretofore considered in studies of seed formation. The first of these is that the incipient embryo is programmed to germinate while still very immature. It does not have the germination mRNA, but removed from the ovular environment it will activate the requisite genes and transcribe and then translate the information for germination. The tendency for young embryos from other species to precociously germinate indicates that this preprogramming is not unique to cotton. The loss of the vascular connection with the vegetative parent signals a large number of developmentally important events. DNA synthesis stops as it does in younger embryo cotyledons when the vascular connection is artifically broken by excision. (It is not known at this point whether the cotyledon cells have

COTYLEDON DEVELOPMENT IN EMBRYOGENESIS

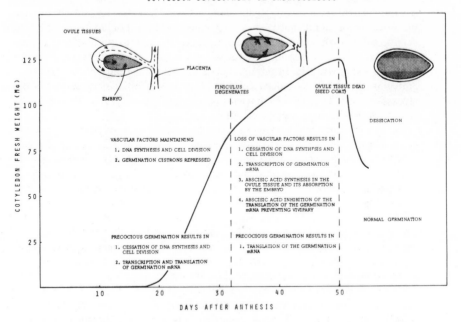

Figure 2   Postulated scheme of developmental events in cotton cotyledon embryogenesis (from Ihle & Dure 37).

become polyploid prior to this time as is the case with legumes.) Prior to this point the cotyledon have been growing as embryonic cotyledons, which suggests that the program for germination is overruled by a program for embryonic growth that is maintained by elements of the vascular flow. Excision causes this later program to fade away.

Secondly, with the removal of regulation by the vegetative plant, the genes for the germination mRNA are activated and their transcription begins but not their subsequent translation. The later process would lead to vivipary—a lethal event—and is prevented in some fashion by the simultaneous appearance of abscisic acid in the tissues of the ovule surrounding the embryo which apparently is taken up by the embryo. Again excision results in precocious germination since the embryos are removed from the source of the abscisic acid, but now precocious germination is insensitive to actinomycin D (in the cotyledons, remember, not in the axis, which must make new cells and hence new RNA). Finally, when the integumentary tissue sclerifies and dies to form the seed coat, the embryo is released from all germination inhibition, but by this time it is desiccated and must await rehydration.

Like most postulates based on scanty data, this raises many more problems than it answers, but hopefully in pursuing these the postulate will get reinforced, modified, or abandoned.

OTHER DICOTS   Castor bean (*Ricinus communis*) embryogenesis and germination is quite different from that of legumes and cotton. The cotyledons do not become the reservoir of stored nutrients during embryogenesis, nor do they appear to develop substantially during germination. Rather the endosperm itself is the tissue site of the polymerization of amino acids and sugars and the depository for the stored oil and phytic acid. In germination metabolism in the endosperm is resumed, protein is hydrolyzed, and gluconeogenesis from lipid effected by the advent of the glyoxylate cycle. The liberated nutrients pass through the cotyledon directly to the growing axis without any polymerization.

Some facets of the biochemistry of castor bean seed formation have been investigated. The accumulation of glycolytic intermediates and of phytic acid, the activities of glycolytic and mitochondrial enzymes, respiration, and RQ values have been measured in whole ovules and in endosperm tissue (54). In general, these enzyme activities increase with the growth of the seed and fall off as desiccation sets in. RQ values above one are reached indicating the transformation of sugars to lipid. Total RNA increases in whole ovules with ovule growth, but then decreases markedly during desiccation (54). An actual decrease in extractable ribosomes has been observed in the endosperm (54, 91) to the point where they appear to have totally disappeared by the time the dry seed is formed. Some must persist, however, or there would be no way of making more of them during germination.

Several other dicot species have had biochemical aspects of their seed formation examined. Oxygen uptake during the development of whole seeds of *Hibiscus esculentus* (25) and poppy (*Papaver somniferum*) (44) has been measured as well as the accumulation of carbohydrate, protein, and nucleic acid in poppy seed (45) and the accumulation of protein at the expense of free amino acid pools in *Brassica napus* (29). Some basic data on embryo development have also been collected for *Datura stramonium* (79). Not enough detailed information on the individual tissues of these seeds exists to perceive a pattern of events as yet.

Monocots

CEREALS   The highly evolved process of seed formation in this group is so similar, and the research performed to date so much in parallel from one species to another, that it is possible to discuss it as a whole. These species make no attempt to store much of the seed reserves in the embryo itself during seed formation other than the small accumulation of oil that takes place in the highly modified single cotyledon, the scutellum. Rather, the carbohydrate energy reserves are polymerized in the endosperm and the protein reserves accumulated in the testa/pericarp cell layer(s) called the aleurone layer(s). The build-up of these storage materials during seed formation has been well documented, beginning with very early work on maize (13). Very comprehensive cataloging of the accumulation of nutrients in the cereal endosperm have appeared for wheat (40–42), rice (1, 55–58, 81), and oats (2), and these processes related to the increase in nucleic acid in maize (38), wheat (41, 82), and oats (82). For example, in maize (38) RNA and DNA increase in a parallel fashion in the endosperm until cell division ceases at about 28 days post anthesis, which is roughly halfway through the time period of seed formation. The concentration of sugars reaches its maximum in this tissue just before the end of cell formation. Starch accumulation begins at the expense of these sugars, reaching its final level at about 46 days. Endosperm protein increases during the cell formation period, but undergoes a second phase of accumulation around 40 days post anthesis, which coincides with the deposition of storage protein in the aleurone layer. Oddly, RNA content decreases in the endosperm during the period of rapid starch synthesis (30–46 days) and both RNA and DNA decrease in the endosperm during desiccation (post 46 days). In contrast, protein, RNA, and DNA build up continuously in the embryo during this 46-day period.

One of the first cell-free amino acid incorporating systems was extracted from maize endosperm and embryo tissue during their development (75), and this study showed that initially the most active system exists in the developing endosperm, but that this activity falls off in later stages of seed formation (perhaps reflecting the loss of RNA and DNA), leaving the embryo with the most active system as desiccation begins.

Recently a possible correlation was explored between the synthesis of cellular enzymic and structural protein, which occurs during the cell proliferation stage, and the synthesis of the storage protein, which occurs later in development, and the population of amino-acyl tRNA synthetases in developing wheat endosperm (68). The overall amino acid makeup of these two classes of protein is different, and this might be reflected in the synthetase levels, but no such correlation was found.

The enzymatic basis for starch synthesis in cereal endosperm has understandably received a great deal of attention. Plant systems may differ somewhat in this synthesis from the classically taught mechanism of glycogen synthesis in mammals(a) by the fact that phosphorylase, generally considered a polymer-shortening enzyme, may produce the 1–4 α-linked primer (acceptor) molecules for starch synthesis (5, 87); (b) UDP-glucose (UDPG) and ADP-glucose (ADPG) can be produced directly from sucrose by the enzyme sucrose-UDP (ADP) glycosyl transferase (27); (c) ADPG as well as UDPG is produced (as mentioned above) and presumably utilized in starch synthesis in endosperm tissue (31, 39, 65, 78). Several detailed developmental studies on the synthesis of starch in developing cereal endosperm from incoming translocated sucrose have appeared recently. Shannon (83, 84) has followed the fate of ^{14}C sucrose produced in vivo from ^{14}C CO_2 in developing maize seeds. Here sucrose is apparently cleaved by invertase as it leaves the phloem at the base of the grain and enters the endosperm. It is re-formed from glucose and fructose as it enters the zone of cells actively synthesizing starch. Here it is converted to starch, apparently from glucose-1-phosphate (G1P) via the UDPG pyrophosphorylase and the UDPG-starch synthetase system, since sucrose-UDP glucosyl transferase cannot be detected during the initial stages of starch synthesis in maize (92). Barley endosperm, on the other hand, presents a more complicated picture based on the extensive enzyme study of Baxter & Duffus (5). Here sucrose-UDP glucosyl transferase, UDPG and ADPG pyrophosphorylases, UDPG- and ADPG-starch synthetases, starch phosphorylase, pyrophosphatase, and nucleoside diphosphokinase levels were all measured during endosperm development. The results suggest that primer molecules for starch synthesis are generated from G1P by starch phosphorylase and that the G1P is produced from sucrose by the sequential action of sucrose-UDP glucosyl transferase (to produce UDPG) and UDPG pyrophosphorylase. Incoming glucose units for primer elongation are provided by UDPG and to a lesser extent ADPG. The fructose moiety of the sucrose molecule is converted to G1P by the obvious enzymes (26). As in the case of peas (95), the activity of ADPG pyrophosphorylase is considerably lower than that of UDPG pyrophosphorylase at all developmental stages and its contribution to starch accumulation remains unclear.

OTHER MONOCOTS   The most basic parameters, dry and fresh weight, carbohydrate, protein, and nucleic acid levels, and oxygen uptake, have been established for *Zephyranthes lancasteri* seeds (46, 47).

EMBRYO CULTURE

This experimental technique has enjoyed a rather long history (for reviews see 67, 76, 77) and its achievements have been remarkable. Very early, intrageneric hybrids were produced from crosses that otherwise fail to produce viable seed (77). Likewise, a hybrid between barley and rye was produced by this technique (20), as were hybrids between *Pinus* species (12, 90). Most embryos, once they have reached a certain point in embryogenesis, precociously germinate when excised. Usually excision stops cell division in cotyledons, giving rise to considerably reduced seedlings (37, 101), which nevertheless can grow into mature, healthy plants. This poses a problem in using embryo culture for studying in vitro the embryogenesis of embryos that have reached this stage. Measurements of RNA and protein synthesis in these excised embryos by isotope incorporation in reality are measurements of germinative synthesis. Abscisic acid has been shown to prevent the precocious germination of excised cotton embryos (35, 37), but it remains to be demonstrated that these embryos continue to develop as they do in vivo.

The culture of very small embryos (pro embryo and heart stage) has met with very limited success. Most of these embryos die or develop abnormalities when cultured, although occasionally normal embryonic development is achieved, followed by normal germination into viable seedlings (12, 59). Success here usually has been the result of manipulating the osmotic pressure, pH, and composition of the nutrient media (60). What is needed is a clearly defined set of conditions that will allow very young embryos to continue embryonic development in culture rather than to slowly begin precocious germination. These conditions quite possibly are related to the biosynthetic capabilities of these very small embryos, which may not have the biosynthetic enzymes necessary to produce the end products of intermediary metabolism (amino acids, nucleotides, etc) from carbohydrate intermediates. However, embryos have been shown capable of activating genes when excised (37). Consequently, what ultimately may be needed to allow for routine test tube embryogenesis is a knowledge of the composition of the incoming vascular flow with respect to regulatory molecules. As we have seen in cotton, once this flow ceases in vivo cotyledon cell division stops and germination mRNA is transcribed (and vivipary avoided by the appearance of abscisic acid). Furthermore, when this flow is prematurely stopped by excision, these events are induced prematurely. Once routine test tube embryogenesis is achieved, then the influence of the regulatory molecules of the milieu can be examined one by one.

Surprisingly, very remarkable success has been realized in the past several years in the culture of still younger seed tissues, as evidenced by the work of Beasley and collaborators. The initial success obtained by Beasley was in culturing cotton ovules excised just after fertilization (6). From the ovule epidermis these ovules developed cotton fibers indistinguishable in number and length from in vivo fibers (8). Occassionally complete embryogenesis was obtained producing normal seedlings and ultimately mature plants (10). Next, the development of fibers from the ovule epidermis was achieved with unfertilized ovules (7, 9), but only when the proper balance of exogenous plant hormones was maintained in the culture medium or when the medium contained extracts from germinating pollen. Unquestionably, exciting and enlightening results can come out of these types of experiments provided the culture systems are used to explore the regulation of developmental processes and not considered merely as ends in themselves.

PROSPECTUS

As can be seen from the foregoing review, some basic information has been established for the development of angiosperm seeds. Most of it constitutes a description of the contents of the various tissues and organs during development. Some of it has broached in a small way the problems of developmental regulation. Many curious and stimulating facets have been uncovered that can now serve as a basis for further investigations. Some of these facets (some few out of very, very many) are listed below in the form of unanswered but certainly approachable questions.

1. What are the biosynthetic capabilities of the developing endosperm and embryo? Do these tissues sequester amino acids, nucleotides, etc from the vascular flow, or do they synthesize them from carbohydrate precursors?

2. In dicots, how is the digestion of the nucellus and endosperm effected? Do they hydrolyze themselves or does the growing embryo secrete the requisite enzymes for the purpose?

3. Are there unusual transport mechanisms involved in the flow of nutrients through the seed tissues to the embryo?

4. What influences cause young embryos to grow and develop as embryos rather than to germinate precociously as they do when excised? Are they related to the composition of the vascular flow? What is the role of vascular phytohormones and endogenously synthesized phytohormones on the developmental processes? Seeds are a rich source of many of these hormones and of their precursors and derivatives. Their interacting influences must regulate much of what occurs in seed formation.

5. How general is the polyploidization of cotyledon cells among dicots? Is it related to the vast accumulation of protein, starch, and/or lipid in this organ? Examples of polyploidization in the animal world are generally restricted to organs that have become terminally differentiated; this is not true of cotyledons which often

undergo much further maturation in germination. Does this polyploidization merely serve to store purines and pyrimidines for the germinating axis?

6. Do embryonic dicot cotyledons really transcribe and store unprocessed mRNA for germination enzymes? If so, is its transcription activated by the loss of regulatory factors emanating from the vascular flow? Does ABA prevent vivipary by maintaining this mRNA unprocessed; and if so, by what process?

7. Does seed development in monocots, especially in the cereals, represent a more advanced form of this process? Does it represent a biological saving in both energy and time? By this we mean, does the use of endosperm as the storage tissue and the reduction of cotyledon development give these plants an advantage irrespective of other differences in fructification characteristics?

8. What is the status of proplastids and their replication in seed tissues during their development? Chloroplast DNA is probably carried by all plant cells, as shown by the demonstrations of totipotency of cultured cells. During embryogenesis does the proplastid population of the cotyledons and axis become different? Are there differences in proplastid population and function in epigynous vs hypogynous cotyledons?

This list can be made substantial indeed, and, depending on one's point of view, other equally provocative questions can be asked of these systems.

Seed development as a system for studying the regulation of development suffers from the lack of a detailed genetics. Only rarely can mutants come to our aid. Yet higher animal developmental systems have the same drawback. On the other hand, the development of seeds has one very nice feature that should be stressed; namely, that of developmental synchrony. It is relatively easy to obtain developing material all of which is at precisely the same developmental stage. All that is necessary is to tag flowers at the time of anthesis and harvest the pods, bolls, etc when necessary. Ovule weight or size can be used to define its developmental stage. As can be seen from the literature, the choice of organisms for study has hindered progress to some extent in that most of the seeds studied are from seasonal, annual plants whose year-around greenhouse propagation would be troublesome. On the other hand, if perennial, bushy species with a neutral photoperiod that flower continuously and that have good greenhouse properties are selected for study, a year-round supply of synchronously developing material is easily established. Obviously, the agronomic importance of the species must be considered also.

To some extent this review is intended to be a thinly veiled plea for more investigators to use the study of seed development as a tool for the study of the regulation of developmental processes. Generally an area of biology moves forward only when the number of laboratories working in the area reaches a critical mass. Only then can investigators build on each other's findings and accumulate enough basic data to allow for truly sophisticated experiments to be carried out. The extant literature in the area of seed development is full of suggested developmental mechanisms, some possibly representing mechanisms used throughout the biological world, some possibly unique to plants, but all of them fascinating, exciting, and important. When this subject is reviewed again, perhaps some of these mechanisms will be known.

Literature Cited

1. Asada, K., Kasai, Z. 1962. *Plant Cell Physiol.* 2:397
2. Ashton, W. M., Williams, P. C. 1958. *J. Sci. Food Agr.* 9:505
3. Bain, J. M., Mercer, F. V. 1966. *Aust. J. Biol. Sci.* 19:49
4. Barker, G. R., Rieber, M. 1967. *Biochem. J.* 105:1195
5. Baxter, E. D., Duffus, C. M. 1973. *Phytochemistry* 12:1923
6. Beasley, C. A. 1971. *BioScience* 21:906
7. Beasley, C. A. 1973. *Science* 179:1003
8. Beasley, C. A., Ting, I. P. 1973. *Am. J. Bot.* 60:130
9. Beasley, C. A., Ting, I. P. 1974. *Am. J. Bot.* 61:188
10. Beasley, C. A., Ting, I. P., Feigen, L. A. 1971. *Calif. Agr.* 25:6
11. Beevers, L., Poulson, R. 1972. *Plant Physiol.* 49:476
12. Berlyn, G. P., Miksche, J. P. 1965. *Am. J. Bot.* 52:730
13. Bernstein, L. 1943. *Am. J. Bot.* 30:517
14. Bhatnagar, S. P., Johri, B. M. 1972. In *Seed Biology,* ed. T. T. Kozlowski, 1:71–149. New York: Academic
15. Bils, R. F., Howell, R. W. 1963. *Crop Sci.* 3:304
16. Bisson, C. S., Jones, H. A. 1932. *Plant Physiol.* 7:91
17. Brady, T. 1973. *Caryologia* 25 suppl.: 233
18. Briarty, L. G., Coult, D. A., Boulter, D. 1969. *J. Exp. Bot.* 20:358
19. Brink, R. A., Cooper, C. D. 1947. *Bot. Rev.* 13:423
20. Brink, R. A., Cooper, C. D., Ausherman, L. E. 1944. *J. Hered.* 35:67
21. Corti, E. F., Sarfatti, G. 1973. *Caryologia* 25 suppl.: 1–314
22. Crocker, W., Barton, L. V. 1953. *The Physiology of Seeds.* Waltham: Chronica Botanica
23. Danielsson, C. E. 1952. *Acta Chem. Scand.* 6:149
24. Danielsson, C. E. 1956. *Physiol. Plant.* 9:212
25. Das, V. S. R., Rao, M. P. 1962. *Indian J. Plant Physiol.* 5:230
26. De Fekete, M. A. R., Cardini, C. E. 1964. *Arch. Biochem. Biophys.* 104:173
27. Delmer, D. P., Albersheim, P. 1970. *Plant Physiol.* 45:788
28. Dure, L. S. III, Waters, L. C. 1965. *Science* 147:410
29. Finlayson, A. J., Christ, C. M. 1971. *Can. J. Bot.* 49:1733
30. Flinn, A. M., Pate, J. S. 1968. *Ann. Bot.* 32:479
31. Frydman, R. B., DeSouza, B. C., Cardini, C. E. 1966. *Biochim. Biophys. Acta* 113:620
32. Galitz, D., Howell, R. W. 1965. *Physiol. Plant.* 18:1018
33. Grzesiwk, S., Mierzwinska, T., Sojka, E. 1962. *Fiziol. Rast.* 9:682
34. Ihle, J. N., Dure, L. S. III 1969. *Biochem. Biophys. Res. Commun.* 36:705
35. Ibid 1970. 38:995
36. Ihle, J. N., Dure, L. S. III 1972. *J. Biol. Chem.* 247:5034
37. Ibid, 5048
38. Ingle, J., Beitz, D., Hageman, R. H. 1965. *Plant Physiol.* 39:735
39. Jenner, C. F. 1968. *Plant Physiol.* 43:41
40. Jennings, A. C., Morton, R. K. 1963. *Aust. J. Biol. Sci.* 16:318
41. Ibid, 332
42. Ibid, 384
43. Johnansen, D. A. 1950. *Plant Embryology.* Waltham: Chronica Botanica
44. Johri, M. M., Maheshwari, S. C. 1965. *Plant Cell Physiol.* 6:61
45. Ibid 1966. 7:35
46. Ibid, 49
47. Ibid, 385
48. Joshi, P. C., Wadhwani, A. M., Johri, B. M. 1967. *Proc. Nat. Acad. Sci. India* 30:37
49. Klein, S., Pollock, B. M. 1968. *Am. J. Bot.* 55:658
50. Kollöffel, C. 1970. *Plant (Berlin)* 91: 321
51. Loewenberg, J. R. 1955. *Plant Physiol.* 30:244
52. Maheshwari, P. 1950. *An Introduction to the Embryology of Angiosperms.* New York: McGraw-Hill
53. Makower, R. U. 1969. *J. Sci. Food Agr.* 20:82
54. Marré, E. 1967. *Current Topics in Developmental Biology* 2:76. New York: Academic
55. Matsushita, S. 1958. *Mem. Res. Inst. Food Sci. Kyoto* 14:24
56. Ibid, 30
57. Ibid 1959. 17:23
58. Ibid. 19:1
59. Mauney, J. R. 1961. *Bot. Gaz.* 122:205
60. Mauney, J. R., Chappell, J., Ward, B. J. 1967. *Bot. Gaz.* 128:198
61. McKee, H. S., Robertson, R. N., Lee, J. B. 1955. *Aust. J. Biol. Sci.* 8:137
62. Merrick, W. C., Dure, L. S. III 1973. *J. Biol. Chem.* 247:7988
63. Millerd, A., Simon, M., Stern, H. 1971. *Plant Physiol.* 48:419

64. Millerd, A., Whitfeld, P. R. 1973. *Plant Physiol.* 51:1005
65. Murata, T., Minamikawa, T., Akazawa, T., Sugiyama, T. 1964. *Arch. Biochem. Biophys.* 106:371
66. Nagle, W. 1962. *Naturwissenschaften* 49:261
67. Narayanaswami, S., Norstog, K. 1964. *Bot. Rev.* 30:587
68. Norris, R. D., Lea, P. J., Fowden, L. 1973. *J. Exp. Bot.* 24:615
69. Ohmura, T., Howell, R. W. 1962. *Plant Physiol.* 15:341
70. Öpik, H.~1965. *Exp. Cell Res.* 38:517
71. Öpik, H. 1968. *J. Exp. Bot.* 19:64
72. Payne, P. I., Boulter, D. 1969. *Planta* 84:263
73. Poulson, R., Beevers, L. 1973. *Biochim. Biophys. Acta* 308:381
74. Raacke, I. D. 1957. *Biochem. J.* 66:110
75. Rabson, R., Mans, R. J., Novelli, G. D. 1961. *Arch. Biochem. Biophys.* 93:555
76. Raghavan, V. 1966. *Biol. Rev.* 41:1
77. Rappaport, J. 1954. *Bot. Rev.* 20:201
78. Recondo, E., Danhut, M., Leloir, L. F. 1963. *Biochem. Biophys. Res. Commun.* 12:204
79. Rietsema, J., Blondel, B. 1959. *The Genus Datura,* ed. G. S. Avery, S. Satina, J. Rietsema, 196–219. New York: Ronald
80. Rowan, K. S., Turner, D. H. 1957. *Aust. J. Biol. Sci.* 10:414
81. Saio, K. 1964. *Plant Cell Physiol. Tokyo* 5:393
82. Semenenko, G. I. 1957. *Fiziol. Rast.* 4:320
83. Shannon, J. C. 1972. *Plant Physiol.* 49:198
84. Shannon, J. C., Dougherty, C. T. 1972. *Plant Physiol.* 49:203
85. Sharpé, A., Van Parijs, R. 1971. *Arch. Int. Physiol. Biochim.* 79:1042
86. Singh, H., Johri, B. M. 1972. *Seed Biology,* ed. T. T. Kozlowski, 1:21–75. New York: Academic
87. Slabnik, E., Frydman, R. B. 1969. *Biochem. Biophys. Res. Commun.* 38:709
88. Smith, D. L. 1971. *Ann. Bot.* 35:511
89. Ibid 1973. 37:795
90. Stone, E. C., Duffield, J. W. 1950. *J. Forest.* 48:200
91. Sturani, E., Cocucci, S. 1965. *Life Sci.* 4:1937
92. Tsai, C. Y., Salamini, F., Nelson, O. E. 1970. *Plant Physiol.* 46:299
93. Turner, D. H., Turner, J. F. 1957. *Aust. J. Biol. Sci.* 10:414
94. Turner, D. H., Turner, J. F., Lee, J. B. 1957. *Aust. J. Biol. Sci.* 10:407
95. Turner, J. F. 1969. *Aust. J. Biol. Sci.* 22:1145
96. Vecher, A. S., Matoshko, I. V. 1965. *Biokhimiya* 30:808 (Eng. transl.)
97. Walbot, V. 1971. *Develop. Biol.* 26:369
98. Walbot, V. 1972. *Planta (Berlin)* 108:161
99. Walbot, V. 1973. *New Phytol.* 72:479
100. Walbot, V., Capdevila, A., Dure, L. S. III 1974. *Biochem. Biophys. Res. Commun.* 60:103
101. Walbot, V., Clutter, M., Sussex, I. 1972. *Phytomorphology* 22:59
102. Walbot, V., Harris, B., Dure, L. S. III 1974. *Developmental Biology of Reproduction,* ed. C. Markert. New York: Academic
103. Wardlaw, C. W. 1955. *Embryogenesis in Plants.* New York: Wiley
104. Waters, L. C., Dure, L. S. III 1966. *J. Mol. Biol.* 19:1
105. Watson, J. D. 1970. *The Molecular Biology of the Gene.* Menlo Park: Benjamin. 2nd ed.
106. Wheeler, C. T., Boulter, D. 1967. *J. Exp. Bot.* 13:229
107. Wollgiehn, R. 1960. *Flora (Jena)* 148:479

Ann. Rev. Plant Physiol. 1975. 26:279–308

GENETIC ANALYSIS AND
PLANT IMPROVEMENT

·❖7592

T. B. Rice and P. S. Carlson

Department of Crop and Soil Sciences, Michigan State University, East Lansing, Michigan 48824

CONTENTS

INTRODUCTION

Genetic manipulation has become a standard tool in the biological analysis of microorganisms. The generation, recovery, and analysis of defined genetic variants in microbial organisms is as common and powerful a technique in the analysis of biological processes as routine biochemical, physiological, and cytological procedures. In most investigations, mutants which block or modify the biological process of interest have become essential for an adequate analysis of the phenomenon.

Genetic manipulations constitute only a single component of the large number of tools a molecular biologist has at his disposal. The fact that the techniques of so many disciplines have become routine in microbiology has released the molecular biologist from the limitations of being a physiologist, a biochemist, or a geneticist, and has permitted him to focus on important questions rather than on a discipline. The plant sciences are still a long way from such a unity of experimental approach, but it is important for individuals in various disciplines to begin to move in that direction. Certainly, major advances will come when the central techniques of a discipline can be used as routine procedures by those in other fields.

This review will attempt to view higher plant genetics as a technique which can contribute to the plant sciences rather than as an end in itself. It will discuss the characterization, analysis, and physiological importance of several classes of genetic variants; attempts to devise microbial-like selective systems with higher plants; and possibilities for extending the range of mutant recovery schemes. This review does not survey the entire literature, and many important experimental systems are not described. The interested reader will find more complete reviews of plant genetics in other *Annual Reviews* (20, 37, 101, 137, 138, 170, 178), in volumes of *Advances in Genetics* (2, 70, 136, 162, 177), in several journals (33, 77, 132), and in a number of textbooks (83, 98, 179).

Though genetic analysis has most often been used to define the organization of metabolic pathways (189), mutant lesions have been used to dissect a number of other biological pathways as well. Phenomena such as transcription and translation (44, 52, 75, 106, 112), enzymology and protein chemistry (31, 173), membranes and cell walls (43, 95, 125), the cell cycle and sporulation (94, 168, 176), and behavior (19, 103) have all been subject to a mutational analysis. The important observation is that genetic variability can be used to probe cellular and developmental phenomena as well as metabolic pathways.

The use of genetics as an analytical tool with higher plants need not proceed in the same direction as it has over the past 25 years with microbial organisms. This is due to several major factors. Our knowledge of basic genetic processes is now rather complete, and current emphasis is focused on understanding genetic components which are unique to eukaryotes. The analysis of metabolic pathways is no longer of paramount importance, and higher order biological processes now challenge our understanding. Finally, modern biochemical technology has removed the necessity for recovering many classes of mutants. These advances include the availability of specific inhibitors (13), microscale biochemical assays (56, 73, 165), and defined in vitro transcriptional and translational systems (79, 80, 133, 147).

This review will place major emphasis on plant breeding, for this is where the fields of genetics and plant physiology have a broad common ground. The plant breeder uses the tools of genetics and existing or induced genetic variability to maximize physiological processes (e.g. response to a given environment, disease resistance, yield). Hence, genetics is not only an analytical tool in plant physiology, but genetic variability is the substratum for maximizing the physiological processes essential in crop production. The relationship between the genotype and the phenotype must become more clearly defined if we are to understand and modify the rate-limiting physiological processes in crop growth and yield.

MUTANT SELECTIVE SCHEMES

The potential use of genetics as an analytical tool with higher plants is limited by the paucity of biochemically defined mutant types. Examination of a broad spectrum of mutants with specific defects is required for the genetic analysis of any process, from a single step in a biochemical pathway to a complex developmental sequence. The majority of mutants known in higher plants, however, are morphological variants. Unfortunately, the phenotypic alterations which allow recognition and isolation of these mutants are separated from the underlying molecular basis by several levels of organization.

The striking success of the genetic method with microorganisms, particularly *Escherichia coli,* is of course not accidental. This organism was chosen for precisely those characteristics that favor the rapid isolation and characterization of defined mutant types (1). *E. coli* has a short generation time, a sexual cycle, and can be grown in a controlled, chemically defined environment. Small pools of intracellular metabolites insure a rapid response to changes in its environment. Finally, its haploid genome allows immediate expression of recessive mutations without recourse to extensive backcrossing. Thus it is a relatively straightforward procedure to mutagenize 10^6 to 10^9 genetically identical organisms. The few mutant individuals are recovered by exposing the entire population to the appropriate selective conditions. The key to success lies in the design of stringent selective conditions to recover defined mutant types specific to the process being analyzed.

The spectrum of selective methods can be classified under two broad headings. There are direct or positive selective procedures in which conditions favor growth of cells with the desired mutant phenotype, and there are enrichment or negative selective procedures in which selection is against the growth of wild type cells (102). In each instance the objective is to give the mutant organisms a selective advantage. The variety of positive selective screens is enormous, but all rely on conditions that discourage the growth of wild type cells. One approach is to look for resistant mutants through the use of toxic agents such as antibiotic drugs (13) or analogs of metabolic compounds (71). An alternative is to select for mutants that have acquired a new metabolic capacity such as the ability to metabolize a substitute carbon source (31, 91). In each case growth depends on acquisition of a mechanism to cope with the altered environmental conditions. Negative selective screens are based on methods that distinguish between growing nonmutant organisms and the nongrowing (frequently auxotrophic) mutants. The most efficient procedures are those in which growth results in an organizational or metabolic imbalance that ends in lethality. This is the principle of the penicillin enrichment method (116), the starvation enrichment method (9, 118, 150), and enrichment by lethal synthesis (18, 155). Other procedures such as filtration enrichment (65), which are similar in principle, have been developed for use with particular organisms. The general approach on entering a new area of mutant selection is to consider the expected properties of the new mutant type and how one of these can be translated into a selective advantage.

The use of selective conditions to recover defined mutant types in higher plants has lagged well behind its use with other organisms. Principally this can be attributed to the basic design of the plant, which has none of the characteristics of microorganisms that favor genetic analysis. Plants are large organisms; they have long generation times and large nutrient pools; they are composed of cells that represent many distinct differentiated states; and genetically they are diploid, or more often polyploid. It is very difficult to obtain a genetically homogeneous population (40, 67). The characteristics of the whole plant, which permit adaptation to a changeable environment, preclude the application of biochemical selection methods (185). Generally only approaches that depend on a visual assay, such as changes in color or morphology, have been successful (42, 139).

Recent years have seen advances in several areas which hold the promise of extending microbial-like selection systems to higher plants. At present somatic cell cultures, which offer many of the advantages of a microbial system, appear to have the most potential. It is possible to grow large, relatively homogeneous populations of cells in a short time (183). Cells can be grown in chemically defined media (183), heterotrophically or photosynthetically (14, 39), and are more responsive to modifications in that environment than are the whole plants. Several procedures exist for obtaining haploid cell lines (113, 184). Cell cultures can be treated with a broad spectrum of physical and chemical mutagens. Thus most of the selective screens devised for microorganisms should be applicable to cultures of somatic cells. Both positive and negative selective procedures have been effective (32, 35). Advances in this area have been reviewed recently (37). In addition to mutation, techniques for directed modification of the genome are being investigated. These possibilities include direct DNA transfer by transformation (99, 117), viral mediated transfer (34, 50), organelle uptake (34, 152, 153), and cell hybridization (36, 108, 111, 154). Development of these approaches also depends on utilizing precisely defined selective conditions. The last feature of this system is the possibility of regenerating an entire plant from undifferentiated cultured cells (158, 190). The totipotency of somatic cells, which has been expressed in several species, has enormous theoretical and practical importance. Regeneration would permit standard genetic and developmental analyses of mutant types selected in vitro. Regeneration is essential for the application of in vitro procedures to the important problems of agricultural species, but unfortunately this has not been reproducibly accomplished for any cultivars of the major crops.

Two other techniques are potentially useful for isolating defined mutant types. These are germination of seeds under controlled conditions (114), and the newly developed technique for generating haploid plantlets from isolated pollen (141). Each technique has inherent advantages and limitations. The serious limitation, which applies equally to both procedures, is that only positive selective conditions will be applicable. Any mutant that blocks either of these developmental processes will be lost. Seed selection has been successfully applied to oat seeds from mutagenized, selfed plants (M_2 seeds). Mutants with increased resistance to Victoria blight were recovered when M_2 seeds were exposed to the *Helminthosporium victoriae* toxin (114). Seed selection has also been the vehicle for obtaining chlorate resistant mutants in *Arabidopsis* (145) and mutants with increased resistance to herbicides in both wheat and tomatoes (149). Certain difficulties are imposed by the large size of the seed and the endosperm, and by the developmental complexity of the seed and the germination process. But these are substantially offset by the fact that the problem of trying to regenerate whole plants is avoided. Moreover, many physiological processes involve interactions between several tissue types, and seed selection probably provides the best method to select for mutations that affect this level of organization.

Pollen culture is a technique with great potential for use in mutant isolation procedures. The haploid pollen grains can be regarded as individual organisms (46, 141, 142, 144, 184); isolated pollen would be mutagenized, allowed to initiate development, and then exposed to selective conditions that favor mutant plantlets. As with seed selection, problems associated with redifferentiation are eliminated. Diploidization to obtain a fertile plant occurs spontaneously (109) and can be induced with colchicine (186). The potential of this method has yet to be realized. At present its application is limited because of the relatively low frequency of plantlet formation, and it has been successful with only a few species.

Thus there are three systems in higher plants that are potentially useful for recovering defined biochemical mutant types. Many of the selective methods devised for bacterial and fungal systems should be applicable for selecting induced variants or for screening existing gene pools for natural variants. As in microorganisms this approach should provide much useful information about the control of metabolic processes. However, the challenges offered by higher plants are the problems associated with multicellular organization. These include the regulatory and developmental mechanisms that control growth and shape morphology, the physiology of hormonal functions, and physiological aspects of yield. In vitro selective techniques may also be applicable to some of these problems. However, the design of selective conditions requires an understanding of the physiology of these processes on which to base hypotheses concerning the extent and direction of genetic involvement. Progress toward an in vitro approach to these difficult problems awaits further elaboration of the physiological principles involved.

AN APPROACH TO ELUCIDATE THE METABOLIC LESION UNDERLYING MORPHOLOGICAL MUTATIONS

Although much basic work in plant genetics has utilized mutations which alter the morphology of an individual, little is known about the underlying metabolic lesions which are responsible for these aberations, or the ways in which these lesions exert an effect on the gross morphology of the individual (33, 42, 137–139, 162). If these morphological variants are to be utilized as tools in the investigation of physiological processes, then it is essential to identify the underlying metabolic deficiency and reconstruct the sequence of events which generates the final phenotype (86, 88). Mutations affecting morphology are generally recovered because they can be easily identified, and not because they demonstrate alterations in any interesting biochemical or physiological characteristics (57). Hence it is not at all clear what area of metabolism is affected by the lesion.

The past literature on plant genetics contains several instances where mutants have functioned as essential components in physiological and biochemical studies, much as they have in work with microbes (138). Hence the microbial approach can be applied to higher plants, but why has it so rarely been utilized? The most obvious explanations are that plants are not amenable to the standard microbial selective systems (121), and that novel selective schemes for higher plants have not been developed. If a consideration of the unique organizational and physiological features of the plant were combined with the principles of microbial selective systems, new techniques for recovering classes of defined mutant types might be developed. In those instances where the genetic approach has proven applicable, a simple program for the visual recovery of mutants gave types in which the nature of the underlying biochemical lesion was clear. For example, in mutants which have altered chlorophyll or other photosynthetic functions, or lack normally occurring flavonoid or carotenoid pigments, the nature of the defect presents a clear indication of the area of metabolism which is altered (92, 119, 151). Mutants which demonstrate a mineral deficiency syndrome on adequate medium or mimic the effects of growth factor or hormone induced abnormalities also provide a clue as to the underlying biochemical lesion (28, 148, 195). By using these morphological variants, only a limited number of metabolic pathways can be attacked, and in each case the mutant phenotype dictates the approach. These biochemical processes are complex and unprecedented in microbial systems. Unfortunately, examples of the successful utilization of mutations in plant physiology are rare, and even those systems with the most potential have not been extensively analyzed. This is due in large part to the difficulties of biochemical analysis in the available systems (81, 92). Thus only those processes which yield distinct visual phenotypes are open to analysis, but these are often not the processes which physiologists and biochemists would choose for study. Extension of the usefulness of morphological mutants requires a methodology to characterize the underlying biochemical alteration in a range of mutant types.

Biochemical genetics and molecular biology have approached the problem of genetic analysis from a different perspective. Mutant individuals are recovered not because of any ill-defined morphological alteration, but because they satisfy certain biochemical and physiological criteria. In Beadle & Tatum's classic work (10), mutations of *Neurospora* were recovered because of their growth characterictics on different, biochemically defined media. Since a mutant phenotype could be reversed by the addition of known low molecular weight compounds, the altered enzymatic reactions could be readily identified. The individual metabolic intermediates or end products required by the mutant organism clearly pinpointed the underlying metabolic lesion. The biochemical characterization of these mutants was made possible by the criteria imposed during mutant selection. This approach has wide application in microorganisms, where a range of metabolic pathways (e.g. amino acid metabolism, sugar catabolism) have been genetically dissected by recovering auxotrophic variants dependent on supplementation with low molecular weight compounds for growth (60, 85, 164, 199).

Several attempts have been made to correct morphological abnormalities in higher plants by supplementing the growth medium with known low molecular weight compounds (121). These attempts have largely failed to show positive effects. While there is little reason to expect that a strict auxotrophic mutant would survive and be recovered as a morphological variant, many morphological mutants may have a partial auxotrophic deficiency; a response to nutritional supplementation has been observed in a few cases (127, 166, 191). The lack of success with the majority of mutants may be a consequence of the fact that the higher plant system is developmentally and physiologically much more complex than is the microbial system. In any presumed auxotrophic mutation the required nutrient would be taken up by the root, and the morphological abnormality is expressed in the shoot. Uptake and permeability barriers, inefficient translocation, or application after a critical developmental stage might block the effect of a required supplement.

Since scoring for variants which require low molecular weight compounds has not proved to be a generally useful tool in the analysis of morphological mutations, another approach which also focuses on small metabolic intermediates may prove useful. The approach concentrates attention on the species and quantities of low molecular weight intermediates found in mutant and wild type individuals. The major assumption behind this approach is that the state of the small molecules reflects the functioning of the macromolecular components (e.g. enzymes) of the organism. If there is an alteration in macromolecular organization, this change should be reflected in the spectrum of low molecular weight components present (85, 164). If, for example, the activity of an enzyme is altered by a mutational event, then the precursor to that activity might be expected to accumulate or the products of that reaction should disappear. Predictable alterations in low molecular weight intermediates have often been observed in known auxotrophic variants; they have also been found in those morphological variants of *Neurospora* where the primary lesion is known (26, 85, 164). This is essentially the approach used by the physician when analyzing blood or urine, and which has yielded knowledge of primary genetic defects in humans (76). The approach assumes that many genetic perturbations will alter the rates of metabolic pathways in definite ways, and that simple enzymatic defects can have striking consequences on morphogenesis. It is a well-established biochemical fact that certain metabolites acting as specific effector molecules (e.g. hormones, cAMP, inducers) are able to alter the spectrum of biochemical constituents present in a cell, and hence its physiological state (5, 6, 115, 163). Presumably the final morphology of the organism is modified as a consequence. Abnormal concentrations of metabolites may also permit nonspecific interactions that result in a metabolic disturbance which has a changed morphology as its final expression (93).

The approach involves extracting and analyzing as many low molecular weight species as possible by a range of different techniques. Differences between the mutant and wild type are noted and then used in an attempt to predict a possible metabolic alteration to explain the shift in the observed pattern. Once a hypothesis based on these observations can be formulated, the limited number of possible enzymatic lesions can be examined. There are certainly drawbacks to such an approach. First, not all small molecules can be detected by routine techniques, and hence it is impossible to survey the entire range of metabolic pathways. Second, the small molecules generated by a block in one pathway may be metabolized by other biochemical routes. Third, the accumulation of metabolic intermediates may be due to the pleiotropic effects of the mutational event and give obscure clues as to the primary lesion. Fourth, the approach can only give information when low molecular weight intermediates are involved in the blocked reaction. In any case, the approach offers a potential way of obtaining hints about the location of the primary biochemical lesion in an otherwise enigmatic situation.

There have been several attempts to correlate an imbalance in low molecular weight intermediates with a distantly removed phenotype. Perhaps the best known metabolic pathway to benefit from this approach is that of aromatic amino acid metabolism in man (76, 104). Several metabolic disorders occur in this pathway, but only two of them will be considered here. In one metabolic defect, alkaptonuria, individuals homozygous for a single recessive gene are unable to degrade 2,5-dihydroxyphenylacetic acid (homogentisic acid), which is a product of tyrosine metabolism. Homogentisic acid accumulates in the blood and is excreted in the urine where it turns black on exposure to air. During development, dark pigment is produced in the cartilage and other parts of the body. As would be expected, the blood of these individuals lacks the enzyme homogentisic oxidase, causing the accumulation of its substrate. Phenylketonuria, another metabolic defect in this pathway, is characterized by mental retardation and decreased body pigmentation (187). Homozygous individuals lack the enzyme phenylalanine hydroxylase, and hence cannot normally metabolize that amino acid. This leads to excess quantities of phenylalanine in the blood and body fluids. Much of the amino acid is converted by an alternative pathway to phenylpyruvic acid which is then excreted in the urine in large amounts. Here the absence of an enzyme leads to an accumulation of its substrate and related intermediates (187). In both of these cases, the necessary information for biochemical analyses could be deduced from alterations in the levels of low molecular weight compounds and not from the final phenotype.

One other system where low molecular weight intermediates have been essential in defining the metabolic lesions underlying morphological variants is *Neurospora*. In one interesting example the morphological mutation *col-2*, which conditions a colonial phenotype, was shown to accumulate glucose-6-phosphate (27). The mutants contained ten times the quantity of this compound present in wild type sibs. This observation was then utilized to demonstrate that the *col-2* mutation alters the structure of glucose-6-phosphate dehydrogenase, leading to an accumulation of its substrate (27). A number of related morphological defects in *Neurospora* have been shown by these techniques to affect carbohydrate or phospholipid metabolism (26, 169). Hence, phenotypically similar morphological variants may alter related or parallel biochemical processes.

After a consideration of how it might be possible to elucidate the underlying biochemical lesion in mutant individuals, it is important to ask if such an undertaking has merit, especially for an individual whose discipline is plant physiology. The use of mutants in a physiological analysis does provide several important experimental advantages. First, the isolation of mutants defective for a given metabolic function demonstrates that a particular in vivo alteration has an effect on the organism: if it is possible to recover mutants blocked in distinct physiological processes, then these processes exist as important biological entities. Second, the phenotypic alteration in the mutant organism is due only to that mutational event, while chemical alterations in phenotype, induced by antibiotics, analogs, or poisons, are often associated with a multiplicity of effects. Hence it is essential to identify the primary genetic lesion of morphological mutations so that the power of the physiological and genetic approaches can be combined in the analysis of interesting biological processes.

QUANTITATIVE INHERITANCE AND HETEROSIS

Individual Mendelian factors which are characterized by distinct phenotypic effects are rarely observed in natural populations (48). Many characters observed outside of the laboratory display continuous phenotypic variation and are inherited as quantitative traits under the control of multiple genes (66, 74, 107). A trait which demonstrates continuous phenotypic variation is termed a metric trait. A polygene is a gene which exerts a small but quantifiable effect on the phenotype of a character, often of a smaller magnitude than environmental influences, and, in concert with other equivalent genes, exerts genetic control over a quantitative trait. Since it is technically difficult to demonstrate positively the existence and character of genes with small phenotypic effects on complex biological characters, no satisfactory biochemical evidence is available to substantiate the existence of a distinct class of genetic units whose effect is minor and whose actions are cumulative. Nor is multiple factor inheritance a mutually exclusive alternative to monogenic inheritance. There are intermediate modes and patterns of inheritance between these two poles; the individual genetic units of a multiple genetic series, such as the inheritance of grain color in several cereals (140), have often been termed polygenes (128). Although a number of lines of direct and indirect experimental evidence show that the genetic principles which govern the inheritance of polygenes are the same as those which apply to Mendelian genes, polygenes remain statistical entities without a known molecular basis for their function (129).

The classic work of East (53) defined the criteria of polygenic segregation that demonstrate its Mendelian mode of inheritance. The properties of nuclear genes behaving in a Mendelian fashion are two: segregation and linkage. Although neither of these characters can be directly demonstrated for polygenes, indirect observations confirm that they do occur. The extent of segregation and linkage are examined using standard genetic methods, with the exception that one examines the statistical aspects of the phenotype of whole populations rather than their individual members. Segregation and linkage have been observed in many studies of quantitatively inherited traits, and a methodology has been generated for describing the behavior of multiple genetic factors (129).

Although statistical techniques have provided breeders with adequate tools to use polygenes in their breeding programs and to predict responses to artificial selection (146, 178), they have not provided the geneticist with any indications of what the mechanism of action of a polygene might be. The definition of a polygene precludes its analysis by standard Mendelian or molecular techniques (129). Statistical techniques in this sense are descriptive and not analytical. The question still remains, what is the molecular basis of a polygene? Polygenes may represent an unexplored class of genes that are involved in minor or "modifier" roles and that are essentially exchangeable units. Alternatively, a polygene may not be a distinct genetic unit. The phenomenology of polygenic action might be explained as pleiotropic effects of other major genes. Indeed, a polygene may be an artificial unit, an artifact of the methods used in its analysis. Viewed as part of a larger biological process, a number of major genes, encoding for distinct biochemical events, would appear to act as polygenes. Thus what is a Mendelian factor at one level of analysis may become part of a polygenic series at another level. Furthermore, the majority of allelic variation observed in natural populations is not of an all-or-nothing dicotomy, but consists of subtle differences in the protein encoded by a gene. These observed biochemical differences, as exemplified by allelic variants in enzymes (isozymes), most often have no detectable effect on the phenotype of the organism (72, 171). The subtle differences in the products of Mendelian genes may account for the genetic phenomena that have been used as a basis to postulate the existence of polygenes or multiple genetic units.

Hence there are several possible ways to view the underlying genetic organization which results in the phenomena of polygenic inheritance. Polygenes may represent a real and discrete class of genetic elements which fit the criteria of the classic definition. This possibility has been discussed above. Alternatively, polygenic inheritance may be due to major Mendelian genes that have minor effects on complex biological phenomena. An examination of this possibility requires consideration of a characterized genetic system and how its expression may fulfill the phenomenology of polygenic inheritance. Finally, polygenes may be an expression of novel genetic elements within the genome; these elements, their organization, and how they might satisfy the observations of quantitative inheritance will be considered.

Consider the *lac* operon and its products as an example of a well-characterized genetic system (11, 159, 198, 199). An examination of its organization may well provide insights into the kinds of genetic alterations which form the basis of polygenic inheritance, and the reductionist approach used in its analysis may provide an experimental means of understanding polygenic action. The *lac* operon is a functional, polarized grouping of adjacent genes involved in a common metabolic pathway, that of lactose utilization. The DNA sequences involved are separated into two regions and constitute about 0.1% of the *E. coli* chromosome. One region contains the controlling elements of the operon and three structural genes, while the second region contains a regulatory gene. The controlling elements consist of a *promoter locus* (*p*), and an *operator locus* (*o*), which are adjacent to the structural genes, and a *regulator gene* (*i*) located near the *o* and *p* loci but not contiguous with them. The *p* locus is the site at which RNA transcription is initiated. The *o* locus lies between the *p* locus and the structural genes, and is the chromosomal site where the *lac* repressor binds, thus inhibiting transcription. The *i* gene encodes for the *lac* repressor protein which is active as a tetramer and which interacts specifically with the *o* locus. Immediately adjacent to the operator are the three structural genes *z, y,* and *a,* which encode for the three enzymes of the operon. β-Galactosidase, encoded by the *z* gene, is active as a tetramer and is composed of identical subunits. β-Galactoside permease or M protein, encoded by the *y* gene, is active as a monomer and is involved in transporting β-Galactosides into the cell. Thiogalactoside transacetylase is encoded by the *a* gene and is active as a dimer of identical subunits.

E. coli cells grown in the absence of a galactoside contain an average of only one molecule of galactosidase per cell. When grown in the presence of an excess of a suitable inducer to the *lac* operon, approximately 1000 enzyme molecules are present per cell. The kinetics of galactosidase appearance are simple. Upon addition of excess inducer to the growing cells, the enzyme activity increases at a rate proportional to the total increase in protein. The three enzymes of the *lac* operon are coordinately induced.

The expression of the *lac* operon is via the polarized sequential synthesis of polycistronic mRNA from the operon DNA. Synthesis of the specific message commences immediately after induction. The level of operon mRNA in fully induced cells is 60–200 times that in repressed cells, and the message has a half life of from 1 to 3 minutes. There does appear to be an inherent polarity in the *lac* operon since the production of monomeric units encoded by the *z* gene is greater than that of the *y* or *a* genes. This phenomenon may be due to ribosomes departing from the message between the gene boundaries during polycistronic translation, or to preferential degradation of the message at the distal end.

The repressor molecule functions to control the expression of the operon by specifically regulating transcription of the *lac* operon DNA. The repressor acts by binding to the *o* locus and thereby preventing transcription. A range of inducer molecules interact with the repressor, consequently reducing its affinity for the *o* locus and permitting transcription of the structural genes. The *o* locus is contiguous with the structural genes it regulates and exerts a *cis*-dominant effect on their regulation.

The *lac* operon is subject to catabolite repression. That is, enzyme induction is inhibited by the presence of glucose or several derivatives of glucose. This effect can be overcome by high levels of cAMP. The catabolite repression effect is closely associated with the *p* locus. In the presence of glucose or other catabolite repressors the cellular level of cAMP is reduced. This reduction leads to the inability of the cells to initiate *lac* mRNA synthesis. Cyclic AMP acts at the promotor locus in conjunction with a protein believed to be an activator for promoting initiation of RNA synthesis. The mechanism of regulation involved in catabolite repression is in contrast to that effected by the repressor protein, since interaction of the activator protein with the *p* locus is essential for transcription of the operon. The *p* locus is recognized efficiently by RNA polymerase only in the presence of the catabolite activator protein and cAMP.

This discussion has focused on the organization and regulation of the *lac* operon, and has not mentioned the numerous genetic variants which have played a central role in its analysis. After examining some of these genetic variants and the classes of phenotypes which they produce, the possible types of known mutational lesions which may produce polygene-like alterations may be considered. It is important to note that polygenic action has been defined as a phenotypic effect and that no molecular mechanisms are implied (129). Hence any measurable trait that displays a degree of continuous variation may be defined as a metric character. In this particular system, several quantitative traits might be considered. For example, the total amount of enzyme protein for any of the three *lac* enzymes, or their sum, could be quantified in absolute terms (or relative to the total protein) and expressed as a metric trait. Alternatively, enzyme activity could be quantified and defined as the interesting trait. A third possibility would be to measure the total protein in a culture or the amount of growth on lactose (or some other carbon source). The important point is that a metric character, in this case response to growth on lactose, can be approached at one of several levels of organization utilizing a number of different measures. Once the genetic system is known and the molecular components can be examined, events that are clearly the effects of major genes at the primary level of analysis may appear to be polygenic effects at another level.

Genetic variants have been found to occur in all of the structural genes and controlling elements of the *lac* operon (11). Several distinct types of mutants have been observed. The first class of mutants includes typical structural gene mutations: (*a*) β-galactosidase mutations ($z^+ \longrightarrow z^-$) expressed as an alteration in the enzymatic activity of the galactosidase protein; (*b*) permease mutations ($y^+ \longrightarrow y^-$) expressed as an alteration in the capacity to concentrate lactose; and (*c*) transacetylase mutations ($a^+ \longrightarrow a^-$) expressed as an alteration in the activity of the thiogalactoside transacetylase protein. In all of these mutations, the protein encoded by the respective structural genes is produced, often at wild type levels; however, the enzymatic activity of the protein has been altered, causing the phenotypic alteration. Mutational changes in genetic units need not be expressed in the presence or absence dichotomy expected of a rigorous molecular definition. Many genetic lesions are certainly less severe and more intermediate in effect. These partially inactivating mutations, termed bradytrophs, are commonly recovered in biochemical genetic studies (64). Formation of an active or partially active tetramer from several inactive polypeptide subunits also occurs. Such intragenic complementation has been observed between different alleles at the *z* gene in merodiploids. The second class of mutations contains those involving the controlling elements of the operon. Several different types of mutations have been observed in the *i* gene. The mutation $i^+ \longrightarrow i^-$ leads to the constitutive synthesis of all of the enzymes encoded in the *lac* operon regardless of the presence or absence of an inducer molecule. As expected, diploids heterozygous at the *i* gene (i^+/i^-) demonstrate normal regulation. Another type of mutational event at the *i* gene is $i^+ \longrightarrow i^s$ (s is for superrepressed). The i^s mutants are dominant to i^+ in diploids (i^+/i^s) and do not permit induction of the *lac* enzymes at normal inducer concentrations. Only when the inducer concentration is 1000-fold its normal level is significant enzyme synthesis observed. Finally, there is the mutation $i^+ \longrightarrow i^q$ (q is for quantity) where the total amount of *i* protein (the *lac* repressor) is altered. Mutations have also been recovered in the *lac* operator. It is important to note that all *o* mutations act in a *cis*-dominant fashion, since they produce no diffusible product. The primary observed mutational lesion is $o^+ \longrightarrow o^c$ (c is for constitutive) which permits enzyme synthesis in the absence of the inducer. The constitutive nature of these mutations in only partial, and full levels of enzyme synthesis are only observed if the inducer is present. Mutations in the *p* locus also display *cis*-dominance, and they prevent or greatly reduce enzyme synthesis. This is presumably due to a decrease in affinity of the promoter region for RNA polymerase. Hence there is sufficient genetic variation to generate a continuously varying phenotype. There are other types of mutational lesions which affect the *lac* operon (e.g. polarity mutations, mutations in catabolite repression), but the mutational lesions which have been discussed proved an adequate basis for investigating how polygenic effects might appear.

Each of these known mutant types may, at some level of analysis, demonstrate a "polygenic effect," that is, a small but quantifiable influence on a metric trait. When the analysis of a mutation occurs close to that gene's primary product (i.e. individual RNA or protein species), then it often demonstrates a distinct phenotype. However, when the analysis of a mutation occurs not at the level of the primary product, but several levels of biological organization removed from that primary product, the phenotype of that lesion may become less distinct. If growth on limiting lactose is the important trait, then a genetic alteration in any component of the operon may exert a polygenic effect. This could involve controlling the level of enzyme production (*i* gene, *p* or *o* locus), or qualitatively altering the activity of an enzyme (*z, y,* or *a* gene) or any combination of these mechanisms. If the activity of the *lac* enzymes per cell is of interest, then an alteration in any genetic component might be involved, either by changing the levels of the proteins (*i* gene, *p* or *o* locus) or by altering the activity of the proteins present (*z, y,* or *a* gene) or both. If the total amount of *lac* enzyme protein is of interest, then alleles in the controlling elements may demonstrate polygenic effects. This would be especially true of constitutive mutations. Mutations that affect the efficiency of complete transcription of the polycistronic message could also be involved.

The important point here is that for this relatively simple, well-characterized system combinations of standard mutational types in defined genes can result in a range of activity levels when the pathway is considered as a whole. These variants fulfill the criteria of polygenic inheritance. It is easy to generalize from this analogy and to visualize how more complex developmental and physiological metric characters, involving tens or hundreds of loci, could be controlled by similar mechanisms. However, this analogy does not demonstrate that these variants are acting as classical polygenes. The reverse approach is actually required. That is, it will be necessary to find a case of polygenic inheritance which fits all of the criteria of the definition, and then to analyze its genetic organization utilizing the techniques of molecular genetics.

The eukaryotic genome contains novel genetic elements, families of genetic loci, which have many of the characteristics of polygenes (29, 58, 100, 130). Although none of these families have been shown to be involved in specifying quantitative traits, a consideration of their genetic organization may help clarify our concepts about polygenes and reveal how their action may account for polygenic inheritance.

A multigenic family is generally a series of genetic units which share extensive nucleotide homology and appear to have similar or overlapping phenotypic functions. They have been found to encode for several central genetic and biochemical functions (29, 58, 100, 130). Their genetic and metabolic organization closely parallel some components expected of classically defined polygenes, and their analysis may provide a clue as to the nature of polygenic action.

Repetitious DNA sequences have been demonstrated in eukaryotes by several different techniques. DNA-DNA hybridization has shown that a significant proportion of the DNA in higher organisms is organized into families of genes ranging from 50 to 10^6 members, all of which share a degree of sequence homology (24, 25, 130). Three classes of DNA sequences have been defined in the eukaryotic genome: highly repetitive sequences ($\geq 10^5$ copies), middle repetitive sequences (10^2–10^4 copies), and unique sequences (one or only a few copies). These classes of DNA sequences occupy different proportions of the genome, but generally highly repetitive sequences constitute about 5–20%, middle repetitive sequences 10–40%, and unique sequences 50–80% of the total DNA. DNA-DNA hybridization studies combined with density gradient centrifugation have characterized a catagory of these sequences which includes the repeating satellite DNA sequences. The function of these sequences is unknown, but they are presumably involved in the structure and functioning of the eukaryote chromosome. These DNA sequences are not transcribed, so their effects on the genome must be a function of their base sequences. (There is no evidence that polygenes are transcribed or translated.) One of the best characterized satellite DNA families is that of the mouse (68). When DNA isolated from *Mus musculus* is sheared into fragments about 400 to 500 nucleotides in length and analyzed in a CsCl gradient, there is a fraction of the DNA which displays an unusually high AT content. The high AT satellite comprises about 10% of the DNA of the mouse, and contains about 10^6 copies of a nucleotide sequence some 300 base pairs in length. These sequences are very similar to each other (93–95% sequence homology) and are not transcribed into RNA. Shorter repeating units of six base pairs have been detected within the 300 base pair structures. Clusters of these satellite genes have been mapped at the centromeric regions of almost all of the mouse chromosomes. Despite the sequence homology in the mouse, these structures bear no resemblance in nucleotide sequence to the satellite genes of other closely related species. In this instance, there are a large number of identical loci scattered throughout the genome. If the function of these loci was of a regulatory nature, then each locus could effect a small alteration in some common quantitative trait.

RNA-DNA hybridization provides another tool which demonstrates the existence of repetitious DNA sequences. The genes encoding for the various classes of histones, the ribosomal RNA, and the transfer RNA species are all present in multiple copies (29, 58, 175). Of the multigenic families the ribosomal genes of *Xenopus laevis* are the best described. They have a common nucleotide sequence and a defined function. Electron micrographs and RNA-DNA hybridization studies have demonstrated that the ribosomal genes have a unit organization of an 18s, 28s, and spacer regions. This unit is tandemly repeated 450 times in a haploid genome at a single genetic locus, the nucleolar organizer (16). The 18s and 28s genes are transcribed into RNA while the major spacer is not (29). All copies of the ribosomal gene appear identical in nucleotide sequence (29). A similar type of genetic organization is found among the approximately 400 histone genes, which show a similar organization in a number of diverse species. The amino acid sequence of a given histone has been highly conserved throughout evolution (58). Also, there is little or no heterogeneity among the DNA sequences of the histone family (194). In this case the identical genetic units are active through the products of their transcription or translation.

As a final example, consider a situation in which similar genetic loci are involved in a common biochemical function. The summation of the information encoded in these loci is a much wider range of functional capacity. Protein sequence data has revealed such a multigenic family involved with storing larger amounts of related genetic information (100). This results from having nonidentical family members with partially overlapping specificity. For example, most antibody families encode for multiple antibody specificities. The antibody molecule is generally composed of identical subunits, each consisting of a light and heavy polypeptide chain, all covalently joined by disulfide bonds (55). In mammals these chains are encoded by three unlinked families of genes, two light chain families, d and k, and the heavy chain family. All antibody chains can be divided into an amino terminal portion, the variable region, and a carboxy terminal portion, the constant region. The variable regions are different from one antibody to the next, while the constant regions are relatively constant for a given class of antibody. The variable regions of each antibody family are encoded by a set of multiple genes, while the constant region is encoded by one or a small number of genes. The number of variable genes ranges from 50 to several thousands. The genes encoding for the variable region are clustered in one chromosomal region and vary markedly in their numbers among different mammals.

Although the categories of multiple gene families discussed above do not completely satisfy the definition of a polygenic system, their characteristics do confirm the existence within the eukaryote genome of the type of genetic organization postulated to underlie polygenic inheritance. Whether polygenic inheritance is actually due to the action of such families of genes remains uncertain.

Analysis of the genetic organization of quantitative inheritance certainly deserves a fresh experimental approach. The proper experimental system will not, it would appear, be found among the many examples of polygenic control of characters such as yield, stature, etc, where the final phenotype is the product of an infinite number of interacting biological processes. It seems better to take the reductionist approach and choose a more simple analytic system. If a well-defined physiological process or even the quantity of an individual protein could be used as a quantitative trait, then the possible number of genetic inputs would be reduced. By bringing the quantitative trait closer to a primary gene product, and removing the possibility of higher order biological interactions, known molecular mechanisms can be analyzed, and polygenic functions might be more completely defined. It appears that this strategy affords perhaps the only route to more fully understanding the components of quantitative inheritance.

It is quite possible to view this discussion and its juxtaposition to polygenic inheritance as a contrived case. However, the definition of a polygene, when viewed from a contemporary genetic perspective, permits few other experimentally verifiable approaches towards understanding the biochemical organization of a polygene. Known molecular mechanisms can account for the phenomenon of polygenic action, and it is not necessary to invoke the action of any new molecular mechanisms.

The ideal experimental system in this case would first satisfy the definition of polygenic inheritance, then would be amenable to the reductionist approach of molecular genetics, and finally would allow the use of biochemical and genetic technology in its analysis.

Heterosis or hybrid vigor (and the converse, inbreeding depression) is an example of another classically defined genetic phenomenon which remains uncharacterized in molecular terms and is described almost entirely by statistical methods (54, 82). The analysis of heterosis suffers from the lack of a defined experimental system. Its analysis, like that of polygenic inheritance, is at the level of final phenotype which is far removed from the primary gene product. Although biochemical mechanisms exist which might account for the phenomenon of heterosis (63), no molecular analysis has been undertaken in a system where heterosis has been first observed. Like polygenic inheritance, heterosis can be observed at several levels of biological organization.

The molecular mechanisms which may provide the basis for heterotic effects (or overdominance if only a single gene is involved) can be examined by returning to a consideration of the *lac* operon. The expression of heterosis requires heterozygosity. Hence genetic phenomena that influence the final level of expression of the *lac* operon in merodiploid *E. coli* should be considered. If total growth of the culture on limiting lactose is the character of interest, then any interaction which allows better utilization of that sugar could be considered a heterotic effect. Since heterosis (or inbreeding depression) occurs following a genetic cross, any genetic or molecular interaction except a *cis*-dominant effect could provide the mechanism. Only the interactions between the *p* and *o* loci and the structural genes are *cis*-dominant. Intergenic complementation (epistasis) could result in an increased rate of lactose utilization. Intragenic complementation could result in the formation of a multimeric protein with altered activity. Such an alteration might result in increased β-galactosidase activity or in decreased operator binding capacity of the repressor. In a regulatory system characterized by positive control of gene activity, an interaction of the activator molecule with the *trans*-receptor locus could result in greater total activity. Such an interaction could occur in the *lac* system during release of catabolite repression. Thus the flow of metabolites along the catabolic pathway might also provide the basis of a heterotic effect. These three possibilities —regulatory interactions, intragenic complementation, and epistasis—all may contribute to the appearance of heterosis. The important point is that there are known molecular mechanisms that can account for the phenomenon of heterosis. There is no evidence that any of these mechanisms are in fact involved in heterosis. Heterosis, like quantitative inheritance, is a phenomenon which is observed primarily in crosses between individuals in natural populations. Since naturally occuring alleles are often cryptic in their phenotypic effects, the analysis of a heterotic interaction is much more difficult and subtle than characterizing the interactions of more severe mutational lesions. What is needed is an experimental system with a clearly defined character that satisfies both the definition of a heterotic effect and the requirements for effective manipulation and analysis of the discrete genetic and biochemical components involved.

In the analysis of both quantitative inheritance and heterosis it is essential that once the definition of the phenomena has been satisfied in a particular experimental system, the causal biological relationships be elucidated using the reductionist approach of molecular genetics. In both instances it is clear that the final phenotypes, so often used as measures of these phenomena, are not an adequate expression for their analysis. It is important to identify, isolate, and characterize the actual rate-limiting physiological processes which are responsible for the change in phenotype. An understanding of these processes should provide the plant breeder with additional tools for the rational design and efficient construction of new strains.

DEVELOPMENTAL GENETICS AND CROP YIELD

The fruit or grain of a crop plant is a differentiated organ, and as such is the endpoint of an integrated developmental process. Because of its agricultural and economic importance, analysis of the fruit has centered on characterizing yield. Yield is a quantitative expression of the production of an end product; the developmental mechanisms involved in maturation are not usually considered or evaluated (123, 196, 197). The developmental geneticist views the process of fruit production and yield from a different perspective. This approach involves a description and analysis of the processes and events whereby a fertilized egg gives rise to a complex of several differentiated cell types. The focus is on dissecting the sequence and timing of the developmental events involved in generating the final phenotype. The developmental geneticist views yield as the summation of all of these separate but integrated biological processes.

When an agronomist or plant breeder discusses yield, he is referring to the property of a community. The variability expressed both within and among populations of individual organisms constitutes higher order fluctuations which the developmental biologist is not equipped to analyze. Hence this discussion will deal with yield as a developmental phenomenon of the individual plant, and with its possible genetic manipulation. An attempt will be made to gain insight into how yield can be subjected to a developmental analysis by analogy with other more characterized systems.

Yield is a character potentially limited by the fluctuations of numerous physical, chemical, and biological factors. Besides such environmental parameters as climate, soil structure, nutrient levels, light quality and intensity which may be rate limiting, and the influence of such managerial practices as the density and spacing of planting, irrigation, fertilization, and pest control (47, 123, 196, 197), the developmental and physiological characteristics of the whole plant play a major role in determining the actual and theoretical yield capacity. General characters important in the growth and metabolism of the whole plant, such as photosynthetic efficiency or nutrient utilization, are factors in determining yield (47, 123, 181, 196). The morphology of the entire plant is also an important component determining yield, as is demonstrated by the effect of characters like dwarf growth habit or leaf shape and attitude (4, 49, 122, 134). Basic physiological decisions such as partitioning of the plant's available resources between reproductive and vegetative structures certainly affect yield (192, 197). However, yield is also dependent on characters which are specific to the seed, and constitute its unique genetically determined developmental and physiological composition. These characters, such as the final size of the seed, the seed's ability to mobilize nutrients, or the spectrum of macromolecules held in storage, have not been well characterized. In most breeding programs seed development and its regulation are taken as given. Since a significant fraction of the carbohydrates, amino acids, and other metabolites required by the seed are derived from vegetative tissues (12, 96, 197), more robust plants may produce a greater quantity of fruit and therefore demonstrate higher yields. This discussion will take the opposite approach, and will attempt to view the seed as the endpoint of a developmental sequence which can be modified genetically for the purpose of increasing yield. That is, in a situation where all other parameters are optimal, what are the rate-limiting factors affecting yield inherent in the seed itself? Which biochemical or developmental processes might be modified genetically to release this limitation? What kind of developmental genetic system is represented by the seed, and what analogies exist which demonstrate the kinds of genetic perturbations which are permissible? Some aspects of plant development and their relationship to yield have been described (96), but few attempts have been made to explore the possibility that modifications of seed development could increase the yield capacity of the major crop species. Because of their agricultural importance, much of the discussion and many examples refer to the cereals.

Since the subject is the developmental genetics of yield, and since the mature seed is the developmental endpoint of interest, it is appropriate to consider the mechanisms involved in seed development. Although seed development of the Gramineae has been well studied (45), it has been difficult to characterize the developmental sequence in terms of other well-known angiosperms (8). In the absence of a uniform development pattern for the cereals, the principal processes involved will be outlined in very general terms.

Subsequently, other comparable developmental systems will be examined. This discussion will rely heavily on *Drosophila* for the following reasons: the level of biological complexity is similar; there are several examples of tissue interactions analogous to those observed in the developing seed; and the genetic analysis of *Drosophila* development has been very productive. The conceptual framework and developmental insights that have evolved in this system should be instructive to a discussion of possible genetic modifications of seed development.

The mature seed is a mosaic structure composed of several differentiated tissue types and two genetically distinct clones of cells (97, 124). At fertilization the female gametophyte contains seven cells derived from three divisions of the haploid macrospore: the egg cell and two synergids lie at the micropylar end while the three antipodal cells lie at the opposite pole. All of these cells are haploid. A large central cell which is the progenitor of the endosperm contains the remaining two haploid nuclei. Four layers of diploid maternal tissue—the nucellus, two integuments, and the ovule wall—surround the gametophyte and complete the structure of the ovule. All of these tissues are involved in the formation of the seed. In the Gramineae as in other angiosperms a double fertilization occurs (124). Two sperm nuclei enter the embryo sac; one of the nuclei fuses with the female pronucleus in the egg cell to form the diploid zygote, and the other enters the central cell, fuses with the two polar nuclei, and forms the primary triploid endosperm nucleus. The snyergids and the antipodal cells remain haploid.

Pollination and fertilization stimulate seed development (124, 143). Initial activity is most intense in the maternal tissues, the diploid integuments, nucellus and ovary wall, and the haploid antipodal cells. The nucellus surrounds the embryo sac and provides a pathway for the movement of nutrients to the embryo and endosperm. Enlargment of the nucellus, which varies in extent with different species, begins soon after pollination and appears to involve both cell division and cell enlargement (124, 143). The integumentary layers generally undergo extensive enlargement early in development, spurred by growth of the nucellus and endosperm. As the ovule matures into a seed, these layers differentiate to form the hard, protective coverings of the seed coat. However, in some species such as maize the integuments have only a minor role (157), and instead the ovary wall differentiates the pericarp to serve the same function. The antipodal cells begin to divide after pollination, and 24 or more are present by fertilization (3, 157). The prominent development of these cells is characteristic of the Gramineae. Activity is also seen in the endosperm shortly after fertilization. The primary endosperm nucleus undergoes synchronous divisions within a syncytium to approximately the 128 to 500 N stage (21, 22, 157), depending on the species. The triploid nuclei then move to the periphery of the central cell where individual endosperm cells are formed. The antipodal cells become quite large and polyploid as the endosperm develops, but they soon become disorganized, and by the time the endosperm has become cellular they have usually disappeared (15, 157). Early growth of the endosperm appears to be at the expense of the nucellus since the endosperm digests away the nucellus as it expands (157). Midway through development, cells in the center of the endosperm cease division and grow by cell enlargement; their ploidy level increases by endomitosis (51). Cells on the periphery continue to divide, and late in development they differentiate to form the aleurone layer (124). The endosperm acts as a buffer between the developing embryo and the maternal tissue, supplying nutrients for the embryo. In the Gramineae the endosperm also serves as a storage tissue and is involved in the synthesis of a unique spectrum of protein, carbohydrate, and lipid storage molecules.

The embryo is the final component of the seed to begin development. The initial cleavage divisions are slow, so that development of the embryo lags well behind the endosperm. Embryogenesis conforms to the Asterad type (45, 124), since each of the first generation blastomeres contributes to development of the embryo and the second cleavage division is longitudinal. Following the early morphogenetic events, the embryo grows rapidly to maturity, drawing nutrients from the surrounding endosperm tissue.

This description of seed development highlights the fact that the seed is primarily composed of two very distinct developmental systems, the embryo and the endosperm, which arise from separate fertilization events and which are derived from different clones of cells. These two components follow independent and autonomous developmental sequences, and interaction between the two systems is not a prerequisite for their normal formation. The embryo, for example, can be excised at an early stage and cultured on a simple defined medium (156, 182). In cases where the development of the endosperm is genetically blocked (e.g. interspecific hybrids), the embryo can often be cultured in vitro and a mature plant can be recovered (23). In fact, a process analogous to embryogenesis can proceed in vitro on chemically defined media from pollen (87) or somatic cells (158) in the complete absence of any endosperm function. Though an endosperm is most often required for normal embryo development, it does not appear to play any unique function or role in the development of the embryo beyond providing a nutritional milieu. Hence modification of the endosperm is not limited by constraints imposed by any special requirements of the embryo. The endosperm is also able to develop and mature in the absence of an embryo (69). Seeds containing only the endosperm or only the embryo have been found in capsules that have developed from incompatible interspecific *Datura* crosses (7). Since no correlation could be established between the size of the seed and the presence of an embryo in some species crosses, it appears that endosperm maturation required no specific information from the embryo. Thus each tissue proceeds as if its developmental fate were determined at an early state. No inductive interactions or other significant exchanges of developmental information are apparent. Besides being the product of an independent fertilization event and having a different ploidy level, the endosperm displays a unique spectrum of gene products which are not expressed in the embryo or the whole plant (105, 167). The literature is full of examples of tissue-specific genetic characters expressed only in the triploid tissue of the seed (138, 139). Examples include defined enzymatic proteins, genes determining pigment production, morphological features, and gross protein, carbohydrate, and lipid molecules.

The triploid tissue of the seed is an excellent material for the developmental geneticist. It is a complex assemblage of tissues whose developmental sequence is distinct, temporally defined, and genetically controlled by a number of loci which are tissue specific. As a single organ its developmental process is more simply manipulated experimentally and intellectually than that of the whole plant. The triploid tissues are organized into a very plastic organ. A wide spectrum of biochemical and morphological alterations have no effect on its development or its function of supplying nutrients for the growing embryo (139). It is, in the final analysis, a largely expendable organ under laboratory conditions since the embryo can be rescued using in vitro culture methods. In several instances experimental systems displaying similar attributes have permitted the recovery and characterization of a wide range of genetic variants. The quality of being an expendable component permits the isolation and analysis of a number of absolute and conditional mutant types, since the lack of functioning does result in an immediate lethal event and since many genetic loci are expressed only in that component. By these criteria the chloroplast in *Chlamydomonas reinhardi* can be considered an expendable organelle, since the organism can live at the expense of an exogenous carbon source in the absence of chloroplast function. The experimental utilization of this fact has permitted an extensive molecular genetic analysis of its organization and functioning (120). In *Drosophila* a similar relationship exists in the larval stage between the larval tissues and the imaginal discs. This relationship, which is described in detail below, has been exploited to examine the genetic control of disc development.

The developmental genetics of *Drosophila* is an active field of research utilizing what appears in many ways to be a parallel genetic system. Since both the genetics and development of *Drosophila melanogaster* have been extensively characterized, it is proving to be a useful organism in which to investigate several fundamental developmental problems. Principally these are the role of the egg cytoplasm in programming embryonic and subsequent development and the process of cell determination. Numerous types of mutants have been induced and recovered in *Drosophila*. These mutants may be utilized by analogy and extrapolation to suggest the type of variants which can be expected to occur in the seed of higher plants. A juxtaposition of the two developmental systems may provide insights into seed biology and some hints as to how a developmental geneticist could approach the character of yield.

The *Drosophila* life cycle consists of two distinct stages: the larval and the adult. During the intervening stages of pupation and metamorphosis, most of the larval structures are histolyzed and a new form of the organism, the adult fly or imago, is constructed from clusters of undifferentiated cells in the larva, called imaginal cells (17). The imaginal cells are set aside early in embryogenesis, perhaps as early as the blastoderm stage (38). The imaginal discs contain the cells which will form the epidermal structures of the adult. Each disc is characterized by its position in the larva and by the part of the adult which it forms (78). The discs attain their final size and characteristic shape during the extended period of larval growth. Mature discs consist of undifferentiated, essentially embryonic cells whose fate is rigorously determined, awaiting the hormonal signal to begin differentiation (78, 89). The early onset of determination and the temporal separation between the determination of imaginal cells and their subsequent differentiation make *Drosophila* an appropriate organism in which to study the process of cellular determination and the possible early involvement of the egg cytoplasm.

Developmentally the relationship between the *Drosophila* larva and the imaginal discs is very similar to that between the endosperm and the embryo in the seed. Although the imaginal discs originate from a relatively large number of cells, the divergence of larval and imaginal cells occurs early in embryogenesis and the subsequent development of each clone follows a distinct and independent pathway. Growth of the larval tissues occurs by an increase in cell size and effective ploidy level through polytenization (17), as occurs in the endosperm of many plant species. In both the larva and the endosperm, the final number of cells is fixed at an early stage. The mode of differentiation increases the synthetic capacity in those cells for the output of specialized products required later in development; these cells have no function beyond this life stage. The imaginal discs, however, fulfill the same role as the plant embryo, that is, they are the embryonic form of the following life stage. Cells of both the plant embryo and the imaginal discs remain rather quiescent until development of the host tissue reaches an advanced stage. Both remain diploid and divide normally during the process of determination and elaboration of the progenitors for different adult structures. As early as the imaginal discs can be recognized, they can be surgically removed from the larva and grown to maturity by culturing them in vivo in the abdomen of an adult female (90). Mature discs differentiate normally when they are removed from their normal location in the larva (or recovered from the abdomen of an adult female) and injected into the abdominal cavity of a host larva prior to metamorphosis (61). Differentiation of mature discs has also been observed under controlled conditions in vitro (126). Most recently it has been demonstrated that normal imaginal structures can be recovered from one or a few cells that have been removed from their normal environment at the earliest cellular stage of embryogenesis (the blastoderm stage in *Drosophila*) and allowed to complete development under in vivo culture conditions (38). Each of these results shows that the imaginal cells have the capacity for autonomous development; there is no evidence for developmental information coming from the larvae. Thus, like the interaction between the plant embryo and endosperm, the demands of the discs on the larva are principally nutritional.

In *Drosophila,* the analysis of developmental mutants has contributed significantly to our understanding of the relationship between the larval and imaginal cell types. Each mature imaginal disc has a characteristic size and morphology and contains a wide variety of determined cell types (78). Several classes of mutants are known which block or modify the development of one or more discs without altering the larval tissues (172, 180). The most severe phenotype is that of the "discless" mutants in which formation of all of the discs is blocked while the structure and function of the larval tissues is normal (172). A second large group of mutants includes those in which the imaginal discs are present but one or more are abnormal. The variety of abnormalities includes discs that are too small or too large, discs with a different shape or texture, and discs that are unable to differentiate. Another class of mutants, the homeotic mutants, affect the determination process directly, since a portion of one disc is altered and gives rise to a structure normally formed by a different imaginal disc (78). Only a few mutants are known which specifically affect the other half of this system, the larval tissues. However, many are sure to be found among the late embryonic and larval lethal mutants now being isolated. All of these mutants, with tissue-specific and stage-specific defects, help to resolve this complex developmental process into its individual steps, just as biochemical lesions helped in the reconstruction of biochemical pathways. These mutants emphasize the fact that critical decisions related to the developmental capacity of a tissue, as well as its final size and morphology, are under precise genetic control.

However, many of the morphological variants found in *D. melanogaster* are not developmental mutants of the type discussed above, but result instead from the secondary effects of mutants which alter general metabolic processes. Thus it is appropriate to consider the features expected of developmental mutants and examine procedures that have been successfully utilized to recover them. The autonomous development of a tissue requires the precise, sequential activation of many genetic functions, including general metabolic processes as well as tissue-specific developmental information. These two genetic components are integrated into the final biological product. Mutations that modify the regulatory processes defining the unique characteristics of a tissue ought to be found among the class of mutants in which the primary defect can be localized to a given stage and tissue. Since the general functions are required by all tissues, a lesion in the structural locus encoding for that information will alter the final phenotype of a number of tissue types. For example, alleles of the *rudimentary* locus, which have a defect in pyrimidine metabolism, can produce an abnormal wing phenotype and reduce female fertility (84). Alleles of the *rosy* locus, which alter a subunit of xanthine dehydrogenase, produce both color and morphological abnormalities in the larval Malpighian tubules and in the adult eye (41). In both cases a metabolic block leads to alterations in several tissues at different stages. Although these mutations are of genetic and biochemical interest, they do not provide new information relevant to the development of the tissues they affect. These mutations provide no insight into the regulatory mechanisms essential to the development of a given tissue. To a developmental geneticist, mutants defective in general metabolic processes and their associated pleiotropic effects constitute developmental noise. They modify developmental events but are not directly involved in defining the unique characteristics of individual tissue types. Their effects can only be secondary and indirect.

There are instances in *Drosophila,* however, of mutants with unique tissue-specific, developmental effects. These include mutants that specifically alter the imaginal discs, such as the mutants described above and certain maternal-effect mutants, like *grandchildless* in *D. subobscura.* Adult females homozygous for the *grandchildless* mutant produce defective eggs that develop normally except that the pole cells, which include the primordial germ cells, do not form (62). Hence the defective eggs produce morphologically normal adult progeny that are sterile (and thus childless) because they possess no germ cells. Thus the mutation blocks the development of a very specific cell type. Recent attempts to isolate additional mutant types with tissue-specific defects emphasize the need for stringent selective conditions which account for the role of a given tissue in the biology of the organism. For example, several criteria were applied in sequence to select for imaginal disc-specific mutants in *D. melanogaster* (172, 180). To recover these mutants, several investigators concentrated first on the analysis of lethal mutants (since few viable mutants with disc-specific abnormalities are known) that die late in development, after the larval stage and before emergence of the adult. Normal functioning of the imaginal discs is essential to the organism during this interval. Previously it had been shown that the late lethal mutant *lethal giant larvae* produces fully grown larvae having severely defective discs (88), suggesting that the discs might not be essential structures during the larval stage. Thus the first requirement for a prospective disc-specific mutant was that development be arrested during the stage when the imaginal discs become essential for formation of the adult fly, and that prior development, when the discs are apparently dispensable structures, produce fully grown larvae. Only the late lethal mutants were subsequently examined for evidence of abnormal disc development. Mutant larvae were dissected and examined microscopically for missing discs or discs with an abnormal morphology. If present, the discs were tested for their developmental capacities using the in vivo or an in vitro assay system to induce differentiation. The late lethal criterion significantly reduced the number of mutants to be analyzed by the second more difficult set of procedures, thus improving the efficiency of the overall selection process. The approach has been remarkably successful. Large numbers of mutants with defective imaginal disc development have been isolated, including many with phenotypes not previously observed (172). This class of mutants appears to provide the best means of dissecting imaginal disc development, the process of cell determination, into its component parts.

The same rigorous approach has been utilized to recover additional maternal-effect mutants for examining the extent to which the egg cytoplasm is involved in the initial determination of different cell types (160, 161). In *Drosophila* the meiotic divisions occur at the end of oogenesis, so the oocyte differentiation is controlled by the diploid, maternal genome. The fact that divergence of the larval and imaginal cell lines may begin as early as the blastoderm stage, prior to detectable zygotic gene activity, suggests that some component of the egg cytoplasm, presumably part of the egg cortex, is involved in the determination process (160). Any dependence of this process on localized ooplasmic substances should be reflected in a class of maternal-effect mutants with tissue-specific defects, such as *grandchildless*. For example, if the distinction between larval and imaginal cells has been programmed into the egg, there should be a class of maternal-effect late larval, lethal mutants. The problem has important theoretical and practical applications to seed development where the divergence of clonal seed components clearly occurs within the embryo sac prior to fertilization, and thus under the direction of the maternal genome, and portions of the seed (seed coats) are actually derived from maternal tissues. The protocol to search for this mutant type follows. The first step was to isolate female-sterile (*fs*) mutants, that is, mutants where homozygous mutant females are unable to produce viable adult progeny. Many *fs* mutants were isolated using standard genetic manipulations. Since reduced female fertility is a phenotype that is part of the pleiotropic syndrome of many morphological mutants, the *fs* mutants were carefully examined for morphological variants, reduced viability during development of homozygotes, and reduced male fertility. Mutants with any abnormality other than female sterility were excluded from further consideration. The determinative events occur at or following the blastema stage. Thus the final selective criterion applied was that the early cleavage events should be normal; this eliminates mutants where the eggs are not fertilized as well as those that are unable to initiate or maintain cleavage divisions. The approach was quite successful. Utilization of these sequential selective steps allowed the recovery of six interesting mutants from nearly 4300 homozygous mutagenized stocks examined (160). Each of the mutants has a remarkably uniform phenotype. Although none of the new maternal-effect mutants is a later-larval lethal, each has a novel phenotype that is certain to improve our understanding of the role of egg components in zygotic development.

Our understanding of seed biology would benefit greatly from the same type of developmental genetic analysis. Several possible mutant types can be predicted based on the developmental and genetic analogies between the seed and the *Drosophila* larva. There should be mutants that specifically block formation of either the embryo or the endosperm; mutants that alter the size or morphology of the embryo or the endosperm; mutants that alter the spectrum of storage molecules in the endosperm; and mutants that block or alter formation of the seed coats. In short, there should be mutants that block or modify each process required to form the seed. Since each of the suggested variants affects the morphology or the composition of the seed, screening procedures to isolate the mutants should be straightforward. It bears repeating, however, that care should be taken to select mutants with tissue-specific defects rather than those with pleiotropic phenotypes. The distinct genetic differences between some components of the seed should facilitate the isolation of different mutant classes: the embryo has a diploid hybrid genome; the endosperm is a triploid hybrid; and the nucellus, integuments, and ovary wall have a diploid maternal genotype. The analysis of developmental mutants will be essential to understanding the different processes involved in seed development, how the separate processes are regulated, which processes are limiting in terms of yield, and to what extent the genetic controls can be modified. This background is essential if an informed approach for modifying the yield characteristics of the seed is to be attempted.

Reduced to its simplest terms, the goal of the developmental geneticist interested in increasing quantitative yield is either to increase the total number of seeds produced by a plant, to increase the average size of the seeds produced, or both. The number of florets and their average fertility are the factors that contribute to determining the number of seeds produced by each plant. Several physiological parameters affect the average fertility of the florets (188, 197). The important developmental character appears to be the duration of the flowering stage. If this stage could be lengthened, then more flowers should be produced, unless the number of potential florets is a fixed genetically determined trait. [Number of rows of kernels in corn is genetically determined (59)]. Whichever process limits the output of each plant, and it may vary with different crop species, modification of the regulatory elements for that developmental process may relieve the genetic constraints and improve the yield capacity of the species. Alternatively, mutants which hasten the onset of flowering, perhaps analogous to homeotic mutants, would increase the proportion of growth directed towards reproduction. Under good filling conditions, the size of the seed is limited by the size of the seed coat in some species (135). The maternally derived integuments produce the seed coats; each is a two-celled layer which increases in size by restricting cell division to a single plane. Thus a fixed size suggests that a fixed number of cell divisions occur. If so, this number is likely to be a genetically controlled and modifiable trait. However, in some varieties of wheat the 1000-grain weight has a yearly variation of as much as 50 percent (135), indicating that it is not the capacity of the seed that is limiting, but the ability to fill the seed completely. Many physiological factors influence this process, but one, the ability to function as a metabolic sink and to mobilize nutrients, appears to be inherent in the seed (12). Since the physiology of this process in not well understood, it is difficult to suggest mechanisms for possible modifications. But variants that draw resources from a larger area or for a longer time should attain the maximum size more frequently.

If the concern is for increasing the qualitative aspect of yield, for example, by increasing the protein content of the seed, then several developmental options are open. Since the proteins in the embryo are nutritionally superior to those in the endosperm (110), one possibility is to attempt to increase the size of the embryo, thereby altering the embryo to endosperm ratio. The existence of the mutant *lethal giant discs* in *D. melanogaster,* in which the imaginal discs of the mature larva are approximately 50 percent larger than normal (30), promotes the feasibility of this approach in the seed. An alternative method of changing the distribution of proteins within the seed would be to find a variant with a multilayered aleurone, since the aleurone also has a more nutritional protein spectrum (110). Without more information on the role of the aleurone during seed development, it is difficult to evaluate this approach. Finally, the geneticist may attempt to alter the spectrum of proteins (or other storage molecules) found in the endosperm itself by selecting for variants such as *opaque-2* in corn or the new high lysine strains of sorghum (131, 174). The sorghum strains were recovered from among several thousand naturally occurring strains examined for the floury endosperm phenotype. Identification of this trait, which is also associated with the *opaque-2* mutation, required processing longitudinal sections of several kernels from each strain. Recent results suggest that the scanning electron microscope may provide a more rapid means of examining samples (193). The effectiveness of any approach to modifying the nutritional character of the endosperm depends on finding more rapid and efficient methods to screen large numbers of potentially mutant seed.

In contemplating the variety of possible genetic modifications of seed development, it is important to realize that modern agricultural species have been released from many of the environmental pressures that resulted in the evolution of their present form. Efforts in mutation breeding should emphasize the importance of the seed as an end product over its reproductive function. Areas of conflict between man's needs and the plant's needs can be circumvented through the use of seed stocks or conditional mutants.

ACKNOWLEDGMENTS

We thank Drs. W. Adams, R. Chaleff, M. Constantin, J. Grafius, M. Lieb, J. Polacco, and A. Srb for their thoughtful criticisms of this review.

Literature Cited

1. Adelberg, E. A., Ed. 1960. *Papers on Bacterial Genetics.* Boston: Little, Brown. 450 pp.
2. Allard, R. W., Jain, S. K., Workman, P. L. 1968. *Advan. Genet.* 14:55–132
3. Anderson, A. M. 1927. *J. Agr. Res.* 34:1001–18
4. Athwal, D. S. 1971. *Quart. Rev. Biol.* 46:1–34
5. Atkinson, D. E. 1966. *Ann. Rev. Biochem.* 35:85–124
6. Audus, L. J. 1972. *Plant Growth Substances.* London: Hill. 553 pp.
7. Avery, A. G., Satina, S., Rietsema, J. 1959. *Blakeslee: The Genus Datura.* New York: Ronald. 289 pp.
8. Batygina, T. B. 1969. *Rev. Cytol. Biol. Veg.* 32:335–41
9. Bauman, N., Davis, B. D. 1957. *Science* 126:170
10. Beadle, G. W., Tatum, E. L. 1941. *Proc. Nat. Acad. Sci. USA* 27:499–506

11. Beckwith, J. R., Zipser, D., Eds. 1970. *The Lactose Operon.* Cold Spring Harbor, NY: Cold Spring Harbor Lab. 437 pp.
12. Beevers, H. 1969. See Ref. 47, 169–80
13. Benveniste, R., Davies, J. 1973. *Ann. Rev. Biochem.* 42:471–506
14. Bergmann, L. 1968. In *Les Cultures de Tissus de Plantes,* 213–22. Paris: C.N.R.S. 381 pp.
15. Bhatnagar, S. P., Johri, B. M. 1972. In *Seed Biology,* ed. T. T. Kozlowski, 1:77–149. New York: Academic. 416 pp.
16. Birnstiel, M. L., Wallace, H., Sirlin, J. L., Fischberg, M. 1966. *Nat. Cancer Inst. Monogr.* 23:431–44
17. Bodenstein, D. 1950. In *Biology of Drosophila,* ed. M. Demerec, 275–367. New York: Hafner. 632 pp.
18. Bonhoeffer, F., Schaller, H. 1965. *Biochem. Biophys. Res. Commun.* 20: 93–97
19. Brenner, S. 1974. *Genetics* 77:71–94
20. Brink, R. A. 1973. *Ann. Rev. Genet.* 7:129–52
21. Brink, R. A., Cooper, D. C. 1947. *Bot. Rev.* 13:423–77
22. Ibid, 479–541
23. Brink, R. A., Cooper, D. C., Ausherman, L. E. 1944. *J. Hered.* 35:67–75
24. Britten, R. J., Davidson, E. H. 1971. *Quart. Rev. Biol.* 46:111–38
25. Britten, R. J., Kohne, D. E. 1966. *Carnegie Inst. Washington Yearb.* 65:78–106
26. Brody, S. 1973. In *Developmental Regulation,* ed. S. J. Coward, 107–53. New York: Academic. 266 pp.
27. Brody, S., Tatum, E. L. 1966. *Proc. Nat. Acad. Sci. USA* 56:1290–97
28. Brown, J. C., Holmes, R. C., Tiffin, L. O. 1958. *Soil Sci.* 86:75–82
29. Brown, D. D., Wensink, P. C., Jordan, E. 1971. *J. Mol. Biol.* 63:57–73
30. Bryant, P. J., Schubiger, G. 1971. *Develop. Biol.* 24:233–63
31. Campbell, J. H., Lengyel, J. A., Langridge, J. 1973. *Proc. Nat. Acad. Sci. USA* 70:1841–45
32. Carlson, P. S. 1970. *Science* 168:487–89
33. Carlson, P. S., Ed. 1973. *Basic Mechanisms in Plant Morphogenesis.* Brookhaven Symp. Biol. Vol. 25. 421 pp.
34. Carlson, P. S. 1973. *Proc. Nat. Acad. Sci. USA* 70:598–602
35. Carlson, P. S. 1973. *Science* 180: 1366–68
36. Carlson, P. S., Smith, H. H., Dearing, R. D. 1972. *Proc. Nat. Acad. Sci. USA* 69:2292–94
37. Chaleff, R. S., Carlson, P. S. 1974. *Ann. Rev. Genet.* 8:267–78
38. Chan, L. N., Gehring, W. 1971. *Proc. Nat. Acad. Sci. USA* 68:2217–21
39. Chandler, M. T., Tandeau de Marsac, N., de Kouchkovsky, Y. 1972. *Can. J. Bot.* 50:2265–70
40. Chase, S. S. 1969. *Bot. Rev.* 35:117–67
41. Chovnick, A. 1966. *Proc. Roy. Soc. B* 164:198–208
42. Clayberg, C. D., Butler, L., Kerr, E. A., Rick, C. M., Robinson, R. W. 1966. *J. Hered.* 57:189–96
43. Davies, D. R., Plaskitt, A. 1971. *Genet. Res.* 17:33–43
44. Davies, J., Nomura, M. 1972. *Ann. Rev. Genet.* 6:203–61
45. Davis, G. L. 1966. *Systematic Embryology of the Angiosperms.* New York: Wiley. 528 pp.
46. Derbergh, P., Nitsch, C. 1973. *C.R. Acad. Sci.* 276:1281–84
47. Dinauer, R. C., Ed. 1969. *Physiological Aspects of Crop Yield.* Madison, Wis.: Am. Soc. Agron., Crop Sci. Soc. Am. 396 pp.
48. Dobzhansky, T. 1951. *Genetics and the Origin of Species.* New York: Columbia Univ. Press. 364 pp. 3rd ed.
49. Donald, C. M. 1968. *Euphytica* 17:385–403
50. Doy, C. H., Gresshoff, P. M., Rolfe, B. G. 1973. *Proc. Nat. Acad. Sci. USA* 70:723–26
51. Duncan, R. E., Ross, J. G. 1950. *J. Hered.* 41:239–68
52. Dunn, J. J., Studier, F. W. 1973. *Proc. Nat. Acad. Sci. USA* 70:3296–3300
53. East, E. M. 1915. *Genetics* 1:164–76
54. East, E. M., Jones, D. F. 1919. *Inbreeding and Outbreeding.* Philadelphia: Lippincott. 285 pp.
55. Edelman, G. M., Gall, W. E. 1969. *Ann. Rev. Biochem.* 38:415–66
56. Egyhazi, E., Daneholt, B., Edström, J. E., Lambert, B., Ringborg, U. 1971. *J. Cell Biol.* 48:120–27
57. Ehrenberg, L. 1971. In *Chemical Mutagens,* ed. A. Hollaender, 2:365–86. New York: Plenum. 610 pp.
58. Elgin, S. C. R., Froehner, S. C., Smart, J. E., Bonner, J. 1971. *Advan. Cell. Mol. Biol.* 1:1–57
59. Emerson, R. A., East, E. M., 1913. *Bull. Agr. Exp. Sta. Nebr.* 2:1–120
60. Englesberg, E. 1971. In *Metabolic Pathways,* ed. J. H. Vogel, 5:257–96. New York: Academic. 576 pp.
61. Ephrussi, B., Beadle, G. W. 1936. *Am. Natur.* 70:218–25

62. Fielding, C. J. 1967. *J. Embryol. Exp. Morphol.* 17:375–84
63. Fincham, J. R. S. 1966. *Genetic Complementation.* New York: Benjamin. 143 pp.
64. Fincham, J. R. S., Day, P. R. 1963. *Fungal Genetics.* Philadelphia: Davis. 326 pp. 2nd ed.
65. Ibid, 55–57
66. Fisher, R. A. 1918. *Trans. Roy. Soc. Edinburgh* 52:399–423
67. Fisher, R. A. 1956. *The Theory of Inbreeding.* New York: Academic. 150 pp.
68. Flam, W. G., Walker, P. M. B., McCallum, M. 1969. *J. Mol. Biol.* 40:423–43
69. Flemion, F., Uhlmann, G. 1946. *Contrib. Boyce Thompson Inst.* 14:283–93
70. Flor, H. H. 1956. *Advan. Genet.* 8:29–54
71. Fowden, L., Lewis, D., Tristram, H. 1967. *Advan. Enzymol.* 29:89–163
72. Gabriel, O. 1971. *Methods Enzymol.* 22:578–604
73. Gall, J. G., Pardue, M. L. 1969. *Proc. Nat. Acad. Sci. USA* 63:378–83
74. Galton, F. 1889. *Natural Inheritance.* London: Macmillan. 259 pp.
75. Garen, A. 1968. *Science* 160:149–59
76. Garrod, A. E. 1923. *Inborn Errors of Metabolism.* London: Frowde, Hodder, Stoughton. 216 pp. 2nd ed.
77. Gaul, H. 1964. *Radiat. Bot.* 4:155–232
78. Gehring, W., Nöthiger, R. 1973. In *Developmental Systems: Insects,* ed. C. H. Waddington, S. J. Counce, 211–90. New York: Academic
79. Gilbert, J. M., Anderson, W. F. 1971. *Methods Enzymol.* 20(C):542–49
80. Gold, L. M., Schweiger, M. 1971. *Methods Enzymol.* 20 (C) 537–42
81. Goodwin, T. W., Ed. 1965. *Chemistry and Biochemistry of Plant Pigments.* New York: Academic. 583 pp.
82. Gowen, J. W., Ed. 1952. *Heterosis.* New York: Hafner. 552 pp.
83. Grant, V. 1964. *The Architecture of the Germplasm.* New York: Wiley. 236 pp.
84. Green, M. M. 1963. *Genetica* 34: 242–53
85. Greenberg, D. M. 1969. In *Metabolic Pathways,* ed. D. M. Greenberg, 3: 238–316. New York: Academic. 622 pp.
86. Grüneberg, H. 1963. *The Pathology of Development.* Oxford: Blackwell. 309 pp.
87. Guha, S., Maheshwari, S. C. 1966. *Nature* 212:97–98
88. Hadorn, E. 1961. *Developmental Genetics and Lethal Factors.* New York: Wiley. 355 pp.
89. Hadorn, E. 1965. *Brookhaven Symp. Biol.* 18:148–59
90. Hadorn, E., Hurlimann, R., Mindek, G., Schubiger, G., Staub, M. 1968. *Rev. Suisse Zool.* 75:557–69
91. Halpern, Y. S., Umbarger, H. E. 1961. *J. Gen. Microbiol.* 26:175–83
92. Harborne, J. B. 1967. *Comparative Biochemistry of the Flavonoids.* New York: Academic. 383 pp.
93. Hartman, P. E., Roth, J. R. 1973. *Advan. Genet.* 17:1–106
94. Hartwell, L. H. 1974. *Bacteriol. Rev.* 38:164–98
95. Henry, S. A., Fogel, S. 1971. *Mol. Gen. Genet.* 113:1–19
96. Heslop-Harrison, J. 1969. See Ref. 47, 291–321
97. Heslop-Harrison, J. 1972. In *Plant Physiology,* ed. F. C. Steward, 6C:133–290. New York: Academic. 450 pp.
98. Hess, D. 1968. *Biochemische Genetik.* New York: Springer-Verlag. 353 pp.
99. Hess, D. 1972. *Naturwissenschaften* 59:348–55
100. Hood, L. 1973. *Stadler Symp.* 5:73–143
101. Hooker, A. L., Saxena, K. M. S. 1971. *Ann. Rev. Genet.* 5:407–24
102. Hopwood, D. A. 1970. In *Methods in Microbiology,* ed. J. R. Norris, D. W. Ribbons, 3A:363–433. London: Academic. 505 pp.
103. Hotta, Y., Benzer, S. 1972. *Nature* 240:527–35
104. Hsia, D. Yi-yung. 1959. *Inborn Errors of Metabolism.* Chicago: Year Book Publ. 358 pp.
105. Inglett, G. E., Ed. 1972. *Symposium: Seed Proteins.* Westport, Conn.: Avi. 320 pp.
106. Jacobson, A., Gillespie, D. 1970. *Cold Spring Harbor Symp. Quant. Biol.* 35:85–94
107. Johannsen, W. 1909. *Elements der exakten Erblichkeitslehre.* Jena: Fisher. 735 pp.
108. Kao, K. N., Michayluk, M. R. 1974. *Planta* 115:355–67
109. Kasperbauer, M. J., Collins, G. B. 1972. *Crop Sci.* 12:98–101
110. Kaul, A. K. 1973. In *Nuclear Techniques for Seed Protein Improvement,* 1–106. Vienna: Int. At. Energy Ag. 422 pp.
111. Keller, W. A., Melchers, G. 1973. *Z. Naturforsch.* 28:737–42
112. Khesin, R. B. et al 1969. *Mol. Gen. Genet.* 105:243–61
113. Kimber, G., Riley, R. 1963. *Bot. Rev.* 29: 480–531

114. Konzak, C. F. 1956. *Brookhaven Symp. Biol.* 9:157–76
115. Koshland, D. E., Neet, K. E. 1968. *Ann. Rev. Biochem.* 37:359–410
116. Lederberg, J., Zinder, N. 1948. *J. Am. Chem. Soc.* 70:4267–68
117. Ledoux, L., Huart, R., Jacobs, M. 1971. In *Informative Molecules in Biological Systems*, ed. L. Ledoux, 159–75. Amsterdam: North-Holland. 466 pp.
118. Lester, H. E., Gross, S. R. 1959. *Science* 129:572
119. Levine, R. P. 1969. *Ann. Rev. Plant Physiol.* 20:523–40
120. Levine, R. P., Goodenough, U. W. 1970. *Ann. Rev. Genet.* 4:397–408
121. Li, S. L., Redéi, G. P., Gowans, C. S. 1967. *Mol. Gen. Genet.* 100:77–83
122. Loomis, R. S., Williams, W. A. 1969. See Ref. 47, 27–47
123. Loomis, R. S., Williams, W. A., Hall, A. E. 1971. *Ann. Rev. Plant Physiol.* 22:431–68
124. Maheshwari, P. 1950. *An Introduction to the Embryology of Angiosperms.* New York: McGraw-Hall. 453 pp.
125. Mäkelä, P. H., Stocker, B. A. D. 1969. *Ann. Rev. Genet.* 3:291–322
126. Mandaron, P. 1970. *Develop. Biol.* 22:298–320
127. Mathan, D. S., Cole, R. D. 1964. *Am. J. Bot.* 51:560–66
128. Mather, K. 1943. *Biol. Rev.* 18:32–64
129. Mather, K., Jinks, J. L. 1971. *Biometrical Genetics.* Ithaca, NY: Cornell Univ. Press. 382 pp.
130. McCarthy, B. J., Farquhar, M. N. 1972. *Brookhaven Symp. Biol.* 23:1–43
131. Mertz, E. T., Bates, L. S., Nelson, O. E. 1964. *Science* 145:279
132. Miflin, B. J. 1973. In *Biosynthesis and its Control in Plants*, 49–68. London: Academic. 364 pp.
133. Miller, O. L. Jr., Beatty, B. R. 1969. *Science* 164:955–57
134. Monteith, J. L. 1969. See Ref. 47, 89–110
135. Murata, Y. 1969. See Ref. 47, 235–59
136. Nayar, N. M. 1973. *Advan. Genet.* 17:153–292
137. Nelson, O. E. 1967. *Ann. Rev. Genet.* 1:245–68
138. Nelson, O. E., Burr, B. 1973. *Ann. Rev. Plant Physiol.* 24:493–518
139. Neuffer, M. G., Jones, L., Zuber, M. S. 1968. *The Mutants of Maize.* Madison, Wis.: Crop Sci. Soc. Am. 74 pp.
140. Nilsson-Ehle, H. 1909. *Lunds Univ. Arrkr.* N. F. 5, 2:1–22
141. Nitsch, C. 1974. *C. R. Acad. Sci.* 278:1031–34
142. Nitsch, C., Norreel, B. 1973. *C. R. Acad. Sci.* 276:303–6
143. Nitsch, J. P. 1971. In *Plant Physiology*, ed. F. C. Steward, 6A:413–501. New York: Academic. 541 pp.
144. Nitsch, J. P. 1972. *Z. Pflanzenzucht.* 67:3–18
145. Oostindiër-Braaksma, F. J., Feenstra, W. J. 1973. *Mutat. Res.* 19:175–85
146. ' Panse, V. G. 1940. *J. Genet.* 40:283–302
147. Paul, J., Gilmour, R. S. 1968. *J. Mol. Biol.* 34:305–16
148. Phinney, B. O., West, C. A. 1960. *Ann. Rev. Plant Physiol.* 11:411–36
149. Pinthus, M. J., Eshel, Y., Shchori, Y. 1972. *Science* 177:715–16
150. Pontecorvo, G., Roper, J. A., Hemmons, L. M., MacDonald, K. D., Bufton, A. W. J. 1953. *Advan. Genet.* 5:141–238
151. Porter, J. W., Anderson, D. G. 1967. *Ann. Rev. Plant Physiol.* 18:197–228
152. Potrykus, I. 1973. *Z. Pflanzenphysiol.* 70:364–66
153. Potrykus, I., Hoffmann, F. 1973. *Z. Pflanzenphysiol.* 69:287–89
154. Power, J. B., Cummins, S. E., Cocking, E. C. 1970. *Nature* 225:1016–18
155. Puck, T. T., Kao, F. 1967. *Proc. Nat. Acad. Sci. USA* 58:1227–34
156. Raghavan, V., Torrey, J. G. 1963. *Am. J. Bot.* 50:540–51
157. Randolph, L. F. 1936. *J. Agr. Res.* 53:881–916
158. Reinert, J. 1973. See Ref. 183, 338–55
159. Reznikoff, W. S. 1972. *Ann. Rev. Genet.* 6:133–56
160. Rice, T. B. 1973. *Isolation and characterization of maternal-effect mutants: An approach to the study of early determination in Drosophila melanogaster.* PhD thesis. Yale Univ., New Haven, Conn. 147 pp.
161. Rice, T. B., Garen, A. 1975. *Develop. Biol.* In press
162. Rick, C. M., Butler, L. 1956. *Advan. Genet.* 8:267–382
163. Robison, G. A., Butcher, R. W., Sutherland, E. W. 1971. *Cyclic AMP.* New York: Academic. 531 pp.
164. Rodwell, D. M. 1969. See Ref. 85, 317–74
165. Rosenberg, M. 1970. *Proc. Nat. Acad. Sci. USA* 67:32–36
166. Rudolph, A., Scholz, G. 1972. *Biochem. Physiol. Pflanzen* 163:156–68
167. Scandalios, J. G. 1969. *Biochem. Genet.* 3:37–79
168. Schaeffer, P. 1969. *Bacteriol. Rev.* 33:48–71

169. Scott, W. A., Mishra, N. C., Tatum, E. L. 1973. *Brookhaven Symp. Biol.* 25: 1–18
170. Sears, E. R. 1969. *Ann. Rev. Genet.* 3:451–68
171. Shannon, L. M. 1968. *Ann. Rev. Plant Physiol.* 19:187–210
172. Shearn, A., Rice, T. B., Garen, A., Gehring, W. 1971. *Proc. Nat. Acad. Sci. USA* 68:2594–98
173. Sherman, F., Stewart, J. W. 1971. *Ann. Rev. Genet.* 5:237–56
174. Singh, R., Axtell, J. D. 1973. *Crop Sci.* 13:535–39
175. Sirlin, J. L. 1972. *Biology of RNA,* 226–30. New York: Academic. 525 pp.
176. Slater, M., Schaechter, M. 1974. *Bacteriol. Rev.* 38:199–221
177. Smith, H. H. 1968. *Advan. Genet.* 14:1–54
178. Sprague, G. F. 1967. *Ann. Rev. Genet.* 1:269–94
179. Stebbins, G. L. 1971. *Chromosomal Evolution in Higher Plants.* Reading, Mass.: Addison-Wesley. 216 pp.
180. Stewart, M., Murphy, C., Fristrom, J. W. 1972. *Develop. Biol.* 27:71–83
181. Stoy, V. 1969. See Ref. 47, 185–202
182. Street, H. E. 1969. In *Plant Physiology,* ed. F. C. Steward, 5B:3–224. New York: Academic. 454 pp.
183. Street, H. E., Ed. 1973. *Plant Tissue and Cell Culture.* Berkeley: Univ. California Press. 503 pp.
184. Sunderland, N. 1973. See Ref. 183, 205–39
185. Sussex, I. M. 1973. See Ref. 33, 145–51
186. Tanaka, M., Nakata, K. 1969. *Jap. J. Genet.* 44:47–54
187. Thompson, J. S., Thompsom, M. W. 1973. *Genetics in Medicine.* Philadelphia: Saunders. 400 pp. 2nd ed.
188. Togari, Y., Kashiwakura, Y. 1958. *Proc. Crop Sci. Soc. Jap.* 27:3–5
189. Umbarger, H. E. 1969. *Ann. Rev. Biochem.* 38:323–70
190. Vasil, I. K., Vasil, V. 1972. *In Vitro* 8:117–27
191. Walles, B. 1963. *Hereditas* 50:317–40
192. Wardlaw, I. F. 1968. *Bot. Rev.* 34:79–105
193. Wassom, C. E., Hoseney, R. C. 1973. *Crop Sci.* 13:462–63
194. Weinberg, E., Birnstiel, M. L., Purdom, I. F., Williamson, R. 1972. *Nature New Biol.* 240:225–28
195. Weiss, M. G. 1943. *Genetics* 28:253–68
196. Wittwer, S. W. 1974. *BioScience* 24: 216–24
197. Yoshida, S. 1972. *Ann. Rev. Plant Physiol.* 23:437–64
198. Zubay, G. 1973. *Ann. Rev. Genet.* 7:267–87
199. Zubay, G., Chambers, D. A. 1971. See Ref. 60, 297–349

Ann. Rev. Plant Physiol. 1975. 26:309–40

STOMATAL ACTION[1]   ❖7593

K. Raschke

MSU/AEC Plant Research Laboratory, Michigan State University, East Lansing, Michigan 48824

CONTENTS

[1]Abbreviations used: ABA (abscisic acid); CAM (Crassulacean acid metabolism); CMU [3-(*p*-chlorophenyl)-1,1-dimethylurea]; DCMU [3-(3,4-dichlorophenyl)-1,1-dimethylurea]; PEP (phosphoenolpyruvate).

Wenn der Mensch ein Loch sieht, hat er das Bestreben, es auszufüllen, dabei fällt er meist hinein.

Kurt Tucholsky

STOMATAL FUNCTION (THE CONTROL TASK)

Most of the basic mechanisms involved in stomatal action are not yet understood. We do not know how plant cells take up ions, how ions are moved between cell compartments, how carbohydrate metabolism is coupled to ion transport, how a loss of water from leaves triggers the synthesis of abscisic acid, how abscisic acid affects ion transport, and so on. Insufficient tools are available to study quantitatively the metabolism of the guard cells. The volume of guard cells constitutes only one-thirtieth or less of the total volume of the epidermis. The lumen of a pair of guard cells, if it is large, has a volume of about 10^{-11} liter. We have been unable so far to separate guard cells in quantities sufficient for biochemical analysis from the tissue in which they are embedded. Yet we try to understand and describe stomatal action because we are fascinated by the ability of the stomata to respond to changes in the environment within minutes, even seconds, and we realize, of course, that virtually all of the carbon assimilated by higher land plants has entered them through the stomata, and we are aware that one-half to six-sevenths of the water precipitated on the land returns to the atmosphere by evapotranspiration from vegetation (42,

185). On the average, transpiration through the stomata is almost three times larger than evaporation from the soil (42, 84).

Land plants are in a dilemma as long as they exist: assimilation of CO_2 from the atmosphere requires an intensive gas exchange; the prevention of excessive water loss demands that gas exchange be kept low. For the solution of this dilemma, higher plants evolved an epidermis covered with a cuticle of low permeability for water vapor and CO_2, and plants evolved the stomatal feedback system for the reconciliation of the two opposing priorities (Figure 1). The apparent diffusion resistance of the cuticle is in general 10 to 100 times higher (e.g. 27, 57) than the lowest resistance of the stomatal pores (around 1 s cm^{-1}). The cuticular resistance increases with desiccation (113), and may in some xerophytes (121) and tree leaves (57) be as high

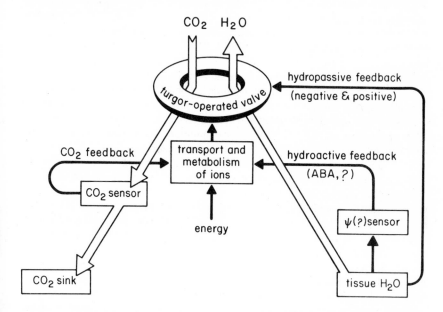

Figure 1 Simplified and hybrid schema of the stomatal feedback system. The guard cells function as a turgor-operated valve modulating uptake of CO_2 from the atmosphere and water loss from the mesophyll to the atmosphere. The mechanism sensing CO_2 is located in the guard cells. The terms "hydropassive" and "hydroactive" were coined by Stålfelt (176) in 1929; they are here used in a broader sense, particularly the term "hydropassive." "Hydropassive" refers to changes in turgor caused by changes in the water potential in the leaf; the solute content of the guard cells remains unaffected. Hydropassive responses include effects of changes in epidermal pressure on stomatal aperture (positive feedback). The hydroactive feedback is probably between mesophyll and the guard cells, with ABA serving as one of possibly several messengers. Not entered are (*a*) the transfer of latent heat through the stomata and the feedback through leaf temperature (p. 329); (*b*) the direct loss of water from the epidermis to the atmosphere (p. 327); (*c*) the possibly existing intracellular hydroactive feedback in the guard cells (pp. 326, 334); and (*d*) the linkage of the stomata to plant hydraulics.

as 360 or 460 s cm^{-1}. Even in mesophytes like *Vicia faba*, the cuticular resistance may reach this magnitude if plants are deprived of their water supply for a few days.

In some species stomatal resistances can be smaller than 1 s cm^{-1} (e.g. 23, 27, 95). In other species gas exchange through the stomata is reduced by plugs of epicuticular wax in the stomatal antechamber or by wax felts covering the pore (71, 157). Under comparable conditions, C$_4$ plants appear to have stomatal diffusion resistances two to three times higher than C$_3$ plants (23). This ratio appears to depend on the age of the plants (169).

Stomatal opening has to be synchronized with the CO$_2$ requirement of the assimilatory tissue. One can think of several ways to achieve this task: (*a*) light starts and maintains photosynthesis in the mesophyll and independently and directly triggers and maintains stomatal opening as well; (*b*) stomatal aperture follows an endogenous diurnal rhythm; (*c*) a messenger, e.g. a product of CO$_2$ assimilation, travels from the mesophyll to the guard cells and signals a demand for CO$_2$; (*d*) stomata recognize the demand for CO$_2$ by sensing the CO$_2$ depletion of the intercellular spaces. The last two possibilities would involve closed feedback loops, the first two would not. As other reviewers have summarized before (78, 103), and as will be further amplified below, stomatal opening is not an obligatory response to light; therefore explanation *a* is unlikely. Turning to *b*, we realize that stomata of several species do open and close in circadian rhythms in continuous light or darkness (44, 103); under these conditions, however, the amplitude of the variation in stomatal aperture is only about one-tenth of that in plants exposed to a daily light-dark cycle (152), and maximal opening occurs in the morning and not at noon (153, 178). In addition, stomatal rhythms usually fade away after the second or third 24 hr period in continuous light or darkness (44, 103, 153). Therefore *b*, acting alone, is insufficient to provide enough CO$_2$ for photosynthesis; an opening caused by an environmental factor must be superimposed (178) in order to increase stomatal conductance sufficiently to supply the photosynthetic apparatus with CO$_2$. With respect to *c*, numerous experiments with isolated epidermes and stomata have demonstrated (e.g. 35, 41, 63, 116) that stomata do not need a product of the photosynthesizing mesophyll as a messenger that commands opening. This conclusion does not rule out the possibility that the mesophyll supplies the guard cells with fuel. There is, however, strong evidence for stomatal opening in response to a depletion of the intercellular spaces of CO$_2$, whether it is caused by photosynthesis or by fixation of CO$_2$ in the dark (79, 80, 119, 122). Stomatal opening in response to a low CO$_2$ concentration ([CO$_2$]) in the leaf provides a simple method for making stomatal conductance proportional to the assimilatory activity of the mesophyll.

There are two ways by which water strain may override an opening caused by CO$_2$ depletion and reduce stomatal aperture. Guard cells may lose water to the tissue surrounding them and as a consequence lose turgor. According to most textbooks, this is the only stomatal response to water stress. But there is a second negative feedback of water loss on guard cells. It involves an "active" reduction of their content of solutes. Stålfelt (176) coined the terms "hydropassive" and "hydroactive" to describe the two kinds of responses to water stress, depending on whether the solute content of the guard cells remains constant or decreases.

Loss of water through the stomata represents a loss of latent heat from the plant. Transpiration lowers leaf temperature and this also provides negative feedback by lowering the water-vapor pressure in the leaf (not shown in Figure 1). Changes in leaf temperature also affect CO_2 assimilation; stomata respond to this change through the feedback of CO_2. Stomatal movement affects transpiration and CO_2 assimilation to different degrees because the temperature dependencies of these two gas exchange processes differ. There is a second reason for a lack of proportionality between the effects of stomatal movement on transpiration and assimilation: transpiration is linearly related to the water-vapor concentration in the intercellular spaces of the leaf but assimilation follows a saturation curve with respect to the intercellular $[CO_2]$. Within certain limits, the temperature dependence of assimilation can be ignored and assimilation can be assumed to depend on $[CO_2]$ linearly. Cowan & Troughton (15) have discussed the effects of stomatal movement on transpiration and photosynthesis, making these simplifying assumptions.

Stomata are components of several feedback loops. If stomata respond to a change in an environmental factor, it is uncertain which feedback loop will determine the ultimate stomatal aperture because CO_2 exchange and water-vapor loss are always simultaneously affected by stomatal movement. Stomata will rarely be in a steady state, either in the field or in the laboratory.

The following two sections summarize observations made on stomata. The third section, containing more speculation than facts, will be on a less solid basis; it is included to stimulate new research. However, a clean separation of facts from hypotheses was not possible in any one of the sections and was therefore not attempted. The fourth section of the review deals briefly with the uncertainty in the interpretation of stomatal behavior, as well as with stomatal behavior with time. Statements without reference numbers refer to the author's own unpublished work.[2]

STOMATAL MECHANICS (THE TURGOR-OPERATED VALVE)

Mechanics of Stomatal Movement

RADIAL MICELLATION   In one way, kidney-shaped guard cells resemble radial tires; the radial orientation of the cellulose microfibrils in the guard-cell walls (168, 204) does not allow the circumference of radial cross sections of the cells to expand during inflation. Guard cells expand by changing shape and by stretching in the direction of their longitudinal, curved axes. Since the micellae in the walls fan out radially from the pore to the periphery of the stomatal apparatus, lengthening of the guard cells is associated with an increase in curvature. In cases of extreme inflation, sometimes occurring in isolated epidermes, the two guard cells together may almost assume the shape of a figure 8 (67, and personal observations).

---

[2]I anticipate that much of this unpublished work, as well as that published provisionally in the annual reports of the MSU/AEC Plant Research Laboratory, will be in press or already published in journals devoted to plant physiology by the time this review appears.

As the guard cells expand, they push each other apart; as the ends of the cells expand, the middle portions of the cells are displaced. In principle, the mechanics of opening is the same in stomata with kidney-shaped guard cells and in the stomata of the Gramineae and Cyperaceae (144, 188). Aylor et al (7) have demonstrated the crucial role of radial micellation with models made from rubber balloons, and have shown that the observed asymmetry in the thickness of the guard cell walls is not essential for the bending of the guard cells during stomatal opening. It is also questionable whether the cuticular ledges assist in stomatal opening. Although the ledges are curved when the stress on them has been released after their enzymatical separation from the epidermis, their main role seems to be preventing liquid water from entering the intercellular spaces of the leaf (162) rather than aiding stomatal opening.

CHANGES IN GUARD-CELL DIMENSIONS DURING OPENING   According to measurements in epidermal strips (144), width and depth of the guard cells and the length of the stomatal apparatus change very little during stomatal opening while the perimeter of the whole stomatal apparatus increases (up to 20% in *Vicia faba*). Much change occurs in the perimeter of the stomatal pore; an increase by almost 60% was measured in *Vicia faba* when the aperture increased from 0 to 13.5 μm. Clearly, the ventral wall cannot be the location of the neutral axis of a bending guard cell, as was assumed by DeMichele & Sharpe (21). When the stoma is closed, the lumen of the middle portion of kidney-shaped guard cells has the familiar triangular cross section, with one corner of the triangle pointing to the pore. In material fixed for sectioning and microscopy, this cross section may sometimes look like a rectangle; this shape, however, is an artefact caused by desiccation of the cell walls. Upon inflation of the guard cells, the triangular cross section rounds up; it becomes elliptical and, at high turgor, circular. The walls surrounding the pore are then almost perpendicular to the tangential plane of the leaf. The outer perimeter of a cross section of a guard cell hardly changes during opening, but the perimeter of the cell lumen increases because the cell walls become thinner. The area of the cross section of the middle portion of the guard cell may increase almost twofold during stomatal opening. The expansion of the polar sections of the guard cells is made possible by a thickening and a fold in the bottom cell wall, both of which expand when the stoma opens (144). Guard cells of *Zea mays* appear to have a similar fold (175, Fig. 13, 14). Optical sectioning of intact stomata in the open and closed states and integration of the cross-sectional areas made it possible to estimate the relationship between stomatal aperture and guard cell volume. It seems this relationship is linear, and the volume of the lumina of guard cells of *V. faba* may almost double when a stoma opens from a closed state to an aperture of 18 μm. The lumina of a pair of guard cells of *V. faba* have a volume of $\sim 10^{-11}$ liter, and the walls of the guard cells occupy about an equal volume. Guard cells vary considerably in size, even within one species.

The walls of guard cells appear to give when turgor increases, i.e. with increasing volume the increment in turgor needed to produce a constant increment in volume decreases until, of course, stiffening occurs (146). This phenomenon may have two causes. One is that some of the volume changes are due to shape changes. The other and perhaps more important cause is the composition of the matrix of the guard cell walls. Walls of guard cells stain deeply with ruthenium red (5, 168, 175) and are digested by snail pectinase (77) and pectinases of fungal origin. Therefore, most probably, the matrix consists of pectins.[3]

The radial micellation of the guard cell walls appears early in the ontogeny as a radial orientation of microtubules (75, 168, 175; B. Palevitz, in preparation). Palevitz also reports that the two guard cells of a developing stomatal complex accumulate K^+ and Cl^- before they separate and form the stomatal pore. Presumably the same mechanism which produces turgor in the functional stoma is already used to exert the stress needed for splitting the cell walls between the two developing guard cells.

THE GRASS STOMA   The stomatal apparatus of the Gramineae and the Cyperaceae can be considered to have reached a higher state of differentiation than the "ordinary" stoma consisting of two kidney-shaped cells. The volume of the middle portions of the guard cells of maize has been reduced to about 0.12 of the total guard cell volume; much less osmotically active material will be needed during stomatal opening to fill the hydraulically inactive middle portion of grass guard cells than that of kidney-shaped guard cells (which occupies from 0.5 to 0.66 of the total cell volume). The middle portion of a grass guard cell, e.g. that of maize, is both narrow (~ 2 μm wide) and shallow (~ 2 μm deep; the bulbous ends are $\sim$12 μm deep; see also the figures in Ref. 175). The middle portion is stiffened like a U-beam by reinforcements of the cell walls (the U opens to the adjoining subsidiary cell). The bulbous ends of the guard cells are radially micellated (204) and act indeed like bellows (166).[4]

SUBSIDIARY CELLS   These specialized cells, found in grasses and other plant families, provide the counterforce to stomatal opening and serve as storage for inorganic ions (148) involved in the stomatal movements. The volume of a pair of subsidiary cells of *Zea mays,* estimated directly and indirectly (148), is a little larger than 10^{-11} liter and thus about four times that of the two guard cells.

The use of the term "subsidiary cell" should remain restricted to epidermal cells which adjoin guard cells and are morphologically different from other epidermal cells (29). The use of the term for any epidermal cell adjacent to a guard cell is erroneous. If necessary, a term like "functional subsidiary cell" could be used for

[3]This fact may explain why my attempts at isolating guard cells from leaf or epidermal tissue by applying pectinases failed; guard cells were usually the first cells to be damaged or killed, due to the known but not yet understood cytotoxic effects of pectin breakdown (e.g. 115).

[4]Aylor et al (7) criticize this view because grass guard cells appear to them like telephone receivers. In my own experience, this impression could be obtained at hydropassive opening or after focussing the microscope on the outer border between subsidiary cell and guard cell in the tangential plane on the epidermis, but not if the microscope is focussed at half depth of the middle portion of the guard cell.

epidermal cells which serve a physiological function similar to that of subsidiary cells but do not differ morphologically from ordinary epidermal cells.

EPIDERMAL PRESSURE   In isolated epidermes, stomatal opening is related mainly to the mechanics of the guard cells, particularly when the ordinary epidermal cells are ruptured. In the intact leaf, the situation is more complicated; the epidermal pressure compresses the stomata and the attached mesophyll may also limit opening (112, 176). The "antagonism" between guard cells and epidermal cells was discussed by von Mohl (111) as early as 1856. He pointed out that the effect of the epidermal pressure on stomatal opening depended on the ratio of areas on which the pressures acted. Recently, DeMichele & Sharpe (21) made the point again. Since epidermal cells have a larger area of attack (because they are deeper) than the guard cells, the epidermal pressure can balance a pressure in the guard cells higher than that in the epidermis. In *Vicia faba*, this "mechanical advantage" of the epidermal cells may vary between 1.2 (wide open stomata) and 2.2 (closed stomata). Epidermal turgor prevents maximal stomatal opening at high water potentials, and changes in epidermal turgor cause transient as well as lasting stomatal responses (139). Epidermal turgor is responsible for positive hydropassive feedback (Figure 1 and pp. 327, 335).

RELATION BETWEEN APERTURE AND PORE AREA   Pore area appears to be a relevant parameter for the evaluation of the stomatal permeability for water vapor and CO_2. Yet stomatal aperture usually is measured rather than pore area. Fortunately, pore area in *Vicia faba* is linearly related to stomatal aperture: area (in μ m^2) = 14.3 X aperture (in μm) (144). Measurements of stomatal width should therefore give suitable estimates for the stomatal conductance for gases. As pointed out in an earlier review (203), stomatal conductance does not follow the perimeter law under natural conditions.

Osmotics of Stomatal Movement

OSMOTIC PRESSURE IN GUARD CELLS   At incipient plasmolysis, osmotic pressures in guard cells may be as high as 100 atm (66). Osmotic pressures in guard cells are usually underestimated because epidermal samples are left on plasmolyzing solutions for too long a time. The half-time for the equilibration of the water potential of the guard cells with that of the solutions is between 6 and 10 s (34, 145). Three half-times should be sufficient to determine plasmolysis. If exposures last longer than 5 min, and if the solution has a water potential < −7 atm, stomata of *V. faba* begin to lose solutes rapidly (145). At water potentials < −70 atm the loss sets in immediately. If, however, plasmolysis is determined after an exposure <2 min, one obtains osmotic pressures for incipient plasmolysis which are as high as 90 atm in *V. faba* (145) for stomata which were wide open. Osmotic pressures of similar magnitude have also been measured after short exposures of open stomata to electrolytic plasmolytica (192). Since the volume of plasmolyzed guard cells is about one-half of the volume of guard cells from wide open stomata (see above), the osmotic pressure of inflated guard cells of *V. faba* is around 45 atm. Such a value agrees well with the cryoscopically determined osmotic pressures in other dicotyledoneous species (30 to 38 atm) (9), particularly if allowances are made for differences in stomatal aperture.

Osmotic pressures of grass guard cells have not yet been determined reliably because plasmolysis is difficult to recognize in these cells.

K^+ SALTS IN GUARD CELLS   Accumulation of K^+ in guard cells was first described as early as 1905 by Macallum (96). Others, like Lloyd (91), have confirmed the observation or, like Iljin (66), have found that stomata could not be plasmolyzed permanently if K^+ salts were used as osmotica. But credit for bringing out the involvement of K^+ in stomatal movement must go to a few Japanese workers and to R. A. Fischer. Imamura's publication of 1943 (68), summarizing 15 years of work on the participation of K^+ and Ca^{++} in stomatal movement, is still a treasure of information. Yamashita (200) demonstrated a good positive correlation between the time course of stomatal movements and the K^+ content of guard cells; starch content of guard cells was sometimes well correlated with aperture (negatively) but at other times considerably out of phase. Ultimately, Fujino (37–39, 41) proposed that the migration of K^+ into guard cells occurs in quantities large enough to increase osmotic pressure and to produce stomatal opening. At about the time Fujino published the results of his work in English, Fischer (32, 35) found that stomata in epidermal strips would open only if they were floating on solutions of K^+ salts. He came to the conclusion that K^+ was the major solute in inflated guard cells. He also showed that CO_2-free air was more effective than light in enhancing K^+ accumulation in guard cells. The quantities of K^+ migrating into guard cells during stomatal opening are sufficient to bring about the observed changes in osmotic pressure. This conclusion was reached after experiments on epidermal strips with radioactive tracers (35) and after elemental analysis of single guard cells with the electron probe microanalyzer (65, 160). Parallel determinations of the elemental content, osmotic pressure, and volume of guard cells of *V. faba* led to the following results (65): K^+ was indeed the only cation taken up in large quantities (4×10^{-12} eq/stoma) by the guard cells of opening stomata; the final concentration reached was 0.9 eq/liter. The contents of the guard cells in Cl, N, P, and S changed little if at all during stomatal opening, indicating that the positive charges of K^+ must have been balanced mainly by organic anions. From striking a balance with respect to osmotica in the guard cells it was concluded that the organic anions were probably divalent. K^+ was also the specific cation transported during stomatal movement of a grass, *Zea mays* (148). The highest K^+ content of the guard cells of maize was 0.4 eq/liter. In this species, chloride turned out to be an important counter ion to K^+; on the average, 40% of the positive charges of K^+ were balanced by Cl^- (148). Histochemical (33) as well as microprobe studies (160, 187) demonstrated a linear dependency of stomatal opening on the K^+ content of the guard cells.

Guard cells can utilize Rb^+ in place of K^+ (35, 63); other alkali ions are taken up only under special circumstances. If, for example, epidermal strips of *Commelina communis* are floated on solutions of salts of K^+ and Na^+, stomata open wider on solutions of Na^+ salts than on K^+ salts (194). Nevertheless, even in this species, K^+ appears to be the cation involved in stomatal opening in the whole leaf; in fact, much of the early evidence for the importance of K^+ uptake by guard cells in stomatal movement was obtained with this species (41, 68). Analyzing stomata from leaves of *C. communis* with the electron microprobe, we found no Na^+ in the guard cells of closed or open stomata. In *Vicia faba,* the specificity for K^+ (and Rb^+) depends on the concentration at which the cations are available. If alkali ions are offered at high enough concentrations, like 100 meq liter $^{-1}$, stomata of *V. faba* are able to utilize all of the alkali species (63). Guard cells of *V. faba* may lose their specificity for K^+ even at low ion concentrations if Ca^{++} is omitted from the bathing solution (126). In the absence of Ca^{++}, guard cells are able to take up large amounts of Na^+ (as measured with the microprobe) (151); the stomata open widely. The role of Ca^{++} in modifying the ion specificity of guard cells is not yet understood, nor is it known why Ca^{++} reduces stomatal opening in some species (41, 68, 195), whether by crosslinking pectins in the cell wall or by inhibiting an ATPase. Under experimental conditions, Tl^+ can serve as analog of K^+. Thallium can be taken up by guard cells and shuttled between guard cells and subsidiary cells; this is my interpretation of observations of Maercker (98).

The involvement of K^+ salts in stomatal movement appears to be general. It has been reported for 50 species (20, 197), including pteridophytes and gymnosperms, and not only for stomatal opening in the light but also for nocturnal opening in crassulacean plants (20).

RATES OF UPTAKE AND LOSS OF K^+ BY GUARD CELLS   Is K^+ transport sufficient to account for the often observed rapid stomatal openings and closures? Early observations of Yamashita would answer this question affirmatively (200). In an investigation specifically directed to this question (148), the movement of K^+ into guard cells of *Zea mays* proceeded at the same relative rate at which stomata moved. Thus even in rapidly responding stomata like those of maize, the rates of K^+ movement are large enough to account for stomatal movements. The K^+ uptake measured over ½ hr was on the average about 10^{-14} eq min^{-1} per stoma or 150 μ mole per gram fresh weight an hour, which is about 12 times the maximal velocity of system 1 uptake by barley roots (28). Over short times (2 min) the velocities of K^+ movement into guard cells can be about 10 times greater, and guard cells of *Zea mays* can lose most of their K^+ within a few minutes (148).

SOURCE OF K^+ In species with slowly opening stomata, like *V. faba,* K^+ seems to come from all over the epidermis (and leaf tissue). The source is difficult to determine because the total volume of the epidermal cells is at least 30 times larger than that of the guard cells. The ion exchange capacity of the cell walls of the epidermis alone may be large enough to store the K^+ needed for stomatal inflation (150). *Zea mays* (148) and many other species (197) possess some epidermal cells which are loaded with K^+ (and Cl^-); however, it has not yet been shown whether these "ion stores" can be utilized by guard cells. However, there is no question that the subsidiary cells of the Gramineae have an important function as reservoirs for K^+ (and Cl^-). Determinations with the microprobe showed that the K^+ (and Cl^-) content of stomatal complexes of *Zea mays* does not change during stomatal movement (148). Rather, the distribution of the ions between guard cells and subsidiary cells is altered. Stomatal movement appears to be based on an ion shuttle between guard cells and subsidiary cells in other species, too (20, 197). In species without morphologically differentiated subsidiary cells one or several epidermal cells adjoining a stoma may fulfill the storage function of a subsidiary cell; these cells often contain large amounts of K^+ (e.g. 33). On the other hand, it is not quite certain whether the subsidiary cells of *Commelina communis* function as ion storages or merely as ion channels to the guard cells (197).

THE ANIONS IN THE GUARD CELLS   Chloride neutralizes on the average about 0.4 of the positive charges of K^+ in the stomata of *Zea mays;* in fact, in a few individual stomata the balance between K^+ and Cl^- was found to be complete (148). In other species, like *Vicia faba,* the participation of Cl^- in stomatal movement was quantitatively quite variable. In epidermal samples taken from illuminated leaves, only 0.05 of the K^+ accumulated by guard cells was compensated by Cl^- (65). The contribution of Cl^- to the anion content of guard cells was larger (about one-third) when epidermal strips of *V. faba* were floated on solutions containing 10 meq Cl^- liter^{-1} (128) which is a relatively high concentration. A reinterpretation of previous histochemical tests (41, 56) involving the reduction of $AgNO_3$ as indicating the presence of halogenides rather than the presence of reducing equivalents (56) or Ca (41) suggests that the participation of Cl^- in stomatal movement may be common; this contention is supported by additional recent work (20). Quantitatively, however, Cl^- does not constitute the most important anion in guard cells; most of the counter charges for K^+ in the guard cells are provided by organic anions. The content of organic acids of epidermal tissue was found to increase with stomatal opening (2, 132). From measured changes in osmotic pressure and K^+ contents of guard cells, it was concluded that the organic anions were divalent (65). Allaway (2) found malate in epidermes of *V. faba* whose stomata were open, but almost none if the stomata were closed. He estimated the change in malate content to be large enough to balance about 0.5 of the K^+. It is likely that aspartate and citrate also provide negative charges (193) in guard cells.

EXCHANGE OF K^+ for H^+ Of the counter ions for K^+, Cl^- accompanies K^+ on its way into guard cells and out of them (148). This seems not to be true for organic anions. When K^+ was offered to guard cells of *V. faba* in combination with presumably nonabsorbable, bulky, and zwitterionic anions (iminodiacetate, benzene sulfonate and 4,4-dimethyl-4,7-diazadecane-1,10-disulfonate), the stomata opened as widely as when K^+ was offered in combination with the absorbable ion Cl^- (150). Organic anions can therefore be produced in the guard cells during stomatal opening. The acidification of the solution on which the samples floated indicated that electroneutrality was maintained by an exchange of K^+ for H^+ (150). The amount of H^+ excreted by guard cells was determined by automatic titration; after correcting for ion exchange with the free space, it was found to be of the same magnitude as the amount of K^+ (or Na^+ for that matter) taken up by the guard cells (150, 151). Subsidiary cells storing K^+ may thus function like exchangers for cations. The identity of the fixed anions is not known; they are not silicates, at least not in maize. Since subsidiary cells also store Cl⁻, they may also have sites able to bind anions.

A discussion of how the transport of inorganic ions and metabolism of organic anions may be linked to each other and to the general metabolism of the guard cells follows later.

CONCLUSION   Turgor (of up to about 50 atm) is produced in the guard cells by the osmotic pressure exerted by imported potassium ions in association with organic anions (which are probably produced in the guard cells during stomatal opening) and with imported chloride ions. Those potassium ions whose charge is balanced by organic ions are taken up in exchange for H^+. Subsidiary cells serve as ion storages, at least in grasses. The relationship between solute content and stomatal opening is linear within the experimental error.

STOMATAL RESPONSES TO CO_2, H_2O, AND TEMPERATURE (THE FEEDBACK LOOPS)

CO_2 Feedback

STOMATAL RESPONSES TO CO_2 Several earlier reviews (e.g. 78, 103) have summarized the evidence that sensitivity to CO_2 is a common feature of stomata. There are exceptions that will be dealt with later.

Stomatal responses to changes in the $[CO_2]$ of the air could be demonstrated strikingly with leaves of Zea mays through which air was forced at constant pressure (136). Any change in $[CO_2]$ was answered by a stomatal adjustment of the air flow through the leaf. Within limits, stomata succeeded in keeping the flow of CO_2 into the leaf constant, probably by keeping the $[CO_2]$ near or in the guard cells constant. Overshoot and oscillations occurred, indicating delayed negative feedback. Stomata began to respond to changes in $[CO_2]$ in a time as short as 3 s (142); the half-time of closing in response to an increase in $[CO_2]$ was 1.8 min in the dark and 1.5 min in the light (at 27° C); if stomatal aperture is expressed in terms of diffusion resistance, the half-time of closing was ∿ 5 min (142). Stomatal opening in response to a reduction of $[CO_2]$ set in equally fast, but the completion of an opening movement took between a few minutes and one hour, depending on the magnitude of the final aperture attained. The velocity of stomatal opening depended more strongly on temperature than did the velocity of closure (141). Stomata of other species like Vicia faba responded more slowly and to a smaller extent to changes in $[CO_2]$ than those of Zea mays.

Experiments with airflow porometers have one flaw: on its passage through the leaf the air has to pass two epidermes, and we are not certain which epidermis exerts control; stomata in the two epidermes of leaves may differ in their behavior with respect to CO_2 (22). But stomata of Zea mays controlled the flow of CO_2 into the leaf also when CO_2 was supplied to the mesophyll by diffusion, as under natural conditions (147). When photosynthesis was changed by changing light intensity, stomatal diffusion resistances varied in a way that the $[CO_2]$ in the leaf hardly changed; it declined only slightly with an increasing rate of CO_2 uptake. We can assume that the turgor of guard cells is controlled by the $[CO_2]$ within these cells (78). The level of the intracellular CO_2 will depend on the distribution of $[CO_2]$ in the immediate vicinity of the guard cells, including the stomatal pore. When light induces an increased uptake of CO_2 by the leaf, the intercellular $[CO_2]$ declines. Stomata tend to keep the intracellular $[CO_2]$ constant by an opening movement, and so the decline in intercellular $[CO_2]$ is smaller than that which would occur if stomata did not respond and their aperture was kept fixed. Similar considerations apply to stomatal responses in the $[CO_2]$ of the air. If external $[CO_2]$ increases, stomata close. Again, due to stomatal reaction, the change in intercellular $[CO_2]$ will be smaller than that which would occur if stomata did not respond. Stomata of *Zea mays* (147) and *Helianthus annuus* (191) closed when external $[CO_2]$ was increased. Intercellular $[CO_2]$ increased linearly with external $[CO_2]$ and photosynthesis was enhanced. In other cases (22, 82) stomata closed in response to increases in external $[CO_2]$ to an extent that less CO_2 was available to the mesophyll than at lower external $[CO_2]$; consequently, photosynthesis declined with increasing $[CO_2]$ in the air surrounding the leaf.

SATURATION KINETICS OF THE STOMATAL RESPONSE TO CO_2 IN THE LIGHT AND IN DARKNESS   In *Zea mays,* the velocity of stomatal closure in response to CO_2 increases with the CO_2 concentration in the air, following saturation kinetics. Half-saturation occurs at approximately 200 μl l^{-1} and saturation between 10^3 and 10^4 μl l^{-1}, both in the light and in darkness (142). A very similar saturation curve is obtained if stomatal conductances of *Xanthium strumarium* are related to the estimated $[CO_2]$ in the stomatal pore (152). Neales (119) has found in *Agave americana,* a plant exhibiting crassulacean acid metabolism (CAM), that during nocturnal CO_2 fixation, reduction of transpiration by CO_2 saturated between 1000 and 1800 μl l^{-1}. Nishida (122) found in other CAM plants a correlation between nocturnal acid accumulation and stomatal opening. Kluge & Fischer (80) demonstrated that this correlation was caused by stomatal responses to a lowering of the $[CO_2]$ in the leaf and not based on a rhythm endogenous to the guard cells.

The stomatal sensor for CO_2 is located in the guard cells themselves, as proven by Mouravieff (116) on isolated individual stomata, and as further indicated by the experiments of Pallaghy (127) with *Zea mays* epidermes in light and darkness. In summary, it can be stated that stomatal responses to CO_2 in the light and in darkness are manifestations of one and the same basic mechanism.

Stomatal sensitivity to CO_2 is specific for this molecule, or for $H CO_3^-$; applications of analogs of CO_2 like ethylene or allene (C_3H_4) (129) or SO_2 at low concentrations (99) do not abolish the effect of CO_2. Stomatal sensitivity to CO_2 does not decline during prolonged exposure to CO_2 (72). Unnaturally high $[CO_2]$ (5%, 20%) applied in darkness may cause opening (36, 92), although this has not been observed in *Zea mays* (142).

SENSITIZATION OF STOMATA TO CO_2 Stomata of some species are insensitive to CO_2 if the water potential of the leaf tissue is high. For instance, stomata of greenhouse grown and well-watered plants of *Xanthium strumarium,* once open, do not respond to CO_2 (143). These stomata however, can, be sensitized to CO_2 by raising the ABA level in the leaf (143). ABA produced in the leaf tissue when transpiration is high (143) or after plants have been chilled (24, in combination with 154) is as effective in sensitizing stomata to CO_2 as exogenous ABA added to the transpiration stream of detached leaves. There are indications that this may be true for other species, too (48, and observations on *Commelina communis*).

The observation of stomatal insensitivity to CO_2 at high water potentials in a few species resolves a difference of opinion between Zelitch (203) and most other stomatal physiologists. Zelitch believes that "in the normal range, CO_2 concentration does not have an important role in the normal opening and closing of stomata." He based his view on his experience with leaf disks floating on water. In my opinion, stomata of these samples had lost their sensitivity to CO_2.

STOMATAL RESPONSES TO LIGHT MEDIATED BY CO_2 Stomata respond to light only indirectly by responding to changes in the $[CO_2]$ brought about in the leaf by light. As early as 1932, Scarth (161) envisaged that stomata are not obligatorily light-operated mechanisms. Freudenberger (36) demonstrated that this was true for green and etiolated leaves brought into the light or kept in darkness. Previous reviewers (78, 103) have summarized the evidence for the role of CO_2 in the mediation of stomatal responses to light. Thus it is necessary here to deal with recent evidence only. In Pallaghy's (127) experiments, stomata in epidermal strips of *Zea mays* opened and the guard cells accumulated K^+ in complete darkness if the air was free of CO_2. Earlier experiments had already shown that stomata of *Zea mays* controlled air flow through the leaf in proportion to photosynthesis in the mesophyll (137), and that stomatal aperture did not depend on the direction of illumination, whether guard cells were illuminated directly or shaded by the mesophyll. Aperture, as indicated by air flow through the leaf, was the same in all cases as long as the amount of light received by the mesophyll remained the same. Further experiments were conducted (46) in which gas exchange was by diffusion and in which intercellular $[CO_2]$ was measured while the leaf samples of *Zea mays* were illuminated with white, red, or blue light. Of the monochromatic light applied, 99.5% was absorbed by the leaf. The epidermis facing the light source thus received 200 times more light than the epidermis away from the light. Nevertheless, stomatal conductances were the same whether an epidermis, upper or lower, was or was not illuminated directly, as long as the quantum flux received by the mesophyll, and consequently the rate of photosynthesis in the mesophyll, remained the same. In further experiments, stomatal conductances were determined while the CO_2 content of the air was varied in the light and darkness. Regardless of the method of altering the intercellular $[CO_2]$, whether by light, darkness, or by the CO_2 concentration in the air surrounding the leaf, one common curve could be fitted to the data points when plotted against the intercellular $[CO_2]$, indicating that stomatal aperture in *Zea mays* was controlled by the $[CO_2]$ in the substomatal cavity. At equal intercellular $[CO_2]$, however, illuminated stomata were open a little wider than the ones in darkness. A direct stomatal response to light, therefore, was superimposed on the curve representing the relationship between stomatal aperture and $[CO_2]$. This direct response can be attributed to the photosynthetic reduction of CO_2 by the chloroplasts within the guard cells (78). In darkness the intracellular $[CO_2]$ is probably above the intercellular $[CO_2]$; in the light it is below. Light controls stomatal opening by reducing the intercellular and intracellular $[CO_2]$. The relative importance of these two processes of CO_2 removal for stomatal behavior will depend on circumstances (such as temperature). Applications of inhibitors of photosynthesis are therefore expected to produce various stomatal responses. DCMU fed to leaves through the transpiration stream caused stomatal closure indirectly by inhibiting photosynthesis in the mesophyll and subsequently increasing intercellular $[CO_2]$ (18, 138). Support for this explanation comes from the observation that stomata retained their ability to open in CO_2-free air after a treatment with CMU (3) (which acts like DCMU). If DCMU is applied to epidermal strips in the light, stomatal aperture is reduced by varying degrees. In some situations, DCMU had a very small effect (64), in others a moderate one, probably increasing with time (128), and there is a report of complete closure caused by $10^{-5}M$ DCMU (86). Inhibition of stomatal opening by DCMU is expected to be most strongly expressed if the intracellular sources of CO_2 are strong and the CO_2-scrubbing action of the guard-cell chloroplasts is important in keeping the intracellular $[CO_2]$ low.

Stomatal opening occurring in the presence of inhibitors of electron transport between photosystems I and II of photosynthesis or in far-red light (64) provides no evidence for the participation of cyclic photophosphorylation in the provision of energy for the stomatal mechanism; neither is a calculation of the feasibility (131) of this assumption any proof. I also doubt my own old hypothesis (137) that the dependence of stomatal opening velocity in *Zea mays* on light intensity in CO_2-containing and CO_2-free air is an indication of a direct utilization of light energy for the opening movement. I rather assume now that this positive correlation was caused by the light-dependent depletion of the intercellular spaces of CO_2; stomatal sensitivity to changes in intercellular $[CO_2]$ is highest at low $[CO_2]$ (142). Stomata of *Xanthium strumarium* open in CO_2-free air in darkness as fast as in light, even after the leaves have been in complete darkness for 60 hr.

If stomata respond to the $[CO_2]$ in the guard cells, they are responding to the balance between (*a*) the intracellular evolution of CO_2; (*b*) diffusion of CO_2 into the guard cells and out of them; and (*c*) CO_2 consumption by photosynthesis and carboxylations in the dark. Relation of stomatal aperture to this interaction rather than just to the intercellular $[CO_2]$ makes stomata more sensitive to a change from light to dark than the saturation kinetics of stomatal closing in response to CO_2 (142) would indicate. Since half saturation occurs near 200 μl l^{-1} (142), one can compute what the relative stomatal aperture would be at $[CO_2]$ expected to be typical for leaves in the dark. For 1200 μl l^{-1}, for instance, I estimate a stomatal aperture which is 0.14 of the maximal. In other words, stomata would not close completely after photosynthesis stops were it not for the intracellularly produced CO_2. Obviously, small differences in the relative magnitudes of the sources and sinks for CO_2 will determine whether stomata are open or closed. In plants well supplied with water, completely closed stomata may be rare even in darkness. A hydroactive response may be necessary to shut the stomata tightly. It is a general experience that often very low light intensities are sufficient to lower the CO_2 level in the guard cells enough to produce opening. Mouravieff has provided an example (117) in which 25 to 30 pE $cm^{-2}s^{-1}$ of red or green light were sufficient to affect the CO_2 exchange of the leaf tissue and to initiate stomatal opening movements.

ACTION SPECTRA AND STOMATAL RESPONSES TO BLUE LIGHT   Considering the previous paragraph, it is not surprising that the action spectra of stomatal responses to light depend on the conditions of the measurement. Kuiper (86) took an action spectrum of maintenance of stomatal opening in epidermal strips of *Senecio odoris*. It resembled that of photosynthesis, with a blue light effect superimposed. Probably photosynthesis by guard cell chloroplasts was needed to keep the intracellular $[CO_2]$ low. As we would predict, Kuiper's material was sensitive to DCMU; application of this inhibitor led to stomatal closure. Karvé's (74) action spectrum of the velocity of stomatal opening in maize and Mansfield & Meidner's (101) determination of the wavelength dependency of stomatal aperture in *Xanthium* leaves agree with Kuiper's results. Further, the enhancement of stomatal opening by blue light was not caused by a lower $[CO_2]$ in the leaf than would be found in red light (101). Different results were obtained by Hsiao et al (62), who not only determined action spectra of stomatal aperture but also action spectra of ^{86}Rb uptake into guard cells of *Vicia faba*. Evidently the stomata of their leaf material were insensitive to CO_2, since the stomata were already open in the dark. As we would expect, only the blue-light effect was conspicuous, with a peak of action extending from 420 to 460 nm and no action > 560 nm (at 1.2 nE $cm^{-2}s^{-1}$). At higher quantum fluxes (6.3 nE $cm^{-2}s^{-1}$) a low peak in the red appeared; possibly photosynthetic reduction of CO_2 in the guard cells assisted opening. The action spectra for ^{86}Rb$^+$ uptake (as tracer for K$^+$) were almost identical with the action spectra for stomatal opening. Blue light is known to enhance respiration (83) and thus probably the energy supply to the turgor mechanism, and blue light probably activates PEP carboxylase (189). Keerberg et al (76) found the effect of blue light on stomata in *Aspidistra* but not in *Phaseolus*. Habermann (45) observed that blue light enhanced opening in wild type but not in the xantha mutant of *Helianthus* (45). Earlier reports of effects of blue light on stomata are summarized in *Physiology of Stomata* (103).

INVOLVEMENT OF PHYTOCHROME?   Transport of K$^+$ is involved in nyctinastic movements of leaves (158, 159) and the state of phytochrome in the motor tissue affects these processes. The similarity with stomatal movements has intrigued some people to postulate a participation of phytochrome also in stomatal movement. There is, however, more evidence against this proposal than for it (74). Evans & Allaway (30) devoted a special study to this question and concluded "phytochrome is involved in the closing reactions of *A. julibrissin* pinnules but apparently not in those of *V. faba* stomata." We (149) observed a transient opening response of maize stomata after exposing them to red light (λ = 660, 681 nm). By simultaneous illumination with light of two wavelengths, a wide range of P_{fr}/P ratios could be established (47), showing that phytochrome was not the pigment involved and that the response was probably mediated by the photosynthetic pigments. Habermann (45) has reported an effect of red and far-red light on stomata, but the observed stomatal apertures and their changes were so small that they could have been caused equally well by phytochrome-mediated changes in the epidermis or the mesophyll. We have therefore no conclusive evidence to date for the participation of phytochrome in the operation of the stomatal mechanism. There is as yet no species known in which presence and action of phytochrome is essential for stomatal action. This conclusion does not rule out the possibility that species may be found in which the state of phytochrome does modify stomatal responses.

CONCLUSION   Much evidence supports the view that the CO_2 concentration in the guard cells determines stomatal aperture; stomata respond to light indirectly by responding to the reductions in the $[CO_2]$ in the mesophyll as well as in the guard cells. These responses close feedback loops which synchronize stomatal opening with the demand of the mesophyll for CO_2.

Open stomata of water-saturated leaf tissue may lose the sensitivity to CO_2, but exposure to ABA sensitizes them again to CO_2. Blue light may enhance stomatal opening but is not essential for opening.

H_2O Feedback

STOMATAL RESPONSES TO DECREASES IN LEAF-WATER POTENTIAL AND THE INVOLVEMENT OF ABA   Almost all textbooks convey the impression that turgor changes produced by changes in the water potential in the leaf are sufficient to explain stomatal closing in a plant suffering water saturation deficits. In general, stomata are insensitive to reductions in water potential until a threshold is passed; then they close rapidly and more or less completely. The position of the threshold can be anywhere between −7 and −18 atm. Hsiao has recently thoroughly reviewed the literature pertaining to this relationship (60).

The switch-like action of the stomata may be inherent in the mechanics of the stomata. A test of this hypothesis on epidermal strips of *Vicia faba* did not confirm this notion. The decline in stomatal aperture with sinking water potential turned out to be gradual and stomata were open at water potentials as low as −60 atm (146). Perhaps the "switch action" of the stomata had been lost when the epidermis was separated from the mesophyll; Stålfelt (176) reported that the mechanical connection between mesophyll and epidermis affected stomatal opening. Our own tests of epidermis with mesophyll attached showed no threshold either, but a threshold appeared when more than a few minutes was allowed for stomatal adjustment (145). This result reminds one of Stålfelt's 1929 reports (176) that water loss from leaves of *V. faba* led to stomatal closure only after delay of about 13 min. This closure could not be reversed instantaneously by resupplying the leaves with water. Stålfelt suspected the involvement of a metabolic response and called this mechanism the hydroactive system. This mechanism evaded identification until very recently, when it was found that ABA was formed rapidly [within 7 min (104)] in wilting leaves (54, 94, 108, 114, 199, 202) and that the synthesis of ABA appeared to be accelerated when the leaf water potential dropped below about -10 atm (201) (this result needs confirmation because not only water potential but also time of exposure varied during the investigation). Another series of studies followed the reports of Little & Eidt in 1968 on the reduction of transpiration by ABA (90) and reports by Mittelheuser & Van Steveninck (106) on stomatal closure by ABA (summary in 60). The experiments of Hiron & Wright (54) impressively demonstrate how ABA is synthesized rapidly in wilting leaves, how stomata close as a consequence, how turgor is regained and ABA synthesis ceases in the recovering turgescent leaf after the regulatory task has been achieved. A doubling of the ABA content of leaves is usually enough to bring about stomatal closure (85, 152). More detailed investigations suggest that in short-term experiments the effect of ABA on stomata depends on the ABA concentration in the transpiration stream and on the strength of the transpiration stream and not on absolute amounts in the leaf (16, 152, 155).

We have no idea whether ABA synthesis is triggered by the reduction in water potential or turgor or the increase in the concentration of a triggering substance in a cell compartment. We assume that ABA is made in the mesophyll, probably in the chloroplasts (104, 134), and travels from there to the stomata where it prevents the uptake of K^+ by guard cells and starch breakdown (59, 100) or leads to a loss of K^+ from guard cells. Stomata respond to ABA within a few minutes (18, 85, 106).

If ABA functions as a messenger of water stress to the stomata, stomata should reopen once the ABA supply is stopped. In short-term experiments this was indeed the case (18, 58), provided the concentrations of ABA applied were not too high. The effect of ABA on stomata is direct and not mediated by an effect on the photosynthetic apparatus (18). Stomata are sensitive to the (+) enantiomer only (19, 85) contrary to what Milborrow stated (104). Trans-ABA is inactive (18, 85). We do not know the fate of ABA during the rapid reversal of stomatal closure after the supply of ABA to detached leaves has been stopped. Cummins (17) showed that the bulk of ABA is not metabolized fast enough to explain stomatal reopening. Is ABA sequestered into a compartment away from its site of action in the guard cells? There is an indication of compartmentation of ABA in the leaf (154). On the other hand, ABA accumulated in pools remote from the guard cells could be responsible for the aftereffects of wilting on stomata (see Hsiao 60) if these pools leak some ABA into the transpiration stream.

The evidence for the involvement of ABA in a hydroactive stomatal feedback system is considerable. Other messengers may also be involved, particularly substances structurally related to ABA, like xanthoxin (104); the search must go on.

HYDROACTIVE FEEDBACK WITHIN GUARD CELLS?   Guard cells rapidly lose solutes if isolated epidermal strips are exposed to water potentials below -7 atm (61, 145). Perhaps a supply of ABA from the mesophyll is not needed for stomatal closure in response to low water potentials and guard cells possess an internal hydroactive system (see below). More work is needed to answer this question.

ABSENCE OF HYDROACTIVE FEEDBACK UNDER CERTAIN CIRCUMSTANCES? When leaves are subjected to very rapid water loss, stomata do not close at a threshold value of leaf water content (66, 139). Does ABA not form rapidly enough?

Or is Iljin's explanation correct that starch is hydrolyzed in rapidly wilting leaves, thus preventing turgor loss, and stomata open widely after rewatering (67)? Kriedemann (personal communication) observed that stomata in leaves of water-stressed *Vitis labruscana* (cv. Concord) opened within a few minutes after rewatering. It is difficult to conceive that catabolism of ABA and K^+ import into guard cells proceeded as fast as leaves regained their turgor. We must rather assume that either the passive, mechanical response to low water potentials was exceptionally effective in regulating water loss in Concord grape vines, or that the effect described by Iljin (67) occurred. Complete absence of stomatal control was reported for one *Acacia* species (49).

POSITIVE HYDROPASSIVE FEEDBACK   While discussing stomatal mechanics (see above) we have seen that changes in epidermal turgor may lead to stomatal responses in the "wrong" direction, e.g. a reduction in water supply leading to reduction in epidermal pressure on the guard cells which in turn leads to a hydropassive stomatal opening and increased transpiration. Similarly, improved water supply often leads to hydropassive closing. These responses can be both transient, as in the Iwanoff effect (69), or lasting (139).

EFFECT OF CUTICULAR ("PERISTOMATAL") TRANSPIRATION ON STOMATAL MOVEMENT   Guard cells gain and lose water by exchange with the neighboring epidermal cells, and can also gain or lose water by evaporation and condensation. Seybold (as discussed in 97) postulated that direct water loss from the guard cells through the cuticle and the cuticular ledges to the atmosphere (peristomatal transpiration) serves as a mechanism sensing the humidity of the air. Maercker (97), using autoradiography after feeding tritiated water to leaves, has provided some evidence for the hydrophilic property of the cuticular ledges and the cuticle above the guard cells. She also observed an accumulation of Tl^+ in guard cells of open stomata and subsidiary cells of closed stomata after adding Tl^+ to the transpiration stream. She interpreted these results as indicating variations in cuticular transpiration from the two cell types. This conclusion is incorrect because Tl^+ is a good tracer for K^+ and therefore unsuitable for the indication of terminal points of the transpiration stream. Nevertheless, the experiments with tritiated water would speak for a stronger cuticular transpiration from guard cells than from ordinary epidermal cells.

Lange et al (88) directed little jets delivering dry or moist air to groups of stomata in isolated epidermes enclosed in little chambers under the microscope. Stomata closed when the humidity of the air flowing over the cuticle was low and opened when the humidity was high. Stomatal aperture depended also on the humidity in the substomatal cavity. That direct evaporation from the epidermis in general and from the guard cells in particular can have a modifying effect on stomatal aperture was also concluded from the observation that detached maize leaves of equal water contents had much higher transpiration resistances in dry air than in moist air (139). Schulze et al (164) found that this was true also in the field. In several species growing in a desert the apparent paradox occurred that in humid air transpiration was higher and leaf water content lower than in dry air. Caldwell (13) reports that stomata of *Rhododendron ferrugineum* are sensitive to wind, while those of *Pinus cembra* are not. Differences in water loss through the cuticle may have caused the difference in stomatal sensitivity to wind. The direct loss of water from the epidermis to the atmosphere may thus exert an open-loop control over stomatal aperture. Under certain circumstances even liquid water can be lost through the cuticle above the guard cells and other epidermal cells if turgor is high. Arens observed this phenomenon (5); pine needles can exude liquid water (163), and I saw "cuticular guttation" in leaf sections of *Zea mays.*

INTERDEPENDENCE OF STOMATAL RESPONSES TO ABA AND CO$_2$ In *Xanthium strumarium,* and possibly in other species also, the presence of ABA is required to make stomata sensitive to CO_2 (143) and stomata of *X. strumarium* respond to ABA only if CO_2 is available (152). A concentration of $10^{-5}M$ ($\pm$) ABA in the transpiration stream has virtually no effect on stomata of *X. strumarium* in CO_2–free air, if the rate of CO_2 evolution by the tissue is not too high. Stomata respond with decreasing delay and increasing velocity of closing to ABA if the [CO_2] in the air is increased. Stålfelt's (177) observation that hydroactive closing was promoted by CO_2 in *Vicia faba* and *Rumex sativa* may refer to the same phenomenon. The enhancement by CO_2 of the effect of ABA on stomata followed saturation kinetics with respect to [CO_2] [apparent $K_m(CO_2) = 170 \ \mu l \ l^{-1}$]; the enhancement by ABA of the stomatal response to CO_2 did not follow saturation kinetics exactly; apparent half-saturation occurred between 10^{-6} and $10^{-5}M$ ($\pm$) ABA at 300 $\mu l \ l^{-1}$ CO_2 (152). The interdependence of stomatal responses to CO_2 and ABA imparts flexibility to the plant with respect to water use efficiency. At high humidity stomata will be completely open if the sun shines. If water deficit develops, some ABA will be formed which sensitizes the stomata to CO_2 (143) and makes stomatal conductance proportional to the requirement for CO_2. Further water stress leads to further production of ABA which overrides the opening response to the depletion of CO_2 of the intercellular spaces by photosynthesis. Abscisic acid appears to act like an endocrine regulating the mutual relationship between the exchange of CO_2 and water vapor by the plant. ABA fed into the transpiration stream leads to an increase in the water use efficiency. In detached leaves of *X. strumarium,* $10^{-5}M$ ($\pm$) ABA reduced the transpiration ratio from 68 to 38 g H_2O/g CO_2, with a reduction of photosynthesis by only 14% (143). The increase in dry weight of young barley plants painted with a 100 μM solution of ABA after 9 days was only 5% less than that of the controls while the transpiration ratio was reduced by 30% (73). The water-saving effect of ABA applications was most striking in the experiments of Mizrahi et al (110). The dry matter of unwatered wheat seedlings, sprayed every 3 days with $3.8 \times 10^{-4}M$ ABA for 32 days, was 47% heavier than that of controls sprayed with water. ABA delayed drought-induced senescence. Reduction of transpiration by ABA may mean survival under conditions of salinity, water-logging, and drought (107, 110).

OTHER EFFECTS OF ABA AND INTERACTION WITH KINETIN   In addition to the direct involvement of ABA in the metabolism of guard cells, there are other effects of ABA which might affect stomatal operation indirectly. ABA given at high concentration increases the water permeability of plant tissues (43) and the exudation from tomato roots (179). There is also a slow induction of stomatal narrowing observed by Tal & Imber (180) in the wilty tomato mutant *flacca* (whose stomata stay open in darkness and light) after eight sprays with ABA solutions during 24 hr. This slow closure is a response different from the rapid stomatal reaction to ABA. Cummins (16) has shown that while stomata in general and stomata of wild type tomato in particular respond to applications of ABA within minutes, stomata of *flacca* do not respond to ABA (or to CO_2 for that matter) in short-term experiments.

In the water relations of plants, kinetin appears to act like an antagonist to ABA (43, 106, 109, 179). Transpiration from detached leaves treated with kinetin is considerably higher than water loss from detached control leaves. Pallas & Box (130) explain this effect as resulting from a hydropassive stomatal opening in the treatments due to a reduction of the solute content of the epidermal cells. It appears at least equally probable that kinetin acts indirectly on stomata by preventing aging. Thus the detached control leaves would be the true "treatments" in which proteolysis proceeded and CO_2 uptake decreased, and the kinetin-treated leaves would be the true "controls" in which original apertures and rates of transpiration were maintained because kinetin checked the acceleration of senescence. Kinetin does not enhance stomatal opening in epidermal strips (gibberellic acid and IAA are ineffective also) (58, 123, 186). Kinetin antagonizes the effect of ABA in detached leaves, but does so only to a small extent (106). The antagonism is pronounced in whole plants (109).

Temperature Feedback

TEMPERATURE DEPENDENCE OF STOMATAL OPENING AND FEEDBACK ON LEAF TEMPERATURE   In general, stomatal opening follows the temperature curve of CO_2 assimilation (50, 140) and the CO_2 feedback functions. According to my interpretation of the work of Neales (120) and Ting et al (184), this hypothesis holds also for the temperature dependence of stomatal opening in CAM plants. Dark fixation of CO_2 decreases in *Agave americana* with increasing night temperature as does stomatal aperture. At temperatures above 35°, stomata of plants well supplied with water become insensitive to CO_2; the stomata are open even if CO_2 assimilation has stopped and CO_2 evolves from the leaf tissue (> 40°C) (140). Schulze et al (165) found that stomata of desert plants open with increasing temperature as long as water stress does not interfere. Loss of stomatal sensitivity to CO_2 at high temperatures can be of great benefit to the plant, as it not only reduces the danger of overheating if sufficient water is available, but also because it keeps the leaf near the optimum temperature for photosynthesis (25, 140). In many species the temperature of brightly illuminated leaves is near air temperature if the latter is ~ 33°C. Below this temperature, leaves tend to be warmer than air; above, they are cooler (89). This peculiarity has two causes: (*a*) the exponential increase in water-vapor pressure with temperature, resulting in an increased negative feedback of transpiration; (*b*) the increase in stomatal aperture with temperature. An analysis of these relationships was made on leaves of *Xanthium strumarium* (25). The crossover temperature was 35°C and stomatal resistance decreased with temperature to a value as low as 0.36 s cm⁻¹ at a leaf temperature of 40°C. The effect was observed in the field, too (25). The largest effect of transpirational cooling on leaf temperature was reported by Lange (87). Temperatures of leaves of *Citrullus colocynthis* growing in a north African oasis were 15° below an air temperature of 50°C.

STOMATAL METABOLISM (CLUES AND CONJECTURE ON THE CONTROLLER)

In this section I will speculate on the relationships between the following processes observed to occur in guard cells during stomatal opening: (*a*) uptake of K^+ into the vacuoles; (*b*) excretion of H^+ from guard cells; (*c*) production of organic acids, particularly malic acid; (*d*) disappearance of starch; (*e*) uptake of Cl^- into the vacuoles. Further, I shall try to explain how CO_2 and ABA may be involved in stopping or reversing these processes.

EXCHANGE OF K^+ FOR H^+ Any one of processes *a, b, c,* or *e* could be the initiator of stomatal opening. Is *a* an active uptake of K^+, responsible for stomatal opening, with all other processes following as consequences? Pallaghy (125) found the electric potential in the guard cells of tobacco to be negative (as in other plant cells) and the potential difference between the cell and the medium to change by about 43 mV for each tenfold change in the concentration of K^+ or Na^+ in the medium. This indicates that K^+ was passively held in the guard cells. (In view of the importance of this conclusion, confirmation and extension of Pallaghy's work is necessary.) The finding that valinomycin caused only a partial loss of K^+ from guard cells of *Zea mays* (6) would support the view that most of the K^+ accumulated in guard cells followed a redistribution of charge. On the other hand, it cannot be ruled out that valinomycin, a hydrophobic substance, was not able to penetrate in large quantity the thick, hydrophilic walls of the guard cells. It is also possible that valinomycin interfered with energy supply and not with K transport into the guard cells. No evidence has been provided yet for the presence of a cation-transporting ATPase in the plasmalemma or tonoplast of the guard cells, and ATPase similar to the one occurring in roots (55). The activity ascribed by Fujino (41) to ATPase in guard cells could have resulted from an acid phosphatase, similar to the ones reported by others (98, 171). Ouabain, an inhibitor blocking the ATPase involved in K^+/Na^+ exchange (53), was shown by Thomas (181, 182) to act on guard cells but no confirmation is available. Recently Turner (187) presented some evidence for an effect of ouabain on stomata in the light. In Mouravieff's (118) experiments ouabain closed stomata but only in CO_2 containing air; in CO_2-free air it was ineffective. Mouravieff suspects that ouabain affects CO_2 uptake and not the stomatal mechanism.

If K^+ is taken up into guard cells following a gradient in electrical potential, an active import of Cl^- (process *e*) could be the driving mechanism. However, since Cl^- rarely balances K^+ completely in guard cells (65, 128, 148), one may conclude that processes *b* and *c*, synthesis of organic acids and excretion of H^+ (150), are the active processes responsible for at least the uptake of the K^+ into the guard cells that is not accompanied by Cl^-. Hydrogen-ion excretion from guard cells is a primary process in stomatal opening, perhaps the most important one. Cations follow the electrical gradient, perhaps by facilitated diffusion. Specificity of ion carriers or channels in the membranes would then account for the stomatal selectivity for K^+. The hypothesis receives support from other work on ion uptake by plants (53, 55, 172, and the papers collected in 4). Transport of Cl^- into the guard cells might be coupled to the electrogenic pump in a way proposed for anion uptake into roots (55) and into cells of *Chara* (170, 173). The presence of a hydrogen ion pump does not exclude altogether an active import of K^+ into guard cells. One could conceive that a K^+ pump aids the stomatal mechanism, particularly at low K^+ concentrations external to the guard cells. The data of Humble & Hsiao (63) on high specificity for K^+ and Rb^+ at low concentrations, but no specificity at high concentrations, could be taken to indicate this.

Expulsion of H^+ from guard cells during stomatal opening would make the cells' interior not only more negative but also more basic. This rise in cytoplasmic pH could trigger the production of organic acids, predominantly malic, as is envisaged to occur in roots (51). These acids provide H^+ for exchange for K^+ and anions to balance the charges of K^+ in the vacuole. The intracellular pH of guard cells is reported to rise during stomatal opening (37, 41, 78). If this phenomenon is not an artefact due to metachromasy of the indicator dyes used (26), then it provides further evidence in favor of the H^+ expulsion hypothesis.

METABOLISM OF ORGANIC ACIDS   Obviously, the well-known disappearance of starch grains in guard cells of opening stomata suggests that the organic acids in the guard cells are derived from starch. Phase shifts between stomatal movement and starch content of guard cells were observed (124, 200) and are not surprising, considering the length of the metabolic paths and the number of pools involved. Direct evidence for the derivation of malate from starch, however, is missing. The ability of guard cells to make starch from glucose-1-phosphate was suggested as evidence for the presence of phosphorylase (124), although the presence of phosphorylase itself was not demonstrated. We have no unequivocal evidence for glycolysis taking place in guard cells, nor for the pentose shunt, nor gluconeogenesis. But we do have evidence supporting the contention that malate is made in the guard cells by carboxylation of phosphoenolpyruvate (PEP). In a pioneering investigation, Willmer et al (198) have demonstrated that the PEP carboxylase activity of epidermal strips is proportional to the number of stomata in the sample. Autoradiographs (122, 167, 198) prove that the ability of the epidermis to fix CO_2 is concentrated in the guard cells. I would estimate that the CO_2 requirement for malate production during stomatal opening is not entirely met by the CO_2 evolved from respiration in guard cells. External CO_2 may be needed as supplement. This requirement for exogenous CO_2 may explain why in some species, like *Xanthium strumarium* (24), sometimes also in *Gossypium hirsutum* and *Zea mays,* maximal stomatal opening does not occur in CO_2-free air but at $[CO_2]$ around 100 μl 1^{-1}.

Participation of glycolic acid metabolism in stomatal action had been proposed by Zelitch (203), but the evidence was indirect. If direct evidence should be presented in the future, one would have to consider the production of part of the stomatal malate from glycolate; guard cells do contain microbodies (168, 183).

ENERGY SUPPLY FOR STOMATAL OPENING   The formation of malate from starch via PEP is exergonic, but transport of malate into the vacuole as well as the expulsion of H^+ and the active uptakes of K^+ and Cl^- (if they are active) will require ATP or reducing equivalents. Expansion and turgor increase of the guard cells (of *Vicia faba*) require at least 0.5 erg per opening movement of a stoma, or the equivalent of the free energy change occurring during the oxidation of about 5×10^{-14} mole glucose (assuming an efficiency of energy utilization of 0.3). Besides this, energy is needed to maintain and turn over enzymes and carriers. Thus various inhibitors are capable of affecting stomatal opening. Some of them are uncouplers, such as DNP, FCCP, and CCCP (39, 128, 196). Others are inhibitors of energy supply or of the transport of phosphorylated compounds like NaN_3 and NaF (39, 41, 128, 190); still others are inhibitors of protein synthesis and ion transport like oligomycin, chloramphenicol, and cycloheximide (128). Detergents like Triton X-100 may also cause stomatal closure effectively, most probably by destroying membranes. The energy required for stomatal movement seems to come from oxidative phosphorylation. Electron micrographs show the presence of many mitochondria in guard cells (105, 168, 175, 183), and activity of dehydrogenases of the tricarboxylic acid cycle has been demonstrated in guard cells (40).

In crassulacean plants, malate formation in the dark by fixation of CO_2 requires oxygen (135). Stomatal movement also requires oxygen, although concentrations of O_2 as low as 50 to 100 microlites/liter are sufficient to release blockage of stomatal movement by anaerobic conditions (93); the oxygen requirement for stomatal movement is fully satisfied at 1.5% (1).

Stomata can open in the dark as rapidly and widely as in the light. It was therefore concluded that light does not directly supply energy to the stomatal mechanism (see above).

STOMATAL CLOSURE   We do not know the metabolic processes going on during stomatal closure. Related observations are: The temperature dependence of closure is smaller than that of opening (141) and seems to indicate a passive or highly catalyzed movement of K^+ (and Cl^-) from the guard cells to the epidermal tissue in general and to the subsidiary cells in particular, if they are present. The assumption that subsidiary cells serve as ion exchangers (see above) however, may be too simple. The subsidiary cells of maize contain numerous mitochondria (175, communication by H. Ziegler, and own observation) which may indicate an active reabsorption of K^+ by subsidiary cells during closure. Strictly anaerobic conditions prevent stomatal closure (1, 93), and NaF (41) as well as NaN_3 (39, 190) are reported to arrest stomata in the open state. Nothing is known about the fate of malate during stomatal closure.

MODULATION OF THE STOMATAL METABOLISM BY CO_2 If CO_2 is required for making malate in opening stomata, how can CO_2 in the vicinity of guard cells lead to stomatal closure? Saturation kinetics of the velocities of stomatal closing are similar to those of PEP carboxylase; it is tempting to speculate that malate formation is not only necessary for stomatal opening but also responsible for stomatal closing. Further investigations will have to show whether the malate level in the cytoplasm of guard cells determines direction and magnitude of ion fluxes into the vacuoles and out of them. The malate level in the cytoplasm reflects the balance between malate formation, malate removal into the vacuole, and deacidification. It is known that high malate concentration inhibits PEP carboxylase and thus its own synthesis. There may also exist a feedback loop through the cytoplasmic pH. In roots, malate formation decreases with decreasing pH, although PEP carboxykinase seems to be involved there (51). Nondissociated aliphatic acids, including malic, cause leakage of ions from roots (70). If the same processes occur in guard cells, high CO_2 would lead to an acidification through malate formation (the changes in carbonic acid levels would produce pH changes of the order of 0.2 pH units only and therefore are too small to be used for the control of the metabolism of the guard cells) (142). This acidification would cause stoppage of formation of further malate and deflation. Removal of CO_2 would lead to a rapid deacidification of the tissue; again analogies can be drawn to occurrences in the root (51, 52). Malic enzyme is present in the epidermis (198). PEP carboxylase and the malic dehydrogenases appear to form a push-pull system which adjusts the intracellular malate level and pH in some relationship to the intercellular $[CO_2]$. This mechanism may serve as the CO_2 sensor of the stomata. It appears worthwhile to investigate whether the differential properties of PEP carboxylase and malate dehydrogenase with respect to temperature (10) are the cause of increasing stomatal opening with temperature. The presence of a thick external cuticle above the guard cells ensures that guard cells respond more to the $[CO_2]$ in the substomatal cavity and the pore than to the atmospheric $[CO_2]$. Guard cells utilizing Cl^- as a major counter ion for K^+ should be more sensitive to CO_2 than the ones employing malate because much malate remains in the cytoplasm and acts as a feedback signal. Stomata of *Zea mays* utilize Cl^- (148) and are indeed highly sensitive to CO_2 (136, 142).

Other authors (62, 128) have tried to explain the CO_2 sensitivity of stomata as resulting from competition for energy between CO_2 fixation and operation of the stomatal mechanism. The effects of CO_2 on ion uptake into leaves of *Elodea* (71a) and cells of *Chara* (173) provide analogies. According to the competition hypothesis, stomatal opening would occur only if energy was left over from CO_2 fixation. I doubt whether basing the function of an important control system like that of the stomata on quite a variable residual term would have survived evolution. Stomatal closure would be the result of a continuous metabolic activity; stomata would open as soon as the postulated fixation of CO_2 stopped, e.g. when the guard cells ran out of PEP, or through product inhibition. There are no observations proving the occurrence of those effects. On the contrary, stomata are known to be able to open in response to CO_2-free air in darkness as fast as in the light, even after they stayed closed for 60 hr in darkness. The competition hypothesis lacks positive evidence.

MODULATION OF STOMATAL METABOLISM BY ABA   The discovery of the simultaneous requirement of CO_2 and ABA for the modulation of stomatal aperture in *Xanthium strumarium* (152) and observations of a similar nature in other species might provide a clue to the mechanism of action of ABA. By elimination of other hypotheses, it appears that ABA reversibly blocks the active excretion of H^+ from guard cells. In the presence of CO_2 this would lead to a rapid acidification of the cytoplasm and to stomatal closure as postulated in the previous paragraph. In many species, the intracellularly evolved CO_2 may suffice to elicit acidification. In species with strong H^+ pumps, malate formation would lead to acidification of the cytoplasm only in the presence of ABA. This would explain sensitization of stomata to CO_2 by ABA (152).

Stomatal opening (and K^+ accumulation in guard cells) can be enhanced by fusicoccin (187); fusicoccin even overcomes the inhibitory action of ABA on stomata (174, 186). This fungal toxin is known to stimulate excretion of H^+ by internodal segments and isolated cotyledons while making the intracellular electrical potential more negative (102). Fusicoccin acts like an additional H^+ pump. Fusicoccin stimulates growth, probably by excretion of H^+ into the cell walls (102). ABA inhibits growth and H^+ secretion of coleoptiles (156). The similarity of the responses of growing systems and guard cells to ABA and fusicoccin is striking and keeps alive the hypothesis that ABA inhibits H^+ expulsion. It was suggested that ABA acts on stomata by inhibiting the formation of α-amylase (104). It seems doubtful that the inhibition of starch breakdown would be fast enough to explain the rapid stomatal response to ABA and that stomatal closure in response to ABA could be explained by an inhibition of α-amylase.

INTRACELLULAR HYDROACTIVE FEEDBACK   It may not be necessary to invoke the participation of ABA in the rapid solute loss from guard cells after exposure to low water potentials (p. 326). Kluge & Heininger (81) have recently shown that vacuoles of CAM plants rapidly begin to lose malate if tissue slices are exposed to water potential <-2 atm, and that this effect saturated at approximately -12 atm. This phenomenon may also be related to the leakiness of roots caused by aliphatic acids (70).

CONCLUSION   Guard cells can adjust their turgor by metabolism of organic acids (mainly malic) and movement of inorganic ions (mainly K^+, H^+, and Cl^-). Direction and magnitude of these processes are possibly controlled by the pH in the cytoplasm, which in turn reflects the relative rates of production and removal of malate. CO_2 is not only a substrate for the acidification of the cytoplasm; it is also needed for the production of osmotica. ABA enhances acidification by inhibiting expulsion of H^+ from guard cells. Malate is required for both opening and closing, however, in different compartments of the guard cell. The open state of the stoma appears as likely as the closed one. Circumstances determine which of the partial mechanisms in the guard cells is active and triggers the other ones. This model of the stomatal metabolism is yet unproven.

DYNAMICS OF STOMATAL ACTION

The stomatal control system is never in equilibrium except during stomatal closure, and even that is not certain. Whenever one feedback loop is "satisfied" another one is not. For example, when CO_2 feedback does not call for further opening, transpiration may deplete the water resources of the plant and hydropassive and hydroactive responses will follow. This results in the well-known morning maximum of stomatal opening in plants growing in the field. Responses to light will depend on epidermal turgor and the moisture content of the air. Therefore, stomata in attached leaves or dry air open faster than in detached leaves or moist air (e.g. 139). Stomatal responses are often ambiguous and therefore difficult to interpret. For example, if leaves of *Zea mays* are darkened for 10 min and then illuminated again, stomata reopen with a rapid movement lasting a few minutes and a slower opening movement follows (137). It has not been possible thus far to find out whether this two-phased response is due to a change of the water relations of the stomata and the leaf tissue or due to a metabolic response in the guard cells. Another example of the difficulty in interpreting stomatal responses is the midday closure of stomata (103). Many reports on stomatal behavior are of little value because the authors have not tried sufficiently to reduce the ambiguity of stomatal responses; quoting examples would serve no useful purpose.

STOMATAL OSCILLATIONS Stomatal responses take time; water moves, pools must be emptied or filled, ions transported, molecules are synthesized or degraded, etc. Transportation lags and times required to metabolize substances in the required amounts introduce delays into the feedback loops, leading to transient responses to perturbations. If a delayed negative feedback signal arrives at the guard cells at a time when it reinforces the initial response instead of reducing it (180° phase shift), oscillations may arise in the feedback loop. Whether oscillations will occur and then die out or will be sustained depends on the relative magnitude of the feedback signal or the degree of amplification of the original disturbance in the feedback loop (as any text on control theory will describe). Any environmental or physiological noise can cause the disturbance. Oscillations with a period of a few minutes occurring after changes in $[CO_2]$ (8) or light intensity (137) have been explained as oscillations caused by a lag in negative feedback in the CO_2 loop. While it is true that these oscillations were triggered by a response of the CO_2 feedback system, it has not yet been proven that the oscillations were maintained by a feedback of CO_2; they could equally well have been maintained by one of the H_2O feedback loops. The ambiguity of stomatal responses holds here too. Farquhar (31) found stomata of cotton got into resonance with sinusoidal perturbations of $[CO_2]$ or humidity at similar frequencies (around 1.5 hr^{-1}).

POSITIVE FEEDBACK Reinforcement of the original signal by a feedback signal does not need any delay if the feedback is positive. Based on observations of other authors, Cowan (14) has designed a simple model of the hydraulic system of the plant in which positive feedback is the essential feature. He can demonstrate with his model the previously made observation that hydropassive positive feedback speeds up stomatal responses. More importantly, he can show that positive feedback can easily give rise to stomatal oscillations. Farquhar (31) described features of Cowan's model in mathematical terms and confirmed it experimentally with whole cotton plants, indicating synchronization of stomatal behavior in the whole plant.

By adjusting the humidity in the air, he kept transpiration constant and thus independent from stomatal movement. As expected, oscillations ceased because a stomatal feedback loop was opened.

Stomatal oscillations need not be triggered by changes in the environment or plant. If the degree of amplification (gain) in one of the feedback loops is high enough and the condition of reinforcement of the feedback signal is met, autonomous oscillations may arise, triggered only by the noise in the environment or plant (14). The gain of a stomatal feedback loop depends not only on plant parameters but also on environmental conditions (31). The occurrence of stomatal oscillations may not be rare. Barrs (8) has surveyed the pertaining literature and discussed earlier models; Johnsson (71b) has contributed more recently. Brogårdh & Johnsson (12) demonstrated that Li^+, given to detached plants, slows oscillations of transpiration and that D_2O either shortens or lengthens the period of the oscillations, depending on how D_2O is administered (11). Penning de Vries (133) presented a model of stomatal functioning which, unlike Cowan's (14), included photosynthesis in the leaf and feedback related to CO_2; I hope it will be used to enhance our understanding of peculiarities in stomatal behavior. In general, models of stomatal behavior have been useful in emphasizing our ignorance. Particular versions of models may be more useful than general ones in that they provide tools to study stomatal behavior with time. One purpose of such models may be to make predictions of the diurnal changes in stomatal conductance in a particular species in a particular, changing environment. Another purpose may be to evaluate the possibility that a certain feature of the stomatal mechanism is responsible for a certain observed phenomenon. We have far too little information, e.g. on time constants of components of the stomatal mechanism, to design predictive models which come anywhere near the truth, but we can use models as analytical tools as Cowan (14) has done. His model has its value even if we realize that it does not include CO_2 feedback and hydroactive feedback, that guard cells and subsidiary cells probably are put in series instead of in parallel, and that the values used for the osmotic pressure in guard cells are too low by an order of magnitude. Models of this kind should, of course, not be used to predict quantitatively stomatal regulation of transpiration and photosynthesis in the field.

POSSIBLE ADVANTAGE OF OSCILLATORY BEHAVIOR OF STOMATA   Stomatal instability has been seen as a result of imperfect design. This need not be true. As Cowan (14) suggested, stomatal oscillations may affect favorably the ratios between CO_2 assimilation and transpiration and may optimize the relationship between assimilation and growth. Stomatal oscillations may also be looked upon as exploratory excursions in the approach to an optimal state (14). For a plant which continuously changes its physiological condition, homeostasis is not necessarily the optimal solution for success in a continuously changing environment.

We need more knowledge not only of the components of stomatal metabolism and plant hydraulics but also of their behavior with time if we want to understand how the stomatal feedback system helps to provide the plant with food while preventing thirst.

ACKNOWLEDGMENTS

I thank my colleagues G. D. Farquhar, A. Lang, M. Pierce, K. L. Poff, C. A. V. K. Richman, and Y. Vaadia for contributing to this review by helpful criticism. The responsibility for the expressed views, however, is solely mine. My unpublished research referred to in the review was supported by the U.S. Atomic Energy Commission under contract No. AT (11-1)-1338.

Literature Cited

1. Akita, S., Moss, D. N. 1973. *Plant Physiol.* 52:601–3
2. Allaway, W. G. 1973. *Planta* 110:63–70
3. Allaway, W. G., Mansfield, T. A. 1967. *New Phytol.* 66:57–63
4. Anderson, W. P., Ed. 1973. *Ion Transport in Plants.* London: Academic. 630 pp.
5. Arens, T. 1968. *Protoplasma* 66:403–11
6. Arntzen, C. J., Haugh, M. F., Bobick, S. 1973. *Plant Physiol.* 52:569–74
7. Aylor, D. E., Parlange, J. Y., Krikorian, A. D. 1973. *Am. J. Bot.* 60:163–71
8. Barrs, H. D. 1971. *Ann. Rev. Plant Physiol.* 22:223–36
9. Bearce, B. C., Kohl, H. C. Jr. 1970. *Plant Physiol.* 46:515–19
10. Brandon, P. C. 1967. *Plant Physiol.* 42:977–84
11. Brogårdh, T., Johnsson, A. 1974. *Physiol. Plant.* 31:112–18
12. Brogårdh, T., Johnsson, A. 1974. *Z. Naturforsch.* 29c:298–300
13. Caldwell, M. M. 1970. *Plant Physiol.* 46:535–37
14. Cowan, I. R. 1972. *Planta* 106:185–219
15. Cowan, I. R., Troughton, J. H. 1971. *Planta* 97:325–36
16. Cummins, W. R. 1971. *On the stomatal response to abscisic acid.* PhD thesis. Michigan State Univ., East Lansing
17. Cummins, W. R. 1973. *Planta* 114:159–67
18. Cummins, W. R., Kende, H., Raschke, K. 1971. *Planta* 99:347–51
19. Cummins, W. R., Sondheimer, E. 1973. *Planta* 111:365–69
20. Dayanandan, P., Kaufman, P. B. *Am. J. Bot.* In press
21. DeMichele, D. W., Sharpe, P. J. H. 1973. *J. Theor. Biol.* 41:77–96
22. Domes, W. 1971. *Planta* 98:186–89
23. Downes, R. W. 1969. *Planta* 88:261–73
24. Drake, B., Raschke, K. 1974. *Plant Physiol.* 53:808–12
25. Drake, B., Raschke, K., Salisbury, F. B. 1970. *Plant Physiol.* 46:324–30

26. Drawert, H. 1968. *Protoplasmologia,* Vol. IID3. Vienna: Springer. 749 pp.
27. Ehrler, W. L., Van Bavel, C. H. M. 1968. *Plant Physiol.* 43:208–14
28. Epstein, E., Rains, D. W., Elzam, O. E. 1963. *Proc. Nat. Acad. Sci. USA* 49:648–92
29. Esau, K. 1965. *Plant Anatomy.* New York: Wiley. 767 pp.
30. Evans, L. T., Allaway, W. G. 1972. *Aust. J. Biol. Sci.* 25:885–93
31. Farquhar, G. D. 1973. *A study of the responses of stomata to perturbations of environment.* PhD thesis. Australian Nat. Univ., Canberra
32. Fischer, R. A. 1968. *Science* 160:784–85
33. Fischer, R. A. 1972. *Aust. J. Biol. Sci.* 25:1107–23
34. Fischer, R. A. 1973. *J. Exp. Bot.* 24:387–99
35. Fischer, R. A., Hsiao, T. C. 1968. *Plant Physiol.* 43:1958–68
36. Freudenberger, H. 1941. *Protoplasma* 35:15–54
37. Fujino, M. 1959. *Kagaku* 29:147–48
38. Ibid, 424–25
39. Ibid, 660–61
40. Ibid 1960. 30:89–90
41. Fujino, M. 1967. *Sci. Bull. Fac. Educ. Nagasaki Univ.* 18:1–47
42. Geiger, R. 1966. *The Climate Near the Ground.* Cambridge, Mass.: Harvard Univ. Press. 611 pp.
43. Glinka, Z., Reinhold, L. 1971. *Plant Physiol.* 48:103–5
44. Haapala, H. 1967. *Aquilo Sec. Bot.* 5:120–32
45. Habermann, H. M. 1973. *Plant Physiol.* 51:543–48
46. Hanebuth, W. F., Raschke, K. 1973. *Plant Research '72, MSU/AEC Plant Res. Lab., Mich. State Univ.,* 139–44
47. Hartmann, K. M. 1966. *Photochem. Photobiol.* 5:349–66
48. Heath, O. V. S., Mansfield, T. A. 1962. *Proc. Roy. Soc. B* 156:1–13
49. Hellmuth, E. O. 1969. *J. Ecol.* 57:613–34

50. Hesketh, J. D., Hofstra, G. 1969. *Can. J. Bot.* 47:1307–10
51. Hiatt, A. J. 1967. *Z. Pflanzenphysiol.* 56:233–45
52. Hiatt, A. J., Hendricks, S. B. 1967. *Z. Pflanzenphysiol.* 56:220–32
53. Higinbotham, N. 1973. *Ann. Rev. Plant Physiol.* 24:25–46
54. Hiron, R. W. P., Wright, S. T. C. 1973. *J. Exp. Bot.* 24:769–81
55. Hodges, T. K. 1973. *Advan. Agron.* 25:163–207
56. Höfler, R. 1939. *Protoplasma* 33: 258–74
57. Holmgren, P., Jarvis, P. G., Jarvis, M. S. 1965. *Physiol. Plant.* 18:557–73
58. Horton, R. F. 1971. *Can. J. Bot.* 49:583–85
59. Horton, R. F., Moran, L. 1972. *Z. Pflanzenphysiol.* 66:193–96
60. Hsiao, T. C. 1973. *Ann. Rev. Plant Physiol.* 24:519–70
61. Hsiao, T. C. 1973. *Plant Physiol.* 51: Suppl. 9 (Abstr.)
62. Hsiao, T. C., Allaway, W. G., Evans, L. T. 1973. *Plant Physiol.* 51:82–88
63. Humble, G. D., Hsiao, T. C. 1969. *Plant Physiol.* 44:230–34
64. Ibid 1970. 46:483–87
65. Humble, G. D., Raschke, K. 1971. *Plant Physiol.* 48:442–53
66. Iljin, W. S. 1915. *Beih. Bot. Zentralbl.* 32:15–35
67. Iljin, W. S. 1932 *Jahrb. Wiss. Bot.* 77:220–51
68. Imamura, S. 1943. *Jap. J. Bot.* 12:251–346
69. Iwanoff, L. 1928. *Ber. Deut. Bot. Ges.* 46:306–10
70. Jackson, P. C., Taylor, J. M. 1970. *Plant Physiol.* 46:538–42
71. Jeffree, C. E., Johnson, R. P. C., Jarvis, P. G. 1971. *Planta* 98:1–10
71a. Jeschke, W. D., Simonis, W. 1969. *Planta* 88:157–71
71b. Johnsson, A. 1973. *Physiol. Plant.* 28:40–50
72. Jones, R. J., Mansfield, T. A. 1970. *J. Exp. Bot.* 21:951–58
73. Jones, R. J., Mansfield, T. A. 1972. *Physiol. Plant.* 26:321–27
74. Karvé, A. 1961. *Z. Bot.* 49:47–72
75. Kaufman, P. B., Petering, L. E., Yocum, C. S., Baic, D. 1970. *Am. J. Bot.* 57:33–49
76. Keerberg, H., Keerberg, O., Pärnik, T., Vill, J., Värk, E. 1971 *Photosynthetica* 5:99–106
77. Kelle, A. 1934. *Zur Physiologie der Nebenzellen des Spaltöffnungsapparates.* PhD thesis. Univ. Münster, Germany
78. Ketellapper, H. J. 1963. *Ann. Rev. Plant Physiol.* 14:249–70
79. Kluge, M. 1968. *Planta* 80:255–63
80. Kluge, M., Fischer, K. 1967. *Planta* 77:212–23
81. Kluge, M., Heininger, B. 1973. *Planta* 113:333–43
82. Koch, W. 1969. *Flora B* 158:402–28
83. Kowallik, W. 1971. *Photosynthesis and Photorespiration,* ed. M. D. Hatch, C. B. Osmond, R. O. Slatyer, 514–22. New York: Wiley-Interscience. 565 pp.
84. Kramer, P. J. 1969. *Plant and Soil Water Relationships.* New York: McGraw-Hill
85. Kriedemann, P. E., Loveys, B. R., Fuller, G. I., Leopold, A. C. 1972. *Plant Physiol.* 49:842–47
86. Kuiper, P. J. C. 1964. *Plant Physiol.* 39:952–55
87. Lange, O. L. 1959. *Flora* 147:595–651
88. Lange, O. L., Lösch, R., Schulze, E. D., Kappen, L. 1971. *Planta* 100:76–86
89. Linacre, E. T. 1964. *Agr. Meteorol.* 1:66–72
90. Little, C. H. A., Eidt, D. C. 1968. *Nature* 220:498–99
91. Lloyd, F. E. 1925. *Flora* 118/119: 369–85
92. Louguet, P. 1965. *Physiol. Vég.* 3:345–53
93. Ibid 1972. 10:515–28
94. Loveys, B. R., Kriedemann, P. E. 1973. *Physiol. Plant.* 28:476–79
95. Ludlow, M. M., Wilson, G. L. 1971. *Aust. J. Biol. Sci.* 24:449–70
96. Macallum, A. B. 1905. *J. Physiol.* 32:95–128
97. Maercker, U. 1965. *Protoplasma* 60: 61–78
98. Ibid, 173–91
99. Majernik, O., Mansfield, T. A. 1972. *Environ. Pollut.* 3:1–7
100. Mansfield, T. A., Jones, R. J. 1971. *Planta* 101:147–58
101. Mansfield, T. A., Meidner, H. 1966. *J. Exp. Bot.* 17:510–21
102. Marrè, E., Lado, P., Ferroni, A., Ballarin Denti, A. 1974. *Plant Sci. Lett.* 2:257–65
103. Meidner, H., Mansfield, T. A. 1968. *Physiology of Stomata.* New York: McGraw-Hill. 179 pp.
104. Milborrow, B. V. 1974. *Ann. Rev. Plant Physiol.* 25:259–307
105. Miroslavov, E. A. 1971. *Bot. Zh.* 56:485–92
106. Mittelheuser, C. J., Van Steveninck, R. F. M. 1969. *Nature* 221:281–82
107. Mizrahi, Y. 1972. *Mechanisms involved in the adaptation of the plant shoot to*

root stress (in Hebrew). PhD thesis. Hebrew Univ., Jerusalem, Israel
108. Mizrahi, Y., Blumenfeld, A., Bittner, S., Richmond, A. E. 1971. *Plant Physiol.* 48:752–55
109. Mizrahi, Y., Blumenfeld, A., Richmond, A. E. 1970. *Plant Physiol.* 46:169–71
110. Mizrahi, Y., Scherings, S. G., Malis Arad, S., Richmond, A. E. 1974. *Physiol. Plant.* 31:44–50
111. Mohl, H. von 1856. *Bot. Ztg.* 14:697–704, 713–21
112. Monzi, M. 1939. *Jap. J. Bot.* 9:373–94
113. Moreshet, S. 1970. *Plant Physiol.* 46:815–18
114. Most, B. H. 1971. *Planta* 101:67–75
115. Mount, M. S., Bateman, D. F., Basham, H. G. 1970. *Phytopathology* 60:924–31
116. Mouravieff, I. 1956. *Le Botaniste* 40:195–212
117. Mouravieff, I. 1963. *Ann. Sci. Nat. Bot.* 12:225–32
118. Mouravieff, I. 1972. *Physiol. Vég.* 10:547–51
119. Neales, T. F. 1970. *Nature* 228:880–82
120. Neales, T. F. 1973. *Aust. J. Biol. Sci.* 26:705–14
121. Neales, T. F., Patterson, A. A., Hartney, V. J. 1968. *Nature* 219:469–72
122. Nishida, K. 1963. *Physiol. Plant.* 16:284–98
123. Ogunkanmi, A. B., Tucker, D. J., Mansfield, T. A. 1973. *New Phytol.* 72:277–82
124. Ono, H. 1953. *Bot. Mag.* 66:182–88
125. Pallaghy, C. K. 1968. *Planta* 80:147–53
126. Pallaghy, C. K. 1970. *Z. Pflanzenphysiol.* 62:58–62
127. Pallaghy, C. K. 1971. *Planta* 101:287–95
128. Pallaghy, C. K., Fischer, R. A. 1974. *Z. Pflanzenphysiol.* 71:332–44
129. Pallaghy, C. K., Raschke, K. 1972. *Plant Physiol.* 49:275–76
130. Pallas, J. E. Jr., Box, J. E. Jr. 1970. *Nature* 227:87–88
131. Pallas, J. E. Jr., Dilley, R. A. 1972. *Plant Physiol.* 59:649–50
132. Pallas, J. E. Jr., Wright, B. G. 1973. *Plant Physiol.* 51:588–90
133. Penning de Vries, F. W. T. 1972. *J. Appl. Ecol.* 9:57–77
134. Railton, I. D., Reid, D. M., Gaskin, P., MacMillan, J. 1974. *Planta* 117:179–82
135. Ranson, S. L., Thomas, M. 1960. *Ann. Rev. Plant Physiol.* 11:81–110
136. Raschke, K. 1965. *Z. Naturforsch.* 20b:1261–70
137. Raschke, K. 1966. *Planta* 68:111–40
138. Raschke, K. 1967. *Ber. Deut. Bot. Ges.* 80:138–44
139. Raschke, K. 1970. *Plant Physiol.* 45:415–23
140. Raschke, K. 1970. *Planta* 91:336–63
141. Ibid. 95:1–17
142. Raschke, K. 1972. *Plant Physiol.* 49:229–34
143. Raschke, K. 1973. *Proc. 8th Int. Conf. Growth Regul.* In press
144. Raschke, K., Dickerson, M. 1973. *Plant Research '72, MSU/AEC Plant Res. Lab., Mich. State Univ.,* 153–54
145. Raschke, K., Dickerson, M., Pierce, M. 1973. *Plant Research '72, MSU/AEC Plant Res. Lab., Mich. State Univ.,* 149–53
146. Ibid, 155–57
147. Raschke, K., Dunn, W. 1971. *Plant Research '70, MSU/AEC Plant Res. Lab., Mich. State Univ.,* 37–41
148. Raschke, K., Fellows, M. P. 1971. *Planta* 101:296–316
149. Raschke, K., Fellows, M. P. 1972. *Plant Research '71, MSU/AEC Plant Res. Lab., Mich. State Univ.,* 38–41
150. Raschke, K., Humble, G. D. 1973. *Planta* 115:47–57
151. Raschke, K., Pierce, M. 1973. *Plant Research '72, MSU/AEC Plant Res. Lab., Mich. State Univ.,* 146–49
152. Ibid 1974. *Plant Research '73,* 25–28
153. Ibid, 29–30
154. Raschke, K., Pierce, M., Popiela, C. C. 1974. *Plant Research '73, MSU/AEC Plant Res. Lab., Mich. State Univ.,* 28
155. Raschke, K., Zeevaart, J. A. D., Pierce, M., Popiela, C. C. 1974. *Plant Research '73, MSU/AEC Plant Res. Lab., Mich. State Univ.,* 28–29
156. Rayle, D. L. 1973. *Planta* 114:63–73
157. Rentschler, I. 1974. *Planta* 117:153–61
158. Satter, R. L., Applewhite, P. B., Galston, A. W. 1974. *Plant Physiol.* 54:280–85
159. Satter, R. L., Galston, A. W. 1971. *Plant Physiol.* 48:740–46
160. Sawhney, B. L., Zelitch, I. 1969. *Plant Physiol.* 44:1350–54
161. Scarth, G. W. 1932. *Plant Physiol.* 7:481–504
162. Schönherr, J., Bukovac, M. J. 1972. *Plant Physiol.* 49:813–19
163. Scholz, F. 1974. *Biochem. Physiol. Pflanz.* 165:253–63
164. Schulze, E. D., Lange, O. L., Buschbom, U., Kappen, L., Evenari, M. 1972. *Planta* 108:259–70
165. Schulze, E. D., Lange, O. L., Kappen,

L., Buschbom, U., Evenari, M. 1973. *Planta* 110:29–42
166. Schwendener, S. 1882. *Monatsber. Berl. Akad. Wiss.* 833–67
167. Shaw, M., Maclachlan, G. A. 1954. *Can. J. Bot.* 32:784–94
168. Singh, A. P., Srivastava, L. M. 1973. *Protoplasma* 76:61–82
169. Slatyer, R. O. 1970. *Planta* 93:175–89
170. Smith, F. A. 1972. *New Phytol.* 71:595–601
171. Sorokin, H. P., Sorokin, S. 1968. *J. Histochem. Cytochem.* 16:791–802
172. Spanswick, R. M. 1972. *Biochim. Biophys. Acta* 288:73–89
173. Ibid 1974. 332:387–98
174. Squire, G. R., Mansfield, T. A. 1972. *Planta* 105:71–78
175. Srivastava, L. M., Singh, A. P. 1972. *J. Ultrastruct. Res.* 39:345–63
176. Stålfelt, M. G. 1929. *Planta* 8:287–340
177. Stålfelt, M. G. 1959. *Physiol. Plant.* 12:691–705
178. Ibid 1965. 18:177–84
179. Tal, M., Imber, D. 1971. *Plant Physiol.* 47:849 –50
180. Tal, M., Imber, D. 1972. *New Phytol.* 71:81–84
181. Thomas, D. A. 1970. *Aust. J. Biol. Sci.* 23:981–89
182. Ibid 1971. 24:689–707
183. Thomson, W. W., De Journett, R. 1970. *Am. J. Bot.* 57:309–16
184. Ting, I. P., Thompson, M. L. D., Dugger, W. M. Jr. 1967. *Am. J. Bot.* 54:245–51
185. Todd, D. K. 1970. *The Water Encyclo-pedia.* Port Washington, NY: Water Inform. Center. 559 pp.
186. Tucker, D. J., Mansfield, T. A. 1971. *Planta* 98:157–63
187. Turner, N. C. 1973. *Am. J. Bot.* 60:717–25
188. Vaihinger, K. 1942. *Protoplasma* 36:430–43
189. Voskresenskaya, N. P. 1972. *Ann. Rev. Plant Physiol.* 23:219–34
190. Walker, D. A., Zelitch, I. 1963. *Plant Physiol.* 38:390–96
191. Whiteman, P. C., Koller, D. 1967. *New Phytol.* 66:463–73
192. Wiggans, R. G. 1921. *Am. J. Bot.* 8:30–40
193. Willmer, C. M., Dittrich, P. 1974. *Planta* 117:123–32
194. Willmer, C. M., Mansfield, T. A. 1969. *Z. Pflanzenphysiol.* 61:398–400
195. Willmer, C. M., Mansfield, T. A. 1969. *New Phytol.* 68:363–75
196. Ibid 1970. 69:639–45
197. Willmer, C. M., Pallas, J. E. Jr. 1973. *Can. J. Bot.* 51:37–42
198. Willmer, C. M., Pallas, J. E. Jr., Black, C. C. Jr. 1973. *Plant Physiol.* 52:448–52
199. Wright, S. T. C. 1972. *Crop Processes in Controlled Environments,* ed. A. R. Rees et al, 349–61. London: Academic
200. Yamashita, T. 1952. *Sieboldia* 1:51–70
201. Zabadal, T. J. 1974. *Plant Physiol.* 53:125–27
202. Zeevaart, J. A. D. 1971. *Plant Phsyiol.* 48:86–90
203. Zelitch, I. 1969. *Ann. Rev. Plant Physiol.* 20:329–50
204. Ziegenspeck, H. 1938. *Bot. Arch.* 268–309, 332–72

Ann. Rev. Plant Physiol. 1975. 26:341–67

APICAL DOMINANCE

♦7594

I. D. J. Phillips

Department of Biological Sciences, University of Exeter, Washington-Singer Laboratories, Exeter, EX4 4QG, Devon, United Kingdom

CONTENTS

INTRODUCTION

Basic physiological systems such as photosynthesis, water and nutrient uptake, cell division and enlargement, respiration, etc are well studied if not all describable in detail at the present time. Even given these, however, there remains the problem of finding out how they are integrated to form a whole plant. As plants typically exhibit open-ended patterns of development, they must possess during the whole of their lives systems for coordinating and regulating the processes of growth and differentiation. In this respect plants offer better subjects for the study of development than do most animals, for the determinate development of the latter largely restricts research on their developmental processes to studies of often small embryos.

One of the ways in which spatial patterns of differentiation arise during the development of multicellular organisms is by a mechanism involving positional information (195, 196). Cells, tissues, and organs interpret positional information from other regions according to their genome and developmental history. Thus positional signaling demands that the positions of cells be specified with respect to boundary regions, and also that there exist a gradient in some parameter. In experiments on animals and cellular slime molds, a distinction has been drawn between models for positional signaling which envisage a propagated (perhaps periodic) signal and those which rely on diffusion of a substance along a concentration gradient between its source and eventual sink in the differentiating cells (36). In studies of animals, emphasis is being placed increasingly upon the importance of mechanisms based on diffusion (35, 36, 197).

Studies of development in higher plants should be helped by their life-long embryonic condition in particular regions, though it cannot be said that understanding of the regulation of developmental processes in plants is any more advanced than that for animals. One characteristic feature of development in many plants is that more shoot apical meristems are initiated than fully develop. This has clear survival value by providing a reservoir of meristems which can replace the apical bud should the latter be damaged or removed by wind or grazing activities of animals. The apical bud meristem is commonly dominant to some degree over the lateral, often axillary, bud meristems. The manner in which the dominance of the apical over lateral buds is achieved has been subject to fairly intensive investigation for almost a hundred years. A primary reason for this interest is that the phenomenon lends itself to ready experimentation into the basis of the spatial organization of developmental activities in plants. Similar to current work on animal differentiation, investigations of apical dominance in plants have tended to stress the role of a diffusible signal. This signal apparently arises in the apical bud, and its effects are expressed in several aspects of shoot morphogenesis, particularly in the correlative inhibition of lateral buds, but also in the regulation of developmental patterns in larger lateral organs such as leaves, branches, rhizomes, and stolons.

THE NATURE OF THE CORRELATIVE SIGNAL

Although it is a commonplace observation that a growing apical bud inhibits the development of other lateral shoot buds, the way in which this influence is transmitted is still a matter of controversy. Early investigators of the problem (see 123) and also some more recent and current workers (52, 57, 79, 94–99) emphasize the importance of the supply of inorganic and organic nutrients, as well as water, in regulating lateral bud development. Much more research, however, has been concentrated upon the possibility that the correlative signal from the apical bud is of a hormonal nature (see 123). It is quite likely that both concepts have validity, and in fact attention has been paid to the regulatory effects of known growth regulators on the distribution of nutrients within the shoot system (41, 123). It is often extremely difficult to distinguish between possible correlative roles of various factors (e.g. nutrients and hormones) as they affect lateral bud outgrowth and their undoubted involvement in biochemical events within the buds themselves and which underly their morphogenetic development.

Hormonal or Nutritional Control?

It may be hoped that some synthesis of accumulated information on the environmental factors or experimental treatments which result in outgrowth or suppression of axillary buds would provide an insight into the nature of correlative forces acting on bud development. The most obvious treatments which allow previously inhibited buds to grow out are those which involve either complete removal of the growing apical bud or prevention of its growth by physical means or by exposure to low temperature (123). Such knowledge does not in itself allow any conclusion to be drawn as to whether the correlative signal is hormonal or purely nutritional, for an active apical bud is both a source of auxin and gibberellin (149) and also a sink for nutrients and water. Stem-girdling treatments, on the other hand, have shown that interruption of the continuum of living cells along the stem (by either ringing or heat-girdling) usually results in outgrowth of lateral buds situated below the girdle, and yet at the same time the upper portion of the shoot continues to grow and consume nutrients and water (61, 156). This suggests that lack of growth by lateral buds is attributable to the downward movement through living tissues of an influence originating in the apical part of the shoot, and not to their starvation of nutrients or water. It can be argued, however, that growth activities of the apical bud above the girdle could be sufficiently diminished to increase the availability of nutrients and/or water to lateral buds, but even so this does not readily explain why only the buds below the girdle grow out. Other evidence against the view that competition for major nutrients provides the physiological basis for correlative bud inhibition includes the absence of stimulatory effect of direct application of nutrients to nondormant potato tuber lateral buds (56), and the observation that NPK levels in *Phaseolus vulgaris* were at least as high in the apical 5 mm of correlatively inhibited axillary buds as in the equivalent growing region of buds released from inhibition by decapitation (121). In pea, inhibited lateral buds contain storage starch, and this together with other evidence led to the conclusion that carbohydrate availability did not limit their growth (180). Similarly, studies of the dominance relationships between unequal buds and shoots revealed that the concentration of amino acids in the inhibited (–) bud was at least equal to that in the dominant (+) bud (100), and that transmission of the inhibitory influence from the + bud to the – bud could occur through a graft union of parenchyma cells, even when nutrient supply conditions were completely equal for both buds (38, 39).

Recognition that the morphological form of lateral shoots in *Solanum andigena,* either as negatively geotropic green leafy shoots or as diageotropic leafless stolons, can be regulated by applications of auxins, gibberellins, and cytokinins (10, 11, 80, 198–200) suggests that the correlative signals involved are hormonal rather than nutritional. Similarly, the angles at which branches and leaves are borne to the stem appear to be regulated by activities of the apical bud or dominant shoot, and here again it has been observed repeatedly that exogenous auxin can in many instances substitute for the apical bud in maintenance of plagiogeotropic or diageotropic behavior by lateral appendages (70, 82, 114, 129, 159, 160, 174).

Considerable evidence thus argues against the view that nutrient availability and supply comprises the basic correlative mechanism regulating lateral bud outgrowth. However, certain studies over the past 20 years (52, 57, 79, 94–99, 176) have been interpreted as demonstrating that nutrient competition and water status are the major factors in the mechanism of apical dominance (see 57, 99). Although it is unquestionable that starvation of nutrients or water would inhibit lateral bud growth, whether the supply of these to the buds serves as the principal mechanism of correlative control of their growth seems doubtful. The main pieces of evidence that have been cited to support the concept of a purely nutritional basis for apical dominance have been records of increased bud or shoot growth in response to either reduction in water stress (96, 97, 99) or increases in inorganic and organic nutrient availability (52, 57, 79, 94–96, 98, 99, 123, 176). Some of these experiments were concerned with the growth of lateral buds in the presence of the apical bud and others with relative growth of + and − paired buds or shoots in decussate species. Nevertheless, it is impossible to conclude on such evidence that the mechanism of correlative inhibition of buds is nutritional rather than hormonal. Measurements of the regulatory effects which nutrient and water supply can have on lateral bud growth have not included determinations of growth in all other parts of the plant. It is therefore possible that an increase in bud growth in response to these factors does not represent a change in the distribution of growth in the plant, but is merely one component of an overall increase with no alteration in the strength of apical dominance. References cited above (100, 121, 180) showed that levels of NPK, amino acids, and carbohydrate were not lower in inhibited buds than in growing buds which had been released from correlative inhibition. On the other hand, data has been presented repeatedly showing that total nutrient contents in buds are, not surprisingly, greater in buds which have escaped from inhibition (57, 94–97, 99). To infer from this that nutrient supply provides the basis for apical dominance does not seem justified. What is more relevant is the concentrations of nutrients in those parts of a bud which are either growing or have the potential for growth, and in *P. vulgaris* and *Pisum sativum* these were found to be no lower in inhibited than in growing buds (100, 121, 180). Further evidence against a purely nutritional basis of correlative bud inhibition is provided by the failure of direct applications of nutrients to buds to abolish or reduce inhibition of one bud by another (38, 39, 56).

It seems likely then that the correlative signal in apical dominance has a hormonal basis. This does not necessarily mean that nutrient and water status are unimportant. Apart from the obvious requirements of growing buds for these substances, it is known that nutrient and water levels can affect hormone levels and distribution in the plant. As long ago as 1937 it was found that inorganic nutrient supply to roots influenced auxin synthesis in the apical bud (7), and more recently cytokinin (66), gibberellin (120, 146), and abscisic acid (63, 64, 150, 201) levels have been found to be markedly affected by water potential in the plant. In general, it is reasonable to conclude that interacting nutritional and hormonal conditions must be appropriate to allow bud outgrowth.

Hormonal Correlative Control

Having already considered some of the bud growth responses to nutritional conditions, it is now appropriate to deal with other factors which appear to affect both endogenous hormonal status and apical dominance with a view to the identification of the primary hormonal signal from the apical bud.

Very soon after it was demonstrated that auxin, probably indole-3-acetic acid (IAA), was synthesized in growing apical buds, it was reported that exogenous IAA could substitute for the apical bud in maintaining inhibition of the axillary buds in bean plants (165, 166). This observation has been confirmed innumerable times in succeeding years for many species. A few exceptions were reported, especially *Coleus,* in which it was found impossible to replace the apical bud with auxin (69). However, recently it has been shown that even in *Coleus* the inhibitory effect of apically applied exogenous IAA is revealed when the plant's normally weak apical dominance is strengthened by reducing its nutritional status (164). Thus there seems little doubt that a primary component of the inhibitory correlative signal is the synthesis of auxin in the shoot tip region, probably the young expanding leaves (188), and its transport down the stem. The recognition that gibberellins are also synthesized in young leaves and exported from them down the stem (72, 73) has led to suggestions that this class of growth hormone also plays a primary regulatory role in correlative bud inhibition (68, 119a, 119b, 131, 142). Treatment with gibberellin of the cut end of a decapitated stem most commonly results in an increased rate of lateral bud outgrowth (15, 74, 122, 124), whereas intact plants frequently show enhanced apical dominance in response to gibberellin treatment (92a, 119a, 119b, 123). It was suggested previously (123) that increased apical dominance in intact plants treated with gibberellin is an indirect effect of the enhancement of growth in the main shoot, and perhaps increased endogenous auxin levels (65, 127, 136). In decapitated *P. coccineus* plants, it was found that gibberellin was antagonistic to auxin on bud outgrowth when both hormones were applied to a fully elongated internode, but that when the application was made to a younger elongating internode, gibberellin gave a slight increase in the inhibitory effect of auxin (125). It was suggested (125) that gibberellin transport through an elongating internode was less than through a mature internode, and that in consequence a smaller quantity of exogenous gibberellin applied to a young internode would reach the lateral buds situated below. This proposal has received support from studies of [^{3}H]-GA_1 transport in *P. coccineus* internodes (126). An alternative explanation put forward by Cutter (38) is that the interacting effects of auxins and gibberellins may differ according to the stage of development of the buds at the time of treatment. It is not possible with available information to decide between the possibilities, although Cutter's suggestion is certainly supported by evidence for sequential roles for different growth regulators during lateral bud outgrowth (3, 29, 134, 135). The control of shoot form in woody species does appear to involve interactions between gibberellins and auxins in lateral branch growth and plagiogeotropism (92a, 119a, 119b, 123).

A further line of evidence that indicates a primary role for auxin in correlative control of development has come from studies of the effects of inhibitors of auxin transport. The auxin transport inhibitors most frequently used in apical dominance studies have been 2,3,5-triiodobenzoic acid (TIBA) (107) and morphactins such as 2,chloro-9-hydroxyfluorene carboxylic acid-(9)-methyl ester (CFM) (78). Treatment of plants with these substances often results in either a reduction or abolition of apical dominance (e.g. 41, 139–141), but their use in research into apical dominance has been criticized recently (192) on the grounds that they not only inhibit auxin transport but also have marked morphological effects on both the main shoot and lateral buds (45, 77, 192, 193). Ethylene, which can also inhibit auxin transport (20, 103), similarly has been found to have effects on correlative inhibition of buds. Continuous exposure of intact plants to ethylene inhibited growth of the main apex but did not bring about outgrowth of lateral buds, whereas a pulse treatment with ethylene for 1–2 days was followed by outgrowth of axillary buds (17, 18, 21, 22). It is possible that the stimulating effect of a brief period of ethylene treatment on lateral bud growth derives from an interruption of auxin transport from the apical bud. This may result from inhibition by ethylene of either polar basipetal auxin transport (20, 103) or of auxin biosynthesis (48), or perhaps both. Why an interruption in auxin transport should result in release of lateral buds from apical dominance is not clear, but the problem is discussed later in connection with the role of vascular development between lateral buds and main stem.

Normally, ethylene does not in itself appear to be concerned as a correlative signal in the regulation of lateral bud inhibition. No diminution of ethylene production was observed in pea plants after their decapitation, and apical dominance was not broken by either hypobaric conditions or increased CO_2 (16, 21), either of which treatments eliminated other symptoms of ethylene action (19). Inhibition by ethylene of lateral bud outgrowth in decapitated plants is probably but one of the many examples of growth inhibition by the gas. The same thing may be said of inhibition by ethylene of sprouting in potato tubers (47). As discussed below, absence of growth in correlatively inhibited buds appears to involve an inability of apical cells to divide. Ethylene is also known to inhibit cell division in apical meristems (4, 5), and it is therefore not surprising that lateral buds do not grow out in the continuous presence of exogenous ethylene. Yet it is interesting to note that cytokinin treatment not only induces growth of lateral buds on intact plants (3, 29, 134, 135, 137), but also overcomes the inhibitory effect of ethylene on lateral bud growth in decapitated plants (21, 22).

There is very little evidence that the other classes of known plant growth hormones, cytokinins and abscisic acid, operate as correlative signals in apical dominance. On the other hand, there are numerous data indicating that cytokinins are essential for lateral bud outgrowth. This has been shown in various ways, but particularly by demonstrations of outgrowth of lateral buds on intact plants following direct applications of cytokinins to the buds themselves (3, 29, 134, 135, 137). It is not known whether the cytokinins required for lateral bud growth have to be supplied from elsewhere in the plant, for example the roots (8, 28, 75), or whether they are synthesized within the bud tissues. If the former is the case, then cytokinin perhaps can be regarded as serving as a correlative signal for bud growth, although even this is doubtful, for the primary correlative factor would be that which determines the supply of cytokinins to the buds. In fact, any concept that the regulation of bud growth resides in the partitioning of a finite and limiting supply of cytokinins between apical and lateral buds infers that cytokinins are no more nor less important than nutrients and water. Recent work by Tucker & Mansfield (168, 169) has indicated that the level of endogenous cytokinins was much higher in inhibited axillary buds of *Xanthium strumarium* than in buds released from inhibition. Consequently, these workers have suggested that correlatively inhibited buds are unable to utilize cytokinins.

A possible role for abscisic acid (ABA) or other growth inhibitory hormones in lateral bud inhibition has been put forward several times (see 123). Dörffling (42–44) found that the levels of several growth inhibitors fell in lateral buds of pea and sycamore following their release from apical dominance. More recently, ABA levels were found to be some 50–250 times greater in nongrowing than growing lateral buds of *Xanthium strumarium* (168, 169). Although the presence of high levels of ABA within lateral buds may play a part in preventing their growth, it is difficult to envisage ABA serving as a correlative signal from the apical bud. This is because most evidence points to mature leaves as the principal sites of ABA synthesis, from where it is translocated acropetally into the apical bud (64, 184). For this very reason it is possible that ABA is concerned in examples of the correlative inhibition of an axillary bud by the mature leaf which subtends it (55). Tucker & Mansfield (168, 169) have suggested that it is the high concentration of ABA in *X. strumarium* lateral buds which prevents them from responding to the high levels of endogenous cytokinins which they measured in inhibited buds.

The primary hormonal correlative signal in correlative inhibition of lateral buds by the apical bud therefore appears to be auxin derived from young growing leaves. The same also seems to be true for other manifestations of apical dominance such as the regulation of orientation and development of branches, leaves, rhizomes, and stolons (70, 82, 114, 123, 129, 159, 160, 174, 198). This does not mean, of course, that other hormonal and also nutritional factors are unimportant. The possible mechanism of auxin action in apical dominance and its interaction with other factors is discussed later.

MORPHOLOGY AND CYTOLOGY OF ARRESTED BUD DEVELOPMENT

All vegetative buds on a plant possess essentially equal developmental potential. The fact that the apical bud is usually more active than the lateral buds suggests that the circumstances of the latter are in some ways less advantageous. The degree of disadvantage experienced by lateral buds clearly varies between species, and within any one species depending on the physiological conditions of individual plants. In all cases it appears that the development of each lateral bud is arrested at some stage. This can occur very early in its ontogeny, so that the bud remains extremely small, perhaps consisting only of a group of organized apical meristematic cells, as for example in the shoot of *Tradescantia* (9, 106) and the rhizomes of *Matteuccia struthiopteris* and *Onoclea sensiblis* (177, 178). In many other species buds normally develop further before apical dominance is clearly imposed, so that morphologically distinct but inhibited buds are present, each possessing short internodes and unex-panded leaves; this is the case in legumes such as pea and bean, which have served as the most popular experimental subjects in studies of correlative inhibition of buds. In yet other plants, such as *Coleus,* lateral bud development normally is not com-pletely arrested, and lateral buds continue growing as leafy shoots, although still subject to an inhibitory influence from the apical bud (69).

Lateral bud development involves many aspects of plant growth and differentia-tion. Thus blockage of one or more of the biochemical events which underly morphogenesis will result in arrested bud development. Particularly relevant are the processes concerned with apical cell division activity, cell extension growth in the internodes, and the initiation and expansion of leaves. Each of these aspects of morphogenesis has received considerable study, and it is clear that they may be regulated by a number of different and interacting factors. In plants such as *Coleus* with normally weak apical dominance, there is no complete block to any of the various morphogenetic processes for lateral shoot growth. On the other hand, in plants which show complete apical dominance, all the processes are completely suppressed, although it may well be that only one of them is directly influenced by the correlative signal from the apical bud. For example, in cases of early cessation of lateral bud development, it would appear that apical meristematic activity is held in check, and in consequence no further formation or development of shoot tissues occurs in the bud. Inhibition by the apical bud of growing lateral shoots, as in *Coleus*, may also primarily involve an effect on the lateral apical meristem, although there is no direct evidence available to support this proposition. Since apically synthesized auxin appears to be the principal correlative signal in examples of both complete apical dominance (where inhibition of lateral bud apical cell division is clear) and incomplete dominance (164), then it is possible that in both cases cell division activity at the lateral apex is suppressed. However, in plants with weak apical dominance like *Coleus*, it can be assumed that growth of the lateral shoots can be stimulated through influences on one or more of the developmental processes already proceeding at less than maximum rates, whereas in species such as *Tradescantia, Matteuccia,* and *Onoclea,* removal of the block to apical cell division activity is an essential prerequisite for outgrowth of lateral buds. Even within one species it is possible that apical dominance is achieved through a number of different factors which attain varying degrees of relative importance during lateral bud ontogeny (38). An extreme example of correlative inhibition of buds is provided by the inhibition by the apex of adventitious bud initiation in the hypocotyl of seedling flax (*Linum usitatissimum*). The adventitious buds are formed from epidermal cells, and auxin can substitute for the apex in suppressing their initiation (30, 59, 85).

A number of investigators have examined the morphology and cytology of lateral buds at varying stages during the imposition of correlative inhibition, and also during the early stages of their outgrowth following experimental removal of apical dominance. Such studies have provided some useful pieces of information about the stages of lateral shoot development which are blocked by correlative forces, and thereby aided the interpretation of experiments involving the effects of exogenous growth regulators on apical dominance. Lateral buds of species with complete apical dominance cease cell division activity very soon after the time of their initiation (9, 37, 38, 106, 177, 178). Decussate species often show anisoclady (86), a condition where the two opposite leaves at one node subtend buds or shoots of unequal size. In some decussate species such as *Hygrophila* and *Alternanthera,* it has been found that paired buds are initiated simultaneously, but that cell division goes on for somewhat longer in the larger bud than in its opposite weaker partner, and that should apical dominance be removed, then the bud with the greatest number of cells develops more vigorously and becomes dominant over its smaller partner (37, 38).

This suggests that dominance of one bud over another can occur very soon after their initiation, and certainly before either has formed leaves. It is difficult to visualize how this occurs in terms of auxin transmission as a correlative signal, for there is much evidence that auxin is synthesized primarily in young developing leaves rather than in the meristematic cells of the shoot apex (149). Even so, grafting experiments on *Hygrophila* led to the conclusion that the larger of each pair of lateral buds exerted dominance over the smaller by hormonal rather than nutritional means (39). In other decussate species which show anisoclady, the dominant bud of a pair appears to be initiated earlier than its partner, and in such cases the correlative relationship between paired buds at a node may be initiated in a similar manner to that between the apical bud and later-formed lateral buds (108).

From the above considerations it is clear that regrettably few studies have been conducted of the morphology and cytology, and virtually none of the physiology, of the changes in lateral buds which occur during the imposition of correlative inhibition. More attention has been paid to events in buds subsequent to their release from inhibition. The physiological aspects of lateral bud outgrowth are considered in the next section of this review, but it is appropriate to deal here with studies of the early morphological and cytological events in buds after their release from correlative inhibition. Several reports demonstrate that no mitoses occur in the apical meristem of a lateral bud on a plant showing complete apical dominance (9, 106, 138, 171, 172). However, there is conflicting evidence on the question as to which stage of the cell division cycle is blocked. For inhibited *Tradescantia* buds it was reported (106) that the apical cells were in the 2C condition (i.e. containing the diploid DNA content), which suggests that DNA synthesis was blocked at the G1 stage of the cell cycle. More recent work on *Cicer arietinum,* on the other hand, has indicated that renewed apical cell division commenced only 1 hr after release from correlative inhibition, which was considered too short a time for significant DNA synthesis to occur (58). Guern & Usciati (58) have suggested that the blockage to cell division in correlatively inhibited buds in *Cicer* occurs at the end of the G2 phase and not in the S phase. Ultrastructural studies of the lateral buds of *Tradescantia paludosa* (9) revealed that inhibited buds possess a characteristic cytohistological zonation, with a "zone of inhibition" at the extreme tip of the bud apex similar to the "meristème d'attente" of the actively growing apices of some species (23). However, actively growing buds of *T. paludosa* do not show a meristème d'attente (9), and therefore it may be important in histochemical studies of nucleic acid synthesis in relation to correlative bud inhibition to carefully distinguish between activities in different parts of the bud apical meristem. In *T. paludosa,* both DNA and histone synthesis were initiated in the "zone of inhibition" by decapitation of the main shoot apex, and cell division activity commenced there in approximately 4 days (9).

It may be concluded that more useful work could be done on the morphological and cytological events during the imposition and removal of correlative inhibition of lateral buds, for such information will aid the interpretation of the many surgical and hormone treatment experiments already performed. One aspect of the morphology of lateral buds in relation to apical dominance which has received increasing attention in recent years concerns the kinetics of the development of vascular tissues between bud and stem. This matter is considered within the next section of this review.

MECHANISM OF ACTION OF THE CORRELATIVE SIGNAL

Inhibition of development in lateral buds subject to the correlative influence exerted by the apical bud, or in one lateral bud inhibited by another, may be envisaged as involving one or more of the following possible mechanisms, each of which has been investigated.

Inhibition by Lack of Essential Factors in Buds?

Early investigators of the problem of correlative inhibition of buds were attracted to an hypothesis involving starvation of buds through the monopolization of nutrients by the apical bud (see 123). This "nutritive theory" has been revived recently as an explanation for apical dominance phenomena in a number of species. Some of the experimental results obtained over the past decade or so which have been interpreted in terms of the nutritive theory (57, 79, 94–97, 99) were discussed above in connection with the problem of the nature of the correlative signal. As already pointed out, increased lateral bud growth in response to raising the nutritional or water status of the plant cannot be taken to necessarily indicate that previous lack of bud growth was a consequence of their starvation of nutrients and water. The major inorganic nutrients (NPK) increase in developing buds after about 12 hr following their release from correlative inhibition (57, 79, 94–97, 99, 100, 121), but it is probable that this is attributable to the increased mass of bud tissue extracted. Only a few experiments have measured levels of nutrients within the apical parts of lateral buds, and these have revealed that concentrations of NPK (121), amino acids (100), and carbohydrates (180) were just as high in inhibited as in growing buds. Should lack of major nutrients be the basis of correlative inhibition, then one may reasonably suppose that direct feeding of inhibited buds with nutrients would induce their development. On the few occasions this has been attempted, no release from inhibition was detected, even under conditions which would appear to have guaranteed entry of nutrients into the buds (38, 39, 56).

The weight of evidence, therefore, does not appear to support the view that lack of major nutrients in buds is the cause of their arrested development. This does not mean that nutrient and water status are unimportant in apical dominance phenomena, but their effects are probably indirect and perhaps in some cases mediated by induced changes in the hormonal status of the plant. Inorganic nutrient deficiency, particularly nitrogen, can both increase the strength of apical dominance (57, 99) and depress endogenous cytokinin levels (175, 200). Certainly one of the more significant advances made in studies of the problem over the past decade has been the discovery that direct applications of synthetic cytokinins to inhibited buds of a range of species can elicit their outgrowth even in intact plants (3, 29, 134, 135, 137). This evidence, together with the well-known requirement for cytokinins in plant cell division processes (60) and knowledge that cell division is blocked in correlatively inhibited buds (9, 58, 106, 123, 138, 171, 172), has led to suggestions that bud inhibition is a consequence of a deficiency in endogenous cytokinins. Technical difficulties apparently have discouraged investigation of the levels of endogenous cytokinins in buds in relation to apical dominance, but it is obviously essential that such data be procured. Tucker & Mansfield (168, 169) nevertheless have recorded a fall in the level of endogenous cytokinins of axillary buds of *Xanthium strumarium* following their release from inhibition, and consequently they suggested that correlatively inhibited buds were unable to utilize cytokinins. This proposal certainly fits the observation that a direct application of kinetin fails to stimulate axillary bud outgrowth in *X. strumarium* (93), but it does not help in understanding the situation in those species whose axillary buds can be prompted to grow by direct cytokinin application. If one assumes that a deficiency in cytokinins is causal in lateral bud inhibition in many species, then the question arises as to whether cytokinins required for initial bud outgrowth are synthesized within the buds themselves or derived from elsewhere in the plant. Sachs & Thimann (135) and Thimann (163) suggested that inhibited buds lack the capacity to synthesize cytokinins, and that removal of the correlative inhibitory signal results in rapid acquisition by the buds of the ability to satisfy their own cytokinin requirements. Thimann and co-workers (80a) very recently demonstrated that lateral buds on decapitated *Pisum sativum* plants were prevented from growing out by their treatment with the antibiotic hadacidin (an inhibitor of adenylosuccinate synthetase). Hadacidin inhibition could be counteracted by application of either N-6-γ-γ-dimethylallylaminopurine or kinetin to the bud, and it was concluded that lateral bud outgrowth is dependent on the synthesis of a cytokinin within buds themselves (80a). Conversely, knowledge that roots synthesize cytokinins which can be exported into the shoot system (8, 28, 75) has led to suggestions that root-derived cytokinins are preferentially transported to the dominant bud or buds (58, 104, 121, 123). Woolley & Wareing (199, 200) in experiments on *Solanum andigena* obtained the clearest indication that lateral bud outgrowth requires a supply of cytokinin from roots. They noted first that either roots or exogenous cytokinin was necessary for development of orthotropic leafy lateral shoots on cuttings. Further, it was observed that there was a residual root effect, in that nodal cuttings taken from plants that had been without roots for 12 days still showed capacity for bud outgrowth, and it was suggested (199) that this was due to "storage" of root-derived cytokinins in shoot tissues. Significantly, it was discovered that enhancement of growth in a rootless apical cutting by pretreatment with gibberellic acid prevented bud outgrowth on subsequent decapitation unless exogenous cytokinin was supplied, suggesting that when "stored" cytokinins are completely exhausted during apical growth it is then possible to demonstrate an absolute dependence of lateral bud outgrowth on an external supply of cytokinin. Impressive as is this evidence, it would be reassuring to see similar experiments done in which growth enhancement to presumably exhaust shoot cytokinins did not involve the addition of exogenous gibberellin, in view of known interactions between gibberellins and cytokinins in bud growth and particularly in the regulation of potato lateral shoot morphogenesis (80, 198–200).

Hypotheses to explain the way in which the distribution of cytokinins between actual and potential growth centers is regulated are essentially identical to those previously put forward for nutrient transport: i.e. the "nutritive theory" and the "nutrient-diversion theory," including hormone-directed transport (104, 123) which is discussed below. Studies on the transport of [^{14}C]-benzylaminopurine from the cut base of *Solanum andigena* decapitated cuttings (199) revealed that cytokinin accumulated in lateral buds prior to their growth and that application of IAA to the apical end of cuttings prevented this, whereas [^{14}C]-adenine (which unlike benzylaminopurine does not stimulate bud outgrowth in this system) did not accumulate in the buds of non-auxin treated cuttings until after bud growth had commenced. However, IAA did not appear to inhibit benzylaminopurine accumulation in buds by diverting cytokinin to the point of IAA application, but it did promote formation in the stem tissues of a metabolite of benzylaminopurine which was not accumulated in the buds (199).

The dramatic demonstrations that direct cytokinin applications to buds can result in lateral bud outgrowth in intact plants (3, 29, 32, 134, 135, 137) and in conversion of horizontal stolons into orthotropic leafy shoots (80, 198–200) are clearly indicative of a fundamental role played by this class of growth regulator in apical dominance phenomena. Much indirect evidence suggests that inhibited buds, or stolon tips, are deficient in cytokinins, but more information is required to establish with certainty (*a*) the levels of endogenous cytokinins in bud apical tissues in relation to correlative inhibition, and (*b*) the sources of endogenous cytokinins apparently required for the initiation of lateral bud outgrowth in many species.

Inhibited buds could conceivably contain suboptimal levels of not only cytokinins but also other growth-promoting hormones. Direct application of either auxins or gibberellins to totally inhibited lateral buds, however, does not induce their development into shoots (29, 123), indicating that any deficiency of these hormones in a bud is not the primary cause of its lack of growth. There seems little doubt, on the other hand, that both auxins and gibberellins are essential for continued development of buds which have been released from apical dominance and are undergoing active cell division in the presence of adequate levels of cytokinins. Thus it has been observed repeatedly that the initial release of buds from inhibition on intact plants by a direct application of cytokinins results in only transitory growth, but that a subsequent direct application of either IAA (3, 135) or GA$_3$ (3, 29) allows development to proceed considerably further, and sometimes to the formation of large leafy lateral shoots. In soybean, enhanced growth of cytokinin-released buds by GA$_3$ was prevented when 5-fluorouracil or 5-fluorodeoxyuridine was applied before but not after the GA$_3$ (3), indicating that the role of cytokinin was to initiate cell division activity and that gibberellin was required for the subsequent enlargement of newly formed cells in the bud. It is clear that various growth regulators must play sequential and interacting roles during release of buds from correlative inhibition (67, 148, 199, 200), and that great care must be exercised not to confuse the possibly different control mechanisms operative during very early and later stages of lateral bud development following their escape from inhibition.

Levels of endogenous auxins and gibberellins in lateral buds in relation to apical dominance have been explored only slightly more than those of cytokinins, and available data cannot be regarded as at all conclusive. One of the earliest hypotheses of correlative inhibition, the "theory of direct-inhibition by auxin" (see 123), presumed that auxin levels were supraoptimal in inhibited buds and would therefore fall when dominance was removed. Slight though it is, experimentally derived data argue that auxin levels in buds rise rather than fall upon their release from inhibition (51, 109, 167). Similar findings have been reported for endogenous gibberellins (144, 167). Together these results suggest that in correlatively inhibited buds the normal capacity for auxin and gibberellin synthesis in young green leaves is absent, but that it appears at some stage after release from inhibition. Why buds released from inhibition not by decapitation but by direct cytokinin treatment usually fail to reach self-sufficiency in auxin and gibberellin is unknown, but is a problem deserving attention.

Inhibition by Presence of Inhibitory Factors in Buds?

Very soon after the original demonstrations that auxin can at least partially substitute for the apical bud in maintaining axillary bud inhibition (165, 166), suggestions were made that although auxin may serve as the correlative signal it does not itself inhibit growth of buds. The "direct theory" of auxin action in apical dominance (see 123, 162) required that apically synthesized auxin be transported basipetally in the main stem, but acropetally into lateral buds to create a supraoptimal concentration there. Although auxin transport in stems generally shows basipetal polarity (54), acropetal auxin transport can be of sufficient magnitude to allow movement into axillary buds (54, 194). Nevertheless, it was found by Snow in a series of ingenious surgical experiments that the inhibitory influence from the apical bud could move to lateral buds along routes which were apparently completely inaccessible to auxin (157; see also 123). Snow was led to propose that auxin from the dominant bud acted indirectly to inhibit lateral bud growth by stimulating in the main stem tissues the formation of a specific hormonal growth inhibitor which moved into lateral buds or shoots. Attempts to extract and quantitatively measure endogenous growth inhibitors were for many years thwarted by lack of effective and convenient fractionation procedures, although as early as 1939 it was reported by Snow (158) that whereas lateral shoots of *Pisum sativum* contained a growth inhibitor the apical bud and main stem did not. Not until the introduction of chromatographic techniques did further useful investigations begin, when Dörffling found declines in levels of several inhibitory fractions from lateral buds of *P. sativum* and *Acer pseudoplatanus* following their release from apical dominance (42–44). A similar relationship has been recorded for potato tubers, which contained a neutral inhibitor that disappeared within 24 hr of removing the actively growing dominant sprouts (56).

Identification of ABA as a natural plant growth inhibitory hormone has led to investigations of its possible role in correlative inhibition of lateral buds. Both GLC and bioassay analyses revealed ABA levels in inhibited buds of *Xanthium strumarium* to be between 50 and 250 times greater than in all other parts of the plant (168, 169). ABA levels within inhibited buds of *X. strumarium* fell from approximately 12,000 to 850 μg.g^{-1} dry weight within 48 hr of decapitation of the main shoot (169). The level of ABA in inhibited axillary buds of *X. strumarium* plants growing in the presence of far-red light was some 60 times greater than in buds on plants not exposed to far-red [absence of far-red light results in axillary bud outgrowth in this species (169)]. These results do indicate a correspondence between axillary bud growth and ABA content in *X. strumarium*, but further work is required on this and other species to establish whether this is a general phenomenon. Even should it be so, it is also necessary to determine the time-course of decrease in ABA level in relation to the kinetics of bud outgrowth.

Experiments involving studies of lateral bud growth in plants treated with exogenous ABA have produced rather ambiguous results to date, perhaps not surprisingly in view of the rather doubtful relevance of the methods of application of the inhibitor. Despite the existence of reports of predominantly basipetal transport of exogenous (±)ABA in excised segments of stems and petioles (see 100a), most available evidence indicates that endogenous ABA is synthesized primarily in mature leaves from where it may be translocated acropetally in the phloem towards the apical bud (64, 184). Nevertheless, reports have appeared of studies in which ABA was applied to the cut surface of the stem of decapitated plants and its effects on axillary bud growth measured. It is difficult to see how the results of such experiments can be interpreted usefully in terms of the natural hormonal regulation of bud development, however, if endogenous ABA is indeed synthesized in mature leaves and not the apical bud. ABA applied apically to decapitated pea seedlings inhibited axillary bud outgrowth as effectively as the apical bud (6), whereas in *Phaseolus vulgaris* ABA on its own slightly enhanced axillary bud growth but in combination with IAA and kinetin served to increase inhibition slightly (62). More thought needs to be given to the possibility of interaction between auxin moving basipetally from the apical bud and ABA being transported from mature leaves. Arney & Mitchell (6) suggested that ABA is synthesized within lateral buds in response to the arrival of auxin from the apical bud, and this proposal, for which there is no evidence either for or against, has been given verbal support by others on the basis of circumstantial evidence only (169).

The Role of Hormone-Directed Metabolite Transport

Considerable evidence indicates that correlative inhibition involves deficiency in lateral buds of some essential factor or factors required for their further development, whether these be nutritional or hormonal. Lack of necessary hormonal stimuli, particularly cytokinins, may in some cases be a result of incapacity for their synthesis or utilization by buds, but it seems more likely that in most plants correlatively inhibited buds are denied an adequate supply of both cytokinins and nutritional substances. The problem thus raised is how the pattern of distribution of essential growth factors is set up and maintained in the shoot. The original nutritive theory (see 123) held that since the apical bud is normally present in the embryo, it will continue to command supplies of nutrients from roots and leaves to the detriment of later-formed lateral buds, by virtue of constituting a metabolic sink towards which solutes move along their concentration gradients. This concept, however, does not adequately encompass several features of apical dominance (123), and partially because of this Went (186, 187) offered the "nutrient-diversion theory," which proposed that metabolites move towards regions of highest auxin concentration. Dominant growing buds and shoots are known to be the principal sites of auxin synthesis and contain the highest levels of auxin (149), and the nutrient-diversion theory holds that metabolites move to these organs in response to a stimulus provided by auxin, rather than as a result of their sink activity. There have been numerous demonstrations that metabolites such as sugars and phosphates, and cytokinins, are indeed translocated to and accumulate in growing buds and other sites of metabolite demand (101, 104). In itself this is unremarkable, but perhaps more significantly it is now well established that metabolites and cytokinins are also mobilized by localized applications of certain growth hormones, particularly auxins which can induce long-distance metabolite transport. The term hormone-directed transport (HDT) has been applied to this phenomenon. The mechanism by which HDT occurs remains obscure (14, 105, 116, 117), but it is essential that it be understood before HDT be confidently incorporated into explanations of correlative growth phenomena such as apical dominance. This is because it is possible that substitution of an excised metabolite-sink, such as an apical bud, by an auxin application to the cut tissues could influence transport processes indirectly by inducing or increasing local metabolite demand. Consequently, attempts have been and are continuing to be made to distinguish between the following possible bases of HDT: (a) a direct effect of the hormone on the phloem transport system, either by influencing the whole pathway of transport or by stimulating a possibly active process of phloem unloading at the point of application (81); and (b) an indirect effect on long-distance transport through activation or maintenance of local sink activity in the hormone-treated tissues.

The first reports of HDT (102, 161) were based on experiments which revealed that nitrogenous compounds and carbohydrates accumulated in auxin-treated stem tissues of bean plants over a period of several days. These experiments did not preclude the possibility that auxin treatment induced the appearance of a new growth center in which metabolites accumulated. Many subsequent experiments on HDT have been designed to exclude as far as possible a localized growth response to auxins or other growth hormones. Booth et al (12) found that ^{14}C-assimilates and [^{14}C]-sucrose accumulated in auxin-treated stem stumps of pea and potato plants within a few hours of auxin application, which was perhaps too short a time for any significant increase in local metabolite demand to appear.

A special problem in studies aimed at establishing whether or not auxins can induce long-distance directed metabolite transport in the absence of a localized enhancement of metabolite demand is that sufficient time must elapse between auxin application and extraction of the treated tissues to allow measurable quantities of metabolite to arrive from a distant source. Unfortunately, this also permits the possibility of generation of sink activity by the hormone, or alternatively it may be that hormone treatment results in maintenance of demand in a decapitated internode over that in the control untreated decapitated internode (i.e. a senescence-delaying effect of the hormone). Several approaches have been adopted to overcome these difficulties in studies of the mechanism of HDT. One has been to shorten the transport period as far as practicable, and also at the same time to utilize internodes which have completed elongation growth. Work on *Solanum andigena* (12), *Pisum sativum* and *Populus robusta* (41), and on *Phaseolus vulgaris* (14, 116, 117) has shown that [14C]-assimilates, [14C]-sucrose, and [32P]-orthophosphate can move towards and accumulate in auxin-treated decapitated nonelongating internodes within a few hours of time of auxin application. Clearly, should sufficient time be allowed, auxin treatment of the cut end of even a nonelongating internode will usually result in initiation of local callus growth which would constitute a sink for carbohydrates and phosphate. In the short-term HDT experiments referred to, there was no visible indication of callus formation during the transport period (usually 3–12 hr). But the question is whether an auxin-induced metabolite sink can appear within about 3–6 hr in nonelongating decapitated internodes of the type used, for there are indications that auxins can stimulate nucleic acid and protein synthesis in growing tissues after about 10 min of treatment (76). However, absence of uracil-2-14C accumulation in stem stumps of *Vicia faba* treated with IAA suggests that exogenous auxin does not necessarily stimulate RNA synthesis and growth in mature internodes (115). On the other hand, even in fully elongated segments of soybean hypocotyl RNA synthesis was enhanced to the extent of 25–30% by IAA and 2,4-D within 3 hr, although at the same time radial growth and fresh weight were also increased (76, 105). Similarly, the sucrose mobilizing effects of auxins in *Hevea brasiliensis* were found to occur only when there was local enhancement of metabolic activities in the auxin-treated region (170). In decapitated *P. vulgaris* plants it was observed that application of a mixture of IAA, GA$_3$, and the cytokinin SD 8339 to the stem stump induced strong HDT of [14C]-assimilates within 3 hr, but that RNA and protein synthesis in the treated internode were also increased after only 2.5 hr of hormone treatment (105). Detailed examination of the metabolic effects of auxin at the site of its application to decapitated second internodes of *P. vulgaris* have indicated that auxin-directed metabolite transport can occur in the absence of local enhancement of growth or metabolism. Thus, in experiments by Patrick & Wareing (116), it was found that 9 hr after decapitation, the top 1 cm portions of IAA-treated internodes possessed no greater capacity for direct [14C]-sucrose uptake or metabolism than did similar but non-auxin treated tissues. Also, neither local respiration rate nor incorporation of [14C]-leucine into protein were enhanced after 9–12 hr of IAA treatment, although protein level was maintained in IAA-treated internodes whereas it fell to about one-third the original value in untreated decapitated internodes (116). These results therefore indicated no overall increase in metabolic rate of IAA treated internodes, and have been confirmed and extended in subsequent work (Patrick and Wareing, unpublished data), but did demonstrate a senescence-delaying effect of auxin (as indicated by retention of protein) which may be of significance in understanding the mechanism of HDT. Even so, further work by the same authors showed that decapitated stems treated with plain lanolin for 3 days retained their responsiveness to IAA in terms of enhanced metabolite transport (117), yet one may have expected on the basis of earlier data (116) that protein loss by the internode would have proceeded to a very marked extent. This perhaps indicates that HDT cannot be explained simply in terms of hormone delayed senescence (1, 92, 151) of internodal tissues. As it has been reported that delaying application of auxin after decapitation reduces or abolishes its inhibitory effect on lateral bud growth (69, 90, 166, 173), then the significance in relation to apical dominance of the observation that auxin-directed transport appeared after even a 3 day delay between decapitation and auxin application (117) is made doubtful, for such an auxin application would not be expected to inhibit lateral bud development.

While a local stimulatory effect of auxin on overall metabolic rate, and consequent sink activity, does not appear to be necessary for its long-distance mobilizing effects on metabolites, it does appear that the mechanism of HDT involves a purely local effect of auxin. It has been suggested that auxin acts in HDT by in some manner activating the phloem along the whole pathway of transport being measured (41). To the extent that HDT obviously does involve long-distance transport of metabolites via the phloem (41, 117), there is no doubt that the phloem translocation mechanism is indeed influenced for some distance beyond the point of auxin application. However, the concept of HDT being dependent on the basipetal movement of auxin in the stem (41) has been shown to be untenable by experiments on bean internodes which revealed IAA-enhanced acropetal movement of labeled metabolites from a source positioned up to 10 cm distant within as short a period as 20 min (personal unpublished data). The velocity of basipetal auxin transport in the internodes approximated only 1 cm·hr^{-1}, and it is clear that any role of auxin in long-distance metabolite transport derives from a local effect in the treated region.

As already considered, available evidence argues against the local auxin effect being one of general enhancement of metabolism, and therefore one is compelled to consider other possible ways in which a localized effect of auxin could result in rapid long-distance activation of translocation in the phloem. One possibility is that auxin may induce certain qualitative changes in metabolism that do not result in an overall alteration in total protein synthesis and/or respiration, but which nevertheless could lead to an increased local demand for particular substrates such as sucrose or phosphate. However, Patrick & Wareing (unpublished data) found that IAA had no effect on the relative levels of stored and recently transported sucrose in the upper ends of decapitated *P. vulgaris* internodes, but did enhance the accumulation of unmetabolised sucrose against a concentration gradient. These results argue against any suggestion of auxin-induced depletion of the pool of sucrose in the upper end of decapitated internodes, and lead one to seek an alternative mechanism to explain auxin-directed transport in the absence of either a general or specific increase in local metabolite demand. Any such proposed mechanism must be concerned with possible effects of hormones directly on the phloem, and it is at this point that one stumbles over the grave deficiencies in our understanding of phloem structure and activity (26, 34, 50, 179). Lack of knowledge of the mechanism of phloem transport does of course allow freedom to hypothesize with relatively little fear of outright refutation, but the value of such exercises is doubtful. It is probably sufficient here merely to suggest the most obvious possible explanations for the effects of auxin in HDT in terms of direct action on the phloem, though emphasizing their conjectural natures. Transport of solutes out of the phloem into the tissues of an internode involves movement across the limiting membranes of sieve-tube elements, and it is at least conceivable that this is an active process (81) which may be stimulated by auxin. Another possibility is that auxin inhibits plugging of the phloem in cut stems, perhaps at least partially through an effect on callose deposition in sieve-tube elements (166a). Both these possibilities envisage that stem sever-ence, a usual procedure in studies of HDT, would be expected to result in a diminution of the functional integrity of the phloem (either at the point of unloading or in longitudinal transport), and that this may be counteracted by exogenous auxin substituting for auxin normally derived from the apical part of the shoot (27). Such ideas are not without attraction, but their evaluation probably will await advances in our general knowledge of phloem functioning. Work on isolated "bark" strips of willow (*Salix viminalis*) showed that IAA treatment of the phloem did stimulate the rate of longitudinal transfer of glucose, but in this case movement away rather than towards the auxin application was increased (83, 84), and consequently the relevance of the finding to HDT is obscure. Nevertheless, it was suggested (84) that IAA stimulated sugar loading into the phloem from stem storage parenchyma, which is of interest here in view of hypotheses for HDT involving effects of auxin on unloading from phloem at the point of auxin application.

It will have been noticed that despite the use of the term hormone-directed transport (HDT), the majority of experiments discussed have been concerned with effects of auxins, particularly IAA. Evidence in the literature on the long-distance mobilizing effects of auxins other than IAA, and of other classes of plant growth regulator, is confusing. Some reports indicate that neither GA_3 nor kinetin, when applied alone, have any effect on long-distance transport, whereas 6-benzylaminopu-rine did enhance ^{32}P transport and accumulation (41, 147). On the other hand, Mullins (105) found that either GA_3 or kinetin, as well as IAA, could stimulate long-distance transport of [^{14}C]-assimilates. Bowen & Wareing (14) recorded con-trasting metabolite mobilizing effects of different auxins (IAA, 1-naphthylacetic acid, 2,4-D, 2,4,5-T) in different species (*Phaseolus vulgaris, Pisum sativum, Coleus blumei, Helianthus annuus*). Furthermore, even in a single species transport of [^{14}C]-sucrose and [^{32}P]-orthophosphate were not similarly influenced by the same growth regulator (14). These results were not at all easy to interpret, but did suggest that HDT is not necessarily a reflection of the appearance of a local nonselective metabolite sink. The differential effects of the various regulators were not explicable in terms of varying rates of their basipetal transport down the stem, which again indicates that HDT results primarily from action of the hormone near to the point of its application (14).

Other types of plant growth regulator may interact with auxins to affect the magnitude of the metabolite mobilizing effect. Although a few reports indicate that either gibberellins or cytokinins on their own can induce HDT (71, 105), more generally it has been found that they have no effect when applied to fully elongated decapitated internodes (41, 147). Unpublished data of Wareing and Johnstone at Aberystwyth have shown that when gibberellins or cytokinins are applied in solu-tion, rather than as dispersions in lanolin, they clearly do induce HDT in certain species (though not in others, such as *Pisum sativum*). Nevertheless, there are well-substantiated cases of gibberellins and cytokinins acting synergistically with auxin in HDT (105, 145, 147, 147a, 185). In view of the well-known interacting effects of stimulatory growth regulators in various aspects of plant development, it may be suspected that synergisms between these substances in HDT indicate that their effects on metabolite transport are indirect through stimulation of metabolism in the treated tissues. However, other explanations have been offered, such as stimulatory effects of gibberellins and cytokinins on basipetal transport of auxin (40, 147, 185). It is not possible to arrive at any firm conclusions on available data as to how synergisms between auxins and other hormones in HDT come about. Evidence has already been discussed which suggested that the action of auxin in HDT is localized, though its effects on metabolite transport are transmitted via the phloem over relatively long distances, and the local effect does not appear to be one of enhancement of metabolite demand.

Vascular Tissue Development in Relation to Correlative Inhibition of Buds

An adequate supply of nutritive and hormonal growth factors, including cytokinins, to a growing lateral bud can be maintained only if the bud is suitably served with a continuous vascular connection with the vascular tissues of the main stem. Suggestions have been made repeatedly that apical dominance involves regulation of the development of vascular connections betweeen buds and stem, by the action of basipetally moving auxin in the stem inhibiting the differentiation of vascular elements of bud traces with consequent starvation of the buds (57, 109, 115, 132, 133, 135, 155). To a certain extent these suggestions have been backed by anatomical investigations of bud vascularization, though conversely there is also considerable factual evidence against the general concept of regulation of bud growth by apical control of vascular tissue differentiation. It was noted by Gregory & Veale (57) that in flax seedlings there appeared to be a correlation between the rapidity of lateral bud outgrowth following decapitation or increased nitrogen status and the amount of vascular tissue present at the base of the bud. However, no really convincing quantitative data were given to provide confidence that lack of bud growth was associated with absence of vascular connections (57). Histological studies of pea seedlings by Sorokin & Thimann (155), however, did indicate to them that the establishment of a connection between basipetally differentiating xylem from a bud with xylem differentiating acropetally from the main stele coincided with marked visible release of the bud from correlative inhibition. Even this cannot be taken as conclusive evidence, however, for their data show that the buds increased in length by 20% within 24 hr, and cell division activity in the buds was detected long before xylem connections occurred some 55–70 hr after decapitation (155). Also, the histological data of Sorokin & Thimann (155) were derived almost exclusively from longitudinal sections of bud-bearing nodes, which allowed at least the possibility of the xylem strand running out of the plane of even a thick section at intervals. Subsequent work on bud vascularization in pea seedlings in relation to apical dominance, using serial transverse sections right along the line of the bud trace, showed no discontinuity in either xylem or phloem even in completely inhibited buds (24). In yet another study of pea plants, it was recorded that lateral bud growth commenced within 4–8 hr of decapitation, yet differentiated xylem did not appear until after 24 hr, and phloem until 4 days had elapsed (180). Similarly, axillary bud outgrowth in decapitated *Helianthus annuus* (91) and in *Bidens pilosus* (31) preceded the appearance of vascular connections between bud and stem, and in *Hygrophila* lateral bud growth occurred at a time when the bud trace contained only procambium (38). For soybean (*Glycine max*), convincing evidence has been presented that correlatively inhibited cotyledonary buds have complete xylem (2) and phloem (118) connections with the vascular system of the main stem. In potato tubers, too, correlatively inhibited buds do not lack vascular connections (53). The correlatively inhibited axillary buds of the primary leaf pair of *Phaseolus vulgaris*, which have served as experimental subjects in numerous studies of apical dominance, also appear to be well "plumbed in" to the main stem (191). Studies of the vascular anatomy of a number of species (*Coleus blumei, P. vulgaris, Pisum sativum, Vicia faba, Helianthus annuus, Lonicera japonica*) by Sachs (133) revealed that growing lateral buds had different patterns of vascularization depending upon whether they were released from correlative inhibition by "natural" means (e.g. spontaneous outgrowth in intact plants) or by decapitation of the main stem. Either the young leaves of the apical bud or apically applied auxin in decapitated plants inhibited the formation of connections between strands of xylem, and whereas a naturally released and growing bud on an intact plant possessed a relatively independent vascular system, a growing bud that had been released by decapitation developed vascular connections with the main stele (133). However, in none of the species or experiments was the vascularization of inhibited buds studied (133), so that although the results provide interesting information concerning the factors regulating patterns of vascularization in the stem, they do not suggest that lack of bud growth was associated with absence of vascular connections with the roots and leaves.

The degree of vascularization of correlatively inhibited buds varies from one species to another (33), from one bud to another in a single plant (57), and also environmental conditions such as those of light (128, 133) and nutrient status (57) influence bud trace development. From the evidence already cited it can be seen that by no means all examples of correlative inhibition are associated with a deficiency in bud-trace development, which indicates that the control of initial bud outgrowth is unlikely to be through apical regulation of vascularization of buds. Of course, once a bud has been released from inhibition and commences growth, then its continued development requires that adequate vascular connections be formed. Stimuli for bud-trace development are likely to arise from the growing lateral bud, as it has been amply demonstrated that active buds do induce differentiation of vascular tissues (25, 189, 190). It can be argued that the developed vascular connections which serve inhibited buds are in some way nonfunctional (109). Even should this be so, and there is no evidence for it, it is salutory to remember that protophloem and protoxylem differentiation from the procambium usually takes place some distance behind the apical meristem or developing leaf primordia of the main shoot (a distance probably equal to or greater than the distance between main stele and apical meristem of an inhibited lateral bud), and that discontinuities in patterns of xylem or phloem differentiation occur normally in the stem, and yet development of the main shoot apex proceeds actively. An example of the apparent deficiency in vascularization of the main shoot apex, which still does not result in arrestment of its growth, is provided by Esau (49) for *Linum,* in which phloem was absent from the 32 youngest leaf primordia and xylem from the 48 youngest primordia.

APICAL CONTROL OF ORIENTATION OF LATERAL ORGANS

Plagiogeotropic leaves and branches are carried at characteristic angles to the stem axis. Similarly, many rhizomes and stolons are diageotropic in that they normally grow at right angles to the longitudinal axis of the plant. The angles of orientation between adaxial surfaces of leaves and branches and their parent stems very commonly increase during their ontogeny, due to epinastic movements (114). Environmental factors, such as light and gravity, play important parts in the regulation of orientation of lateral appendages, but it is clear that apical dominance is also involved.

Decapitation of shoots frequently results in upward, hyponastic movements of leaves, lateral shoots, and even woody branches. Application of auxins can counteract the hyponastic tendency in decapitated plants and induce epinasty in intact shoots (70, 82, 92a, 112–114, 119a, 119b, 143). The mechanisms by which these movements come about are not yet understood, but they can concern differential cell elongation in adaxial and abaxial sides of the organ or formation of tension or compression wood in woody branches. The involvement of gibberellins and ethylene in addition to auxin has been demonstrated in petiole epinasty of *Helianthus annuus* (112, 113) and of gibberellins in lateral branch orientation in conifers (92a, 119a, 119b, 123). The orientation of lateral roots in pea seedlings is also subject to regulatory influences by auxin and ethylene (89). It is clear, nevertheless, that considerable work remains to be done on the hormonal basis of apical control of plagiogeotropism in leaves, branches, and lateral roots.

Orientation of essentially diageotropic organs similarly appears to be regulated through a mechanism which includes synthesis and export of growth hormones from apical regions. Palmer found that a light stimulus to the shoot of *Agropyron repens* induced positive geotropic responses in the underground rhizomes (110), and in several tropical grasses a similar stimulus to the shoot evoked positive geotropism in the otherwise negatively geotropic stolons (111). Separation of a plagiogeotropic rhizome from the parent plant can result in upturning of the rhizome tip (181). In wild potato (*Solanum andigena*), decapitation of the shoot and removal of axillary buds leads to negative geotropism in the tips of previously diageotropic stolons (10). Evidence has been presented that lateral shoot growth and orientation in *S. andigena* involves a triple interaction between auxin and gibberellin from the shoot apical buds and cytokinin from the root tips (11, 46, 80, 199, 200).

ENVIRONMENTAL EFFECTS IN APICAL DOMINANCE

Relationships between mineral nutrition, water availability, and photosynthesis on the one hand, and lateral bud development on the other, have been considered above and discussed in more detail previously (123). Nitrogen (114) and phosphorus (46) supplies have also been observed to influence leaf orientation. It is probable that effects of mineral element supply on apical dominance phenomena are mediated through influences on both hormonal and nutritional status of the plant (7, 123, 130, 175, 200). Correlative inhibition of buds is generally enhanced at low light intensity (see 123), perhaps by lowering photoassimilate and increasing auxin levels, though on the other hand, hyponasty of leaves and branches is favored by low light intensity (13) and apparently also by low auxin level (13, 70, 82, 112–114). Photoperiod also affects apical dominance, in that short days (SD) can reduce and long days increase correlative inhibition (123). These photoperiodic effects also may be mediated through changes in hormonal balance (123), and especially interesting in this connection, considering the importance of cytokinins in lateral bud outgrowth, is the observation that SD treatment of *Perilla frutescens* (a SD plant) markedly increased the flux of cytokinins from root to shoot (8).

Gravity plays a very important role in the regulation of bud growth and also in lateral organ orientation. The term gravimorphism has been applied to the responses of plants to gravity involving changes in the pattern of morphogenesis (182). Placement of a normally negatively geotropic shoot in a horizontal position usually reduces correlative inhibition of buds by the main apex (152–154, 182, 183) and increases epinasty of leaves and branches (87, 88, 112). It is possible that inhibition of basipetal polar transport of apically synthesized auxin in horizontally positioned orthogeotropic shoots may play a part in the effects of gravity on apical dominance. While that may be so, there remains a very fundamental problem of apical dominance which has been emphasized by studies of gravimorphic phenomena (152–154, 182, 183). This is that it is usually the physically highest and upwardly directed bud which achieves dominance over other buds, regardless of time of onset of growth of the buds. When buds are at equal heights, as can be arranged by positioning the stem horizontally, for example, it is the upwardly directed bud nearest the root system which becomes dominant. Wareing & Nasr (183) suggested that nutrients are diverted to the highest upwardly directed meristem, perhaps as a result of hormone-directed transport, but that other factors being equal, proximity to roots lends advantage to a bud. Subsequent research emphasizing the probable importance of the supply of root-synthesized cytokinins for bud growth leads one to speculate that it is the diversion of these (rather than major nutrients) to the upwardly directed bud which is of significance in gravimorphic phenomena. The results of Smith & Wareing (153) showed very clearly that the stimulatory effects of roots on bud outgrowth in *Salix viminalis* occurred even when the roots were supplied only with deionized water. Thus studies of gravimorphism, as well as of other aspects of apical dominance, have indicated that a very fundamental component of correlative growth regulation in plants lies in the mechanism controlling the distribution of cytokinins, probably from the roots, among meristems in the shoot.

CONCLUDING COMMENTS

Progress has been made during the past decade or so in studies of apical dominance, but it is only realistic to recognize that our understanding of the mechanisms by which developmental activities in plants are coordinated remains hazy. Of special significance has been the realization that in addition to auxin other growth regulators, notably cytokinins, are concerned in the regulation of lateral bud outgrowth. Also important has been the gradual erosion of tenability of the purely nutritive hypothesis and the straightforward nutrient-diversion theory of correlative inhibition. Similarly, whereas vascular connections between bud and main stele are undoubtedly essential for continued growth of a lateral, much evidence has accumulated that leads one to doubt that the regulation of vascular tissue differentiation plays a primary part in the initiation of growth activities in a lateral bud. One is left with a rather general concept that auxin from the young leaves of the apical bud serves as the principal correlative signal, influencing either cytokinin synthesis or utilization within lateral buds, or the distribution of root-synthesized cytokinins between meristems in the shoot; in either case it is assumed that the inhibited condition of lateral buds subject to correlative inhibition is a consequence of deficiency in cytokinin. This view of the matter appears sound enough on available evidence, but it must also be kept in mind that not only cytokinins but also other growth regulators (including auxin and gibberellin) plus nutrients and water are required for full outgrowth of laterals, and that the capacity of a previously inhibited bud to synthesize the different hormonal factors may be acquired only gradually and sequentially. Future work can, on the basis of the general auxin/cytokinin concept of correlative inhibition, best be directed toward answering a number of the questions discussed in this review.

First, examination of physiological and cytohistological processes involved in the arrest or suppression of lateral bud development, which occurs after their initiation and partial development, may be of value. This is an aspect of correlative inhibition of buds which has received very little attention in comparison with that directed to understanding the mechanism of release of buds from inhibition.

Also needed are more detailed kinetic data on cytohistological and hormonal changes during the very early and later stages of bud growth following their release from apical dominance, in order to distinguish between factors concerned in the initiation of outgrowth and those involved in subsequent bud tissue development. In this connection, a clearer picture of the possibly sequential roles of different growth regulators during bud outgrowth would be helpful. Much more information is in fact required about the actual levels of individual endogenous growth regulators (particularly cytokinins, auxins, and gibberellins, but also ABA and other growth inhibitors) in buds in relation to correlative inhibition.

In view of the possibility that cytokinin supply to lateral buds may be regulated by auxin transmitted from the apical bud, the mechanism of hormone-directed transport must be fully elucidated, taking into account observations that auxin can influence metabolism of cytokinins with the formation of derivatives which have different transport behavior to that of the parent cytokinin. Further evidence is also necessary on the sites of synthesis of the cytokinins apparently essential to the initial stages of release from correlative inhibition, for the concept of auxin-directed transport of cytokinins in apical dominance presupposes that inhibited buds fail to develop further because of a deficiency in their capacity to synthesize cytokinins. Studies of cytokinin transport in relation to the action of auxin as a correlative signal may also be expected to help answer the fundamental question of why it is the highest upwardly directed apical meristem which normally achieves dominance over others in the shoot system.

ACKNOWLEDGMENTS

I thank Dr. J. R. Hillman, Dr. R. P. Pharis, and Professor P. F. Wareing for kindly providing information and unpublished data to be incorporated into this review.

Literature Cited

1. Addicott, F. T. 1970. *Biol. Rev.* 45:485–524
2. Ali, A. A., Fletcher, R. A. 1970. *Can. J. Bot.* 48:1139–40
3. Ibid 1971. 49:1727–31
4. Apfelbaum, A., Burg, S. P. 1972. *Plant Physiol.* 50:117–24
5. Ibid, 125–31
6. Arney, S. E., Mitchell, D. L. 1969. *New Phytol.* 68:1001–15
7. Avery, G. S., Burkholder, P. R., Creighton, H. B. 1937. *Am. J. Bot.* 24:553–57
8. Beever, J. E., Woolhouse, H. W. 1973. *Nature New Biol.* 246:31–32
9. Booker, C. E., Dwivedi, R. S. 1972. *Exp. Cell Res.* 82:255–61
10. Booth, A. 1959. *J. Linn. Soc.* 56:166–69
11. Booth, A. 1963. *The Growth of the Potato*, ed. J. D. Ivins, F. L. Milthorpe, 99–113. London: Butterworths
12. Booth, A., Moorby, J., Davies, C. R., Jones, H., Wareing, P. F. 1962. *Nature (London)* 194:204–5
13. Bottelier, H. P. 1954. *Ann. Bogor.* 1:185–200
14. Bowen, M. R., Wareing, P. F. 1971. *Planta* 99:120–32
15. Brian, P. W., Hemming, H. G., Radley, M. 1955. *Physiol. Plant.* 8:899–912
16. Burg, S. P. 1972. *Hormonal Regulation in Plant Growth and Development*, ed. H. Kaldewey, Y. Vardar, 397. Weinheim: Verlag Chemie
17. Burg, S. P. 1973. *Proc. Nat. Acad. Sci. USA* 70:591–97
18. Burg, S. P., Apfelbaum, A., Kang, K. G. 1972. See Ref. 16, 263–79
19. Burg, S. P., Burg, E. A. 1967. *Plant Physiol.* 42:144–52
20. Ibid, 1224–28
21. Ibid 1968. 43:1069–74
22. Burg, S. P., Burg, E. A. 1968. *Biochemistry and Physiology of Plant Growth Substances*, ed. F. Wightman, G. Setterfield, 1275–94. Ottawa: Runge
23. Buvat, R. 1952. *Ann. Sci. Nat. Bot. Ser.* 11 13:199–300
24. Caldeira, G. C. N. 1970. *Apical dominance in Pisum sativum.* PhD thesis. Univ. London, U.K. 287 pp.
25. Camus, G. 1943. *C. R. Acad. Sci. Paris* 137:184–85
26. Canny, M. J. 1971. *Ann. Rev. Plant Physiol.* 22:237–60
27. Carr, D. J. 1966. *Trends in Plant Morphogenesis*, ed. E. G. Cutler, 272–73. London: Longmans, Green
28. Carr, D. J., Burrows, W. J. 1966. *Life Sci.* 5:2061–77
29. Catalano, M., Hill, T. A. 1969. *Nature London* 222:985–86
30. Champagnat, M., Culem, C., Quiquempois, J. 1963. *Mémoires Publiés par la Société Botanique de France, Colloque de Morphologie (Tératologie) à Strasbourg*
31. Champagnat, M., Desbiez, M. O., Delaunay, M. 1973. *Ann. Sci. Nat.-Bot. Biol. Veg.* 14:71–86
32. Chvojka, L., Vereš, K., Kozel, J. 1961. *Biol. Plant.* 3:140–47
33. Clowes, F. A. L. 1960. *Apical Meristems.* Oxford: Blackwell
34. Crafts, A. S., Crisp, C. E. 1971. *Phloem Transport in Plants.* San Francisco: Freeman. 481 pp.
35. Crick, F. H. C. 1970. *Nature* 225:420–22
36. Crick, F. H. C. 1971. *Symp. Soc. Exp. Biol.* 25:429–38
37. Cutter, E. G. 1972. *Ann. Bot. London* 36:207–20
38. Cutter, E. G. 1972. *The Dynamics of Meristem Cell Populations*, ed. M. M. Miller, C. C. Kuehnert, 51–73. New York: Plenum
39. Cutter, E. G., Chin Hung-Woon 1972. *Ann. Bot. London* 36:221–28
40. Davies, C. R., Seth, A. K., Wareing, P. F. 1967. *Science* 151:468–69
41. Davies, C. R., Wareing, P. F. 1965. *Planta* 65:139–56
42. Dörffling, K. 1964. *Planta* 60:413–33
43. Dörffling, K. 1965. *Ber. Deut. Bot. Ges.* 78:122–28

44. Dörffling, K. 1966. *Planta* 70:257–74
45. Dostál, R. 1972. *Biol. Plant.* 14:177–85
46. Eckerson, S. H. 1931. *Contrib. Boyce Thompson Inst.* 3:197–217
47. Elmer, O. H. 1932. *Science* 75:193
48. Ernest, L. C., Valdovinos, J. G. 1971. *Plant Physiol.* 48:402–6
49. Esau, K. 1943. *Am. J. Bot.* 30:248–55
50. Eschrich, W. 1970. *Ann. Rev. Plant Physiol.* 21:193–214
51. Ferman, J. H. G. 1938. *Rec. Trav. Bot. Neer.* 35:177–287
52. Fletcher, G. M., Dale, J. E. 1974. *Ann. Bot.* 38:63–76
53. Goodwin, P. B. 1967. *J. Exp. Bot.* 18:78–99
54. Goldsmith, M. H. M. 1969. *The Physiology of Plant Growth and Development*, ed. M. B. Wilkins, 124–62. London: McGraw-Hill
55. Goebel, K. 1880. *Bot. Ztg.* 38:800
56. Goodwin, P. B., Cansfield, P. E. 1967. *J. Exp. Bot.* 18:297–307
57. Gregory, F. G., Veale, J. A. 1957. *Symp. Soc. Exp. Biol.* 11:1–20
58. Guern, J., Usciati, M. 1972. See Ref. 16, 383–400
59. Gulline, H. 1960. *Aust. J. Bot.* 8:1–10
60. Hall, R. H. 1973. *Ann. Rev. Plant Physiol.* 24:415–44
61. Harvey, E. N. 1920. *Am. Natur.* 54:362–67
62. Hillman, J. 1970. *Planta* 90:222–29
63. Hiron, R. W. P., Wright, S. T. C. 1973. *J. Exp. Bot.* 24:769–81
64. Hoad, G. V. 1973. *Planta* 113:367–72
65. Hayashi, T., Murakami, Y. 1953. *J. Agr. Chem. Soc. Jap.* 27:675
66. Itai, C., Vaadia, Y. 1971. *Plant Physiol.* 47:87–90
67. Jackson, D. I., Field, R. J. 1972. *Ann. Bot.* 36:525–32
68. Jacobs, W. P., Case, D. B. 1965. *Science* 148:1729–31
69. Jacobs, W. P., Danielson, J., Hurst, V., Adams, P. 1959. *Develop. Biol.* 1:534–54
70. Jankiewicz, L. S., Szpunar, H., Baranska, R., Rumplowa, R., Fiutowska, K. 1961. *Acta Agrobot.* 10:151
71. Jeffcoat, B., Harris, G. P. 1972. *Ann. Bot.* 36:353–61
72. Jones, R. L., Phillips, I. D. J. 1966. *Plant Physiol.* 41:1381–86
73. Jones, R. L., Phillips, I. D. J. 1967. *Planta* 72:53–59
74. Kato, J. 1953. *Physiol. Plant.* 11:10–15
75. Kende, H. 1965. *Proc. Nat. Acad. Sci. USA* 53:1302–7
76. Key, J. L. 1969. *Ann. Rev. Plant Physiol.* 20:449–74
77. Kraus, E. J., Mitchell, J. W. 1947. *Bot. Gaz.* 108:301–50
78. Krelle, E., Libbert, E. 1968. *Planta* 80:317–20
79. Kulasegaram, S., Kathiravetpillai, A. 1972. *Tea Quart.* 43:180–94
80. Kumar, D., Wareing, P. F. 1972. *New Phytol.* 71:639–48
80a. Kung-Woo Lee, P., Kessler, B., Thimann, K. V. 1974. *Physiol. Plant.* 31:11–14
81. Kursanov, A. L. 1963. *Advan. Bot. Res.* 1:209–78
82. Leike, H., von Guttenberg, H. 1961. *Naturwissenschaften* 48:604–5
83. Lepp, N. W., Peel, A. J. 1970. *Planta* 90:230–35
84. Ibid 1971. 97:50–61
85. Link, G. K. K., Eggers, V. 1946. *Bot. Gaz.* 108:114–29
86. Loiseau, J. E. 1965. *Travaux dédiés à Lucien Plantefol*, 367–89. Paris: Masson et Cie
87. Lyon, C. J. 1963. *Plant Physiol.* 38:145–52
88. Ibid, 567–74
89. Ibid 1972. 50:417–20
90. MacQuarrie, I. G. 1965. *Can. J. Bot.* 43:29–38
91. Marr, C., Blaser, H. W. 1967. *Am. J. Bot.* 54:498–504
92. Martin, C., Thimann, K. V. 1972. *Plant Physiol.* 49:64–71
92a. McGraw, D. C. 1973. *An investigation into the role of gibberellins in apical dominance in the Cupressaceae.* MSc thesis. Univ. Calgary, Canada
93. McIlrath, W. J., Bogorad, L. 1960. *Plant Physiol.* 39:328–31
94. McIntyre, G. I. 1964. *Nature* 203:1190–91
95. McIntyre, G. I. 1968. *Can. J. Bot.* 46:147–55
96. Ibid 1971. 49:99–109
97. McIntyre, G. I. 1971. *Nature New Biol.* 230:87–88
98. McIntyre, G. I. 1972. *Can. J. Bot.* 50:393–401
99. Ibid 1973. 51:293–99
100. McKee, N. D. 1968. *Correlative inhibition between axillary shoots of decapitated seedlings of Pisum sativum.* Hons. BSc thesis. Queens Univ. Belfast. Cited by Cutter (37)
100a. Milborrow, B. V. 1974. *Ann. Rev. Plant Physiol.* 25:259–307
101. Milthorpe, F. L., Moorby, J. 1969. *Ann. Rev. Plant Physiol.* 20:117–38
102. Mitchell, J. W., Martin, W. E. 1937. *Bot. Gaz.* 99:171–83

103. Morgan, P. W., Gausman, H. W. 1966. *Plant Physiol.* 41:45–52
104. Morris, D. A., Winfield, P. J. 1972. *J. Exp. Bot.* 23:346–55
105. Mullins, M. G. 1970. *Ann. Bot.* 34:897–909
106. Naylor, J. M. 1958. *Can. J. Bot.* 36:211–32
107. Niedergang-Kamien, E., Skoog, F. 1956. *Physiol. Plant.* 11:60–73
108. Nolan, J. R. 1969. *Am. J. Bot.* 56:603–9
109. Overbeek, J. van 1938. *Bot. Gaz.* 100:133–66
110. Palmer, J. H. 1955. *An investigation into the behaviour of the rhizome of Agropyron repens Beauv., with special reference to the effect of light.* PhD thesis. Sheffield Univ., U.K.
111. Palmer, J. H. 1956. *New Phytol.* 55:346–55
112. Palmer, J. H. 1964. *Planta* 61:283–97
113. Palmer, J. H. 1972. *J. Exp. Bot.* 23:733–43
114. Palmer, J. H., Phillips, I. D. J. 1963. *Physiol. Plant.* 16:572–84
115. Panigrahi, B. M., Audus, L. J. 1966. *Ann. Bot.* 30:457–73
116. Patrick, J. W., Wareing, P. F. 1970. *Proc. 7th Int. Conf. Plant Growth Substances Canberra,* ed. D. J. Carr, 695–700
117. Patrick, J. W., Woolley, D. J. 1973. *J. Exp. Bot.* 24:949–57
118. Peterson, C. A., Fletcher, R. A. 1973. *J. Exp. Bot.* 24:97–103
119a. Pharis, R. P., Kuo, C. C., Glenn, J. L. 1970. *Plant Growth Substances,* ed. D. J. Carr, 441–48. New York: Springer-Verlag
119b. Pharis, R. P., Ruddat, M., Phillips, C., Heftmann, E. 1965. *Naturwissenschaften* 52:88–89
120. Phillips, I. D. J. 1964. *Ann. Bot. London* 28:17–35
121. Phillips, I. D. J. 1968. *J. Exp. Bot.* 19:617–27
122. Phillips, I. D. J. 1969. *Planta* 86:315–23
123. Phillips, I. D. J. 1969. See Ref. 54, 163–202
124. Phillips, I. D. J. 1971. *Planta* 96:27–34
125. Phillips, I. D. J. 1971. *J. Exp. Bot.* 22:465–71
126. Phillips, I. D. J., Hartung, W. 1974. *Planta* 116:109–21
127. Phillips, I. D. J., Vlitos, A. J., Cutler, H. 1959. *Contrib. Boyce Thompson Inst.* 20:111–20
128. Poppei, R. W., Scott, T. K. 1968. *Am. J. Bot.* 55:707
129. Preston, A. P., Barlow, H. W. B. 1950. *Rep. East Malling Res. Sta. No. 76*
130. Reid, D. M., Crozier, A., Harvey, B. M. R. 1969. *Planta* 89:376–79
131. Ruddat, M., Pharis, R. P. 1966. *Planta* 71:222–28
132. Sachs, T. 1969. *Ann. Bot.* 33:263–75
133. Sachs, T. 1970. *Isr. J. Bot.* 19:484–98
134. Sachs, T., Thimann, K. V. 1964. *Nature (London)* 201:939–40
135. Sachs, T., Thimann, K. V. 1967. *Am. J. Bot.* 54:136–44
136. Sastry, K. S. K., Muir, R. M. 1965. *Plant Physiol.* 40:294–98
137. Schaeffer, G. W., Sharpe, F. T. 1969. *Bot. Gaz.* 130:107–10
138. Schaeffer, G. W., Sharpe, F. T. 1970. *Ann. Bot. London* 34:707–19
139. Schneider, G. 1969. *Ber. Deut. Bot. Ges., Vortr. Gesamtgeb. Bot.* 3:19–42
140. Schneider, G. 1970. *Ann. Rev. Plant Physiol.* 21:499–536
141. Schneider, G. 1972. See Ref. 16, 317–31
142. Scott, T. K., Case, D. B., Jacobs, W. P. 1967. *Plant Physiol.* 42:1329–33
143. Scurfield, G. 1973. *Science* 179:647–55
144. Šebánek, J. 1965. *Biol. Plant.* 7:194–98
145. Ibid 1966. 8:213–19
146. Selman, I. W., Sandanam, S. 1972. *Ann. Bot. London* 36:837–48
147. Seth, A. K., Wareing, P. F. 1964. *Life Sci.* 3:1483–85
147a. Seth, A. K., Wareing, P. F. 1967. *J. Exp. Bot.* 18:65–77
148. Shein, T., Jackson, D. I. 1971. *Ann. Bot.* 35:555–64
149. Sheldrake, A. R. 1973. *Biol. Rev.* 48:509–59
150. Simpson, G. M., Saunders, P. F. 1972. *Planta* 102:272–76
151. Skoog, F., Armstrong, D. J. 1970. *Ann. Rev. Plant Physiol.* 21:359–84
152. Smith, H., Wareing, P. F. 1964. *Ann. Bot. N. S.* 28:283–95
153. Ibid, 297–309
154. Smith, H., Wareing, P. F. 1966. *Planta* 70:87–94
155. Sorokin, H. P., Thimann, K. V. 1964. *Protoplasma* 59:326–50
156. Snow, R. 1925. *Ann. Bot. London* 39:841–59
157. Snow, R. 1937. *New Phytol.* 36:283–300
158. Snow, R. 1939. *Nature (London)* 144:906
159. Snow, R. 1945. *New Phytol.* 44:110–17
160. Ibid 1947. 46:254–57
161. Stuart, N. W. 1938. *Bot. Gaz.* 100:298–311
162. Thimann, K. V. 1937. *Am. J. Bot.* 24:407–12
163. Thimann, K. V. 1972. See Ref. 16, 397
164. Thimann, K. V., Sachs, T., Mathur, S. N. 1971. *Physiol. Plant.* 24:68–72

APICAL DOMINANCE 367

165. Thimann, K. V., Skoog, F. 1933. *Proc. Nat. Acad. Sci. USA* 19:714–16
166. Thimann, K. V., Skoog, F. 1934. *Proc. Roy. Soc. B* 114:317–39
166a. Thomas, B., Hall, M. A. 1974. *Plant Sci. Lett.* In press
167. Thomas, T. H. 1972. *J. Exp. Bot.* 23:294–301
168. Tucker, D. J., Mansfield, T. A. 1972. *Planta* 102:140–51
169. Tucker, D. J., Mansfield, T. A. 1973. *J. Exp. Bot.* 24:731–40
170. Tupy, J. 1973. *Physiol. Veg.* 11:13–23
171. Usciati, M., Codaccioni, M., Guern, J. 1972. *J. Exp. Bot.* 23:1009–20
172. Usciati, M., Codaccioni, M., Rautureau, C., Berreur, P. 1970. *C. R. H. Acad. Sci. Paris* 270:1796–99
173. Vardar, Y., Kaldewey, H. 1972. *Hormonal Regulation in Plant Growth and Development*, ed. H. Kaldewey, Y. Vardar, 401–11. Weinheim: Verlag Chemie
174. Verner, L. 1955. *Res. Bull. Idaho Agr. Exp. Sta.* No. 28
175. Wagner, H., Michael, G. 1971. *Biochem. Physiol. Pflanzen* 162:147
176. Wakhloo, J. L. 1970. *Planta* 91:190–94
177. Wardlaw, C. W. 1943. *Ann. Bot. London* 7:171–84
178. Ibid, 357–77
179. Wardlaw, I. F. 1974. *Ann. Rev. Plant Physiol.* 25:515–39
180. Wardlaw, I. F., Mortimer, D. C. 1970. *Can. J. Bot.* 48:229–37
181. Wareing, P. F. 1964. *Proc. 7th Brit. Weed Control Conf.* 3:1020–30
182. Wareing, P. F., Nasr, T. A. A. 1958. *Nature (London)* 182:379–81
183. Wareing, P. F., Nasr, T. A. A. 1961. *Ann. Bot. N. S.* 25:321–40
184. Wareing, P. F., Saunders, P. F. 1971. *Ann. Rev. Plant Physiol.* 22:261–88
185. Wareing, P. F., Seth, A. K. 1967. *Symp. Soc. Exp. Biol.* 21:543–58
186. Went, F. 1936. *Biol. Zentralbl.* 56:449–63
187. Went, F. 1939. *Am. J. Bot.* 26:109–17
188. Wetmore, R. H., Jacobs, W. P. 1953. *Am. J. Bot.* 40:272–76
189. Wetmore, R. H., Ricer, J. P. 1963. *Am. J. Bot.* 50:418–30
190. Wetmore, R. H., Sorokin, S. 1955. *J. Arnold Arbor.* 36:305–17
191. White, J. C. 1973. *Apical dominance in leguminous plants.* PhD thesis. Glasgow Univ., Scotland. 219 pp.
192. White, J. C., Hillman, J. R. 1972. *Planta* 107:257–60
193. Whiting, A. G., Murray, M. A. 1948. *Bot. Gaz.* 109:447–73
194. Wickson, M. E., Thimann, K. V. 1960. *Physiol. Plant.* 13:539–44
195. Wolpert, L. 1969. *J. Theor. Biol.* 25:1–47
196. Wolpert, L. 1971. *Advan. Morphog.* 6:183
197. Wolpert, L. 1972. *Nature New Biol.* 239:101–5
198. Woolley, D. J., Wareing, P. F. 1972. *Planta* 105:33–42
199. Woolley, D. J., Wareing, P. F. 1972. *New Phytol.* 71:781–93
200. Ibid, 1015–25
201. Wright, S. T. C., Hiron, R. W. P. 1969. *Nature (London)* 224:719–20

Ann. Rev. Plant Physiol. 1975. 26:369–401

PHYCOBILIPROTEINS AND COMPLEMENTARY CHROMATIC ADAPTATION

❖7595

Lawrence Bogorad

The Biological Laboratories, Harvard University, Cambridge, Massachusetts 02138

CONTENTS

INTRODUCTION

Phycobiliproteins (PBPs) are the principal photoreceptors for photosynthesis in blue-green and red as well as some other types of algae. These proteins with covalently linked bile pigment chromophores may comprise more than 60% of the total soluble cellular protein, or almost 20% of the total dry weight of one alga (2, 3); the amounts in other blue-green and red algae are probably comparable; thus PBPs are major metabolic products of these organisms. There are three major categories of PBPs: the blue phycocyanins (PCs) and allophycocyanins (APC) plus the red phycoerythrins (PEs). Some algae contain only species-specific APC + PC, but others contain a PE as well; the ratio of one PBP to another and/or the ratio of PBP to chlorophyll varies in some algae depending upon the color of light in which they are growing. The adaptation is complementary—algae growing in red light tend to form more red light-absorbing pigments and thus to look blue; the converse behavior is exhibited by cells grown in green light.

Interactions of chromophores and polypeptides which make PBPs especially suitable for energy conversion and transfer, the supramolecular organization of biliproteins in the aqueous environment adjacent to photosynthetic membranes, and mechanisms for funneling energy to the chlorophyll-containing reaction centers within the membranes are aspects which may claim the attention of cell and structural biologists, protein chemists, photochemists, and photobiologists, including students of photosynthesis. Complementary chromatic adaptation, on the other hand, is an interesting situation for studying photodifferentiation and developmental biology because both the environmental effectors and the products are well defined. Furthermore, PBPs are produced in massive amounts—they are major products of the cell's protein synthetic machinery easily obtained pure in large quantities.

The last time this subject was examined in the *Annual Review of Plant Physiology,* Ó hEocha (130) enumerated the spectral properties of algal bile pigment-protein complexes and examined the distribution of phycoerythrins, phycocyanins and allophycocyanins. He also reviewed information then available on the amino acid composition of the apoproteins and made some guesses about the structures of the bile pigment chromophores.

During the last decade there has been a great deal of progress attributable to increased experience and skill with these compounds as well as to the development of new methods which simplify the solution of difficult structural problems in organic chemistry. The work of the past 10 years has resulted in the resolution of the structures of the chromophores of phycobiliproteins (although a few doubts persist). Important and interesting discoveries about the structures of the polypeptides with which the chromophores are associated have begun to appear since 1971.

This review will deal with the present information on the composition and organization of the polypeptides of which PBPs are comprised. Next, the structures of the chromophores will be discussed, and some guesses, because no solid information is available, will be advanced about the course of their synthesis. Another major advance in this past decade has been the recognition that in many algae PBPs are organized into bodies of definite structure—the phycobilisomes (PBsomes). The organization of the PBsomes will be reviewed, together with a brief examination of the functions of phycobiliproteins in algae. One of the most fascinating phenomena in the biology of some phycobiliprotein-containing organisms is the regulation by light of the synthesis and accumulation of specific protein-pigment complexes. Complementary chromatic adaptation and how it may work is the next item discussed in this review.

Phytochrome, the photomorphogenic pigment of higher plants, is also a biliprotein but it is not discussed here; it was reviewed comprehensively in this series in 1972 by Briggs & Rice (18). A number of reviews have been published since 1965 dealing with specialized or general aspects of PBPs and their metabolism (6, 14, 23, 131–133, 144, 145, 152).

PROPERTIES OF PHYCOBILIPROTEINS

Absorption and Fluorescence of Phycobiliproteins

PBPs are photosynthetic accessory pigments. They absorb light and the energy is transferred to chlorophyll (43, 49, 115). A PC- and APC-containing chlorophyll-less mutant of *Cyanidium caldarium* cannot carry on photosynthesis (120–122).

Absorption and fluorescence maxima of PBPs are given in Table 1. The prefixes "C-", "R-" and "B-" before the name of the PC or PE designates the algal group, i.e. Cyanophyceae (blue-green algae), Rhodophyceae (red algae), and Bangiales (red algae of this order), from which the PBP with this particular set of absorption characteristics was first isolated, although B-PEs, for example, are found in many groups of red algae. The several cryptomonad PCs are distinguished from one another by their absorption maxima; the same system is used to differentiate one type of cryptomonad PE from another.

Light absorbed by PE is efficiently transferred to PC, then to APC, and finally to chlorophyll. APC is the "chlorophyll *b* of blue-green and red algae" in absorbing at about 650 nm and fluorescing at about 660 nm. However, the APC is coupled to a complex light-harvesting system covering a broad spectral range which, in many algae, can be altered to accommodate to different conditions of illumination. PCs in cryptomonads have evolved to take over the function of APC. R-PCs absorb strongly both at about 550 nm, like C-PE, and 615 nm, like C-PC.

Frackowiak & Grabowski (47, 48, 81, 82) have determined the fluorescence emission spectra of R-PE and C-PC at 90°K. They have also studied the transfer of excitation energy between PBPs and chlorophyllide *a* in vitro. Low temperature absorption and fluorescence spectroscopy of *Anacystis nidulans* has been reported by Cho & Govindjee (29).

Table 1 Absorption and fluorescence maxima of phycobiliproteins[a]

Type	Absorption maxima (nm)				Fluorescence maxima (nm)	
Allophycocyanin			650			660
C-Phycocyanin		615				647
R-Phycocyanin	553	615			565	637
Cryptomonad PC–615		588	615			637
PC–630		588		630		
PC–645		583			645	660
C-Phycoerythrin		565				575
R-Phycoerythrin	498 540	568				578
B-Phycoerythrin		546	565			578
Cryptomonad PE–544		544				
PE–555		555				580
PE–568			568			

[a] Data taken from summaries by Chapman (23), O hEocha (130, 133), plus other information in (75, 135, 169). The absorption maxima given are not universal. PBPs from various species differ from one another in this regard (see examples of variations and discussions 89, 130, 133, 169).

The distributions of PBPs among algal groups are given by Chapman (23) and by Bogorad (10).

Polypeptide Subunits

Every PC and PE examined has been found to be composed of two different kinds of polypeptide subunits. Analyses of denatured polypeptides by electrophoresis on sodium dodecyl sulfate-containing polyacrylamide gels and by other methods are enumerated in Table 2. The molecular masses for the small (α) subunits of phycocyanin range from 10 to 19.7 X 10^3 daltons. The larger (β) subunits range in size from about 14 to 22.2 X 10^3 daltons. Variations within this range are found among the subunits of PEs. The precise molecular weights obtained by SDS-PAGE can be affected by the markers which are used and by the care exercised in carrying out the electrophoresis. Some of the variation may be attributable to such technical problems but, on the other hand, some differences in molecular weight of subunits from a single type of PBP but from different organisms have been observed even in the same laboratory. Most if not all the reported differences are probably real. In the case of *Mastigocladus laminosus* PC, a single band appears on SDS gel electrophoresis at the 14,000 dalton position, but after treatment with 8 molar urea and 2-mercaptoethanol, separation into two components by gel electrophoresis was achieved (8). In this case the two polypeptides have the same molecular weight but differ in charge. Variable results have been obtained with APC from different sources. SDS-PAGE (sodium dodecyl sulfate-polyacrylamide gel electrophoresis) of APCs from *Fremyella diplosiphon* and *Plectonema boryanum* revealed a single band at about the 15,000 to 16,000 dalton position (3, 142), but two polypeptides were observed in similar analyses of APCs from two other blue-green algae (74). It is possible that, as in the case of *M. laminosus* PC, the single band seen on SDS-PAGE of some APCs may contain two differently charged polypeptides of the same mass.

It is unlikely that all of the variations in molecular sizes of subunits can be attributed to the action of proteolytic enzymes, at least in the case of *Fremyella*. In that case, the polypeptides were detected in crude extracts of algae prepared in several different ways (2, 3).

The α subunit of *Fremyella* PE contains one chromophore and the β subunit two (2, 3). PC from a number of organisms follows the same pattern as that of PE, i.e. one chromophore is present on the α subunit and two on the β subunit (77). Each polypeptide chain of APC bears one chromophore (77). Another type of difference is seen in cryptomonad PC. The 10,000 dalton α subunit of *Chroomonas* PC is reported to have an absorption maximum at 560 nm while the β subunit absorbs at 640 nm (109).

Some of the molecular weights given in Table 2 are based on amino acid composition. Calculations are made on the assumption that there is one chromophore per polypeptide chain. If the distribution of chromophores per chain in PCs and PEs is universal, the molecular masses of the β subunits calculated on the basis of the above assumption are about 800 daltons too low.

The column headed "Aggregate size" shows molecular weights observed in aqueous solutions in the absence of detergents or other disaggregating agents. The largest

Table 2 Molecular weights of polypeptide subunits of phycobiliproteins ($\times$ 10^3)

Organism	Subunits[a] Light (α), Heavy (β)		Method[b]	Aggregate size[i]	Ref.
Phycocyanins					
Fremyella diplosiphon	16.3K	17.6K	SDS-PAGE		3
Synechococcus sp. (6301)					
(*A. nidulans*)	15.9	19.1	SDS-PAGE	224K ± 6K	74
Aphanocapsa sp. (6701)	16.6	20.2	SDS-PAGE		74
Anabaena sp. (6411)	17.1	19	SDS-PAGE		74
Nostoc punctiformis					
Anacystis nidulans					
and other species	18.5	20.5	SDS-PAGE	340K	126
Prophyridium cruentum	16[g]	18.4[g]	SDS-PAGE	127K	71
Mastigocladus laminosus	14[c]	14[c]	SDS-PAGE		8
Phormidium luridum	11.9	18.5	SDS, aa		94
Oscillatoria agardhi	12.1	14.1	aa		160
Plectonema boryanum	15.1	17.2	SDS-PAGE		142
Chroomonas sp.[d]	10	16	SDS-PAGE	50K	109
Not given R-PC	18.5[e]	20.5[f]	SDS-PAGE		125
Not given C-PC	19.7	22	SDS-PAGE		125
Phycoerythrins					
Fremyella diplosiphon	18.3	20	SDS-PAGE	180–210K	2,3
Aphanocapsa sp. (6701)	20	22	SDS-PAGE		74
Phormidium persicinum	19.7	22	SDS-PAGE		126
Various red algae PE I	19.8	22.5	SDS-PAGE	250–299K	169
PE II	19.8	22.5	SDS-PAGE	43.5–44.3K	
Porphyridium cruentum-B	17.7[h]	30[h]	SDS-PAGE	280K	71
-b		17.2		110–55K	
Cryptomonas sp. PE-566	11.8	19		35K ± 2K	75,123
Rhodomonas sp. PE-542	11	17.7	SDS-PAGE		21
Allophycocyanins					
Fremyella diplosiphon	16		SDS-PAGE		2,3
Plectonema boryanum	15.3		SDS-PAGE		142
Synechococcus sp. (6301)					
(*A. nidulans*)	15.2	17.2	SDS-PAGE	96K ± 3K	74
Aphanocapsa sp. (6701)	16	17;17.9	SDS-PAGE		74
Porphyridium cruentum	14.6		SDS-PAGE	120K	71

[a] Where reported, variation ranges from about ±0.8 to ±5% of the averaged value.

[b] Abbreviations: SDS-PAGE – polyacrylamide gel electrophoresis in sodium dodecyl sulfate.

         aa – values calculated from amino acid analyses.

[c] Distinguishable by electrophoretic mobility in urea-PAGE.

[d] Absorption maximum of α at 560 nm; of β at 640 nm.

[e] Blue.

[f] Red.

[g] α is blue; β is pink.

[h] α absorption maximum at about 525 nm; β at about 500 nm. *P. cruentum* PE-B = PE I of other red algae; PE-b – PE II. Gantt & Lipschultz (71) suggest that van der Velde's (169) preparations of B-PE II (or b-PE) are contaminated with PE-I to explain the differences in subunit compositions observed by the two groups; species-to-species differences might also account for the data (71). Others have also reported observing multiple forms of R-PE (114, 123).

[i] See text for discussion of factors which influence aggregation.

molecular sizes represent $\alpha_6\beta_6$ aggregates. These are generally in equilibrium with $\alpha_3\beta_3$ forms. Ó hEocha (130) has already discussed some of the aggregation-disaggregation phenomena in the PBPs, but to summarize for the present discussion: The dodecamer containing six α and six β subunits predominates in the pH range of 5 to 6, at ionic strengths above 0.1 and at concentrations higher than 1 mg/ml. At higher or lower pH values, at lower ionic strengths, or at lower protein concentrations the dodecamer dissociates to give the aggregate $\alpha_3\beta_3$ (6, 85, 110, 118). In addition, PBPs dissociate irreversibly into 2–4S units at pH 11 or greater (e.g. 87).

As shown in Table 2, cryptomonad PBPs deviate from the $\alpha_6\beta_6 \rightleftharpoons \alpha_3\beta_3$ rule; where analyzed (75, 122), the maximum aggregate size has been found to be 35,000–40,000 daltons, i.e. $\alpha_1\beta_1$.

The amino acid compositions of separated α and β chains of PC have been determined for *Oscillatoria agardhii* (160), *Mastigocladus laminosus* (8), *Phormidium luridum* (94), and of *Synechococcus sp.*, *Aphanocapsa sp.* and *Anabaena sp.* (78). The α and β subunits have been separated by PAGE in urea (8, 9), by gel filtration in Sephadex G-100 in the presence of SDS (94), and by stepwise elution from Bio-Rex 70 with increasing concentrations of acid urea (78). Most authors comment on the similarity of the amino acid compositions of the two subunits. Binder et al (8) show about 19 amino acid residue differences between the α and β subunits of *M. laminosus*. In all of the cases studied except PC from *O. agardhii*, histidine is lacking from one subunit but is present in the other. It is generally absent from the β subunit where these two are distinguishable on the basis of size. Even in the case of PC from *O. agardhii*, there are differences in the histidyl residues between the two subunits (160). [Binder et al (8) pointed out that the difference in content of any amino acid between the α and β subunits is not greater than three residues and that the overall compositions of the two subunits do not differ significantly.]

Methionine is the only N-terminal amino acid of PCs of *Plectonema colothricoides, Phormidium luridum,* and *Synechococcus lividus* (35). In this work the two subunits of the PCs were not resolved. Methionine has also been found to be the N-terminal amino acid in both chains of PC from *O. agardhii* (160) and *C. caldarium* (165), but the amino-terminal residue in the α subunit of *M. laminosus* PC is alanine while that of the β subunit is valine (8). Threonine was found to be the sole N-terminal residue in *Nostoc muscorum* C-PC (124). Both threonine and methionine were found to be N-terminal residues in *P. laciniata* R-PC which had not been resolved into subunits (124). The differences among PCs with regard to methionine versus other amino acids at the N terminus may reflect differences in the processing of completed polypeptide chains in different algae.

Ó Carra (124) and Ó Carra & Ó hEocha (127) found only methionine as the N-terminal residue of all the R-, B-, and C-PEs they studied. PEs were obtained from the red algae *Porphyra laciniata, Ceramium rubrum,* and *Rhodochorton floridulum.* This work was done before PEs were known to consist of two kinds of polypeptide chains, but the presence of only a single N-terminal amino acid suggests that the two chains are the same in this regard.

Serine has been identified as the only carboxy-terminal amino acid residue of *Nostoc muscorum* C-PC (141). Alanine has been identified as the only carboxy-terminal amino acid residue of *Ceramium rubrum* R-PE which had not been resolved into two subunits before analysis.

The two subunits of C-PE from *Fremyella diplosiphon* have been eluted from Bio-Rex 70 with urea by Takemoto & Bogorad (153), using a slight modification of the method described by Glazer et al (78). The amino acid compositions of the two subunits are distinctive but not greatly different. Histidine is present in both. The α subunit yields 12 tryptic peptides; the β subunit 13. At least eight of the peptides are common to the two chains.

The amino acid sequences have been determined for the first 27 amino acid residues from the N termini of the α and β subunits of *Cyanidium caldarium* C-PC (165). All but three residues between 10 and 27 are identical. Overall, 60% homology was found. Extensive homologies have also been found in the first 15 to 18 residues of the α and β subunits of *Synechococcus* sp. (*Anacystis nidulans*) (173). On the other hand, only two of seven amino acid residues are at the same positions in chromopeptides of the α and β chains prepared from C-PC of *Mastigocladus laminosus* (22). If C-PC is like other PCs in carrying two chromophores on the β chain and only one on the α subunit, perhaps only one of the β chromopeptides has been recovered and studied in this research. If there is another, does it resemble the α chromopeptide?

Similarities or identities of some segments of α and β subunits suggest that the gene for one might have been derived from the other and some features were conserved during subsequent evolution, although differences between subunits, e.g. the number of chromophores per polypeptide, have arisen perhaps by reduction as well as substitution. Selection pressure to retain the chromophores obviously would be very high, but pressure to conserve features required for aggregation would be equally strong. The arguments for the latter view are presented later in this review. The ability of PC_α subunits from one species of blue-green alga to form hybrids with PC_β subunits of other species (78) shows that subunit-subunit recognition sites have been retained during evolution of the species as well as within a single species.

The Bile Pigment Chromophores of Phycobiliproteins

From the initial recognition that bile pigments were the chromophores of phycobiliproteins in 1930 (104, 105) until the late 1960s, the structure of these open-chain tetrapyrroles was unclear. Prior to 1962, phycobilins were separated from the proteins by hot or cold strong acids, and the relation of these products to the native bile pigments was hotly argued. In 1962 Fujita & Hattori (56) found that a blue pigment could be extracted from cells of *Tolypothrix tenuis* and *Anabaena cylindrica* with 90% methanol containing 1% ascorbic acid at 60°C. They considered this a new phycobilin or possibly a precursor of the PC chromophore. A purple bile pigment was obtained in the same way from PE-rich cells of *T. tenuis* (59). It was later found (31, 36, 128) that the bile pigments could be removed directly from purified and denatured PBPs using methanol alone. This was the mildest procedure known for separating the chromophores from the proteins. Because the most thoroughly studied algal bile pigments had been obtained by acid treatment of PBPs, some workers considered the compounds obtained in the latter ways as the authentic native chromophores and those obtained by methanol treatment to be artifacts (128). By dissolving the methanol-released bile pigments in 10N HCl, the purple and blue pigments could be converted to the forms removed directly from PBPs with acid. It was shown by three groups about simultaneously (31, 36, 146, 147) that the product of methanol treatment of PC is compound (c) or (d) shown in Figure 1. An alternative structure has been suggested (149) and this possibility has been evaluated by Chapman (23). His extensive discussion of the arguments against the alternative structure are convincing. The structure shown in Figure 1 was established by mass spectroscopy and NMR as well as by chromatography of chromate oxidation products of the compound obtained by treatment of PC with ethanol. A spectrally and chromatographically identical compound is released into the medium by cells of *Cyanidium caldarium* grown in the presence of δ-aminolevulinic acid (162, 163, 167).

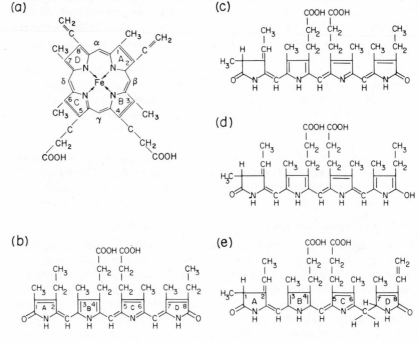

Figure 1   (*a*) Fe protoporphyrin IX. (*b*) Mesobiliverdin IXα (formed by the loss of the α carbon atom from Fe protoporphyrin IX). (*c*) Phycocyanobilin according to Cole et al (31) and Rüdiger et al (147). (*d*) Phycocyanobilin according to Crespi et al (36). (*e*) Phycoery-throbilin (25, 37, 147).

Methanol hydrolysis of PEs yields compound (e) shown in Figure 1 (23, 25, 36, 152). This structure is based on mass spectroscopy and NMR analyses. A similar structure was proposed on the basis of the chromate degradation method (146, 147).

The chromophore of C-PC is designated phycocyanobilin (PCB) and that from C-PE as phycoerythrobilin (PEB). These names have also been applied to the acid-cleavage products of the chromophores from their proteins. It is now clear that the acid-cleavage products are altered forms of the native bile pigments. The practice of designating a particular compound as PCB or PEB, followed by the wavelength of its absorption maximum, is useful to designate various forms released by different methods or produced by alterations after release. Structure (c) in Figure 1 is PCB-630; Structure (e) is PEB-590.

Also included in Figure 1 is the structure of mesobiliverdin. This compound would be formally derived from protoporphyrin IX by removal of the α-bridge carbon and its release as carbon monoxide, plus the reduction of the two vinyl side chains on Rings A and B to ethyl side chains. A hydrogen atom at position 1 of ring A and the ethylidene group, in place of an ethyl group at position 2 of ring A, distinguishes PCB-630 from mesobiliverdin. The bridge carbon between rings C and D is reduced and a vinyl group replaces an ethyl group at position 8 in PEB-590, but otherwise it is identical to PCB-630. This accounts for absorption at a shorter wavelength. Both PEB and PCB yield mesobiliverdin IXα upon refluxing in 1 N KOH in methanol (25, 32–35, 152).

PCB-630 is also obtained by methanol hydrolysis of APC (24). Similar treatment of R-PCs yields both PCB-630 and PEB-590 (24). (For other reports dealing with cleavage of bilins from PBPs and the structure of the chromophores, see 26, 27, 32, 39, 143, 152.)

C- and B-PEs have absorption maxima at 565 nm. Some cryptomonad PEs absorb at 568 nm and others at about 544–555 nm. Of all the PEs, only type R- has absorption maxima at 500 nm and 565 nm as well as a shoulder at 545 nm. O Carra et al (129, 134) attribute the absorption at about 500 nm to phycourobilin, an open-chain tetrapyrrole in which only rings B and C are in conjugation (Figure 1, structure d). Urobilinoid compounds are formed upon the reduction of the three carbon bridges and reoxidation of PEB (33, 129). Ó Carra et al (129) detected the urobilin after strong acid treatment of PEs. On the other hand, Chapman et al (28) were able to detect only PEB after cleavage of bilins from R-, B-, and Rhodomonas PEs with methanol; they could not find urobilins. Chapman et al (28) attribute the absorbance at about 500 nm to an interaction between PE and the protein.

Pecci & Fujimori have also obtained data which argue against the existence of a separate urobilinoid compound (137). B-PE from Porphyridium cruentum has absorption maxima at 565 and 545 nm plus a shoulder at 500 nm. When the B-Pe is treated with p-mercuribenzoate at pH 7 the absorption at 565 nm is lost and a sedimentable purple material with an absorption maximum at about 500 nm forms. They believe that this is one subunit of B-PE. The solution remains red in color and has an absorption maximum at about 545 nm; this is thought to be attributable to the other subunit. The 565 nm band partially reappears after the mercurial is removed from the soluble subunits with mercaptoethanol. Pecci & Fujimori obtained the same bile pigment chromophore, i.e. PEB, from both the purple and red materials. Thus absorbance at either about 500 or 545 nm can be exhibited by PEB-protein complexes.

This problem is not resolved. It may be unraveled as the chromophores of individual subunits of R-PE are studied and as the amino acid sequences which influence the absorption of the chromophores are understood.

Various cryptomonad PCs have absorption maxima at about 580–590 nm plus an additonal band at 615, 630, or 645 nm. The α subunit of *Chroomonas* PC (109) is reported to have an absorption maximum at 560 nm, while the β subunit has an absorption maximum at 640 nm. On the other hand, Chapman et al (28) have reported that PCB-630 is the only bile pigment obtained by methanolysis of *Chroomonas* PC. The recovery of bile pigments after methanolysis is not quantitative; it is easily possible that some might be lost selectively.

Chromophore-Polypeptide Linkages

A number of investigations of the structure of chromopeptides, segments of the polypeptide chain associated with the bile pigments, were carried out before it was recognized that PBPs are generally composed of more than one kind of polypeptide chain. It is possible that the interaction between the chromophore and the polypeptide chain is always through the same amino acids by the same types of bonds and that the amino acid sequences immediately adjacent to the chromophores are always identical, but there is no direct evidence to support such a view at the present time.

Groups on the bile pigments most likely to interact with the protein are (*a*) one of the terminal oxygen atoms (convertible to -OH groups); (*b*) the carboxyl group at the terminus of each propionic acid side chain on rings B and C; (*c*) the ethylidene group on ring A; (*d*) possibly the vinyl or ethyl groups on ring D of PCB or PEB, respectively; and (*e*) pyrrole nitrogen atoms.

Rüdiger (143, 146) has devised a procedure for separating the imides produced by oxidation of bile pigments and identifying them chromatographically. Oxidation of all PEs and PCs studied (146) demonstrated that oxidation with 2N sulfuric acid at 20°C releases ring D and one of the middle rings. Oxidation with this acid at 100° for one hour is required to release ring A and the middle rings. No dipyrrolic fragments are released during the second stage of oxidation, so it is concluded that rings A and C are involved in linkages of the bile pigment with the protein. Originally it was suggested in this work (146) that the propionic acid side chain was joined to an amino acid residue through an ester link. Crespi & Smith (38) have pointed out that the bile pigment is released primarily as the free acid after refluxing PBPs with methanol. In a revised view based in part on work with PE from the cryptomonad *Rhodomonas* (20), it has been suggested that the linkage through the propionic acid side chain may be as an amide involving glutamic acid. One of the other conditions that must be recognized in interpreting linkage data is that PEB and PCB are also liberated as free diacids on treatment of PCs and PEs with the proteolytic enzyme Nagarse (151). All of the bonds must be susceptible to cleavage by this enzyme, or some bonds must be labile enough to break after this enzyme has acted on other bonds.

Several different types of linkages of amino acid residues to substituents on ring A have been suggested. These include a thioether linkage to the proximal carbon of the ethylidene group and esterification by an aspartyl residue of the lactim form of the oxygen on that ring (38), linkage through the pyrrolenine nitrogen with a carboxyl group from the protein [this has been discarded in subsequent discussions from the same laboratory (146; also see discussion in 23)], and a seryl ether of the proximal carbon of the ethylidene (23). Some less specific suggestions have been made regarding seryl esters (93, 125) and the possible involvement of glutamic acid with ring A (125).

Finally, one case (125) of possible linkage through the two-carbon residue on ring D has been suggested in the linkage of a urobilin which is believed by some to be one of the chromophores of R-PE. A thioether linkage is suggested.

The implication that an ether or a thioether linkage involving the two carbon side chains on ring A is the native condition (23, 38) argues, as Chapman (23) points out, that the ethylidene group may be an artifact derived from cleavage, similar but not entirely comparable to the production of hematoporphyrin by the cleavage of heme from cytochrome c.

All of the work discussed so far in this section has been based on analyses of PBPs which have not been resolved into subunits. Work to date has not revealed any general rules for the attachment of the bile pigments to proteins, although Chapman (23) has suggested that a serine ether linkage to ring A and a glutamic acid imide linkage to ring C may be universal.

The only report in which the two subunits have been separated and then subjected to proteolytic enzymes to seek chromopeptides is by Byfield & Zuber (22), who separated the two polypeptide chains of C-PC from *Mastigocladus laminosus* by preparative gel electrophoresis. After pepsin digestion, they obtained a 20-peptide long chain containing the chromophore from the α subunit and a 15-carbon atom long chromopeptide from the α chain. Both chromophore-containing peptides include a blocked half cysteine, and it is considered clearly established by Byfield & Zuber that the chromophore is the blocking group on the α subunit. They assume that this is also the case for the β chain. It is conceivable that a bond involving one of the propionic acid side chains was cleaved in these experiments. No proposal was made as to which part of the PCB is associated with the cysteine of the peptic peptides, but the two-carbon substituent on ring A seems a reasonable candidate. The amino acid sequence near the chromophore in the α chain is Lys-Leu-Ile-Cys [PCB]-Gly-Ala-Ala, and the β chain around the chromophore has the sequence Lys-Ser-Lys-Cys [PCB]-Arg-Asp-Ile-. This does not encourage the belief that linkages in other PBPs will necessarily be the same on both chains, but perhaps only one of two β-chain chromopeptides has been recovered and analyzed.

Conspicuous problems about chromophore structure and chromophore-polypeptide interactions which remain unsolved are: how can absorbance at 500 nm by R-PE be accounted for; what are the covalent linkages between chromophores and polypeptide chains; what kinds of interactions occur between chromophores and the folded protein chains; and how do the chromophores in a single PBP aggregate interact? The next section has some bearing on the last question.

Spectrally Distinguishable Chromophores in Some Single Phycobiliproteins

R-PE has absorption bands at about 500, 545, and 565 nm. That extracted from *Antithammion* sp. absorbs at 497, 538, and 566 nm and fluoresces mainly at about 578 nm (111). But R-PE denatured in 8 M urea shows only 64%, 32%, and 22% of the absorption of the native R-PE at 497, 538, and 566 nm, respectively. The fluorescence of the denatured R-PE is weak and the emission maximum varies with the wavelength of incident light. MacDowell et al (111) found that in light of 479 nm denatured R-PE yielded a fluorescence spectrum with a maximum at 519 and a minor peak at 585 nm. Irradiation at 538 nm resulted in fluorescence emission at 550 and 585 nm. When the denatured PE was illuminated at 566 nm, fluorescence was detected at 578 nm. MacDowell et al (111) conclude that the chromophores are separated and fluoresce independently after the PE is denatured, but energy is absorbed by some chromophores and transferred to other chromophores which fluoresce in the native molecule.

Teale & Dale (154; see also 40) determined absorption, fluorescence-emission, fluorescence-excitation, and fluorescence-excitation polarization spectra of several PBPs. They then studied changes in the spectra as functions of ionic strength and biliprotein concentrations. They concluded from the complexity of the polarization spectra that a single phycobiliprotein contains "s" (sensitizing) and "f" (fluorescing) chromophores. They made the following assignments: PE-s absorption maximum at 540–555 nm; PE-f chromophores absorb at 656–670 nm; PC-s absorption maxima at 600–615 nm; and PC-f chromophores absorb at 630–635 nm.

Glazer et al (79) have determined that the α subunit of *Synechococcus* PC (which bears a single PCB) absorbs at 620 nm, and the β subunit has an absorption maximum at 608 nm. These do not match precisely the values for f- and s-type chromophores as determined by Teale & Dale (154), but Glazer et al conclude that the β subunit carries the s chromophores, whereas the α subunit carries the fluorescing chromophores.

Vernotte (171) has concluded that there is one s chromophore and one f chromophore per 30,000 daltons of C-PC from *Phormidium luridum*. These indirect calculations would appear to be in conflict with the more direct observations of Glazer et al (79).

At higher states of aggregation, i.e. $\alpha_6\beta_6$ rather than $\alpha_3\beta_3$, the long wavelength absorption maximum shifts toward the red, the absorption coefficient increases (85), and marked changes in the ORD and CD spectra occur (17, 136). The absorption spectra of PEs have also been observed to vary with these conditions (3, 50–53).

Immunochemical Relations Among Phycobiliproteins

Bogorad & Walbridge (11, 172) prepared rabbit antibody against PC from the unicellular alga *Cyanidium caldarium* and found it to cross-react with C-PC from four genera of blue-green algae as well as against the R-PC of *Porphyra laciniata*.

Although this wide range of cross-reactions with PCs from eukaryotic and prokaryotic organisms was observed, the antibody did not precipitate C- or R-PEs from a number of algae. Neither did the antibody against *C. caldarium* PC react with APC from *C. caldarium*. Thus an antibody against C-PC precipitated R- and C-PCs from a number of organisms but interacted neither with APC from the homologous species nor with PEs from any one of a number of sources.

Vaughan (170) showed that an antibody prepared against PE from *Ceramium rubrum* or from *Porphyra miniata* precipitated PEs from a large number of different species of red algae. It did not precipitate PE from a cryptomonad.

Berns (5) confirmed and extended the previous observations on the cross-reactivity of antibodies against R-PEs with C-PEs (and the converse) and the cross-reactivity of antibodies against C-PCs with R-PCs (and the converse). He also observed some cross-reactivity of anti-PC with APC. Although in the discussion of his paper he suggests this might be due to some cross-contamination, the point is included in the summary and has been picked up and widely disseminated. The failure of antibody against PC to react with APC had already been noted (11, 172) and was subsequently confirmed (4, 76). Berns (5) also confirmed Vaughn's observation that antibody against PEs from two red algae did not cross-react with PE from a cryptomonad. However, Berns did observe that antibody against PE from the cryptomonad *Rhodomonas lens* did react slightly with PE from the rhodophyte *Porphyridium cruentum*. It is not clear from the summary or text of the paper (5), but both the illustrations and Table I of that report show that PC from the cryptomonad *Chroomonas* does not react with antibody against PCs from various cyanophytes or rhodophytes.

General patterns which have been observed and confirmed (4, 11, 75, 76, 170, 172, and with exceptions, 5) are: (*a*) antibodies against a C- or R-PC from one organism react with R- or C-PCs from any other organisms except perhaps crytomonads; (*b*) the same relationship holds between antibodies and phycobiliproteins of the PE type; (*c*) no cross-reactions have been observed between antibodies against any PC with any PE; (*d*) antibodies against either PE or PC do not cross-react with APC. Thus all R- and C-PCs share some common antigenic determinant(s) which are lacking from PEs or APCs. R- and C-PEs share at least one antigenic determinant which is missing from all other classes of PBPs. APCs do not share antigenic determinants with other classes of PBPs even though the same chromophore is present on APCs and PCs. Furthermore, PBPs from cryptomonads cross-react not at all or only very slightly with antibodies against PBPs from other groups of algae.

It has been found recently (153) that antibody prepared against the α subunit of C-PE from *Fremyella diplosiphon* precipitates the β subunits. Conversely, antibody against β-subunit reacts with the α subunit. These two subunits are now known to have at least eight tryptic peptides in common.

The immunochemical relationship between PBPs in blue-green and red algae has sometimes been taken as evidence that plastids of red algae are derived from cyanophytes. The immunological evidence regarding differences between cryptomonad and other PBPs is in accord with the indications that these pigments have gone in a somewhat different direction evolutionarily; striking differences between cryptomonads and other algae in aggregation patterns and location of PBPs in relation to thylakoids are discussed elsewhere in this review.

Functional Sites and Forms of Phycobiliproteins

In 1956 Myers et al (116) observed 22 nm granules between the thylakoids in chloroplasts of the red alga *Griffithsia flosculosa*. They also observed that the plastid thylakoid membranes themselves might be made up of granular material of about the same dimension. The possibility that some of the granules might be aggregates of phycobiliproteins was mentioned, but this study was primarily concerned with the structure of the cell walls.

Thylakoids of higher plants are frequently closely appressed to one another, but in many algae, particularly those containing PBPs, sets (pairs) of thylakoids are approximately 30–40 nm from one another. The basis for the more or less uniform separation is not understood, except for the observation of Myers et al (116), which was generally disregarded until the work of Gantt & Conti in the mid-1960s.

Using glutaraldehyde-fixation followed by osmium tetraoxide postfixation staining (63–65), beautiful electron micrographs were obtained of sections of the single-celled red alga *Porphyridium cruentum* showing a single line of granules about 32–35 nm in diameter spaced about 40–50 nm center-to-center between thylakoid membranes in the plastid. The first evidence that these bodies might be composed of PBPs came from the observation that they were not disturbed by treatment with methanol-acetone, a solvent which would remove lipid components, but were digested by proteolytic enzymes. Such bodies, now called phycobilisomes (PBsomes), have been isolated and purified (64, 69) on sucrose gradients from glutaraldehyde-fixed and from unfixed *P. cruentum* cells. Samples taken from sucrose gradient preparations of unfixed cells show granules with dimensions 30–32 nm by 45–50 nm which in general resemble PBsomes in situ. PBsomes have also been isolated from three blue-green algae: *Nostoc* sp., *Anacystis nidulans,* and *Agmanellum quadruplicatum* (83).

Antibodies against PE precipitate detached *P. cruentum* PBsomes from suspensions. This indicates that every PBsome of *P. cruentum* contains this PBP. On the other hand, with antibodies against PC some precipitation seems to occur when PBsomes are not completely dissociated, but definitely occurs in dissociating conditions (E. Gantt, personal communication). It is possible that this suggests that at least the antigenic determinants on PC, if not the entire PC molecules, may be buried inside PBsomes. It has also been shown that energy absorbed by PE is transferred to APC by measuring the fluorescence emission spectrum in preparations of PBsomes (70). Illumination of a PBsome preparation with 435 nm light resulted in fluorescence emission at 575 nm (probably attributable to PE) and at 675–680 nm (attributed to APC). [The absorbance and fluorescence emission maxima for the pigments in *P. cruentum* are: B-PE 545, 575; b-PE 545, 570; R-PC 555 and 617, 636; APC 650, 660.] When the PBsome preparation was illuminated at 545 nm, fluorescence was emitted at 575 and at 675–680 nm, attributable to PE and APC, respectively. Illumination with 545 nm light would excite both the R-PE and R-PC present in the preparation. As the phycobilisomes dissociated, emission at 575 increased while that at 675–680 fell, suggesting that energy transfer was failing. The experimental data do not exclude the possibility that some PBsomes contain only R-PC and APC, but this seems unlikely from earlier considerations about energy transfer in PBP-containing organisms (43, 103).

Phycobilisomes have also been observed in situ in *Nostoc muscorum:* 40 nm hemispheres (83); *Synechococcus lividus:* 32–35 nm diameter, 6–7 nm thick discs (44); *Tolypothrix tenuis:* 20–35 nm granules (66); *Fremyella diplosiphon:* approximating the form of hemispheres (66); *Porphyridium aeruginum:* 40 nm discs (67); *Synechococcus sp.* strain 6301: granules about 35 nm in diameter (46); the red alga *Batrachospermum virgatum:* 25–30 nm diameter cylinders which are very long and lie parallel to each other in plane (108). All of these observations are based on thin sections of algae. PBsomes have also been observed in various red and blue-green algae, using the freeze-etch technique (102a, 119, 150).

PC from *Phormidium calothricodes* in the $\alpha_6\beta_6$ form consists of six granules about 3.5 nm in diameter arranged in a circle with a total diameter of 13 nm (7). B-PE from *P. cruentum* in the approximately 300,000 dalton form has the shape of discs about 10 nm in diameter and 5.4 nm thick (62). Crude broken cell preparations of *Synechococcus sp.* 6301 contain discs about 12 nm in diameter which are $\alpha_6\beta_6$ aggregates of C-PC; these algae contain 90–94% of their PBP as C-PC and 6–10% as APC. At pH 8 the $\alpha_6\beta_6$ units seem to disintegrate or at least loosen, for electronmicrographs show amorphous aggregates. The 12 nm diameter discs are re-formed by mixing purified α and β subunits under the proper conditions. On the other hand, if crude preparations at pH 5.2 are allowed to stand for 24 to 72 hours, rows of aggregated phycobiliproteins (RAPs) are built up. These have a diameter of 12 nm, in accord with the expected disc diameter, a strong 5 nm periodicity, and a fainter 3 nm periodicity (45). It is proposed that this periodicity in the negatively stained RAPs shows that pairs of discs form and that these are then aligned into a RAP. The RAPs themselves line up with periodicities in exact register showing a very strong side-by-side interaction. The long, cylindrical RAPs are reminiscent of the observations with *B. virgatum* in which long PBsome cylinders 25–30 nm in diameter were seen (108). Each of the latter could be composed of a circle of six or seven 12 nm diameter RAPs. Thirty-five nm granules are seen in situ in *Synechococcus* sp. 6301 (46).

The alignment of three 12 nm $\alpha_6\beta_6$ subunits edge to edge would make a structure about 36 nm high. Building up a circle of six 12 nm RAPs, perhaps with one in the center, very similar to the suggestion of Edwards & Gantt (44), would fit very well.

PBPs are arranged in PBsomes to absorb and transfer energy to APC. The APC molecules must be sufficiently close to the thylakoids to permit energy transfer to chlorophyll *a*. [Cohen-Bazire & Lefort-Tran (30) have shown that treatment of *Porphyridium cruentum* or *Glococapsa alpicola* with glutaraldehyde has only a slight effect on the absorption spectra of the cells, but a large fraction of the PBPs sediment with thylakoids when cells are broken and centrifuged. Cells of *Anacystis nidulans, Chlorella pyrenoidosa,* and *Porphyridium cruentum* perform photosystem I and II reactions after fixation with formaldehyde and glutaraldehyde (84).] The arrangement of PBPs into rod-shaped PBsomes in *B. virgatum* (108) suggests that the long axis of RAPs can lie parallel to thylakoids. If each RAP must consist of only a single type of PBP, i.e. of PC, PE, or APC, the APC-RAPs must be in contact with the abutting thylakoids. On the other hand, if RAPs can be made up of trapping APCs among PCs and trapping PCs among PEs, arrangements in which the long axes of RAPs are perpendicular to thylakoids—with APCs at the contact points—would seem equally plausible. Additional in vitro assembly of RAPs plus the use of antibodies against PBPs labeled for electron microscopy should help resolve these questions.

Dodge (42) noted that the space inside thylakoids of the cryptomonad *Chroomonas mesostigmatica* contained a large amount of dense staining material. He suggested that the PBPs are probably located within the thylakoids in this alga. This suggestion was taken up by Gantt et al (68). They examined three Cryptophycean algae: *Chroomonas sp., Rhodomonas lens,* and *Cryptomonas ovata* var. *palustris.* Electron microscopy confirmed for each of these three algae the observations of Dodge (42) with *Chroomonas.* By using the same kind of techniques which had been initially employed to determine the composition of the PBsomes in *P. cruentum,* they showed that the dense staining material within the thylakoids is destroyed by proteolytic enzymes but not by lipid solvents. Also, the PBPs were found to leak out of thylakoid preparations in the course of treatment of the latter with proteolytic enzymes.

The entire PBP accessory pigment system appears to have evolved along different lines in Cryptomonads than in red or blue-green algae. The cryptomonad PBPs are immunochemically distinct from PBPs of other algae and they do not form large aggregates. The maximum aggregate size is about 35,000 (75,123), i.e. $\alpha_1\beta_1$ rather than $2\alpha_3\beta_3 \rightleftharpoons \alpha_6\beta_6$, or, for *Chroomonas* PC (109) $\alpha_2\beta_2$ (about 50,000 daltons). The concentration of chromophores in PBsomes is about 0.05–0.1 M assuming 18 bilins per $\alpha_6\beta_6$ in a disc about 10–12 nm diameter and about 3–5 nm thick. Cryptomonad PBPs do not aggregate into PBsomes but are concentrated by confinement within the thylakoids.

The large similarity in stretches of amino acid sequences in α and β subunits of PEs (153) and PCs (165, 173) in blue-green and red algae has been noted above. It seems reasonable to guess that, e.g. one of the genes for apoproteins PE_α and PE_β is derived from the other or that they both arose from a common ancestor in each group of algae. But why have the similarities been maintained? What has been the pressure to prevent wider evolutionary divergence? Transfer of energy from one PBP to another in the sequence PE $\longrightarrow$ PC $\longrightarrow$ APC and from APC to chlorophyll is most likely by the mechanism of inductive resonance of Foerster (43). The efficiency of transfer of energy from one pigment molecule to another by the Foerster mechanism falls off with increasing distance between donor and acceptor. Thus PBPs are effective in energy transfer only when concentrated; a mutation which interferes with aggregation or which places chromophores too far from one another would be disadvantageous. A constraint of this sort could severely limit change. It will be interesting to see if the APC, PC, and PE of a single species have common sequences. If PBPs evolved after chlorophyll *a,* APC is likely to have been the first PBP.

Whether PBsomes are primitive and the cryptomonad intra-thylakoid arrangement derived or the reverse, the evolution of cryptomonad PBP polypeptides would not be restricted by the need to aggregate. On the other hand, no ribosomes have been shown to occur within the vesicles, so some special selective transport arrangement which recognizes each or all of the PBP polypeptides is probably in the membrane. This is a fascinating problem of general biological interest.

The PBP accessory pigment system of Cryptomonads also differs from that in other algae by lacking APC; however, cryptomonad PC-645 fluoresces at the same wavelength as APC. Also, at least one *Cryptomonas* species lacks PE (75).

Biosynthesis of Bile Pigments

The bile pigment biliverdin IX α can be formally derived from protoporphyrin IX by the removal of the α-bridge carbon atom. The production of biliverdin from iron protoporphyrin IX (protoheme) in vitro and the liberation of the α carbon atom of this metallo-porphyrin in CO during synthesis of biliverdin in vivo in mammals have both been examined extensively (see review by Bogorad & Troxler 14). Some probable intermediates have been synthesized (e.g. 90).

In the early and middle 1960s Nakajima and co-workers described a system for the oxidation of pyridine-hemochromogens by preparations of guinea-pig liver. The reality of this as an enzymatic system has been questioned (e.g. 34, 106, 107), although it has been shown (117) that the enzyme degrades pyridine hemochromogen into formylbiliverdin by breaking the porphyrin ring exclusively at the α-methene bridge. The nonenzymatic coupled oxidation of hemochromogens, in contrast, leads to a mixture of biliverdin isomers (138) — i.e. forms in which bridges other than α have been oxidized. The biological significance of the Nakajima system is uncertain.

More recently, Marver and his co-workers described systems not requiring the presence of pyridine to yield biliverdin from protohemin (156–158).

A microsomal fraction from liver or spleen can convert protohemin to biliverdin plus CO. Molecular oxygen is required for the reaction catalyzed by this microsomal heme oxygenase. The reaction is inhibited by carbon monoxide. Protohemin IX is the best substrate, although some other hemins are also effective. Protoporphyrin IX and other metal-free porphyrins do not serve as substrates for this enzyme. When methemoglobin or the α and β chains of hemoglobin serve as substrates, the activities are about one-quarter to one-half those observed with protohemin. Iron protoporphyrin IX in its oxidized form, protohemin, is easily detached from globin (155–158). Godnev et al (80) have reported that PC production by *Anacystis nidulans* is elevated by the administration of protoporphyrin or hematin.

Another enzyme present in the microsomal fraction converts biliverdin to bilirubin (156–158). Biliverdin reductase is an NADPH-dependent enzyme which catalyzes the reduction of the carbon bridge between rings B and C of biliverdin. The product has an absorption maximum at about 500 nm. In mammals, bilirubin glucuronyl transferase, present in liver parenchyma cells, catalyzes the condensation of bilirubin with UDP-glucuronic acid to make the soluble biliverdin-glucuronide. The latter easily passes into the intestines and is excreted. Some steps in bile pigment formation are shown in Figure 2. Hemes and chlorophylls are produced from δ-aminolevulinic acid (for review see Bogorad 12).

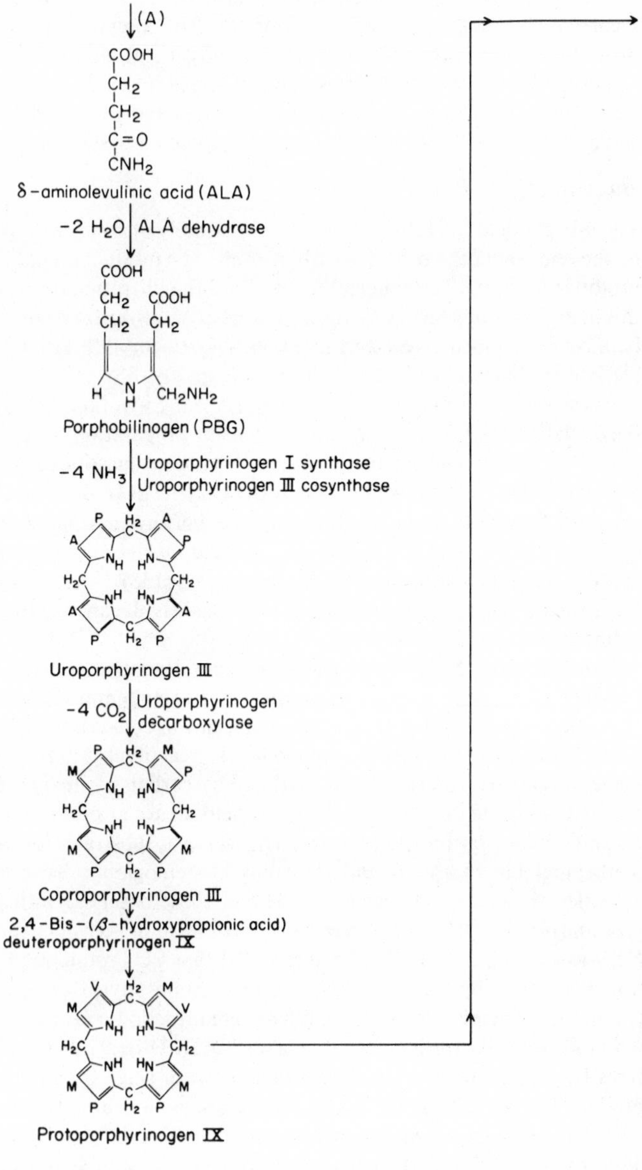
(A)
COOH
CH2
CH2
C=O
CNH2
δ-aminolevulinic acid (ALA)
-2 H2O
ALA dehydrase
COOH
CH2
COOH
CH2
CH2
H
N
H
CH2NH2
Porphobilinogen (PBG)
-4 NH3
Uroporphyrinogen I synthase
Uroporphyrinogen III cosynthase
Uroporphyrinogen III
-4 CO2
Uroporphyrinogen
decarboxylase
Coproporphyrinogen III
2,4-Bis-(β-hydroxypropionic acid)
deuteroporphyrinogen IX
Protoporphyrinogen IX

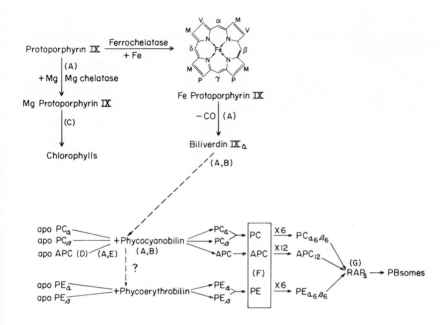

Figure 2 The biosynthetic path to phycobiliproteins [also see Bogorad (12) for details to chlorophylls].

A = -CH$_2$-COOH; M = -CH$_3$; P = -CH$_2$-CH$_2$-COOH; V = -CH = CH$_2$

(A) This step not shown in green plants or PBP-containing algae.

(B) Alternative: PEB ⟶ Biliverdin ⟶ PCB

(C) Photooxidation of Mg protoporphyrin has been suggested as an intermediate step (1) in phycobilin biosynthesis.

(D) Some or all APCs may have α and β subunits.

(E) It has not been determined when the chromophores are joined to the apoproteins.

(F) In cryptomonads PBP aggregation stops at $\alpha_1\beta_1$.

(G) It is not known whether each RAP (row of aggregated phycobiliproteins) is composed of one or more types of phycobiliproteins.

Wild-type and mutant cells of *Cyanidium caldarium* form free PCB and excrete it into the medium when they are supplied with ALA (162, 163). This shows that PCB is formed from ALA. A biliverdin-synthesizing system similar to that described in liver has not been shown to exist in algae (or other plants); however, Troxler et al (161, 166, 168) have shown that carbon monoxide is formed during the production of PCB by cells of *C. caldarium* supplied with ALA. As pointed out earlier, the PCB made by *C. caldarium* from ALA is chromatographically indistinguishable from the methanol cleavage product prepared from *Phormidium luridum* PC (166). (If in fact the ethylidene group is also present on the *Cyanidium* product, this strongly supports the view that the ethylidene group is not an artifact of hydrolysis when the bile pigment is liberated from the protein.)

The biosynthetic relationship between biliverdin, PCB, and PEB is not known. However, all algae which contain PEB also contain PCB. It would seem then that either PCB is an intermediate in the biosynthesis of PEB or that PCB and PEB are formed after a fork in the pathway beyond biliverdin.

THE REGULATION OF PHYCOBILIPROTEIN SYNTHESIS

The influence of light on PBP production has been studied more intensively in the three following cases than in any other algae: *(a) Cyanidium caldarium; (b) Tolypothrix tenuis;* and *(c) Fremyella diplosiphon.*

In *C. caldarium,* PBPs form in response to illumination. The alga can be grown on glucose in darkness, but under these circumstances PBPs are barely detectable. The other two algae exhibit complementary chromatic adaption. This phenomenon was first described by Gaidukov (61), who observed that *Oscillatoria sancta* was red in color after growth under green light, but assumed a blue-green tint when grown under orange light. The color taken on by the alga was complementary to the incident illumination. Kylin (96, 97) and Boresch (15, 16) perceived that the differences in color of the blue-green alga resulted from the presence of different proportions of the red and blue PBPs. The pigments which are most effective in absorbing the wavelength of light with which the algae are illuminated are produced in greater amounts. For example, about 45% of the soluble protein in cells of *Fremyella diplosiphon* grown under fluorescent light is PBP. PE comprises 21% of the total soluble protein, PC 14%, and APC about 10%. On the other hand, when this alga is grown at wavelengths beyond about 600 nm, no PE is detectable, but 47% of the soluble protein can be accounted for as PC and about 16% as APC. There are also differences in the fraction of the total cellular dry weight accounted for by chlorophyll *a* under the two light conditions (4).

The regulation of PBP production in the three algae mentioned is sufficiently distinctive to warrant a separate discussion of each. The photoreceptive pigments through which PBP production is regulated are named "adaptochromes" here. The adaptochrome of *C. caldarium* is further specified by the designation C.c., that of *T. tenuis* by T.t., and that (or those) of *F. diplosiphon* by F.d.

Cyanidium caldarium: This single-celled acidophilic eukaryote normally grows in sulfur hot springs at temperatures below about 55°C and pHs below 4 or 5. Its taxonomy has been in question, but the immunological behavior of its PBPs suggests that it is not a Cryptomonad (11, 172), a conclusion which can also be reached from an analysis of its fine structure (13, 113), and PBsomes occur between its thylakoids (150). It is most likely a Rhodophyte. This alga normally accumulates PC, APC, and chlorophyll *a* when grown in the light, but it can be grown heterotrophically in darkness on glucose. Under the latter conditions wild-type cells produce only traces of chlorophyll, but some mutant strains also accumulate traces of PC and APC (120–122). *Cyanidium* mutants which lack PC and APC or chlorophyll *a*, or both chlorophyll and PBPs, have been isolated (120–122). Light of 420 nm and wavelengths between 550 and 595 nm are most effective in promoting PBP formation in a chlorophyll-less mutant of this alga (120, 122). The general features of the response curve led to the suggestion that a hemoprotein might be the photoreceptor (11, 122), although none has yet been isolated. The production of PBPs requires continuous illumination; it stops when the algae are returned to darkness. The realization that bile pigments are formed from heme in other systems and must certainly come at least via protoporphyrin IX in algae led to the suggestion (11, 122) that photooxidation of a hemoprotein might be a step in the biosynthesis of PBP chromophores. That is, that light might be required for a substrate-level oxidation.

It was found later (162) that dark-grown wild-type cells cultured in 7×10^{-3} M ALA formed uroporphyrin III, coproporphyrin III, and protoporphyrin, as well as large enough amounts of PCB-630 to make the medium visibly blue (162, 164). Inasmuch as PCB could be formed in darkness, the idea of a substrate-level photooxidation was no longer acceptable. [However, Barrett (1) later suggested that phycobilins might arise by photooxidation of magnesium porphyrins.]

These experiments also demonstrated that light has two effects on PBP production in *Cyanidium: (a)* the stimulation of ALA synthesis; *(b)* a direct effect on the production of PBP apoproteins. However, the latter could conceivably promote ALA synthesis. It is also possible that adaptochrome-C.c. regulates chloroplast development, and thus PCB and APC- and PC-apoprotein production are only a few of many manifestations of its action; e.g. it is known that dark-grown *C. caldarium* cells contain small amounts of cytochrome c_{553} but a four- or fivefold increase occurs during greening while the cytochrome c_{550} content does not change (95).

If ALA were produced normally in darkness in *C. caldarium,* the complement of porphyrins and bile pigments observed when ALA was fed should always occur; it does not (162, 163). If protein were present but the chromophore were lacking, administration of ALA and the consequent production of PCB should have permitted the formation of completed PC and APC; it did not (162, 163). It might be argued that the apoprotein is present in darkness, but that an enzyme required to link the apoprotein to the bile pigment requires illumination for activity. This point could not be answered directly, but apoprotein could not be detected in dark-grown *Cyanidium* cells by the use of an antibody against PC (163). This is not conclusive but is the best evidence available now.

A summary scheme for one of the possibilities for the regulation by light of PBP production in *Cyanidium* is shown in Figure 3.

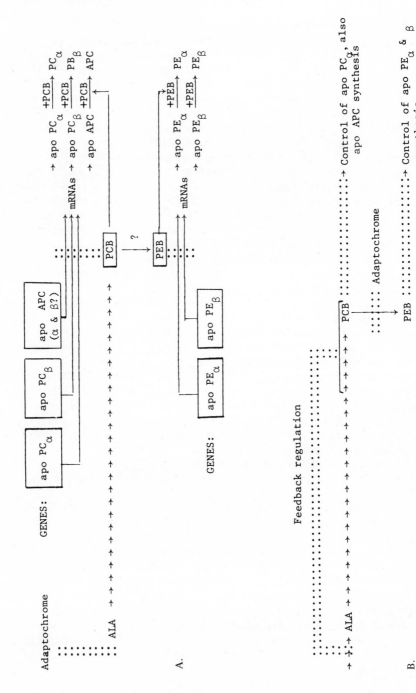

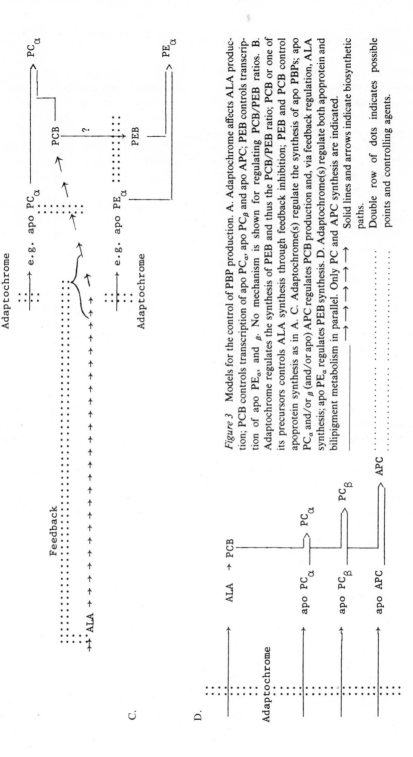

Figure 3 Models for the control of PBP production. A. Adaptochrome affects ALA production; PCB controls transcription of apo PC$_\alpha$, apo PC$_\beta$ and apo APC; PEB controls transcription of apo PE$_\alpha$, and $_\beta$. No mechanism is shown for regulating PCB/PEB ratios. B. Adaptochrome regulates the synthesis of PEB and thus the PCB/PEB ratio; PCB or one of its precursors controls ALA synthesis through feedback inhibition; PEB and PCB control apoprotein synthesis as in A. C. Adaptochrome(s) regulate the synthesis of apo PBPs; apo PC$_\alpha$ and/or PC$_\beta$ (and/or apo) APC regulates PCB production and, via feedback regulation, ALA synthesis; apo PE$_\alpha$ regulates PEB synthesis. D. Adaptochrome(s) regulate both apoprotein and bilipigment metabolism in parallel. Only PC and APC synthesis are indicated.

→ → → → → → → → →   Solid lines and arrows indicate biosynthetic paths.

·············· ··············   Double row of dots indicates possible points and controlling agents.

Two normally PBP-less mutants form PEB-630 and various porphyrins but not PC or APC when grown with ALA in darkness. This suggests that the apoproteins are not accumulated by these mutants. The apoprotein is also not detectable immunochemically in these cases. However, both wild-type and PBP-less mutants form more pyrrolic compounds from ALA if they have been illuminated prior to administration of ALA than if cells had been maintained completely in darkness. The latter suggests that light regulates not only ALA production but affects the activity of enzymes of ALA utilization and PCB production in some additional manner as well (121, 122, 162, 163).

Mutants which produce chlorophyll a but neither PC nor APC can be imagined to be blocked at some step in PCB synthesis beyond protoporphyrin IX, e.g. at biliverdin formation. However, their production of PCB-630 from exogenous ALA prompts three alternative interpretations: (a) the enzyme of PCB synthesis is altered to have a much lower than normal affinity for its substrate, and administration of ALA results in a higher substrate concentration; (b) the primary lesion is in APC and PC apoprotein formation—these apoproteins normally regulate ALA synthesis; (c) separate ALA-synthesizing systems are associated with two (or three) chloroplast multienzyme complexes—one complex includes enzymes for chlorophyll a biosynthesis, another for PCB-630 synthesis—and the primary lesion is in ALA synthesis in PCB-forming complexes.

Tolypothrix tenuis: Fujita & Hattori have described the very interesting chromatic adaptation system in *Tolypothrix tenuis* (54, 55, 57, 58, 60, 86, 88). This filamentous blue-green alga accumulates C-PE, C-PC, and APC. It exhibits classical complementary adaption: red light at wavelengths beyond 600 nm promotes the production of PC and APC but not of PE; 500–600 nm light promotes PE formation (86). In more detailed analyses (55) the action spectrum for PE production was shown to have a maximum at 541 nm. PC and APC production by *T. tenuis* was most strongly promoted by 641 nm light. Wavelengths from 400 to 450 nm were ineffective (41, 55, 58, 86). Thus the system was found to be driven toward PE production by 541 nm light and toward PC production by 641 nm illumination.

Unlike *C. caldarium, T. tenuis* can form PBPs after transfer from light to darkness under some conditions (88). In these experiments, cells in nitrate-free medium were exposed to high light intensities (20,000 lux) for about 24 hr to reduce their content of PBPs. If nitrate was then supplied within the first 6 hr of the dark period, PBP production started about 5 hr after nitrate addition (54, 57, 88). Cells illuminated with fluorescent lamps (relatively rich in wavelengths below about 700 nm) form PE during subsequent dark incubation; cells preilluminated with red light produce PC but not PE in darkness. It was then found (55, 58) that cells which had been illuminated with red light during the bleaching phase (and thus would normally produce PC but no PE during subsequent dark incubation with nitrate) formed PE instead of PC if they were illuminated with green (541 nm) light immediately prior to addition of nitrate and incubation in darkness. Exposure for 6 minutes to 4×10^3 erg/cm^2/sec 541 nm light reversed the effect of the red preillumination

(58). The effect of exposure to green light, in turn, could be reversed with equal energy of 641 nm light. This led Fujita & Hattori (58) to suggest that precursors of PE might be converted to PC plus APC by 541 nm light, and the reverse conversion was driven by 641 nm light. For various reasons the interconversion of precursors seems unlikely (14), but the reversible effects of 541 and 641 nm illumination on PE and PC production were clearly established. Scheibe (148) has prepared an extract of *T. tenuis* in which the absorbance at 650 nm increases upon illumination with green light and decreases upon illumination with red light. Changes in the opposite direction occur at 520 nm, but it is not understood why these are quantitatively about one order of magnitude smaller.

Diakoff & Scheibe (41) have reanalyzed the action spectra and found PC production favored by bands with maxima at 550 and 350 nm (effectiveness ratio $=$ 1:0.08) and PE favored by bands with peaks at 660 and 360 nm (effectiveness ratio $=$ 1:0.25). These observations are in general agreement with those of Fujita & Hattori except that more information is provided about the near-uv. The locations of the action maxima and the ratios for responses to 350 or 360 nm vs 550 or 660 nm light strongly suggests that the chromophore(s) is a bile pigment.

The blue-green alga *Nostoc muscorum* cultivated in darkness with glucose grows as a mass of large, undifferentiated cells within an outer envelope (73, 102). Exposure to small amounts of light promotes formation of filaments during 4 ensuing days in darkness. The action spectrum for induction of filamentation is sharp and shows a maximum at about 650 nm. It is reversed by illumination across a broad band of wavelengths ranging from about 500–600 nm, with a maximum at about 525 nm (99–101). Scheibe (148) has suggested that the action spectrum for reversibility in PE-PC production in *T. tenuis* and for morphology of *N. muscorum* have common features. However, the shape of the absorption curve shown by Scheibe at about 650 nm does not resemble the action spectrum obtained by Lazaroff & Schiff (101) for photomorphogenic effects on *N. muscorum*.

Cells and filaments of *T. tenuis* do exhibit a morphological response to illumination, but it is much different from that of *Nostoc*. Like *Fremyella diplosiphon* (2,4), filaments of *T. tenuis* are longer in cultures grown under fluorescent lamps than in red light (J. Haury & L. Bogorad, unpublished).

As discussed, *T. tenuis* cultures form PBPs in darkness if nitrate is absent from the medium at the beginning of the preillumination period, but produce very little, if any, pigment if nitrate is present at that time (88). However, cells incubated in darkness with nitrate produce PBPs when they are supplied with 7×10^{-3} M ALA (112). PBP production by illuminated cultures is also increased by exogenous ALA (112). The difference between the effect of the administration of ALA on *T. tenuis* and *C. caldarium* is striking: *C. caldarium* produces more bile pigment, while *T. tenuis* produces additional PBPs in light or in darkness. In neither case is the growth rate stimulated by the presence of ALA. ALA, or PCB and PEB or some intermediate could stimulate and regulate the production of the apoproteins. On the other hand, the newly made bile pigments might be joined to preexisting protein. The only experimental evidence available is that chloramphenicol, an inhibitor of protein synthesis in blue-green algae and bacteria, stops ALA-stimulated PBP production

(112), but this antibiotic also slows down the utilization of ALA by *Cyanidium* (162). A possible interpretation of the data available on PBP production and chromatic adaptation in *T. tenuis* is summarized in Figure 3.

Under culture conditions which lead to PE photodestruction in the red alga *Hypnes musciformis,* kinetin reduces the loss or may promote net accumulation (91). Administration of 0.5 mM ALA elicited a similar response. Jennings et al (91) suggest that kinetin, which also promotes chlorophyll *a* formation, may act by controlling ALA formation. The effect of 0.5 mM ALA on the chlorophyll level was not reported.

Fremyella diplosiphon: When grown at wavelengths beyond about 600 nm, this alga forms no PE but contains APC and large amounts of PC. In cultures grown under red light, PBPs constitute 63% of the total soluble protein and almost 20% of the total dry weight (2, 4). Upon transfer to fluorescent illumination, PC production continues for a while and then stops, while the production of PE continues rapidly (about 5×10^8 molecules of PE/mg cells/sec) for at least 3 days. In the converse experiment, about a 20% drop in the total PE per ml of cell culture occurs within about 12 hours after transfer to red light, but then the level remains the same. There is less than 5%, if any, turnover of PE during steady-state growth (4) under fluorescent lamps. Of the PE present at the time the cells are transferred from fluorescent to incandescent lamps, 80% becomes undetectable gradually by dilution. Complementary chromatic adaptation in *F. diplosiphon* involves a fairly rapid cessation of production of one PBP and the commencement or increase in rate of synthesis of the others.

PE production in *F. diplosiphon* is promoted by light at 460 and 550 nm; the ratios of effectiveness are about 7:1. The action spectrum for PC production has peaks at 460 and 600 nm; the effectiveness per quantum is about ten times greater for the shorter wavelength than for the longer one (J. Haury & L. Bogorad, unpublished). It is not known whether two pigments or a single reversible adaptochrome-F.d. controls adaptation in both directions. The action spectrum for PE production by *F. diplosiphon* resembles that for PC and APC synthesis by *C. caldarium* more than that for chromatic adaptation in *T. tenuis.*

The mean filament length of PE-rich *F. diplosiphon* grown under fluorescent light is about 450 μm and the ratio of cell length to diameter is about 2. Cells in red light-illuminated cultures form filaments only 50 μm long and the cells themselves are spherical. The 20% drop in PE content, which occurs about 12 hours after fluorescent light-grown cultures have been transferred to red light, comes at the same time as the death of about 20% of the cells and the breakage of the filaments. The cells which die are called necridia (98). In one experiment the mean filament length dropped from about 550 to 100 μm during this period (4). It has not yet been firmly established that the action spectrum for necridia formation is different from that for promotion of PC formation, but preliminary experiments suggest that different absorbers may be involved (J. Haury, S. Miller, J. Takemoto & L. Bogorad, unpublished). The photomorphogenic pigment in filamentous blue-green algae is here designated "phycomorphochrome."

Mutant strains of *F. diplosiphon* have been obtained which are resistant to rifamycin, a potent inhibitor of the DNA-dependent RNA polymerase of *F. diplosiphon*. The mutants are about normal with regard to both photomorphogenesis and complementary chromatic adaptation. When grown in the presence of rifamycin, some strains retain the capacity for complementary chromatic adaptation but are abnormal with regard to photomorphogenic changes (J. Raiss, S. Miller, J. Takemoto & L. Bogorad, unpublished). Filaments tend to be long in either red or fluorescent light. In a sense, necridia formation is inhibited. RNA polymerase activity in preparations from strains of the latter type is only about 40–60%, rather than 100%, sensitive to rifamycin (S. Miller & L. Bogorad, unpublished). This suggests that two DNA-dependent RNA polymerases or two forms of the same polymerase may be present in the cell. The observed behavior is consistent with mutation to rifamycin resistance of an RNA polymerase (or a polymerase protein factor) required for vital functioning of the cell and complementary chromatic adaptation, while another polymerase, or another factor for the same polymerase, involved in the formation of necridia and not essential for the cell's survival remained sensitive.

To summarize this discussion: There may be separate photoreceptive systems in *F. diplosiphon* for complementary chromatic adaptation (adaptochrome(s)-F.d.) and the photoregulation of cell and filament morphology (morphochrome). This is based on a firm knowledge of the action spectrum for complementary chromatic adaptation (J. Haury & L. Bogorad, unpublished) and preliminary evidence regarding the wavelengths which affect photomorphogenesis (J. Haury, S. Miller & L. Bogorad, unpublished). Furthermore, if the interpretation of the rifamycin-resistant mutants given above is correct, adaptochrome(s) and morphochrome of *F. diplosiphon* regulate transcription but via different DNA-dependent RNA polymerases.

Adaptochromes

Adaptochromes have been defined here as photoreceptive pigments for the regulation of PBP metabolism in complementary chromatic adaptation, as in *F. diplosiphon* or *T. tenuis*, or in the photoregulation of PBP synthesis, as in *C. caldarium*. The action spectra are different in each of the three cases. Thus there is no evidence that the photoreceptive pigment system is the same in these algae although, especially in the blue region, masking by other pigments could lead to erroneous conclusions about the absorption spectrum of the photoreceptor based on action spectra. But, taking the action spectra data at face value, adaptochromes-C.c. and -F.d. are likely to be hemoproteins while adaptochrome-T.t. could be one or two biliproteins. Unlike the situation in *C. caldarium* (122) or *F. diplosiphon* (J. Haury & L. Bogorad, unpublished), blue light has little or no effect on PBP metabolism in *T. tenuis* (41, 55, 60). In the latter, 540–550 nm light promotes PE production and 641–660 nm stimulates PC production (41, 55, 60); responses 1/12 to 1/4 as strong are elicited by 350 and 360 nm light (41).

As already pointed out, the action spectrum for PBP formation in *C. caldarium* has a strong maximum at about 420 nm and lower response at 550–590 nm. This suggests that adaptochrome-C.c. is probably a hemoprotein (122). Poff et al (139, 140) have recently isolated a photoresponsive pigment from *Dictyostelium dis-*

coideum which shows a relatively strong absorption band at 430 and two bands of lesser extinction in the 550–590 nm region. They suggest that this absorber is the photoreceptive pigment for phototaxis in pseudoplasmodia of *Dictyostelium discoideum*. The pigment was solubilized by sonication of the mitochondrial fraction and, based on its absorption spectrum, is judged to be a cytochrome. Perhaps the *Dictyostelium* pigment and adaptochrome-C.c. are similar?

Adaptochrome-F.d. is stimulated strongly by light of about 460 nm and shows a weaker response for PE production at 550 nm; a similarly weaker peak for PC and APC production is at about 600 nm (J. Haury & L. Bogorad, unpublished). It is entirely possible that there are two photoreceptive pigments, both of which have absorption maxima within the 460 nm envelope and separate maxima in the green and red regions of the spectrum. Alternatively, a single photoreversible adaptochrome may be present. Some additional work seeking reciprocal absorbance changes in situ must be carried out. Poff & Butler (139) have studied a blue-photoreceptor pigment in *Phycomyces blakesleeanus* which controls the phototropic response of sporangiophores and the initiation of sporangiophores. They have observed that irradiation with 470 nm actinic light results in bleaching at 445 nm and increased absorption at about 430 nm. The 470 nm light is most effective for eliciting these absorbance changes. Wavelengths around 470 nm are also the most effective for the physiological blue light response. From other data (139) they suggest that in *D. discoideum* there may be a complex of a 465 nm absorber—possibly a flavin —together with a *b*-type cytochrome. The latter is clearly labeled as a speculation which remains to be confirmed. It is possible that the adaptochrome-F.d. system is similar, although its properties could be accounted for by hemoproteins alone.

Zoospore production by the green alga *Protosiphon botryoides klebs* is regulated by a yellow-blue (584–429 nm) photoreversible system (159). It is conceivable, though not very likely, that there are two absorbers with peaks within the "blue envelope" of this system. The spectral characteristics of this system again suggest a hemoprotein photoreceptor and at least to this extent some similarity to the adaptochrome-C.c. and -F.d. systems.

The absorption properties of only three adaptochromes are known, and this information is primarily from action spectra. As more cases are studied, classes of pigments may be identified. At present the two general forms are the presumptive hemoprotein and biliprotein types.

How directly do adaptochromes control PBP formation? Or is PBP metabolism all they control? As already pointed out, adaptochrome-C.c. may affect cytochrome c_{553} production (95), and the chlorophyll as well as PE/PC content of *F. diplosiphon* varies with the color of light under which it is grown (2, 4). The chlorophyll and PBP contents of the red alga *P. cruentum* (19) and the blue-green alga *A. nidulans* (72, 92) also vary with the wavelengths in which the organisms are grown. In all of these species adaptochromes may be controlling not only PBP production but the metabolism of another pyrrolic component of the photosynthetic apparatus as well.

In *T. tenuis*, unlike *F. diplosiphon* where both the photoreceptors and RNA polymerases for PBP synthesis and morphogenesis are different, a single photoreceptive system may control morphogenesis and complementary chromatic adaptation

—i.e. a single photoreceptive system may play the role of phycomorphochrome and adaptochrome. Further fine analyses of action spectra for the various processes need to be made, but it would not be surprising if a change in PBP metabolism is only one of a group of processes controlled by the adaptochrome system in an organism. For example, as suggested above for *C. caldarium,* adaptochrome-C.c. may regulate many features of plastid development. After raising this question, the following enumeration of possibilities can be taken only as offered—as a systemizing summary.

These are some possibilities for the control of production of a particular PBP as well as for integration of apoprotein and chromophore synthesis:

1. The prosthetic group regulates the production of apoprotein [e.g. possible in *T. tenuis* (112) and *Hypnes musciformis* (91).]

2. The apoprotein regulates the formation of a specific bile pigment (e.g. one possible interpretation of the behavior of *C. caldarium*).

3. The production of apoproteins and chromophores are regulated in parallel (e.g. another possible interpretation of the *C. caldarium* experiments).

Then some possible roles for adaptochromes may be:

1. The control of gene transcription.

2. The control of translation of preexisting but previously unread messages.

3. The control of chromophore production, e.g. shifting the enzymatic synthesis of bile pigments from PCB to PEB.

An overwhelming number of combinations of components of these two intersecting sets can be assembled easily. Some are shown in Figure 3. Discovering which are used will be a much more trying task, but these investigations should bring solutions to some general problems in biology.

ACKNOWLEDGMENTS

I am grateful to many of the authors of work cited for discussing their research with me and for providing manuscripts in advance of publication. I am also indebted to my colleagues Jon Takemoto, John Haury, and Stephan Miller for reading and discussing with me all or parts of this review. Part of the preparation of this review was carried on while I was a guest in the Department of Botany at Cambridge University; the kind hospitality provided by Dr. Patrick Echlin and by Professor P. W. Brian and his staff was invaluable and, again, I am grateful.

The work from my laboratory and the preparation of this manuscript was made possible by research support from the National Science Foundation and the Maria Moors Cabot Foundation of Harvard University.

Literature Cited

1. Barrett, J. 1967. *Nature* 215:733–35
2. Bennett, A. 1972. *Metabolic and structural investigations of Fremyella diplosiphon phycobiliproteins.* PhD thesis. Harvard Univ., Cambridge, Mass. 153 pp.
3. Bennett, A., Bogorad, L. 1971. *Biochemistry* 10:3625–34
4. Bennett, A., Bogorad, L. 1973. *J. Cell Biol.* 58:419–35
5. Berns, D. S. 1967. *Plant Physiol.* 42:1569–87
6. Berns, D. S. 1971. *Subunits in Biological Systems, Part A,* ed. S. N. Timasheff, G. D. Fasman, 105. New York: Marcel Dekhov
7. Berns, D. S., Edwards, M. R. 1965. *Arch. Biochem. Biophys.* 110:511–16
8. Binder, A., Wilson, K., Zuber, H. 1972. *FEBS Lett.* 20:111–16
9. Bloomfield, V. A., Jennings, B. R. 1969. *Biopolymers* 8:297–99
10. Bogorad, L. 1962. *Physiology and Biochemistry of Algae,* ed. R. A. Lewin, 385–408. New York: Academic. 929 pp.
11. Bogorad, L. 1965. *Rec. Chem. Progr.* 26:1–12
12. Bogorad, L. 1975. *Chemistry and Biochemistry of Plant Pigments,* ed. T. W. Goodwin. New York: Academic. In press. 2nd ed.
13. Bogorad, L., Mercer, F. V., Mullens, R. 1963. *Photosynthetic Mechanisms of Green Plants,* Publ. 1145, 560–70. Nat. Acad. Sci.-NRC. 766 pp.
14. Bogorad, L., Troxler, R. F. 1967. *Biogenesis of Natural Compounds,* ed. P. Bernstein, 247–313. London: Pergamon. 1209 pp. 2nd ed.
15. Boresch, K. 1919. *Ber. Deut. Bot. Ges.* 37:25–39
16. Boresch, K. 1919. *Arch. Protistenk.* 44:1–70
17. Boucher, L. J., Crespi, H. L., Katz, J. J. 1966. *Biochemistry* 5:3796–3802
18. Briggs, W. R., Rice, H. 1972. *Ann. Rev. Plant Physiol.* 23:293–334
19. Brody, M., Emerson, E. 1959. *Am. J. Bot.* 46:433–40
20. Brooks, C., Chapman, D. J. 1972. *Phytochemistry* 11:2663–70
21. Brooks, C., Gantt, E. 1973. *Arch. Mikrobiol.* 88:193–204
22. Byfield, P. G. H., Zuber, H. 1972. *FEBS Lett.* 28:36–40
23. Chapman, D. J. 1973. *The Biology of Blue-Green Algae,* ed. N. G. Carr, B. A. Whitton, 162–85. Oxford: Blackwell. 676 pp.
24. Chapman, D. J., Cole, W. J., Siegelman, H. W. 1967. *Biochem. J.* 105:903–5
25. Chapman, D. J., Cole, W. J., Siegelman, H. W. 1967. *J. Am. Chem. Soc.* 89:5976–77
26. Chapman, D. J., Cole, W. J., Siegelman, H. W. 1968. *Am. J. Bot.* 55:314–15
27. Chapman, D. J., Cole, W. J., Siegelman, H. W. 1968. *Biochim. Biophys. Acta* 153:692–98
28. Chapman, D. J., Cole, W. J., Siegelman, H. W. 1968. *Phytochemistry* 7:1831–35
29. Cho, F., Govindjee 1970. *Biochim. Biophys. Acta* 216:151–61
30. Cohen-Bazire, G., Lefort-Tran, M. 1970. *Arch. Mikrobiol.* 71:245–57
31. Cole, W. J., Chapman, D. J., Siegelman, H. W. 1967. *J. Am. Chem. Soc.* 89:3643–45
32. Cole, W. J., Chapman, D. J., Siegelman, H. W. 1968. *Biochemistry* 7:2929–35
33. Cole, W. J., Ó hEocha, C., Moscowitz, A., Krueger, W. R. 1967. *J. Biochem.* 3:202–7
34. Colleran, E., Ó Carra, P. 1970. *Biochem. J.* 119:905–11
35. Cope, B. T., Smith, U., Crespi, H. L., Katz, J. J. 1967. *Biochim. Biophys. Acta* 133:446–53
36. Crespi, H. L., Boucher, G. D., Norman, G. D., Katz, J. J., Dougherty, R. C. 1967. *J. Am. Chem. Soc.* 89:3642–43
37. Crespi, H. L., Katz, J. J. 1969. *Phytochemistry* 8:759–61
38. Crespi, H. L., Smith, U. H. 1970. *Phytochemistry* 9:205–12
39. Crespi, H. L., Smith, U. H., Katz, J. J. 1968. *Biochemistry* 7:2232–42
40. Dale, R. E., Teale, F. W. J. 1970. *Photochem. Photobiol.* 12:99–117
41. Diakoff, S., Scheibe, J. 1973. *Plant Physiol.* 51:382–85
42. Dodge, J. 1969. *Arch. Mikrobiol.* 69:266–80
43. Duysens, L. N. M. 1952. PhD thesis. Univ. Utrecht, Belgium
44. Edwards, M. R., Gantt, E. 1971. *J. Cell Biol.* 50:896–99
45. Eiserling, F. A., Glazer, A. N. 1974. *J. Ultrastruct. Res.* 47:16–25
46. Evans, E. L., Allen, M. M. 1973. *J. Bacteriol.* 113:403–8
47. Frackowiak, D., Grabowski, J. 1970. *Photosynthetica* 4:236–42
48. Ibid 1971. 5:146–52
49. French, C. S., Young, V. K. 1952. *J. Gen. Physiol.* 35:873–90
50. Fujimori, E. 1964. *Nature* 204:1091–92

51. Fujimori, E., Pecci, J. 1967. *Arch. Biochem. Biophys.* 118:448–55
52. Fujimori, E., Pecci, J. 1970. *Biochim. Biophys. Acta* 207:259–62
53. Ibid. 221:132–34
54. Fujita, Y., Hattori, A. 1960. *Plant Cell Physiol.* 1:281–92
55. Ibid, 293–303
56. Fujita, Y., Hattori, A. 1962. *J. Biochem. Tokyo* 51:89–91
57. Ibid. 52:38–42
58. Fujita, Y., Hattori, A. 1962. *Plant Cell Physiol.* 3:209–20
59. Fujita, Y., Hattori, A. 1963. *J. Gen. Appl. Microbiol.* 9:253–55
60. Fujita, Y., Hattori, A. 1963. *Microalgae Photosyn. Bact.*, 431–40
61. Gaidukov, N. 1902. *Abh. Preuss Akad. Wiss.* 5
62. Gantt, E. 1969. *Plant Physiol.* 44:1629–38
63. Gantt, E. Conti, S. F. 1965. *J. Cell Biol.* 26:365–81
64. Ibid 1966. 29:423–34
65. Gantt, E., Conti, S. F. 1966. *Energy Conversion by the Photosynthetic Apparatus, Brookhaven Symp. Biol.* 19:393–405
66. Gantt, E., Conti, S. F. 1969. *J. Bacteriol.* 97:1486–93
67. Gantt, E., Edwards, M. R., Conti, S. F. 1968. *J. Phycol.* 4:65–71
68. Gantt, E., Edwards, M. R., Provasoli, L. 1971. *J. Cell Biol.* 48:280–90
69. Gantt, E., Lipschultz, C. A. 1972. *J. Cell Biol.* 54:313–24
70. Gantt, E., Lipschultz, C. A. 1973. *Biochim. Biophys. Acta* 292:858–61
71. Gantt, E., Lipschultz, C. A. 1974. *Biochemistry* 13:2960–66
72. Ghosh, A. K., Govindjee 1966. *Biophys. J.* 6:611–17
73. Ginsburg, R., Lazaroff, N. 1973. *J. Gen. Microbiol.* 75:1–9
74. Glazer, A. N., Cohen-Bazire, G. 1971. *Proc. Nat. Acad. Sci. USA* 68:1398–1401
75. Glazer, A. N., Cohen-Bazire, G., Stanier, R. 1971. *Arch. Mikrobiol.* 80:1–18
76. Glazer, A. N., Cohen-Bazire, G., Stanier, R. Y. 1971. *Proc. Nat. Acad. Sci. USA* 68:3005–8
77. Glazer, A. N., Fang, S. 1973. *J. Biol. Chem.* 248:659–62
78. Ibid, 663–71
79. Glazer, A. N., Fang, S., Brown, D. M. 1973. *J. Biol. Chem.* 248:5679–85
80. Godnev, T. N., Rotfarb, R. M., Gvardiyan, V. N. 1966. *Dokl. Akad. Nauk SSSR* 169:1191–94
81. Grabowski, J. 1972. *Photosynthetica* 6:291–97
82. Grabowski, J., Frackowiak, D. 1972. *Photosynthetica* 6:142–49
83. Gray, B. H., Lipschultz, C. A., Gantt, E. 1973. *J. Bacteriol.* 116:471–78
84. Hallier, U. W., Park, R. B. 1969. *Plant Physiol.* 44:535–40
85. Hattori, A., Crespi, H. L., Katz, J. J. 1965. *Biochemistry* 4:1213–24
86. Hattori, A., Fujita, Y. 1959. *J. Biochem.* 46:521–24
87. Ibid, 633–44
88. Ibid, 1259–61
89. Haxo, F., Ó hEocha, C., Norris, P. 1955. *Arch. Biochem. Biophys.* 54:162–73
90. Jackson, A. H., Kenner, G. W. 1968. *Biochem. Symp.* 28:3–18
91. Jennings, R. C., Broughton, W. J., McComb, A. J. 1972. *Phytochemistry* 11:1937–44
92. Jones, L. W., Myers, J. 1965. *J. Phycol.* 1:7–14
93. Killilea, S. D., Ó Carra, P. 1968. *Biochem. J.* 110:14–15
94. Kobayashi, Y., Siegelman, H. W., Hirs, C. H. W. 1972. *Arch. Biochem. Biophys.* 152:187–98
95. Kupelian, R. H. 1964. *The development and characteristics of two water-soluble cytochromes from the alga Cyanidium caldarium.* PhD thesis. Univ. Chicago, Chicago, Ill. 35 pp.
96. Kylin, H. 1910. *Hoppe-Seyler's Z. Physiol. Chem.* 69:169–239
97. Kylin, H. 1912. *Sv. Bot. Tidskr.* 5:531–44
98. Lamont, H. C. 1969. *Arch. Mikrobiol.* 69:237–59
99. Lazaroff, N. 1966. *J. Phycol.* 2:7–17
100. Lazaroff, N. 1973. See Ref. 23, 162–85
101. Lazaroff, N., Schiff, J. 1962. *Science* 137:603–4
102. Lazaroff, N., Vishniac, W. 1961. *J. Gen. Microbiol.* 25:365–74
102a. Lefort-Tran, M., Cohen-Bazire, G., Pouphile, M. 1973. *J. Ultrastruct. Res.* 44:199–209
103. Lemasson, C., Tandeau de Marsac, N., Cohen-Bazire, G. 1973. *Proc. Nat. Acad. Sci. USA* 70:3130–33
104. Lemberg, R. 1930 *Liebig's Ann. Chem.* 477:195–245
105. Lemberg, R., Bader, G. 1933. *Liebig's Ann. Chem.* 505:151–77
106. Levin, E. Y. 1966. *Biochemistry* 5:2845–52
107. Levin, E. Y. 1967. *Biochim. Biophys. Acta* 136:155–58

108. Lichtlé, C., Giraud, G. 1970. *J. Phycol.* 6:281–89
109. MacColl, R., Habig, W., Berns, D. S. 1973. *J. Biol. Chem.* 248:7080–86
110. MacColl, R., Lee, J. J., Berns, D. S. 1971. *Biochem. J.* 122:421–26
111. Macdowall, F. D. H., Bednar, T., Rosenberg, A. 1968. *Proc. Nat. Acad. Sci. USA* 59:1356–63
112. Maldonado, A. A. 1967. *Effect of δ-aminolevulinic acid on the biliprotein content of Tolypothrix tenuis.* MS thesis. Univ. Chicago, Chicago, Ill. 35 pp.
113. Mercer, F. V., Bogorad, L., Mullens, R. 1962. *J. Cell Biol.* 13:393–403
114. Mieras, G. A., Wall, R. A. 1968. *Biochem. J.* 107:127–28
115. Mohanty, P., Braun, B. Z., Govindjee, Thornber, J. P. 1972. *Plant Cell Physiol.* 13:81–91
116. Myers, A., Preston, R. D., Ripley, G. W. 1956. *Proc. Roy. Soc. B* 144:450–59
117. Nakajima, O., Gray, C. H. 1967. *Biochem. J.* 104:20–22
118. Neufeld, G. J., Riggs, A. F. 1969. *Biochim. Biophys. Acta* 181:234–43
119. Neushal, M. 1970. *Am. J. Bot.* 57:1231–39
120. Nichols, K. E. 1962. *Action spectra studies of the formation and photosynthetic participation of phycocyanin in wild-type and mutant-type cells of Cyanidium caldarium.* PhD thesis. Univ. Chicago, Chicago, Ill. 51 pp.
121. Nichols, K. E., Bogorad, L. 1960. *Nature* 188:870–72
122. Nichols, K. E., Bogorad, L. 1962. *Bot. Gaz.* 124:85–93
123. Nolan, D. N., Ó hEocha, C. 1967. *Biochem. J.* 103:39–40
124. Ó Carra, P. 1965. *Biochem. J.* 94:171–74
125. Ibid 1970. 119:2–3
126. Ó Carra, P., Killilea, S. D. 1971. *Biochem. Biophys. Res. Commun.* 45:1192–97
127. Ó Carra, P., Ó hEocha, C. 1962. *Nature (London)* 195:173–74
128. Ó Carra, P., Ó hEocha, C. 1966. *Phytochemistry* 5:993–97
129. Ó Carra, P., Ó hEocha, C., Carroll, D. M. 1964. *Biochemistry* 3:1343–50
130. Ó hEocha, C. 1965. *Ann. Rev. Plant Physiol.* 16:415–34
131. Ó hEocha, C. 1966. *Biochemistry of Chloroplasts,* ed. T. W. Goodwin, 1:407–21. London:Academic. 476 pp.
132. Ó hEocha, C. 1968. *Biochem. Soc. Symp. No. 28: Porphyrins and Related Compounds,* 91–105
133. Ó hEocha, C. 1971. *Oceanogr. Mar. Biol. Ann. Rev.* 8:61–82
134. Ó hEocha, C., Ó Carra, P. 1961. *J. Am. Chem. Soc.* 83:1091
135. Ó hEocha, C., Ó Carra, P., Mitchell, D. 1964. *Proc. Roy. Irish Acad.* 63:191–200
136. Pecci, J., Fujimori, E. 1969. *Biochim. Biophys. Acta* 188:230–36
137. Pecci, J., Fujimori, E. 1970. *Phytochemistry* 9:637–40
138. Petryka, Z., Nicholson, D. C., Gray, C. H. 1962. *Nature* 194:1047–49
139. Poff, K. L., Butler, W. L. 1974. *Nature* 248:799–801
140. Poff, K. L., Loomis, W. F., Butler, W. L. 1974. *J. Biol. Chem.* 249:2164–67
141. Raftery, M. A., Ó hEocha, C. 1965. *Biochem. J.* 94:166–70
142. Rice, H. 1971. *Purification and partial characterization of oat and rye phytochrome.* PhD thesis. Harvard Univ., Cambridge, Mass.
143. Rüdiger, W. 1968. *Biochem. Soc. Symp. No. 28: Porphyrins and Related Compounds,* 121–30
144. Rüdiger, W. 1970. *Angew. Chem. Int. Ed.* 9:473–80
145. Rüdiger, W. 1971. *Fortschr. Chem. Org. Naturst.* 29:128–39
146. Rüdiger, W., Ó Carra, P. 1969. *Eur. J. Biochem.* 7:509–16
147. Rüdiger, W., Ó Carra, P., Ó hEocha, C. 1967. *Nature (London)* 215:1477–78
148. Scheibe, J. 1972. *Science* 176:1037–39
149. Schram, B. L., Kroes, H. H. 1971. *Eur. J. Biochem.* 19:581–94
150. Seckbach, J. 1972. *Microbios* 5:133–42
151. Siegelman, H. W., Chapman, D. J., Cole, W. J. 1967. *Arch. Biochem. Biophys.* 122:261
152. Siegelman, H. W., Chapman, D. J., Cole, W. J. 1968. *Biochem. Soc. Symp. No. 28: Porphyrins and Related Compounds,* 107–20
153. Takemoto, J., Bogorad, L. 1975. *Biochemistry.* In press
154. Teale, F. W. J., Dale, R. E. 1970. *Biochem. J.* 116:161–69
155. Tenhunen, R. et al 1972. *Biochemistry* 11:1716–20
156. Tenhunen, R., Marver, H. S., Schmid, R. 1968. *Proc. Nat. Acad. Sci. USA* 61:748–56
157. Tenhunen, R., Marver, H. S., Schmid, R. 1969. *J. Biol. Chem.* 244:6388–94
158. Tenhunen, R., Ross, M. E., Marver, H. S., Schmid, R. 1970. *Biochemistry* 9:298–303
159. Thomas, J. P., O'Kelley, J. C. 1973. *Photochem. Photobiol.* 17:469–72

160. Torjesen, P., Sletten, K. 1972. *Biochim. Biophys. Acta* 263:258–71
161. Troxler, R. F. 1972. *Biochemistry* 11:4235–42
162. Troxler, R. F., Bogorad, L. 1966. *Plant Physiol.* 41:491–99
163. Troxler, R. F., Bogorad, L. 1966. *Biochemistry of Chloroplasts,* ed. T. W. Goodwin, 2:421–26
164. Troxler, R. F., Brown, A. 1970. *Biochim. Biophys. Acta* 215:503–11
165. Troxler, R. F., Brown, A., Foster, J. A., Franzblau, C. 1974. *Fed. Proc.* 33:1446
166. Troxler, R. F., Dokos, J. M. 1973. *Plant Physiol.* 51:72–75
167. Troxler, R. F., Lester, R. 1967. *Biochemistry* 6:3840–46
168. Troxler, R. F., Lester, R., Brown, A.,

White, P. 1970. *Science* 167:192–93
169. van der Velde, H. H. 1973. *Biochim. Biophys. Acta* 303:247–57
170. Vaughan, M. H. Jr. 1964. *Structural and comparative studies of the algal protein phycoerythrin.* PhD thesis. Massachusetts Inst. Technol., Cambridge, Mass.
171. Vernotte, C. 1971. *Photochem. Photobiol.* 14:163–73
172. Walbridge, C. T. Jr. 1963. *Immunochemical relations of various phycocyanins with serum immune to Cyanidium caldarium phycocyanin.* MS thesis. Univ. Chicago, Chicago, Ill. 26 pp.
173. Williams, V. P., Freidenreich, P., Glazer, A. N. 1974. *Biochem. Biophys. Res. Commun.* 59:462–66

Ann. Rev. Plant Physiol. 1975. 26:403–25

INCOMPATIBILITY AND THE POLLEN-STIGMA INTERACTION[1]

❖7596

J. Heslop-Harrison

Cell Physiology Laboratory, Royal Botanic Gardens, Kew, Richmond, Surrey, England

CONTENTS

INTRODUCTION

Incompatibility in the sense of this review refers to the partial or complete incapacity of a pollen-fertile hermaphrodite or monoecious angiosperm to set viable seed on

[1]The survey of literature pertaining to this review was completed in June 1974. The review does not attempt to cover the rich genetical literature on the subject of angiosperm incompatibility systems in any detail, and the papers selected for attention have been chosen because of their possible relevance to physiological mechanisms.

self-pollination. In this usage, the term is roughly equivalent to Darwin's "self-sterility" (18), but admits the possibility of varying degrees of self-infertility. Self-incompatibility (SI) systems, together with other devices promoting cross-pollination such as dichogamy, dioecism, and various specializations of flower structure, regulate the breeding system by ensuring a greater or lesser amount of outcrossing in a population. They are complemented by isolation mechanisms which act to set the limits to gene exchange, marking the boundaries, so to speak, of biological species. The overall effect is to establish a regulated degree of heterozygosity in the population, a major factor in determining the capacity for response to selection and so the evolutionary potential of a species (93). SI is extremely widespread among the flowering plants (12, 23, 84) and is evidently much more important as an outbreeding device in the group than sexual polymorphism, of which dioecism, the simplest version, is the preferred system among animals (36).

Whitehouse (119) has suggested that the early adoption of SI as an outbreeding mechanism played an important part in the evolutionary success of the angiosperms. This seems entirely probable, but it is by no means clear how the various systems are themselves related in an evolutionary sense, the more so because the genetical evidence is currently pointing to a greater degree of diversity in the systems present among extant flowering plant species than has hitherto been suspected (88). SI mechanisms are of course necessarily subject to selection, but like all components of the breeding system, in a retrospective mode. It is conceivable that each of the principal systems known today has had a long evolutionary history, representing among them the most successful trends established during an early and fluid period of angiosperm evolution selected because of the adaptive flexibility they conferred upon the lineages in which they prevailed. As noted by Bateman (9), a polyphyletic history of this kind is suggested by the fact that the main SI systems of today are confined to particular families.

Inter- and Intraspecific Incompatibility

Fertility barriers between species take a great many forms, but where the relationship is close it is sometimes found that they are phenomenologically analogous with intraspecific SI systems in that they act in the stigma or style to prevent pollen germination or to frustrate pollen tube growth. The obvious distinction is to be seen in the effects: while interspecific systems preclude too remote a union, intraspecific SI systems prevent one that is too close. Many authors from Darwin onwards have noted that barriers to free interbreeding may arise through the gradual decay of co-adaptation when allopatric populations undergo evolutionary divergence. Any of the manifold interactions vital to a biparental mating system could be affected in this way, including those between pollen and pollen tube and stigma and style. Maladjustments of this kind contributing to an interspecific breeding barrier are in no real sense comparable with SI systems, and it would be a mistake to refer to them in terms that might suggest homology.

The operation of intraspecific compatibility controls may, however, transgress species boundaries, for in some genera the pollen behavior in artificial interspecific cross-pollinations is unquestionably related in some sense to the SI systems effective within the species concerned. Phenomena of this type were first systematically investigated by Lewis & Crowe (74). It may be noted that pollen-pistil interactions between closely related species could be affected by the prevailing SI systems without the fact having any bearing whatever on the nature of the isolating mechanism keeping the species apart. This would be expected were the physiological controls in the pistils of the species concerned identical, representing the legacy of a common ancestor; an analogy would be two separate communication systems operating with the same coding. It is pertinent, nevertheless, to ask whether there might truly be interspecific barriers dependent upon physiological controls similar to those operating in SI systems—similar, that is, in their biochemical basis, although not necessarily sharing any elements of genetical control. Some recent experiments on interspecific hybridization (58, 59) have offered some evidence in this connection, but within the limitations of this article this will not be considered.

The General Nature of Interactions in SI systems

In an efficient SI system, a stigma receiving self-pollen should not be impaired in consequence for the acceptance of appropriate nonself pollen, nor should the operation of the controls entail a loss of potential female fertility by the wholesale sacrifice of ovules. These conditions are fulfilled by prefertilization controls in which the interactions between the partners are restricted to the vicinity of the inhibited pollen grain or pollen tube. The possible levels of interaction are shown in Table 1, which summarizes the cardinal events leading up to fertilization in a typical angiosperm (38). In steps 1 to 5, the interactions are likely to be between the male gametophyte, together with any surface materials it may bear, and the sporophytic parent of the female gametophyte that would be its prospective mate. Only in step 6 can there be direct interaction between the two gametophytes. This analysis has genetical implications. On the female side, compatibility control would lie with the diploid sporophyte in steps 1 to 5, and with the haploid gametophyte in step 6. On the male side, control in steps 1 to 4 could depend upon the haploid gametophyte, or could be exerted by its diploid parent through the agency of sporophytically synthesized materials carried with it. From step 5 onwards, control on the male side might be expected to lie entirely with the male gametophyte.

Table 1   Levels of interaction in SI systems[a]

Fate of the male gametophyte:	Interaction with the sporophyte involves:
1. Capture	Stigma surface materials
2. Hydration	
3. Germination	Stigma surface and the underlying papillae
4. Tube penetration	
5. Tube growth through the style	Transmitting tissue of the style
6. Entry into the female gametophyte and gamete discharge	None: interactions are with the female gametophyte

[a] From Heslop-Harrison et al (42).

We shall see in a later section how these expectations have been justified in genetical studies. Here we may note another point concerning the "information" content needed in the operation of SI systems. The rejection of "self" in SI, like the rejection of "nonself" in vertebrate immune systems, implies a recognition event. In contradistinction to the behavior in immune systems, however, the recognition is not conditioned by encounters with foreign molecules, but is constitutive in the sense that the recognizing female tissues are preprogrammed to identify and react against another cell produced by the same individual and carrying a sample of the same genome. Obviously, in seeking an explanation of the self-rejection, there can be no appeal to any principle of differentiation. No doubt the factors that provide a background to the response are specific to style and stigma on the one hand and pollen on the other, representing different pathways of differentiation in the two tissue lineages, but it is not easy to see how the uniqueness of the self-rejection response can arise without the transcription at some level of the genes conferring specificity that is necessarily held in common. Correspondingly, the acceptance of compatible pollen must result from the transcription of key alleles *not* held in common. In practice, the genetical analysis of the two most important SI systems has shown that control resides in one, or in some cases two, loci, and that the presence of a wide range of compatible genotypes within a given population is guaranteed by multiple allelism—in some instances such as clover with a very long series of alleles indeed. Contemplation of these facts shows that no *simple* system of complementarity or lack of complementarity—enzyme and co-factor, for example—can be expected to explain the phenomenon. Because of the numerous discriminations inherent in SI systems, we are compelled to postulate the participation of information-carrying molecules, and the currently most popular theories of the physiology of the response cast proteins in the executive role. There is now a body of evidence supporting this attribution for one important SI system, and a review of this occupies a considerable part of this article.

We begin, however, with a summary of the genetics of SI systems and a survey of some of the characteristics of pollen and stigma, separately and interaction, that must necessarily play a part in regulation of germination and pollen tube growth.

GENETICS OF INCOMPATIBILITY SYSTEMS

Darwin (18) and Correns (17) provided some evidence bearing upon the inheritance of self-sterility, but the credit for the first comprehensive analysis of the genetics of such a system rests with East & Mangelsdorf (24). Their work on *Nicotiana* led to the discovery of one of the principal SI systems, now known to be characteristic of the Solanaceae and to occur also in many other families. The "oppositional factor" hypothesis of East & Mangelsdorf held that any pollen grain carrying an *S* (incompatibility) allele identical with one present in the pistil would be rejected. The control on the male side is therefore with the haploid gametophyte. In the classical gametophytic incompatibility (GSI) system, control is given by a single locus with many alleles, and there is no structural differentiation of the flower in the different incompatibility classes. The type may be characterized as gametophytic, monofactorial, and homomorphic. A bifactorial gametophytic system was discovered in the Gramineae by Lundquist (87). In this, each locus, S and Z, is polyallelic. Pollen is rejected when the alleles at the two loci are both represented also in the stigma and style, and not otherwise. Bifactorial gametophytic systems have now been reported in other families including dicotyledons, suggesting that they may be widespread among angiosperms (88).

Sporophytic control of pollen behavior in SI systems was suspected first by Correns (17) from observations on the Cruciferae, and the genetics of this type of control were worked out for this family by Bateman (10, 11). A similar system had already been described for the Compositae by Hughes & Babcock (48). In the sporophytic incompatibility system (SSI) the behavior of the pollen is determined by the genotype of its diploid parent, and again a single locus with an allelic series is concerned, with no structural differentiation of the flower in different compatibility classes. The system is sporophytic, monofactorial, and homomorphic.

Heteromorphic SI systems have been known in various families since Darwin's work (18), the best characterized being in the Primulaceae. Here the pin-thrum difference in the flowers is associated with differences in pollen and stigma, and pollen behavior is strictly related to the parental genotype; the control is therefore sporophytic. The genetic mechanism is an S-gene complex with elements controlling each of the structural and physiological features of the system (28, 94). With respect to inheritance, the situation is quite similar to that seen in animal sex determination, where the switch depends on the segregation of an XY chromosome pair (68). Heteromorphic systems will not be considered in this review.

The genetical situations summarized above are shown in Table 2, expanded from Lewis (70). Continued experimental breeding work with many genera has exposed various complexities, the details of which are not of any special concern for this article. Recent reviews of the genetics of incompatibility have been given by Arasu (1), Townsend (114), and de Nettancourt (100), and reference should be made to these for access to the more modern literature of the subject.

Table 2 Morphology and genetics of angiosperm incompatibility systems[a]

Morphology	Genetics			
	Control		No. of loci	No. of alleles
	Pollen	Stigma/Style		
Homomorphic	Sporophytic; independent gene action, or various dominance relationships	As pollen	1	Many
	Gametophytic; independent gene action in each haploid grain	Independent gene action in each diploid cell	1,2	Many
Heteromorphic	Sporophytic, with dominance	Dominance	1	2

[a] Based on Lewis (70).

CYTOLOGICAL AND PHYSIOLOGICAL CORRELATES OF SI SYSTEMS

In 1957 Brewbaker (12) pointed out a number of correlations between type of SI system and characteristics of the pollen and pollen tube behavior. In general, GSI pollen is binucleate (one vegetative cell nucleus, one generative cell nucleus), readily germinates in vitro, and is relatively long-lived. The inhibition of incompatible male gametophytes takes place after germination and while the tube is growing in the style. In contrast, SSI pollen is trinucleate (one vegetative cell nucleus, two gamete nuclei), does not easily germinate in vitro, and is short-lived. The inhibition of incompatible grains takes place on the stigma surface; either germination is inhibited, or the tube formed grows abnormally without penetrating the stigma papillae. More recently, further correlates have been noted. Hoekstra (43) compared four binucleate pollens, including some from genera with known GSI systems, with four trinucleate pollens, three from composites with the SSI system. The respiratory rates of the latter were greater by factors of 2 to 3 than the former when compared on a gross weight basis, and 5 to 10 on a protein basis. The pollen was stored in controlled 97% RH, and the time for respiration to fall to 50% of the initial rate was measured. The Compositae (SSI, trinucleate pollen) reached the 50% level in 2–3.75 hr, while *Nicotiana alata* (GSI, binucleate pollen) did not reach this level until after 15 hr.

As Brewbaker himself noted, the correlations pointed out (12) are not perfect. They do not apply in heteromorphic SI systems, nor do they extend to the Gramineae, where a bifactorial GSI system prevails with pollen that is trinucleate, has a high respiration rate, and is short-lived. Nevertheless, there can be little doubt of their significance in relation to the incompatibility response, and we shall pursue some of the implications in later sections. Meanwhile, we may note that the type of incompatibility control is reflected also in characteristics of the stigma. The families known to have SSI systems have papillate stigmas and no copious stigma exudate, while in most—although not all—genera with GSI systems the stigma is smooth, or has low papillae, and bears a plentiful fluid secretion (38).

THE POLLEN GRAIN IN THE INCOMPATIBILITY RESPONSE

General Features

The difference in the genetic control of pollen behavior in GSI and SSI systems has obvious implications in relation to the time and site of S-gene action on the male side. In monofactorial GSI systems of the kind first investigated in the Solanaceae, all plants rising from legitimate pollinations are heterozygous at the S locus. Segregation takes place at meiosis to give two classes of spores; these give rise to male gametophytes that will function only in pistils whose diploid cells carry another S allele at each locus. Thus an $S_1 S_2$ plant produces S_1 and S_2 pollen; S_1 pollen will function in an $S_2 S_3$ style but not in any style with S_1 at one locus, and so on. These facts show that S-allele transcription cannot take place earlier than the first cleavage in the meiocyte after the telophase of the first meiotic division, and the likelihood is that it follows upon the conclusion of the second division, after the cleavage that isolates the four spores. The synthesis of the incompatibility substances could occur in the spores, or in the gametophytes arising from them, or possibly in both.

The SSI system exemplified in the Cruciferae and Compositae evidently is also monofactorial. Various dominance relationships have been demonstrated in the investigated species (106, 113) but as we have seen, in all cases the genetical results show that pollen behavior is determined by the diploid parental genotype. This means that the effective gene products carried to the stigma must be synthesized by diploid tissue. Pandey (102, 103) argued that in SSI the S-alleles must act before meiosis (heteromorphic system) or at the latest after meiosis and before the cleavage of the mother cell cytoplasm (homomorphic system). By this means the equal distribution of the diplophase synthesized incompatibility substances to the haploid spores of the tetrad would be ensured. Heslop-Harrison (33, 34) offered an alternative explanation for the working of the SSI systems, namely that the synthesis of the incompatibility substances is not a function of the meiocyte itself, but of the associated sporophytic "nurse" tissue of the anther, the tapetum. According to this view, S-gene products accumulating in the tapetum are transferred upon the dissolution of the tissue to the pollen exine during the final stages of pollen maturation, to be conveyed to the stigma in the cavities or embayments of the outer sculptured layer.

It is now known that fractions carried in the pollen wall do play a part in incompatibility responses. The wall conveys both sporophytically and gametophytically synthesized proteins, and we will now examine their origins and fates.

The Pollen Wall

Most angiospermic pollens have a compound wall, with an inner pectocellulosic layer and an outer layer of sporopollenin, a material characterized by Shaw and co-workers (13) as an oxidative polymer of carotenoids and carotenoid esters.

The inner layer, the intine, is produced within the exine by the protoplast of the haploid spore and later, after pollen mitosis, by the vegetative cell of the male gametophyte. The synthesis of the polysaccharide moiety involves dictyosome activity in the cortical layers of the cytoplasm (19, 35), much as in the growth of the primary wall of a somatic cell. However, there is a complication: during the development of this layer, protein-containing tubules, leaflets, or plates are incorporated, sometimes over the whole of the surface (51) and sometimes principally in the vicinity of germination apertures where these are present (53). Several acid hydrolases are present among the included proteins, but immunological and other evidence indicates that the enzymes account for but a small proportion of the total protein load of the intine (40, 52–55). In the Compositae the synthesis of the intine proteins is evidently conducted in stratified, ribosomal endoplasmic reticulum in the outer cytoplasm, and the insertion into the wall involves a separation of successive layers at the plasmalemma in a manner not previously recorded in plant cells (53). In other species the proteinaceous inclusions in the apertural intine and elsewhere are in the form of tubules, and these have the appearance of irregular, finger-like extensions from the plasmalemma during development (51). Eventually they are cut off by the formation of an unpenetrated cellulosic layer before the cessation of intine growth.

The outer sporopollenin layer is the exine. At maturity it usually shows a distinct stratification, with an inner nonsculptured part, the nexine, and an outer sculptured sexine (27). A feature present in most angiosperm families is a columnar layer formed of numerous radially directed rods rising from the nexine. This layer may be roofed over to give the tectate exine, or the rods may stand free, or be connected above by the fusion of their swollen heads to create walls. The rods (technically, bacula) are generally distributed in a patterned manner, a common arrangement both in monocotyledons and dicotyledons being such that the walls form raised polygons over the pollen grain surface. If there is no tectum, the cavities so formed are open to the outside but sealed within by the nexine; this is the state in the Cruciferae. If a tectum is present, this is invariably interrupted by perforations or micropores, giving communication from the outside of the grain to the continuous space between the columns. This type of architecture within the exine has been aptly likened to the crypt of a cathedral; to complete the image one must suppose the roof of the crypt to be perforated by roughly circular holes at fairly regular intervals, corresponding to the micropores.

It will be seen that the structural features of the exine are admirably adapted to receive and hold externally derived materials. Indeed, it is now possible to understand the intricate sculpturing of this part of the pollen wall—long of interest to palynologists as a source of taxonomic differentiae—as an adaptive system concerned with the transport of products of central importance in the physiology of pollen.

The transfer of tapetum-derived materials to the exine has been described in detail for several families. These accounts suggest that the phenomenon is likely to be a very general one among angiosperms, although there may well be exceptional groups such as those with water-distributed pollen, for example, and families without a sculptured exine, like the Cannaceae (110). Typically the tapetum synthesizes protein which is held in cisternae of the endoplasmic reticulum until the dissolution of the tissue. The released proteins are then injected through the micropores of the tectum and into the baculate layer in tectate pollen (40), or deposited in the surface depressions in nontectate grains (21, 41) such as those of the Cruciferae. Lipids are also often synthesized abundantly in the later life of the tapetum, and these are also transferred to the pollen surface, together with the carotenoids that give the yellow or orange color so characteristic of many pollens (34). The lipids generally form an outer sealing and binding layer, constituting the *Pollenkitt* of German authors of the 1930s (50).

Using cytochemical methods it is possible to establish some properties of the exine fractions in situ (40). In most of the species so far studied, little enzymic activity is associated with the exine compared with the intine inclusions, but nonspecific esterase is usually detectable. A positive PAS reaction suggests the presence of glycoproteins, confirmed by tests of extracted material.

From the above account, it will be seen that the pollen wall at maturity holds mobile fractions derived both from the haploid gametophyte and the diploid (parental) sporophyte.

Emission of Pollen Wall Proteins

It is well known that proteins pass out freely from moistened pollen, even from intact cells where free passage through the plasmalemma seems improbable (89, 111). The mobile fractions are in fact those housed in the wall layers—derived, as we have seen, partly from the haploid gametophyte and partly from its diploid parent. Evidence for this has come from the localization by immunofluorescence of the origins of the antigens present in leachates from intact pollen (46, 51, 52, 54, 55, 57) and from a pollen print method which allows the precise localization of the sites of emission of wall-held constituents (40). A by-product of this work has been a demonstration that the principal hay fever allergen of *Ambrosia* (ragweed) pollen, Antigen E, is held in the interbacular cavities of the exine and also in the intine (46).

Pollen-print methods allow the kinetics of protein release to be followed. Grains brought into contact with hydrated agarose films take up water as on a stigma, and the wall-held fractions pass out into the film, where they can be identified by staining properties, enzymic activity, or by immunofluorescence using antibodies to known pollen antigens (42). The exine-held (sporophytic) fractions are transferred from the surface embayments, or pass out through the micropores of the tectum, within seconds after the beginning of hydration. The intine-held (gametophytic) fractions pass out from the germination apertures where these are well developed, but may also diffuse through the exine when this does not present an impermeable layer, as, for example, in the Iridaceae (51). The intine release does not normally begin for some minutes after hydration.

Characterization of the Emitted Fractions

The wall-held materials can be obtained without severe contamination from the vegetative cells of pollen grains by leaching in osmotically balanced media under conditions which preserve the integrity of the cell membranes. Knox and co-workers have shown that short-term diffusates of the pollen of *Brassica oleracea* (Cruciferae, SSI) prepared in this manner contain both proteins and glycoproteins (56). Thin-layer gel chromatography, polyacrylamide gel electrophoresis, and gradient diffusion tests against rabbit antibodies prepared against total leachates suggest that the principal constituent is a protein of approximately 10,000 daltons (Fraction A), which probably represents some 70% of the total protein present. Five other fractions have been detected: Fractions B, D, and E, glycoproteins of approximately 16,000, 35,000, and 45,000 daltons respectively, and Fractions C and F, as yet ill defined. Immunoelectrophoretic separation of the antigens using antisera to the total leachate gave six precipitin arcs, of which the principal one is likely to correspond to Fraction A of 10,000 daltons. In similar studies of the wall proteins of *Cosmos* (Compositae, SSI), 9 protein bands were detected by polyacrylamide gel electrophoresis (47), two likely to be glycoproteins. The four major fractions have molecular weights in the range of 10,000 to 40,000. Immunological analysis of the same pollen diffusates showed some 6 antigens.

So far, work on the composition of the pollen wall diffusates has been restricted mainly to species of families with SSI systems. Similar diffusates can be obtained from GSI species, although these are yet to be fractionated and characterized.

THE STIGMA SURFACE

Recent work on the physiology of incompatibility responses has redirected attention to the stigma surface, particularly in SSI systems, since it is here that the inhibition occurs in an incompatible pollination (112), but also in certain plants with GSI (20, 60). There have been no serious attempts to classify stigmas in a physiological sense, but certain differences in the receptive surfaces are readily recognized (38). The surface cells of the pollen-accepting areas are always more or less of a glandular type, and are frequently elongated into unicellular or multicellular papillae. A notable distinction is between the "dry" stigma type, where the papillae bear no free fluid surface, and the "wet" type, where there is a more or less copious secretion.

Wet stigmas are found in many GSI families, including Solanaceae, Liliaceae, Rosaceae, and Onagraceae. The secretion accumulates throughout the life of the stigma and forms the medium in which the captured pollen germinates. The stigma exudate is frequently emulsified, with dispersed lipid droplets. In some genera there is an overlying lipid layer, forming the "liquid cuticle" described by Konar & Linskens (60) in *Petunia,* a genus with a *Nicotiana* type of GSI system. Martin (91, 92) found 8 fatty acids after esterification of the *Petunia* stigma lipids, with chain lengths of 11–20 C atoms. Sugars (60) and phenolic compounds (92) occur in stigmatic exudates, and mucopolysaccharides contribute to its viscosity (66, 67). Free amino acids have been reported. In some GSI species, there is little protein in the readily removed fraction of the exudate [e.g. *Petunia* (61)], although in *Lilium* 7% dry weight is represented by protein (66). Notwithstanding the variation in protein content, it is usually possible to detect enzyme activity in the exudate of GSI stigmas by cytochemical methods. Nonspecific esterase activity has been reported, sometimes dispersed in the aqueous phase and sometimes in a layer in close proximity to the underlying stigma surface (20, 38).

"Dry" stigma surfaces occur in the two main SSI families, Cruciferae and Compositae, and in various other families, including the Gramineae with a bifactorial GSI system. Such stigmas are dry only in a relative sense, since the papillae are now known to bear a hydrated proteinaceous pellicle (95). This layer, which can be observed with both optical and electron microscopes, is disrupted by pronase digestion, but not by lipase (38, 95). It shows intense nonspecific esterase activity with various substrates, but in the Compositae, Cruciferae, and Caryophyllaceae has not been found to have any other enzymic properties. The pellicle overlies the cuticle of the stigma papillae, and the cuticle itself has unusual features, being made up of short radially directed rodlets with intervening discontinuities. This structure provides for the passage of water into the pellicle from the protoplast of the papilla.

The development of the proteinaceous layer of the stigma surface has been followed in species of the Caryophyllaceae, Cruciferae, and Malvaceae (38, 42). In all cases, the superficial esterase activity appears first in the bud, and increases during maturation, even for a period after anthesis. From electron microscopic observations on *Silene,* it seems that the proteins are synthesized in microbodies in the cortical cytoplasm of the papilla. These discharge at the plasmalemma, and the proteins diffuse thence across the pectocellulosic part of the wall and through the cuticular rodlets to aggregate on the outer surface. This mode of origin suggests that the pellicle is to be viewed as a kind of dried-down secretion.

The pellicle proteins can be released into sodium dodecyl sulphate medium roughly balanced to the tonicity of the vacuolar contents with suitable osmoticum, and under these conditions it seems that no appreciable intracellular material is extracted. In the Cruciferae, the leachates produced give satisfactorily clean protein absorption spectra (38). Polyacrylamide gel electrophoresis of the pellicle material from *Hibiscus* (Malvaceae) showed some seven bands, with most of the protein concentrated into three. Two of these bands showed nonspecific esterase activity in the gels.

The stigma surface proteins probably are among the antigens studied by Nasrallah and co-workers (96–98, 118) from the stigmas of *Brassica oleracea*. This work is reviewed in a later paragraph.

THE SPOROPHYTIC SELF-INCOMPATIBILITY SYSTEM

The Pollen-Stigma Interaction

As we have noted, the papillate stigmas of the two principal SSI families, Compositae and Cruciferae, are of the "dry" type, without a copious fluid secretion but with a hydrated overlying proteinaceous pellicle. The transfer of water begins as soon as contact is established, and the emission of wall materials seems to proceed as rapidly on the natural stigma as upon gel films (38, 41, 42, 52, 54). In the Cruciferae, the pollen wall proteins bind to the pellicle within 10–15 min, and thereafter cannot readily be released by leaching (41).

The stigma surface pellicle in this way forms a receptor site for the sporophytic wall proteins. The binding is associated with an increase in cytochemically detectable esterase activity (42), and the erosion of the cuticle of the stigma papilla begins beneath the emerging pollen tube tip. In families with cuticularized stigmatic surfaces the local dissolution of the cuticle is an essential preliminary to the penetration of the pollen tube. This process was first reported in an important paper by Christ (16), who found that in the Cruciferae the pollen tube after entry through the cuticle pursues its way through the wall of the papilla to enter the style. Christ's optical microscopic observations have been confirmed by electron microscopy (21, 61) and generalized to other families with cuticularized stigma papillae.

Christ's observations (16) led him to suggest that the cuticularized surface of the stigma papilla might form the actual incompatibility barrier in the Cruciferae, since he found that tubes from self pollen failed to penetrate and begin their growth through the wall. Assuming that this indicated the absence of an active cutinase, he proposed two possible explanations: *(a)* that cutinases carried by the pollen are inhibited on an incompatible stigma, or *(b)* that pollen-borne enzyme precursors require specific activators found only on compatible stigmas. Cutinases were detected in germinating pollen of Cruciferae by Linskens & Heinen (32, 83), and the "cutinase hypothesis" has gained some support from other authors (63, 107). However, recent electron microscopic observations on cruciferous species with well-developed SSI systems (21, 49) have shown that the cuticle is eroded in both compatible and incompatible pollinations, so that Christ's hypothesis now seems rather less attractive. Nevertheless, it is by no means ripe for abandonment, and is discussed further in a later section.

Characteristics of the Rejection Reaction in an Incompatible Pollination

In the SSI system, the rejection of an unacceptable grain is accompanied by distinctive events in both pollen and stigma papilla. The reactions are cell by cell, very often between one alighting pollen grain and one stigma papilla; the rejection reaction is not shared throughout the stigma. On the pollen side, an individual grain may *(a)* fail to germinate, without phenotypic evidence of abnormality; *(b)* begin germination, but produce a very short outgrowth that soon becomes occluded by callose; *(c)* produce a tube that grows in an abnormal manner without penetration, ultimately once more producing callose; or *(d)* produce a tube that attempts penetration but is frustrated immediately thereafter, again with callose deposition. On the stigma side, the contact with an incompatible pollen grain is followed by an enhancement of cyclosis and permeability and thereafter the rapid deposition of callose. This usually takes place in a site immediately adjoining the contact surface with an incompatible grain. These effects, familiar in a general sense to earlier workers, have recently been described in precise detail from cytochemical and electron microscopic study of various crucifers (21, 22, 41), and parallels have been recorded for the Compositae (52).

The part played by the β-1,3-glucan, callose, in the rejection reaction is a notable one. In both SSI and GSI systems, incompatible pollination is associated with shifts in the carbohydrate metabolism of the stigma and style (84, 108, 115), but callose deposition in the stigma papillae seems to be an exclusive feature of SSI systems. Here the reaction is so consistent that it can be used as a bioassay for incompatibility (39).

Exine-borne Fractions and the Incompatibility Response

Using the callose rejection reaction in the stigmatic papillae as a test of activity, proof of the participation of exine-borne materials in the incompatibility response has now been obtained for various species of Cruciferae. The simplest demonstration is provided by the use of agarose "pollen prints" prepared as described above (39, 41, 42). Gel fragments carrying the diffusate from exines of *Iberis semperflorens* provoke the characteristic callose deposition when placed on the stigma of a genotype of the species incompatible with the pollen parent, while those with diffusate from a compatible genotype do not. A similar demonstration has been given for *Raphanus sativus,* in this case using cellulose acetate membranes into which the exine-borne materials have been withdrawn by suction (22). The successful pioneer experiments of Kroh (84) have thus received their explanation.

The first attempts have been made to identify the precise constituents of the exine-borne materials responsible for the rejection reaction (41, 56). Concentrated diffusates from pollen of *Brassica oleracea* from parents homozygous for a known S-allele were subjected to thin-layer gel chromatography and the gels assayed directly by the implantation of stigmas of the same and different S-genotypes. The stigmas were then tested for the presence of callose in the papillae. The rejection reaction appeared in incompatible but not in compatible combinations, distributed somewhat irregularly throughout the zone of the chromatograms carrying wall proteins and glycoproteins. Evidently the technique at this stage is too crude to allow attribution of activity to any particular fraction, but it seems assured that the effective molecules are indeed proteins or glycoproteins.

Since the pollen wall materials used in these experiments are derived principally from the exine, they are in fact products of the tapetum of the parental anther. The sporophytic control of the incompatibility reaction in the SSI system is thus neatly explained.

Specificity of Pollen and Stigma Proteins

The foregoing results point to the conclusion that in the SSI system the critical recognitions are effected on the surface of the stigma papillae, when exine-borne fractions from the pollen grain pass out onto the pellicle. The pellicle is the obvious candidate for the role of receptor site, and from analogy with other recognition systems, more particularly the lymphocyte, one might hypothesize that the immediate consequence of binding is the transmission of stimuli that initiate further responses according to the genotype combination.

One implication of this hypothesis is that the specificities expressed in the SSI response are carried in the wall proteins on the pollen side and the stigma surface proteins on the pistil side. How then do they express the genetic identity that a self-rejection system must in some way incorporate? Disappointingly, attempts to detect differences in the pollen proteins of SSI species that might reflect incompatibility genotype have been unsuccessful. Immunoelectrophoretic separation of the antigens of wall diffusates of different S-gene homozygotes of Brassica oleracea gave essentially identical patterns (42), and similarly no differences have been detected in immunological studies of whole pollen extracts (96–98, 109). It may be noted that this situation is different from that found in the GSI species Oenothera organensis, reviewed further below, where S-genotype differences are detectable in the antigenic constituents of pollen (69).

In contrast with the results from pollen, striking differences in the patterns of stigma antigens related to incompatibility genotype have been found in SSI species. Nasrallah and co-workers (96–98, 118) raised antisera in rabbits to total saline extracts of macerated Brassica oleracea stigmas from plants of known S-genotypes, and made comparisons by standard immunodiffusion techniques. Tests using total extracts from three homozygous genotypes revealed the presence of a unique antigen for each S-allele, and in extracts from heterozygotes, the two predictable parental antigens were present. These important results suggest that the S-gene specified antigens are present in considerable amounts in the stigma. Moreover, perhaps most pertinently of all, the antigens could be detected in the diffusates from intact stigmas under conditions which would prevent the interference of material passing out from cut cell surfaces. This last result strongly suggests that the S-gene related fractions are on or very near the surfaces of the stigma papillæ; there is an obvious possibility that they are species of proteins present in the pellicle. Immunofluorescence studies indicate that the bulk of the rapidly released stigma antigens of *Brassica* are in fact superficial (38). In further work, Nasrallah and co-workers (96) have shown that different banding patterns appear when the protein complements of *Brassica olera-cea* stigmas bearing different *S*-allele combinations are separated by polyacrylamide gel electrophoresis. In the extracts of six homozygous and two heterozygous *S*-allele genotypes, unique bands were observed for each allele. Diffusion tests against the appropriate antiserum for one of the homozygous *S*-allele extracts suggested that the unique protein band observed in the gels represented the unique antigen detected immunologically, and the finding can probably be generalized for all others.

The matter currently rests at this point, with the enigmatic but seemingly well-supported finding that in the SSI system of the Cruciferae differences in the proteins of the stigma related to the *S*-genotype can be detected by various means, while no differences have yet been detected in the proteins of the pollen, even though these are known from experiment to be involved in the SI system.

THE GAMETOPHYTIC SELF-INCOMPATIBILITY SYSTEM

Physiological work on GSI systems has been concentrated mainly upon Onagraceae (*Oenothera*), Liliaceae (*Lilium*), Solanaceae (*Petunia, Nicotinia, Lycopersicum*), and Leguminosae (*Trifolium*). Much of the research on *Trifolium* has been concerned with temperature and other environmental effects on the SI response; it is of interest mainly from the point of view of controlling SI for the purposes of breeding, and so far it has yielded little of importance in relation to the biochemistry of incompatibility. Since the work in question has been fully reviewed recently by Townsend (114), it will not be treated further here.

The other GSI families listed above show a considerable variation in the phenomenology of the SI response, and they will be considered separately.

Onagraceae: the Genus Oenothera

As we have seen, the sites of inhibition in SSI and GSI systems tend to be different, with the block in SSI families at the stigma surface and in GSI, lower in the style. The genus *Oenothera* with an undoubted GSI system has a somewhat exceptional behavior. Emerson (25) reported that incompatible pollen of *Oenothera organensis* either failed to germinate on the stigma or ceased growth with no appreciable penetration. Hecht (31) found that incompatible pollen of this species produced tubes that showed some growth into the style at lower temperatures, but at the temperature where the reaction was most positive, 27°C, the inhibition was almost complete in the stigma. Moreover, Hecht's elegant grafting experiments (8, 30) proved that there was no more than a slight difference in the rate of growth of compatible and incompatible tubes in the style once the stigmatic barrier was passed. Fine-structural observations by Dickinson & Lawson (20) show that in *O. organensis* the outer layer of stigmatic cells and the secretion they carry is indeed the site of inhibition of incompatible tubes. The surface cells are in fact dead or at least moribund at the time when the filter operates; this suggests that the discrimination is effected by products synthesized before the stigma becomes receptive.

Lewis, who was responsible for the introduction of serological methods into research on incompatibility, published a highly informative pioneering paper in 1952 (69) showing that antigens were present in pollen extracts of *O. organensis* that were specific to the *S*-alleles represented. In later work (90), antisera were raised to pollen of uniform genotype from homozygous parents, and using this the specificity of the pollen-borne antigens for the *S*-genotype was further confirmed, with the additional observation that the *S*-specified antigens, presumed to be proteins, diffused freely from intact pollen. Using this fact, Lewis et al (73) were able to discriminate between individual pollen grains of different *S*-genotypes on agar films loaded with appropriate antiserum. Confirmation of the view that the *S*-gene specified products were proteins came from partial fractionation by gel electrophoresis. Although it has not been proved that the antigens studied by Lewis and collaborators are derived from intine sites, there seems every likelihood that this will prove to be the case. We have already noted that no *S*-allele related differences in the antigen complements of pollen have yet been demonstrated for SSI species.

Lewis and co-workers (73, 90) concluded from their work on the pollen antigens of *O. organensis* and the kinetics of the pollen-stigma interaction that there could be no induced response of the immune type lying behind the SI reaction. The fine-structural observations of Dickinson & Lawson (20) mentioned above support this conclusion. The control is exerted very rapidly, while the tubes are in a region of dead cells and their products. Evidently then, in the GSI system of *Oenothera*, the pistil partner must be preprogrammed to cope with incompatible pollen, just as in the SSI system of the Cruciferae and Compositae. The specificity of the response suggests that the active constituents are proteins, comparable with those postulated to play a corresponding role in the stigma pellicle in SSI plants. Kumar & Hecht (64) found that the incompatibility response in *Oenothera* could be eliminated by immersing the stigmas in water at 50°C for 5 min; this evidence of heat lability can be taken to support the proposition that the specificity resides in proteins (65).

Liliaceae: the Genus Lilium

In the SI species of *Lilium*, the inhibition of incompatible tubes occurs after they have progressed a considerable distance through the style. The style is hollow and lined with secretory cells forming a transmitting tissue; the tube progresses over and in the secretion products of the cells and not through a solid tract of tissue (105). There is no indication of a stigmatic control of the SI response, and tests of stigmatic exudate in vitro do not suggest that it contains any materials acting differentially on compatible and incompatible tubes (105).

The conclusion therefore must be that products of the cells lining the stylar canal must be responsible for the SI reaction. There have been several reports of treatments affecting the rejection response, which in *L. longiflorum* decays in any event some 5 days after anthesis (7). Heat treatment at 50°C before pollination is effective (44), and also X-radiation (45). It has been reported that the growth substance naphthalene acetamide removed the SI response in *L. longiflorum* when applied to the cut surface of a tepal before pollination (26). Because the transmitting tissue is readily accessible in the hollow style, the effects of exogenous chemical treatments can be followed in a direct manner, and this opportunity has been exploited especially by Ascher, who has explored the responses to various inhibitors and extracts. In agreement with results reported by Rosen (105) from tests in vitro, Ascher & Drewlow (6) found no indication of any effect on the SI reaction when stigmatic fluid from different genotypes was injected into the stylar canal. The growth of both compatible and incompatible tubes was promoted by this treatment irrespective of the source of the fluid, but the differences in growth rate attributable to the SI system were not masked in any genotype combination tested. That the dilution of the canal secretions by added fluid did not modify the response is itself significant, for it points again to a very intimate and local interaction between pollen tube and transmitting cells.

Ascher (4, 5) has also examined the effects of various metabolic inhibitors. Injection of 6-methylpurine, an inhibitor of RNA synthesis, before, at, or shortly after anthesis was found to permit incompatible tubes to grow at the same rate as compatible; that is to say, the inhibitor eliminated the discrimination attributable to the working of the SI system without impairing pollen tube growth. Ascher interpreted this result as meaning that RNA synthesis was needed for the production of a tube-inhibiting factor. However, puromycin, an inhibitor of protein synthesis, was found to be effective in modifying the SI response only when injected into the style up to midday on the day of anthesis, and not thereafter (5). This result is not readily reconciled with the idea that a postanthesis production of a messenger-RNA fraction is required to direct the synthesis of inhibiting proteins, and Ascher leaves open the possibility that the incompatibility factors produced by the style may be themselves RNA species. This has received some support from further experiments in which crude nucleic acid extracts from virgin styles of different genotypes were injected into the stylar canals at the time of pollination. According to Campbell & Ascher (15), pollen tube growth was reduced whenever the genotype providing the extract matched that of the tube, with no response when there was no match. It will be important now to follow the effects of purified fractions.

Solanaceae

The three genera of this family most used in work on the physiology of SI appear not to differ in any significant feature, and accordingly they will be treated together.

The solanaceous stigma bears a copious secretion fluid, and the pollen germinates in this (60). The style is solid, and the tubes grow in a matrix material between the widely separated and elongated stylar cells. Tracing pollen tube growth in styles of this kind has been greatly facilitated by the decolorized-aniline blue test for callose (2), introduced for pollen tube studies by Linskens & Esser (82). The tubes are made visible in stylar tissues by the fluorescence of an inner callosic wall layer, present in both compatible and incompatible tubes through much of their length.

The earliest electron microscopic studies of tube growth in the Solanaceae were by van der Pluijm & Linskens (117), who concluded that the matrix material of the style in *Petunia hybrida* is akin to a very thick middle lamella and principally composed of pectins. Later work has shown that apart from pectic substances, acidic low molecular weight carbohydrates are present in extracts (63). Incompatible tubes are inhibited at some point during their passage, and such tubes are characteristically thick-walled, with numerous vacuoles towards the tip. In this work on *Petunia hybrida*, $KMnO_4$ fixation was used, permitting little interpretation of cytological detail, but this has now been given for *Lycopersicum peruvianum* by de Nettancourt et al (101). In this species, incompatible tubes penetrate through the first third of the length of the style and then commonly burst at the tip. Compatible tubes have a wall of two layers, and the authors consider that the growth of the inner nonfibrillar and possibly callosic layer is brought about by the accretion of vesicles originating in the cytoplasm of the tip region of the tube. They conclude that the bursting of incompatible tubes results from an active destruction of the inner wall layer at the tip rather than from a simple growth inhibition, and they support the view that the tube wall is the site of action of the incompatibility factors. The bursting at the tip is accompanied by the release of numerous particles, also seen within the tubes, and the inhibited tubes are distinguishable through the presence in the cytoplasm of concentric whorls of ribosomal endoplasmic reticulum. De Nettancourt et al (101) develop the idea that the bursting of incompatible tubes in the style is akin to the bursting of compatible tubes at the time of discharge of the gametes into the embryo sac—an active process, that is, requiring a specific "discharge" signal.

Since 1953, Linskens and his collaborators have been responsible for a long series of contributions on the SI response in *Petunia hybrida* (81, 84). Serological work on the conducting tissue of the style and on pollen produced two important pieces of evidence: (*a*) that style and pollen each contained antigens that were apparently specific to the *S*-genotype, and (*b*) that these antigens were identical in pollen and style (79). This finding provides the only evidence at present for the identical character of *S*-allele products in pollen and pistil for any SI plant. The likelihood is that the antigen variation found by Linskens in *Petunia* pollen reflects unlike protein spectra. Differences in protein complements of the pollens of two solanaceous species, *Nicotiana alata* and *N. langsdorfii*, related to *S*-genotype were reported by Pandey (104), who found unlike banding patterns of peroxidase isozymes when extracts were subjected to polyacrylamide gel electrophoresis.

Linskens' earlier work had shown considerable differences in the metabolism and constitution of self- and cross-pollinated styles. With both self- and cross-pollination, respiratory rate increased initially, but whereas the rate fell again in selfed styles after 12 hr when tube growth would be halted, there was a continuing rise in cross-pollinated styles (75, 76). Furthermore, the protein content of the styles underwent characteristic changes (76, 86). Linskens recorded that after a compatible pollination a single new glycoprotein fraction made its appearance in the style, but after selfing, two such fractions with different electrophoretic mobilities appeared. These new glycoprotein complexes, referred to by Linskens as "ward bodies," seemed to incorporate contributions from both pollen and stylar partners, judged from the results of tracer experiments (77, 78). Linskens & Tupy, reporting on the effects of self- and cross-pollination on the amino acid pools of the style, recorded a fall in protein content in the first 24 hr after pollination, a fall that was more marked following selfing. Since the free amino acid pool did not rise correspondingly, they concluded that the amino acids released from protein breakdown were lost through entering a respiratory pathway (86).

In recent work by Linskens' group, changes in glycosidase activity have been noted following upon pollination (85). The activity of N-acetyl-β-glucosaminase, β-galactosidase, and α-mannosidase showed a two to fivefold increase in cross-pollinated styles, an increase not found in selfed styles. Evidence has also been given of enhanced synthesis of RNA in the style of *Petunia* after pollination, presumably attributable to activity both in pollen tubes and stylar tissue (116, but see also 29). The RNA produced in compatible pollinations is said to be messenger-like, and to promote the incorporation of labelled amino acids into acid precipitable material (implied protein) when injected into oocytes of *Xenopus laevis* (116). This work is of the greatest interest, and the publication of the full results could well contribute massively to our understanding of postpollination events in the *Petunia* style.

CONSPECTUS

SI systems have provided a fertile field for speculation, and many theoretical schemes have been offered to explain the observed responses (e.g. 3, 72, 80, 90). These will not be reviewed in any detail here, nor will any addition be made to them. Instead we will seek to draw some conclusions from the evidence as it stands and to point out certain lines along which progress may now be expected.

Reference has already been made to the high information content required in SI systems to explain the large numbers of discriminations that have to be made. We have also noted the special circumstances implied by the central phenomenon, the rejection of "self." There seems no alternative but to assume that the S-alleles in the two reacting partners are transcribed at some point in time to give information-carrying products which later come into interaction, either directly or through some more or less remote secondary agency still bearing the imprint of specificity. This interaction, the primary recognition event, marks the point from which the later acceptance or rejection responses are entrained. These consequent responses are obviously of interest in their own right, but some may be quite remote, metabolically speaking, and not necessarily relevant to the recognition system itself.

The primary recognitions in the SSI system and in the GSI system of *Oenothera organensis* must be quite rapid—completed within minutes of pollination. From the timing of the recognition reaction and the circumstances in which it occurs, we may suppose that the events are local and between pre-formed systems. There is no induction phase. What then of the GSI systems of *Lilium* and the various solanaceous genera? When the rejection reaction is of the nature of a certation phenomenon occurring throughout the length of the style and over a period even of hours, it obviously is not feasible to separate "recognition" and "response" phases. There is likely to be a continuing dialogue from the moment when the interacting cells make first contact and then throughout growth until the accumulated "response" brings growth to a stop in an incompatible tube. Further, it is not possible to conclude that the reactions are between pre-formed systems. The synthesis and release of S-gene products could be initiated at the time of pollination and continue throughout the growth phase. On the stylar side, the stimulus could arise from pollination. In orchids, the secretory activity of the pollen-tube transmitting tissue of the style and ovary is induced by the pollination stimulus (36), and the same could be true in some limited degree for the corresponding tissues in the style of *Lilium*.

To be set against this is the demonstration by Linskens (79) that in *Petunia* antigens specific to the *S*-alleles are present in identical form in both pollen and virgin stylar tissue. Taken at its face value, this must mean that the *S*-genes are transcribed in both as a normal part of prepollination differentiation. The fact would not, of course, preclude further transcription after pollination and the gradual build-up then to effective levels for the working of the SI system (33).

Lewis (71) has argued that the evidence from mutation in the GSI system indicates that one and the same cistron produces both pollen and stylar proteins and that these must be identical. The general proposition that self-rejection depends on interaction of identical proteins offers conceptual difficulties, and Lewis (72) has proposed a "dimer" hypothesis to explain how such a system might be worked at the molecular level, as follows:

(1) The *S* gene complex produces a polypeptide whose specificity, determined by the primary structure, is different for each allele. Each allelic polypeptide is an identical molecule in pollen and style. (2) The polypeptide polymerises into a dimer in both pollen and style. (3) The first step in the incompatible reaction is that the same dimers in pollen and style, and only the same, combine to form a tetramer with the aid of an allosteric molecule which may be glucose, a protein or one of many small molecules. (4) The second step in the incompatible reaction is that the tetramer acts as a genic regulator either to induce synthesis of an inhibitor or to repress the synthesis of an auxin of pollen tube growth.

As Lewis noted, this theory has implications that could be tested experimentally. At present, Linskens's evidence that in *Petunia* antigens specific to the *S*-alleles exist in identical form in both pollen and style is the total available bearing upon the hypothesis, and there is clearly a need now for the *Petunia* results to be checked in other GSI species.

It may be noted that the observations on the SSI system of *Brassica* present a severe problem, for here immunological methods certainly as sensitive as those used by Linskens (79) and Lewis (69) reveal *S*-allele related differences in the antigens of the stigma, but not among those of the pollen. There is a danger in arguing from a negative, but it must at least be accepted as a possibility that immunological methods do not detect the differences certainly present among the exine proteins of different *S*-genotypes of *Brassica* because the determinants are masked in some manner so that the differentials are not recognized by the rabbit immune system. If this be so, then the idea that the self-recognition event in the SSI system involves an interaction of *identical* pollen and stylar fractions requires revision.

A promising line of thought here is perhaps offered by the proposition that the *S*-allele specified moieties are parts of molecules which differ otherwise in pollen and pistil in the presence of organ-specific parts. Again, mutational evidence is suggestive (71, 100). Mutations destroying the SI system may affect pollen and pistil independently thus indicating some element of organ specificity (71); but at the same time, new S-gene specificities when they arise affect both pollen and pistil, demonstrating the common element of activity (99). On the basis of such evidence, Lewis has concluded that the S-locus in GSI systems is tripartite, with a stylar part, a pollen part, and the S part proper. Transcription of such a compound locus with allowance for a differential control in pollen and pistil of the readout of the organ-specific parts could produce different protein species in the two partners still carrying the S-imposed specificity.

Among theories based upon the idea of interaction between molecules with common and differential parts from pollen and pistil, that of Sampson (107) is most noteworthy in that it suggests a link between inter- and intraspecific incompatibility systems. Sampson suggested that "(a) the incompatible reactions . . . depend upon the combination at complementary sites of pollen and stigmatic molecules, and (b) that the degree to which the combining surfaces of these incompatibility substances are complementary increases with the closeness of genetic relationship." He referred to the two sites as "S-allele" area and "species" area. Two species will be compatible only if combination takes place at the species area. Within a species, incompatibility will be caused by a combination at the S-allele area resulting from S-allele identity; compatibility will depend upon the absence of S-allele area combination. As Sampson puts it, "Incompatibility results from combination at both areas or none."

Incompatibility phenomena in plants recently have been catching the attention of immunologists (14), and it is interesting that Macfarlane Burnet in a seminar given in the Australian National University in 1972 produced a hypothesis with distinct similarities to that of Sampson. This postulated a mechanism "based upon something equivalent to the V and C ("variable" and "constant") segments of the immunoglobulin molecule." The C segment is presumed to differ from species to species and so is equivalent to Sampson's "species area"; the V segment varies within the species, corresponding thus to the "S-allele area." Correct C-C recognition provides for compatibility, supposing only that no V-V association occurred. Absence of C-C contact results in incompatibility, as does V-V contact.

Theories like the foregoing readily account for differences in the immunological specificities of S-complex proteins in pistil and pollen, but they do not give any obvious explanation for a situation such as that in *Brassica,* where S-specified variation is detectable in the stigma but not in the pollen. We are driven back to the proposition considered earlier, namely that in the S-complex of the pollen the S-specified variation is invisible to the immune system.

Turning to the pollen-stigma interaction in the SSI system, we may note that there is now a perfectly good physiological explanation for one of the correlations noted by Brewbaker (12). The inhibition of incompatible tubes is on the surface of the stigma because this is as far as the pollen grain can carry the wall-held sporophytic message that identifies its parentage. Is the response to this message then to lift a barrier for a compatible grain, or to set up one for an incompatible grain? In various terms these alternatives have been featured in most discussions of the SI mechanism (70). Certainly an incompatible pollination in the Cruciferae does induce changes in contiguous papillae which, if not truly a setting up of a barrier to the unwanted tube, provide a good simulation of such. What then of the cutinase hypothesis of Christ (16)? The electron microscopic observations of Dickinson & Lewis (21) indicate that incompatible tubes of *Raphanus* ultimately do breach the cuticle of the stigma papilla, but the authors report that they make varying amounts of growth over the surface first, suggesting that there are access problems. These problems seem to be more marked still in *Brassica* (11). Suggestively, in certain Caryophyllaceae the entry of the tube in compatible pollinations is much delayed when the proteinaceous pellicle of the stigma is removed enzymically (37). This suggests that the pellicle carries a factor, possibly a peptide, which enhances the activity of a pollen-borne cutinase, probably located in the intine.

We may designate the control over stigma penetration as System I. In families like the Cruciferae and Compositae with cutinized, dry stigmas it could play a part in governing interspecific compatibility, excluding the pollen tube from the style when the mismatch of cutinase precursor and activator is too extreme. System II would then be that in which *S*-allele specificity is expressed in SSI families, the primary recognition taking place on the stigma surface, the reactants being the tapetum-derived proteins of the pollen exine and the corresponding *S*-allele related stigma proteins (42). The probable existence of two such independent control systems will need to be taken into account in future experimental work since their effects could well be confused. System I should yield readily enough to the techniques of enzyme chemistry. System II undoubtedly will prove more complex, but the analogy of the lymphocyte may be a useful one to pursue. For example, the possibility that the cyclic nucleotides may be involved in the secondary responses, including the callose rejection reaction, should now be examined.

Literature Cited

1. Arasu, N. N. 1968. *Genetica* 39:1–24
2. Arens, K. 1949. *Lilloa* 18:71–75
3. Ascher, P. D. 1966. *Euphytica* 15:179–83
4. Ascher, P. D. 1971. *Theor. Appl. Genet.* 41:75–78
5. Ascher, P. D. 1974. *Incompat. Newslett.* 4:57–60
6. Ascher, P. D., Drewlow, L. W. 1971. *Pollen: Development and Physiology,* ed. J. Heslop-Harrison, 267–72. London: Butterworths
7. Ascher, P. D., Peloquin, S. J. 1966. *Am. J. Bot.* 53:99–102
8. Bali, P. N., Hecht, A. 1965. *Genetica* 36:159–71
9. Bateman, A. J. 1952. *Heredity* 6:285–310
10. Ibid 1954. 8:305–32
11. Ibid 1955. 9:53–58
12. Brewbaker, J. L. 1957. *J. Hered.* 48:271–77
13. Brooks, J., Shaw, G. 1971. See Ref. 6, 99–114
14. Burnet, F. M. 1971. *Nature* 232:230–34
15. Campbell, R. J., Ascher, P. D. 1974. *Incompat. Newslett.* 4:33–37
16. Christ, B. 1959. *Z. Bot.* 47:88–112
17. Correns, C. 1913. *Biol. Zentralbl.* 33:389–423
18. Darwin, C. 1876. *The Effects of Cross and Self Fertilisation in the Vegetable Kingdom.* London: Murray
19. Dickinson, H. G., Heslop-Harrison, J. 1971. *Cytobios* 4:233–43
20. Dickinson, H. G., Lawson, J. 1975. *Proc. Roy. Soc. B* 188:327–44
21. Dickinson, H. G., Lewis, D. 1973. *Proc. Roy. Soc. B* 183:21–38
22. Ibid. 184:149–165
23. East, E. M. 1940. *Proc. Am. Phil. Soc.* 82:449–518
24. East, E. M., Mangelsdorf, A. J. 1925. *Proc. Nat. Acad. Sci. USA* 11:166–71
25. Emerson, S. 1940. *Bot. Gaz.* 101:890–911
26. Emsweller, S. L., Uhring, J. 1965. *Hereditas* 52:295–306
27. Erdtman, G. 1952. *Pollen Morphology*

and Plant Taxonomy. Waltham, Mass.: Chronica Botanica
28. Ernst, A. 1936. *Z. Indukt. Abst. Vererb.* 71:156–230
29. Godfrey, C. A., Linskens, H. F. 1968. *Planta* 80:185–90
30. Hecht, A. 1960. *Am. J. Bot.* 47:32–36
31. Hecht, A. 1964. *Pollen Physiology and Fertilisation,* ed. H. F. Linskens, 237–43. Amsterdam: North Holland
32. Heinen, W., Linskens, H. F. 1961. *Nature* 191:1416
33. Heslop-Harrison, J. 1968. *Nature* 218:90–91
34. Heslop-Harrison, J. 1968. *New Phytol.* 67:779–86
35. Heslop-Harrison, J. 1968. *Port. Acta Biol.* 10:235–40
36. Heslop-Harrison, J. 1972. *Plant Physiology: A Treatise,* ed. F. C. Steward, Chap. 9, 133–289. New York: Academic
37. Heslop-Harrison, J., Heslop-Harrison, Y. 1975. *Ann. Bot.* 39:163–65
38. Heslop-Harrison, J., Heslop-Harrison, Y., Barber, J. 1975. *Proc. Roy. Soc. B* 188:287–97
39. Heslop-Harrison, J., Heslop-Harrison, Y., Knox, R. B. 1973. *Incompat. Newslett.* 3:75–76
40. Heslop-Harrison, J., Heslop-Harrison, Y., Knox, R. B., Howlett, B. J. 1973. *Ann. Bot.* 37:403–12
41. Heslop-Harrison, J., Knox, R. B., Heslop-Harrison, Y. 1974. *Theor. Appl. Genet.* 44:133–37
42. Heslop-Harrison, J., Knox, R. B., Heslop-Harrison, Y. 1975. *Proc. Linn. Soc. Bot.* In press
43. Hoekstra, F. A. 1974. *Incompat. Newslett.* 3:52–54
44. Hopper, J. E., Ascher, P. D., Peloquin, S. J. 1967. *Euphytica* 16:215–20
45. Hopper, J. E., Peloquin, S. J. 1968. *Can. J. Genet. Cytol.* 10:941–44
46. Howlett, B. J., Knox, R. B., Heslop-Harrison, J. 1973. *J. Cell Sci.* 13:603–19
47. Howlett, B. J., Knox, R. B., Paxton, J. D., Heslop-Harrison, J. 1975. *Proc. Roy. Soc. B* 188:167–82
48. Hughes, M. B., Babcock, E. B. 1950. *Genetics* 35:570–88
49. Kanno, T., Hinata, K. 1969. *Plant Cell Physiol.* 10:213–16
50. Knoll, K. 1930. *Z. Bot.* 23:609–18
51. Knox, R. B. 1971. *J. Cell Sci.* 9:209–37
52. Ibid 1973. 12:421–44
53. Knox, R. B., Heslop-Harrison, J. 1970. *J. Cell Sci.* 6:1–27
54. Ibid 1971. 9:239–51

55. Knox, R. B., Heslop-Harrison, J. 1971. *Cytobios* 4:49–54
56. Knox, R. B., Heslop-Harrison, J., Heslop-Harrison, Y. 1975. *Proc. Linn. Soc. Bot.* In press
57. Knox, R. B., Heslop-Harrison, J., Reed, C. 1970. *Nature* 225:1066–68
58. Knox, R. B., Willing, R. R., Ashford, A. E. 1972. *Nature* 237:381–83
59. Knox, R. B., Willing, R. R., Prior, L. D. 1972. *Silvae Genet.* 21:65–69
60. Konar, R. N., Linskens, H. F. 1966. *Planta* 71:372–87
61. Kroh, M. 1964. See Ref. 31, 221–24
62. Kroh, M. 1967. *Züchter* 36:185–89
63. Kroh, M. 1973. *Biogenesis of Plant Cell Wall Polysaccharides,* ed. F. Loewus, 195–205. New York: Academic
64. Kumar, S., Hecht, A. 1965. *Naturwissenschaften* 52:398–99
65. Kwack, B. H. 1965. *Physiol. Plant.* 18:297–305
66. Labarca, C., Kroh, M., Chen, M., Loewus, F. 1969. *Plant Physiol.* suppl. 2
67. Labarca, C., Loewus, F. 1973. *Plant Physiol.* 52:87–93
68. Lewis, D. 1949. *Biol. Rev.* 24: 472–96
69. Lewis, D. 1952. *Proc. Roy. Soc. B* 140:127–35
70. Lewis, D. 1954. *Advan. Genet.* 6: 235–95
71. Lewis, D. 1960. *Proc. Roy. Soc. B* 151:468–77
72. Lewis, D. 1965. *Genet. Today* 3:657–63
73. Lewis, D., Burrage, S., Walls, D. 1966. *J. Exp. Bot.* 18:371–78
74. Lewis, D., Crowe, L. K. 1958. *Heredity* 12:233–56
75. Linskens, H. F. 1953. *Naturwissenschaften* 40:28–29
76. Linskens, H. F. 1955. *Z. Bot.* 43:1–44
77. Linskens, H. F. 1958. *Ber. Deut. Bot. Ges.* 71:3–10
78. Ibid 1959. 72:84–92
79. Linskens, H. F. 1960. *Z. Bot.* 48:126–35
80. Linskens, H. F. 1965. *Genet. Today* 3:629–35
81. Linskens, H. F. 1975. *Proc. Linn. Soc. Bot.* In press
82. Linskens, H. F., Esser, K. 1957. *Naturwissenschaften* 44:16
83. Linskens, H. F., Heinen, W. 1962. *Z. Bot.* 50:338–47
84. Linskens, H. F., Kroh, M. 1967. *Encycl. Plant Physiol.* 18:506–30
85. Linskens, H. F. et al 1969. *C. R. Acad. Sci. Paris B* 269:1855–57
86. Linskens, H. F., Tupy, J. 1968. *Züchter* 36:151–58

87. Lundquist, A. 1955. *Hereditas* 40: 278–94
88. Lundquist, A. 1975. *Proc. Roy. Soc. B* 188:235–45
89. Mäkinen, Y. L. A., Brewbaker, J. L. 1967. *Physiol. Plant.* 20:477–82
90. Mäkinen, Y. L. A., Lewis, D. 1962. *Genet. Res.* 3:352–63
91. Martin, F. W. 1969. *Am. J. Bot.* 56:1023–27
92. Martin, F. W. 1971. See Ref. 6, 262–66
93. Mather, K. 1943. *Biol. Rev.* 18:32–64
94. Mather, K., De Winter, D. 1941. *Ann. Bot.* 5:297–311
95. Mattsson, O., Knox, R. B., Heslop-Harrison, J., Heslop-Harrison, Y. 1974. *Nature* 247:298–300
96. Nasrallah, M. E., Barber, J., Wallace, D. H. 1969. *Heredity* 24:23–27
97. Nasrallah, M. E., Wallace, D. H. 1967. *Nature* 213:700–1
98. Nasrallah, M. E., Wallace, D. H. 1967. *Heredity* 22:519–27
99. Nettancourt, D. de 1969. *Theor. Appl. Genet.* 39:187–96
100. Nettancourt, D. de 1972. *Genet. Agr.* 26:163–216
101. Nettancourt, D. de et al 1973. *J. Cell Sci.* 12:403–19
102. Pandey, K. K. 1958. *Nature* 181: 1220–21

103. Pandey, K. K. 1960. *Evolution* 14:98–115
104. Pandey, K. K. 1967. *Nature* 213:669
105. Rosen, W. G. 1971. See Ref. 6, 239–54
106. Sampson, D. R. 1960. *Am. Natur.* 94:283–292
107. Sampson, D. R. 1962. *Can. J. Genet. Cytol.* 4:38–49
108. Schlösser, K. 1961. *Z. Bot.* 49:266–88
109. Sedgley, M. 1973. *Incompat. Newslett.* 3:39–40
110. Skvarla, J. 1970. *Am. J. Bot.* 57:519–29
111. Stanley, R. G., Linskens, H. F. 1965. *Physiol. Plant.* 18:47–53
112. Stout, A. B. 1931. *Am. J. Bot.* 18: 686–95
113. Thompson, K. F. 1972. *Heredity* 28: 1–7
114. Townsend, C. E. 1971. See Ref. 6, 281–309
115. Tupy, J. 1959. *Biol. Plant. (Praha)* 1:192–98
116. Van der Donk, J. A. W. M. 1974. *Incompat. Newslett.* 4:30–32
117. Van der Pluijm, J. E., Linskens, H. F. 1966. *Züchter* 36:220–24
118. Wallace, D. H., Nasrallah, M. E. 1968. *Cornell Univ. Agr. Exp. Sta. Mem.* 406:1–23
119. Whitehouse, H. K. L. 1950. *Ann. Bot.* 14:199–216

Ann. Rev. Plant Physiol. 1975. 26: 427–39

GAS VESICLES   ❖7597

A. E. Walsby

Marine Science Laboratories, University College of North Wales, Menai Bridge, Anglesey, Wales

CONTENTS

INTRODUCTION

The gas vesicle is probably the only stable gas-filled structure found inside the protoplasm of living cells. A macrocrystalline organelle of beautiful simplicity, it is potentially of interest in the study of a number of physiological phenomena. However, those who study the physiology of higher plants are possibly unaware of its existence as it occurs exclusively in the lowest forms of plant life, the prokaryotic blue-green algae and bacteria. This short review provides an introduction to the gas vesicle, first presenting the better-established information on its structure and properties, the evidence for which I have recently reviewed in detail elsewhere (46), then discussing current investigations on gas vesicle ultrastructure, and finally considering those developments from studies on gas vesicles which may be of particular interest to physiologists.

Gas Vesicle Structure

Gas vesicles are the minute, hollow structures which make up the gas vacuoles of certain prokaryotic organisms (2). They have been found in most truly planktonic

427

blue-green algae (e.g. 15, 19, 33, 51) and in diverse bacterial groups (e.g. 4, 35, 39, 49, 55). With few exceptions (22) they are restricted to aquatic microorganisms (4).

Gas vesicles simply comprise a thin, continuous wall or membrane which entirely encloses a hollow space (18). Their usual shape is a cylinder closed by conical ends (2, 15; see Figure 1a). In all blue-green algae the cylinder is of constant width (e.g. 2, 15, 16, 33), about 70 nm, while its length varies, usually with a mean value of 300 nm (16, 44) and a typical maximum of 1 μm, though rarely to 2μm. The end caps are right cones with an end angle of 70° to 96° (16), depending on the method of measurement, and an altitude of about 50 nm (46). Gas vesicles of identical form are found in some bacteria (28, 55), though in most they are somewhat shorter and perhaps wider than the algal vesicles (46). The gas vesicles of halobacteria are much wider (23, 36, 44), up to 300 nm, and usually are without the central cylinder.

In all cases gas vesicles show the same distinctive ultrastructure, establishing their homology in the different groups. The wall, only about 2 nm in thickness, is composed of ribs running with a periodicity of 4.5 nm (15, 46) at right angles to the main axes of the cylinder and cones, probably in the form of a continuous helix (16), though the alternative model of a series of stacked hoops is not definitely ruled out (46). At the central region a more prominent rib is usually seen (40; see Figure 1b) which has been interpreted as the place where the two identical halves of the structure are joined back-to-back (46). The gas vesicle is formed by a continuous growth process over a number of hours (25, 40). Small double-coned structures are first produced. They enlarge to the diagnostic width and then continue to extend as cylinders.

CHEMISTRY   The gas vesicle wall, though often referred to as a membrane, has no lipid (or carbohydrate) component (14, 21, 50). Analysis of gas vesicles isolated from two species of blue-green algae shows them to be made up entirely of protein of rather similar composition (6, 13). Fifteen amino acids are present with alanine, valine, glutamic acid, isoleucine, leucine, and serine together accounting for about 70% of the total. The absence of cysteine precludes disulphide linkages. The gas vesicles of *Halobacteria* are also proteinaceous (21, 36) but show differences in amino acid composition, presumably connected with the need for protein stability in a highly saline environment (3).

Circular dichroism studies (14) and X-ray diffraction analyses (Blaurock et al, unpublished; see 46) indicate the importance of β-pleated structure in the protein.

Figure 1 (*opposite*) (*a*) Freeze-fracture through a cell of the heterotrophic bacterium *Prosthecomicrobium pneumaticum* showing the rows of gas vesicles (Electron micrograph X 60,000). (*b*) One of the gas vesicles from *a*, enlarged, showing the 5 nm ribs. Note the more prominent central rib (X 180,000). (*c*) Freeze-fracture through a cell of *P. pneumaticum* which has been exposed to a pressure of 1 MPa showing the collapsed, flattened vesicles, the ribs still visible (X 60,000). Micrographs *a-c* by D. Branton and A. E. Walsby. (*d*) Phase contrast light micrograph of the blue-green alga *Oscillatoria agardhii* showing gas vacuoles, aggregations of gas vesicles, which appear as bright areas (X 1100). (*e*) The same algal filament as in *d* (photographed by method of Walsby, ref. 49) after exposure to pressure which has collapsed the gas vesicles. Note decrease in filament width (X 1100).

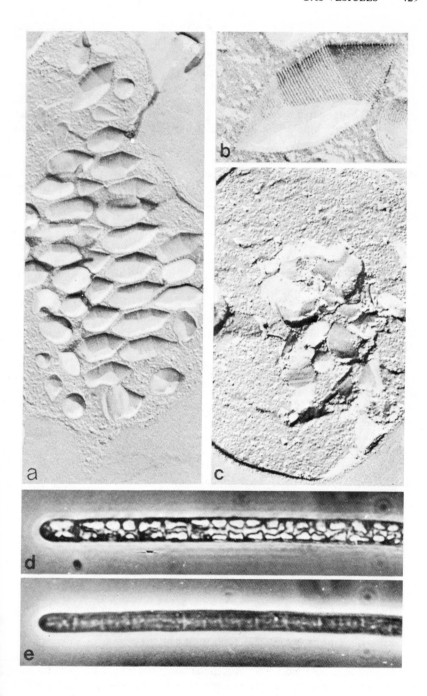

In the algal vesicles only one species of protein may be present, but the evidence for this crucial feature requires careful consideration (see below).

The density of collapsed gas vesicles, determined by density gradient centrifugation, is about 1.3 g cm^{-3} (13, 34), typical of protein. That of the intact structure has been calculated to be about 0.13 g cm^{-3} in the blue-green algae, and 0.08 g cm^{-3} in the halobacteria (46), the differences being due to differences in shape. Independent estimates (17, 46) put the weight of a single vesicle at about 0.5 x 10^{-16} g, which corresponds nicely with the volume of membrane protein calculated from electron microscope measurements.

PHYSICAL PROPERTIES   The hollow space inside a gas vesicle is freely available to gases by diffusion (42). The gas vesicle membrane has been shown to be permeable to O_2, N_2, Ar (42), CO_2, CO, H_2, and CH_4 (44). The vacuole gas is therefore in constant, dynamic equilibrium with the gas dissolved in the surrounding medium or cytoplasm. Usually it will be air modified by the metabolic processes of the cell. It is emphasized, however, that while gases normally will be present in the structure, they should not be regarded as being contained by it.

Almost certainly, the gas vesicle membrane is permeable to water vapor, as it is to other gases. Liquid water probably is prevented from condensing or creeping inside the structure by the hydrophobic nature of its inner surface (44), though this property has not been established by direct investigation as yet. The outer surface of the gas vesicle, by contrast, is hydrophilic (44). This presumably minimizes interfacial tension which would tend to collapse the structure.

A gas vesicle is a rigid but brittle structure. It collapses suddenly and irreversibly when a certain differential pressure known as the "critical pressure" is established across its wall (44). However, it will withstand prolonged and repeated applications of subcritical pressures indefinitely without significant elastic compression (44). This critical pressure varies considerably even for gas vesicles within the same cell, ranging from 40 to 180 kPa (1 atm equals $\sim$ 100 kPa) in halobacteria, 200 to 700 kPa in certain blue-green algae (44), and up to 1100 kPa in others (31, 52).

Chemical and physical conditions which affect proteins in general influence the strength of gas vesicles (3, 12). Their critical pressure falls (in extreme cases to zero) when they are exposed to temperatures above 40°C, pH's above 10 and below 5, the action of proteolytic enzymes, and agents which compete for hydrogen bonds and destroy hydrophobic bonds (3). The vesicle flattens on collapse (2; see Figure 1c). The central 70 nm diameter cylinder deforms to a 110 nm (= ½ πd)-wide envelope (16, 51) from which the flattened conical ends tear away (16, 44). The flattened cylinder may also split into equal lengths (50), perhaps separating along the prominent central rib. The gas which was present escapes by diffusion into the surrounding medium (44).

If special precautions are taken to avoid generating excessive pressure (50), gas vesicles can be isolated in an intact state from bacteria and blue-green algae without distortion of shape or size (36, 51). Cells must be lysed by osmotic shrinkage (50) or using conditions which weaken the cell wall (13, 23, 36). The released vesicles can be separated from most other subcellular components by accelerated flotation in a centrifuge [again taking care to avoid excessive pressures being generated by the accelerated water columns (50)]. In combination with techniques of membrane filtration or molecular sieving, preparations better than 97% pure have been obtained (3, 13). Suspensions of gas vesicles are extraordinarily turbid. Very dilute suspensions held in a light beam appear bluish normal to the beam and reddish by the transmitted light; spectrophotometric measurements indicate an approximation to Rayleigh scattering where the intensity of scattering is proportional to λ^{-4} (43). Nearly all of the scattering comes from the hollow (gas-filled) spaces and disappears instantaneously on collapsing the structures. The abrupt clearing of concentrated gas vesicle suspensions from a milky-white to a watery transparency is quite dramatic. Measurements of this pressure-sensitive scattering have been used in making quantitive measurements of gas vesicles (44, 47).

GAS VACUOLES

Gas vacuoles are simply aggregations of gas vesicles. They are visible under the light microscope as areas showing different refractility to the surrounding cytoplasm (see Figure 1d), usually of irregular outline and, under high magnifications, tinged pink in color (7), perhaps due to differential scattering. In the blue-green algae (2, 19) and in *Methanosarcina* (54), most of the vesicles occur in these aggregates and are packed in close hexagonal stacking like the cells of a honeycomb. Their alignment may be a consequence of their high axial ratio (46), as it is less marked in bacteria which produce shorter vesicles. In certain blue-green algae there may be 10,000 vesicles in a single cell (24), arranged in a small number of vacuoles. The individual vesicles collapse independently of one another as the pressure is raised (44), but if a large pressure is suddenly applied, they all collapse together and the vacuoles instantaneously disappear (see Figure 1e). This phenomenon serves to distinguish gas vacuoles unambiguously from other refractile granules in cells (20, 49). While gas vacuoles may be surrounded by photosynthetic lamellae (33), they are not enclosed by any limiting membrane, and on the cell lysing or swelling into a spheroplast, the vesicles immediately move free of one another, showing active Brownian movement. For the most part then, the properties of the gas vacuole are those of the constituent vesicles.

For certain purposes, however, it is necessary to consider the properties of gas vacuoles in toto. The efficacy of gas vacuoles in providing buoyancy depends not only on the density of the constituent vesicles, but on how closely they pack together. The interstitial space between the cylindrical vesicles of blue-green algae occupies $1 - \pi/6\sqrt{1/3}$ or about 9.3% of the space in the vacuole. If this is filled with cytoplasm of density, say 1.05 g cm^{-3}, the overall density of the vacuole is about 0.22, compared with the 0.13 g cm^{-3} of the vesicle. A density of this order for gas vacuoles is also indicated by interference light microscope measurements (8).

The interactions of gas vacuoles with light are considerably more complex than those of isolated gas vesicles. Spectrophotometric measurements indicate that the simple Rayleigh-scattering relationship no longer holds (32) and interference, reflection, and refraction have been implicated (see 46). In some algae, such as *Trichodesmium* (38) and *Nostoc* (41), the gas vacuoles occur in a distinctive layer at the cell periphery, which may have light-shielding significance. In other organisms, e.g.

Oscillatoria redekei (53) and certain bacteria (22, 49), they occur only at the ends of the cell, though it is not clear why.

THE PROTEIN SUBUNIT

The value of the gas vesicle as a model for the study of certain biologically important phenomena lies in its great simplicity, putatively a paracrystalline structure built from a single protein macromolecule into a simple geometrical form. This concept of the gas vesicle, suggested principally by its resemblance to certain viral structures (34), even before its chemistry had been properly determined, now requires rigorous examination. The existence of the single protein subunit cannot yet be regarded as definitely proved, but it is strongly suggested by the following, independent lines of investigation.

ELECTROPHORESIS   The protein of algal vesicles dissolved in acid and phenolic systems produces a single mobile band on gel electrophoresis, though a substantial residue is left at the origin (6, 14, 46). This residue furnishes more of the same mobile band on further extraction (14), though the possibility that it also contains other nonextractable proteins has not been completely eliminated, and further quantitative investigation is still required. The mobile protein has the same travel constant as known proteins of 14,000 daltons (14), though in the system used, mobility may vary with both charge density and molecular weight. Attempts to solubilize the algal vesicle protein in sodium dodecylsulphate, in which the charge density is uniform, have not been successful (6, 14).

N—TERMINAL ANALYSIS   Dansylation of the *Microcystis* gas vesicle protein produced only one labeled amino acid derivative, dansyl-alanine (14). This is consistent with there being but a single protein, while not eliminating the possibility of two proteins with the same N-terminal amino acid. Quantitative estimation of the N-terminal derivative could be used to determine the molecular weight of the parent protein.

AMINO ACID ANALYSIS   Quantitative analysis of amino acids in the gas vesicle protein from *Anabaena* shows that most of the amino acids occur in integer ratios which suggest an empirical formula with a molecular weight of 15,000 daltons (6). Again, these results theoretically do not rule out two molecules of similar composition,or whose molecular weight sum to 15,000, or the presence of a small proportion of a second protein.

In summary, the results obtained with *Microcystis* and *Anabaena* gas vesicles support the single protein model, though more precise data are clearly needed. We should be cautious also in extrapolating these results to other vesicles of other organisms; preliminary results (P. Falkenberg, personal communication) suggest that there are two proteins in the *Halobacterium* vesicles, though these are, of course, also atypical in other respects.

Arrangement of the Protein Subunits

If there is a single protein subunit, then it should be possible to define the physical properties of the gas vesicle in terms of the chemistry of the protein macromolecules and their mutual interactions. There is a paucity of published information in these areas at present, and one looks forward to the results of sequencing studies being undertaken by M. Jost and X-ray diffraction studies by A. E. Blaurock.

A single protein molecule will easily span the 2 nm width of the membrane (13, 17). Presumably, amino acids with hydrophobic side chains will predominate at the inner surface, and those with hydrophilic side chains at the outer surface (46); electron density profiles of collapsed membrane stacks support this arrangement (Blaurock et al; see 46). Gas molecules presumably permeate the structure via gaps between adjacent proteins in the membrane. The gap sizes might be investigated with gases of larger molecular size using methods previously described (42, 44). The rigidity of the membrane will reside in the mutual interactions of the individual molecules. Electron microscopy seems to indicate particles arranged along the membrane ribs (15, 16, 51) like beads on a string, though the repeating interval is not clearly defined either by direct observation (16) or by optical averaging techniques (see 46). The observation (16) that collapsed vesicle sheets often fold at an angle of 60° to the ribs may indicate that this is the angular displacement of particles from one rib to the next; alternatively, it may represent the orientation of primary peptide chains, perhaps that of the cross-β structure. Attempts have been made to relate the various periodic features seen along the ribs by electron microscopic and X-ray diffraction techniques to the component protein molecules (13, 46), but these must be regarded as speculation until we have definite information on the molecular weight of the protein.

FORMATION OF THE GAS VESICLE

At present we know very little about gas vesicle formation though it seems an exciting field to study. First, the process is a slow and continuous one (40) in which it is possible to follow initiation and growth from changes in the numbers and length distribution of vesicles (25). Second, gas vesicle formation has been found to proceed for a short time in cell-free systems (26), suggesting that defined in vitro systems can be developed to study all the factors required for assembly. Third, it should be possible to resolve the separate steps of vesicle protein synthesis [inhibited by antibiotics (24)] and assembly [prevented by pressure (unpublished data)]. The suggestion that gas vesicles grow from a central point (40) in two directions is a good one because it does not require the reorganization of previously made sections; presumably it could be tested by pulse-labeling experiments (40). The hollow space must be a product of the subunit assembly process (42).

Paradoxically, we know more about the regulation of gas vesicle collapse than formation at present, and both processes contribute to the control of cell gas-vacuolation. While it is impossible for collapsed vesicles to be re-erected by inflation (42), the possibility that particles from collapsed membranes are used in assembling new ones is untested. Gas vesicles are constitutively present in many organisms, though they are apparently inducible systems in others (39–41); the basis of control has not yet been studied, however.

INTERRELATIONS OF SHAPE, SIZE, STRESS, AND FUNCTION

Presumably, within the limits of material available, evolution tends to produce a biological structure with the shape best suited to the fulfillment of its function. The gas vesicle, with its simple, regular shape, provides a good model for discussing this principle. The gas vesicle provides buoyancy (see below) by forming a hollow structure which withstands certain pressures. [The necessity for a rigid rather than a flexible structure is discussed in detail elsewhere (45).] The shape giving the lowest density is a hollow sphere with the maximum ratio of diameter to wall thickness. However, the critical collapse pressure of such a structure varies inversely as this ratio. Consequently, for a given wall thickness, an upper size limit is imposed by the pressures encountered. The sphere is also the best shape for withstanding pressures, but only if its wall is of homogeneous construction, with its components having the same contact relationships in all directions. This requirement probably cannot be satisfied by protein molecules with their inherent asymmetry. The asymmetry of the gas vesicle protein is manifest in the ribbed appearance of the membrane. It has been argued (44) that the strength of the gas vesicle resides in its individual ribs, there being little lateral support between one rib and the next. This being so, the most efficient shape now becomes a cylinder of stacked ribs whose maximum diameter is limited by the same pressure considerations.

This theory still requires experimental verification but does receive support from the observation that the narrow vesicles of blue-green algae are considerably stronger than the much wider ones of halobacteria; intermediate cases of size and strength are observed in other bacteria. In each case (44), the strength of the gas vesicles is correlated with the cell turgor pressure to which they are exposed (see below). Perhaps the diameter determines the maximum critical collapse pressure of a gas vesicle which, as mentioned above, shows wide variation. This variation is itself an important component of a response which controls the level of gas-vacuolation in cells (see below).

Being built on a molecular scale, the gas vesicle is very small compared with the cell and large numbers are needed to provide buoyancy. Here again, the cylindrical vesicle is more efficient than a spherical one in that it can be stacked with less intervening space and gives a gas vacuole of lower density.

TURGOR PRESSURE MEASUREMENT

Gas vesicles can be used for measuring the turgor pressure of cells which possess them. The critical-pressure distribution of gas vesicles is usually found to be significantly lower in turgid cells, suspended in their culture medium or in water, than in otherwise identical cells suspended in a hypertonic sucrose solution. The difference between the two critical pressure measurements is equal to the turgor pressure of the cells (44).

In practice, the measurements of gas vesicle collapse with pressure are made by light-scattering methods (44, 47), the two sets of pressure-collapse curves are plotted (with and without sucrose) and the turgor pressure calculated from the separation between them (see Figure 2). Some uncertainties arise when the two curves are not parallel; when they are closer together at the top the turgor pressure probably varies in different cells (44), and when they are closer at the bottom there is possibly loss of turgor as the vesicles collapse (47).

Following the manometric experiments of Green & Stanton (10), the gas vesicle method provides the only other direct confirmation of the classical equation between turgor pressure and the difference between internal and external osmotic pressure of cells (44). It has also provided the first reliable estimates of turgor pressure in prokaryotic organisms, where the plasmolytic method is rendered ineffective by the reluctance of the plasmalemma and cell wall to separate. The turgor pressure of blue-green algal cells usually lies within the range of 200 to 500 kPa (about 2 to 5 atm), while that of various bacteria is usually somewhat lower (5, 44, 52). The finding that the extremely halophilic *Halobacteria* have no cell turgor pressure (44) is in keeping with the remarkable discovery that they contain no free water (9).

The gas vesicle method has been used to follow changes in cell turgor pressure implicated in an important behavioral response of planktonic blue-green algae, in which they may lose their buoyancy when exposed to an increased light intensity (42). The stages of the response appear to be as follows: (*a*) an increase in photosynthetic rate with increased light intensity, sustaining (*b*) higher levels of low molecular-weight photosynthetic products, generating (*c*) increased turgor pressure,

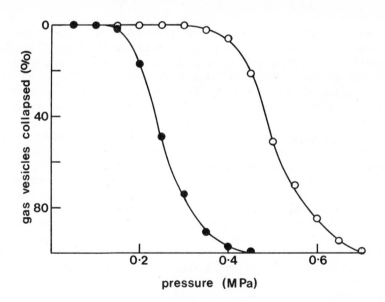

Figure 2   Collapse of gas vesicles with pressure in the blue-green alga *Anabaena flos-aquae*. ● Alga suspended in culture medium; ○ alga suspended in hypertonic sucrose solution (0.4 *M*). The turgor pressure of the cells in the culture medium is equal to the mean separation of the two curves (see 44).

causing (*d*) collapse of the weakest gas vesicles, and thereby (*e*) a decrease in buoyancy (5, 44). The significance of this buoyancy-regulating response to light intensity is discussed below.

FUNCTIONS

There is no evidence of metabolic activity associated with the gas vesicle protein, and any function that it performs therefore must be executed by the structure it makes. Presumably the functions are performed by the intact, hollow structure rather than the collapsed, flattened vesicle, and in studying proposed functions, experimental control systems are conveniently provided by the simple application of pressure. Alternatively, gas-vacuoleless mutants, obtained from a number of organisms (21, 23, 48), can be employed.

Gas vacuoles cannot store gas on account of their permeability (42), but it is suggested that they might provide channels for its diffusion to and from metabolic sites (37). It seems doubtful, however, that over the dimensions of the prokaryotic cell, diffusion rates through cytoplasm become limiting. Another dubious suggestion is that gas vesicles, by increasing the ratio of cell surface to cytoplasm volume (11), might stimulate increased rates of nutrient uptake, something that might be achieved with more economy by having smaller cells (46). Neither of these suggestions has been tested experimentally.

LIGHT SHIELDING   The light-scattering properties of intact gas vesicles have prompted discussion of their possible light-shielding role. Using very dilute suspensions of *Anabaena floss-aquae,* it has been found that the rate of photosynthesis increases only marginally (by 4%) when the gas vacuoles are collapsed, indicating only a small degree of light shielding within the same cell (32). Parallel experiments investigating the minimum lethal dose of ultraviolet radiation show that gas vacuoles offer no useful protection from these damaging wavelengths (32). The effects of the presence of gas vacuoles on photosynthesis in *Microcystis aeruginosa* are very similar at low cell concentrations (M. Jost, personal communication), but in very dense suspensions the gas vacuoles do reduce the useful light penetrating a culture by up to 29% (29), so that in a sense there is sacrificial shielding of cells deep in the culture (see 46). This may be of significance in surface blooms, but whether it determines the survival of the alga remains to be demonstrated. In a third alga, *Nostoc muscorum,* spectrophotometric data suggest that the gas vacuoles, which are oriented at the cell periphery in this species, do provide significant shading of the photosynthetic pigments in individual cells (41), though for a quantitative estimate of this shading effect, measurements by the direct methods described above (32) are required.

BUOYANCY   That gas vacuoles can and do provide buoyancy is not disputed. They have a lower density than any known solid or liquid, and they are accumulated by cells of various aquatic microorganisms (e.g. 2, 20, 28, 48) in sufficient proportions to provide positive buoyancy. This has been demonstrated by the classical experiment in which the buoyancy is lost upon destroying the vacuoles by pressure (20).

The buoyancy conferred by gas vacuoles does not, of course, have any direct effect on cell metabolism, and to appreciate the ensuing advantages the physiological requirements of the organism must be related to its ecological distribution. The majority of gas-vacuolate microorganisms are aquatic and inhabit nonturbulent waters (lakes and ponds rather than rivers or seas). In these waters the physical and chemical conditions (such as light, temperature, and concentrations of oxygen and nutrients) show great variation with depth. Gas vacuoles, which mostly occur in nonmotile species (4, 27), potentially provide an organism with the means of selecting a position in the vertical column where these conditions best suit its growth requirements.

The simplest case is when optimum conditions prevail at the water surface, where gas vacuoles enable the organism to float. For example, aerobic heterotrophs such as *Prosthecomicrobium* (35) may derive advantages from the high oxygen tensions at the meniscus. Evidence for this has been provided by the observation that a gas-vacuoleless mutant of *P. pneumaticum,* which sinks in static liquid culture, grows more slowly than the vacuolate wild type which floats. In shaken culture, both forms grow at the same rate. Experiments conducted in special culture vessels closed by gas-permeable membranes at top and bottom indicate that the differences observed in static culture are due to the oxygen diffusion gradient (48). The aerobic halobacteria may enjoy the same benefits at the surface of brine pools in which the solubility of oxygen is low.

The case of organisms which stratify at a certain depth below the water surface is more complex; they must continually regulate their gas vacuole content to prevent their sinking down or floating up from the preferred level. Investigations on a population of *Oscillatoria agardhii* forming a discrete layer at a depth of 5 meters in Deming Lake, Minnesota (52), suggest that the alga maintains its position by using the buoyancy-regulation response to light intensity described above and thereby avoids damaging light intensities near the surface and limiting intensities below the photic zone. By parallel responses, gas-vacuolate photosynthetic bacteria may come to occupy the upper layers of the hypolimnion where their special requirements for low light intensity and anaerobic conditions (27) are met. This region is often also rich in diverse other gas-vacuolate bacteria (39, 49) whose metabolic requirements have, for the most part, yet to be defined.

While planktonic blue-green algae may often form metalimnetic populations in stratified lakes (see 1, 52), they are more obvious, and hence better known, in forming surface waterblooms. The advantages of buoyancy are less obvious here, with the high light intensity and local depletion of nutrients perhaps leading to death of the alga. However, the studies of Reynolds (30, 31) suggest that the surface aggregation results from the alga's slow buoyancy-regulating response being unable to accomodate rapid changes in turbulence of the water mass. In turbulent periods, the alga is mixed through a large depth and, experiencing a low mean light intensity, becomes overbuoyant. In ensuing periods of calm it floats rapidly to the surface where light inhibition of its photosynthesis may cause it to become trapped. The alga's rapid movement to the surface is a consequence of its large colony size, which, under normal conditions of turbulence in relatively shallow, unstable water, may provide hydrodynamic advantages.

CONCLUSIONS

In a sense, the terms "gas vesicle" and "gas vacuole" are misnomers since the properties, functions, and formation of the structures they describe depend on the hollow spaces they enclose rather than the gas which happens to diffuse into them (42). Winogradsky (54), who in 1888 was the first to demonstrate the structures, showed singular prescience in describing them as *"Hohlungen"* or *"Hohlräumen"* (hollow cavities or spaces), a terminology that unfortunately was lost in ensuing discussions on the nature of the vacuole gas (20). In the recent era, electron microscopy has uncovered the structure of gas vesicles (2, 16) which has facilitated the interpretation of their physical properties (42, 44). Current studies on the gas vesicle cross the borders of physics, biochemistry, molecular biology, physiology, and ecology, but fortunately this complexity is offset by the simplicity of its structure, permitting a broad view of its biological implications.

This simplicity at the same time commends the study of gas vesicles to biologists not specifically interested in prokaryotic organisms. The structure may provide new methods for studying cell turgor pressure and a model system for investigating such unrelated topics as the mutual interactions of proteins in membranes, the evolution of shape in response to stress, organelle formation at the molecular level, and the chain reactions of simple events interpreted as behavioral responses.

Literature Cited

1. Baker, A. L., Brook, A. J. 1969. *Arch. Hydrobiol.* 69:214–33
2. Bowen, C. C., Jensen, T. E. 1965. *Science* 147:1460–62
3. Buckland, B., Walsby, A. E. 1971. *Arch. Mikrobiol.* 79:327–37
4. Cohen-Bazire, G., Kunisawa, R., Pfennig, N. 1969. *J. Bacteriol.* 100:1049–61
5. Dinsdale, M. T., Walsby, A. E. 1972. *J. Exp. Bot.* 23:561–70
6. Falkenberg, P., Buckland, B., Walsby, A. E. 1972. *Arch. Mikrobiol.* 85:304–9
7. Fogg, G. E. 1941. *Biol. Rev. Cambridge Phil. Soc.* 16:205–17
8. Fuhs, G. W. 1969. *Österr. Bot. Z.* 116:411–22
9. Ginzburg, M., Ginzburg, B. 1973. *Proc. 1st Int. Congr. Bacteriol. Symp.* 1:123
10. Green, P. B., Stanton, F. W. 1967. *Science* 155:1675–76
11. Houwink, A. L. 1956. *J. Gen. Microbiol.* 15:146–50
12. Jones, D. D., Haug, A., Jost, M., Graber, D. R. 1969. *Arch. Biochem. Biophys.* 135:296–303
13. Jones, D. D., Jost, M. 1970. *Arch. Mikrobiol.* 70:43–64
14. Jones, D. D., Jost, M. 1971. *Planta* 100:277–87
15. Jost, M. 1965. *Arch. Mikrobiol.* 50:211–45
16. Jost, M., Jones, D. D. 1970. *Can. J. Microbiol.* 16:159–64
17. Jost, M., Jones, D. D., Weathers, P. J. 1971. *Protoplasma* 73:329–35
18. Jost, M., Matile, P. 1966. *Arch. Mikrobiol.* 53:50–58
19. Jost, M., Zehnder, A. 1966. *Schweiz. Z. Hydrol.* 28:1–3
20. Klebahn, H. 1922. *Jahrb. Wiss. Bot.* 61:535–89
21. Krantz, M. J., Ballou, C. 1973. *J. Bacteriol.* 114:1058–67
22. Krasil'nikov, N. A., Duda, V. I., Pivovarov, G. E. 1971. *Microbiology* 40:592–97
23. Larsen, H., Omang, S., Steensland, H. 1967. *Arch. Mikrobiol.* 59:197–203
24. Lehmann, H. 1971. *Ber. Deut. Bot. Ges.* 84:651–27
25. Lehmann, H., Jost, M. 1971. *Arch. Mikrobiol.* 79:59–68
26. Ibid 1972. 81:100–102
27. Pfennig, N. 1967. *Ann. Rev. Microbiol.* 21:285–324
28. Pfennig, N., Cohen-Bazire, G. 1967. *Arch. Mikrobiol.* 59:226–36
29. Porter, J., Jost, M. 1973. *J. Gen. Microbiol.* 75:xxii
30. Reynolds, C. S. 1967. *Shropshire Conserv. Trust Bull.* 10:9–14

31. Reynolds, C. S. 1973. *Proc. Roy. Soc. London B* 184:29–50
32. Shear, H., Walsby, A. E. 1975. *Brit. Phycol. J.* In press
33. Smith, R. V., Peat, A. 1967. *Arch. Microbiol.* 57:111–22
34. Smith, R. V., Peat, A., Bailey, C. J. 1969. *Arch. Mikrobiol.* 65:87–97
35. Staley, J. T. 1968. *J. Bacteriol.* 95:1921–42
36. Stoeckenius, W., Kunau, W. H. 1968. *J. Cell Biol.* 38:337–57
37. Taylor, B. F., Lee, C. C., Bunt, J. S. 1973. *Arch. Mikrobiol.* 88:205–12
38. Van Baalen, C., Brown, R. M. 1969. *Arch. Mikrobiol.* 69:79–91
39. Van Ert, M., Staley, J. T. 1971. *J. Bacteriol.* 108:236–40
40. Waaland, J. R., Branton, D. 1969 *Science* 163:1339–41
41. Waaland, J. R., Waaland, S. D., Branton, D. 1971. *J. Cell Biol.* 48:212–15
42. Walsby, A. E. 1969. *Proc. Roy. Soc. London B* 173:235–55
43. Walsby, A. E. 1970. *Water Treat. Exam.* 19:359–73
44. Walsby, A. E. 1971. *Proc. Roy. Soc. London B* 178:301–26
45. Walsby, A. E. 1972. *Symp. Soc. Exp. Biol.* 26:233–50
46. Walsby, A. E. 1972. *Bacteriol. Rev.* 36:1–32
47. Walsby, A. E. 1973. *Limnol. Oceanogr.* 18:653–58
48. Walsby, A. E. 1973. *Proc. 1st Int. Congr. Bacteriol. Symp.* 1:218–58
49. Walsby, A. E. 1974. *Microbiol Ecol.* 1:51–61
50. Walsby, A. E., Buckland, B. 1969. *Nature* 224:716–17
51. Walsby, A. E., Eichelberger, H. H. 1968. *Arch. Mikrobiol.* 60:76–83
52. Walsby, A. E., Klemer, A. R. 1974. *Arch. Hydrobiol.* 74:375–92
53. Whitton, B. A., Peat, A. 1969. *Arch. Mikrobiol.* 68:362–76
54. Winogradsky, S. 1888. *Beiträge zur Morphologie und Physiologie der Schwefel Bacterien.* 1. Leipzig: Felix. 120 pp.
55. Zhilina, T. N. 1971. *Microbiology* 40:587–91

Ann. Rev. Plant Physiol. 1975. 26:441–81

MEMBRANE BIOGENESIS[1]   ❖7598

D. James Morré

Departments of Botany and Plant Pathology and Biological Sciences, Purdue University, West Lafayette, Indiana 47907

CONTENTS

[1]Certain of the studies reported here were supported by grants from the National Institutes of Health CA 13145 and HD 06624. Purdue University AES Journal Paper No. 5785.

441

INTRODUCTION

This review will consider the biogenesis of membrane constituents and the assembly of these constituents into a biological membrane. Emphasis will be on what constituents must be synthesized, the mechanisms involved, and when and where the events occur within the cell. The question of how—the biochemical pathways of lipid, protein, and carbohydrate synthesis—has been the subject of reviews (22, 40, 69, 123, 126, 149, 156, 243, 250, 391, 416, 418, 432, 464) and will be summarized here only to emphasize recent findings.

As essential parts of all living cells, membranes form a major component of the boundary between protoplasm and the environment (57, 59, 60, 346). In addition to this most obvious role of controlling influx and efflux of cellular metabolites (87) and environmental signals (62, 345), membranes serve a variety of other functions in secretion of macromolecules (20, 66, 274, 275, 324, 346), energy transduction (30, 294), and in assuring the efficient operation of multienzyme systems (56, 59). Finally, many processes may be regulated at the membrane level such as the action of hormones (62, 142), the replication of DNA [at least in bacteria; (see 108 and 155 for references)], immunological surveillance and other surface phenomena (57, 140, 198), and in the manifestation of normal and abnormal growth and development (60, 345, 346).

Membranes occur both as surface and internal components of protoplasm. Some of the internal membranes function as discrete organelles, e.g. chloroplasts and mitochondria (30). These internal membranes provide a vastly increased surface area within the cytoplasm for a variety of metabolic activities (346). Most, if not all, biological membranes have enzymes on their surfaces; some enzymes operate efficiently only when associated with a membrane surface (56, 59).

Thus membranes emerge as complex and specific cell components of considerable structural and functional diversity. Questions of how such structures are formed within the cell must be asked in the context of what constituents must be assembled to ensure membrane diversity as well as the functional specificity characteristic of living cells.

Except for chloroplasts and microbodies (glyoxysomes), the vast majority of information on biogenesis of eukaryotic membranes is derived from mammalian cells and tissues rather than from plants. To the extent possible in this review, information from plants will be integrated into a more general framework derived from a variety of eukaryotic sources.

COMPOSITION AND DIVERSITY OF MEMBRANES

Membranes are composed primarily of lipids and proteins with lesser amounts of asymmetrically distributed carbohydrates and other prosthetic groups (45, 59, 346).

With most eukaryotic membranes, including those of higher plants (142), lipids and proteins are present in approximately equal amounts. The precise proportions vary with the type of membrane and its degree of specialization (191).

Membrane lipids are principally glycerophosphatides (250) combined with sterols, neutral glycerides, sphingolipids, and glycolipids (140, 142, 187, 190, 191). They are heterogeneous with respect to fatty acid composition (224, 301), although the distribution is controlled with considerable precision (156, 250). Among the more detailed analyses of phospholipids is that of Kuksis & Marai (203), who identified 38 molecular species of avian egg phosphatidylcholine. Similar studies have not been carried out with purified plant phospholipids. Except for a paucity of sphingolipids, the same general classes of phospholipids are found in plant and animal membranes (158, 187, 190, 191, 243, 250, 266).

The protein composition of mammalian membranes is wide and varied as shown by polyacrylamide gel electrophoresis (45, 59, 91, 103, 268, 362, 463, 470). A spectrum of proteins is indicated also for plant membranes (461; W. N. Yunghans, unpublished observations). Yet most of the major protein components of membranes remain unidentified and uncharacterized or incompletely resolved in published gels. Certainly there are many more enzymatic activities associated with endoplasmic reticulum or plasma membranes than those represented by the 20 or 30 electrophoretic bands commonly reported.

Membrane proteins are classified variously as either catalytic (enzymes) or structural (no known catalytic function); extrinsic (near the membrane surface and more easily dissociated from the membrane) or intrinsic (deeper within the membrane interior); transmembrane (through the membrane) or amphiatic (with both hydrophobic and hydrophilic regions) (45, 335, 379, 382). Amphiatic properties of membrane proteins imply varying degrees of association with the lipid portions of the membrane and a propensity toward self-assembly (221, 335, 382). For example, cytochrome b_5 and NADH cytochrome b_5 reductase of mammalian endoplasmic reticulum are amphiatic proteins containing a hydrophobic peptide segment comprising 25% of the molecular weight (335). These hydrophobic sequences provide for attachment to the membrane while the hydrophilic/catalytic segments extend away from the membrane. A catalytically active fragment lacking only the hydrophobic peptide cannot bind to the membrane.

The precise organization of proteins and lipids within any membrane is unknown, although several reasonable models have been proposed (59, 379). Much of the lipid exhibits bilayer characteristics and at least some of the proteins are located within the membrane while other proteins are located exterior to the lipid (45, 379). The 75 Å intramembranous particles observed by electron microscopy in freeze-fracture preparations have been suggested as proteins or glycoproteins inserted into the lipid layer (306, 405), but their identification with enzymes or enzyme complexes known to be associated with different membranes is generally inconclusive (404).

Not all membranes are alike. The semiautonomous organelles (chloroplasts and mitochondria) have definite inner and outer membrane systems, each with a characteristic structure, function, and chemical composition (9, 227, 266, 362). These organelles contain DNA and complete transcription and translation systems and are characterized by enzyme systems for generating ATP through respiratory- or photosynthetic-chain-linked phosphorylation (30, 197). A second category of membranes comprises the endomembrane system [the nuclear envelope-endoplasmic reticulum-Golgi apparatus-vesicle-(vacuole-lysosome-outer organelle membrane-plasma membrane) complex] which exists as a functional and structural continuum (271, 274, 275). Structural and biochemical comparisons show many similarities as well as important differences among endomembrane components. Major phospholipids, enzymes, and intrinsic membrane proteins are found in all endomembrane components but in differing proportions (24, 108, 187, 190, 191, 226, 243, 250, 255, 268, 271, 274, 470). Other enzymatic activities and carbohydrate and prosthetic groups are detected only in specific portions of the endomembrane system (60, 91, 103, 186, 192, 268, 274, 336, 355).

Even within a single morphologically identifiable membrane type such as endoplasmic reticulum or Golgi apparatus, not all regions are alike (89, 154, 274). Within systems of interassociated membrane, patterns of membrane characteristics emerge as exemplified by measurements of membrane thickness (134, 274, 275, 414). Tonoplast and plasma membranes are usually the thickest of the cellular membranes in electron micrographs and have been identified as end products of the activities of the endomembrane system (274). Nuclear envelope and rough endoplasmic reticulum are usually the thinnest of the endomembranes and have been identified as potential sites of membrane biogenesis. Other types of membranes which lack ribosomes and have biochemical or structural characteristics intermediate between the generating elements and end products have been grouped into the broad functional category of transitional elements (274, 275). Additionally, membranes are asymmetrically organized with carbohydrate prosthetic groups of glycoproteins and glycolipids (6, 19, 154, 201, 336, 426), phospholipids and sterols (45), and proteins (379, 407).

Any consideration of membrane biogenesis must include similarities as well as differences among cellular endomembranes along with functional, kinetic, and spatial considerations commensurate with known biochemical pathways and current structural concepts (277). To facilitate consideration of these many facets of membrane biogenesis, the remainder of this review is subdivided to consider first the available biochemical pathways for biogenesis of membrane constituents; known or potential sites for synthesis of membrane constituents; temporal, kinetic, and regulatory aspects of membrane biogenesis; and finally, a consideration of the biogenesis and assembly of specific membrane types. In this way, both generalities and exceptions may be gleaned from what is known about the properties of biological membranes.

BIOSYNTHESIS OF MEMBRANE CONSTITUENTS

With intact organs, tissue slices, cells, or organisms, various low molecular weight precursors are rapidly incorporated into all classes of cellular membranes and membrane constituents (11, 12, 14–16, 22, 60, 69, 88, 92, 96, 97, 126, 149, 181, 235, 243, 250, 267, 300, 332, 339, 350, 391, 418, 449). Individual enzymatic steps are known for some membrane constituents but not for others. Evidence is most complete for membrane lipids (158, 243, 250, 416, 418). In contrast, much less is known about details of the mechanisms of synthesis of membrane proteins.

Phospholipids and Glycerides

The main biosynthetic pathways summarized in Figure 1 were initially derived from studies with animal cells or microorganisms (35, 69, 168, 184, 194, 243, 250, 350, 416, 418), although most have now been demonstrated for higher plants as well.

From in vivo results with tomato root, Willemot & Boll (449) suggested that phosphatidylethanolamine (PE) was synthesized by decarboxylation of phosphatidylserine (PS) and phosphatidylcholine (PC) by the methylation of PE. The methylation pathway for biosynthesis of PC has been established in vivo, as well as direct methylation of phosphatidyl-N-methylethanolamine to phosphatidyl-N,N-

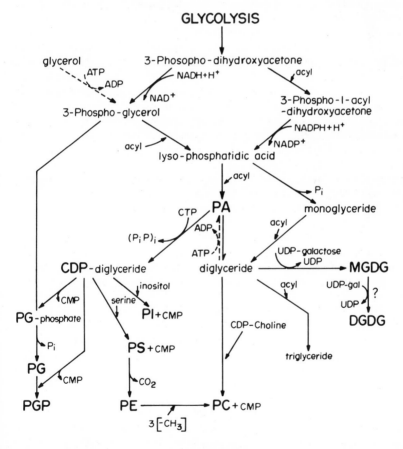

Figure 1   Biosynthetic pathways for glycerolipids. Various base exchange, acylation-deacylation, and degradative reactions are not shown. Adapted from Mazliak (243).

dimethylethanolamine and of the latter to PC by S-adenosylmethionine (235). In these studies, neither the direct methylation of PE nor synthesis of PS by exchange reactions with serine could be demonstrated.

Morré et al (276) demonstrated the complete cytidine nucleotide pathway for PC biosynthesis in onion stem; work of Devor & Mudd (76), Tanaka et al (398), and Marshall & Kates (235) established this pathway for spinach leaves. The cytidyl transferase has been verified also for barley endosperm (173) and castor bean cotyledon (218), but unequivocal evidence for synthesis of PE by a similar pathway is lacking. Macher & Mudd (225) suggest that the ethanolamine and choline transfers are catalyzed by the same enzyme, and they summarize evidence favoring decarboxylation of PS as the predominant pathway of PE biosynthesis. They were unable to detect the formation of CDP-ethanolamine from phosphorylethanolamine. Biosynthesis of PE by decarboxylation of PS has been described for tomato root (449), pea seedlings (422), and spinach leaves (235). Additionally, PE and PS may be formed by various exchange reactions (422).

Analogous nucleotide pathways have been demonstrated for phosphatidylinositol (PI), PS, and phosphatidylglycerol (PG) (234, 259, 386). The results of Marshall & Kates (235) show that biosynthesis of PS in spinach leaves may occur by the same pathway as that operating in bacteria (184), namely by a phosphatidyltransferase reaction between CDP-diglyceride and L-serine, a major contrast with animal tissues where the exchange pathway is the only pathway known (26, 35, 183, 309). Moore et al (260) present evidence for an exchange reaction catalyzed by endoplasmic reticulum of castor bean endosperm for synthesis of PS. This pathway is a potentially major pathway also in rat liver (26).

Synthesis of PG from CDP-glyceride and glycerol-phosphate has been described for spinach leaves (234), cauliflower inflorescence (81), and castor bean endosperm (259). The PG may serve as a substrate for synthesizing cardiolipin (diphosphatidylglycerol), a distinctive phospholipid component of mitochondria (81, 165, 250, 371). Diacylglycerols are synthesized by the monoacylglycerol and α-glycerophosphate pathways, both of which contribute to formation of triglycerides (168, 174).

The fatty acid portions of phospholipids and glycerides are synthesized by the enzyme system fatty acid synthetase (FAS) from acetyl-coenzyme A (CoA) and malonyl CoA (223, 229, 243, 391, 415, 440). In most bacteria and plants, the components of FAS, the biosynthetic enzymes and acyl carrier proteins (ACP) are found as discrete proteins (229, 391, 432) but in yeast and animals, these proteins are integrated into a stable multifunctional complex (47, 223, 341, 415, 440). The ACP of plants and of most bacteria functions as the acyl group carrier in all reactions of the sequence. Pathways and subcellular distributions of fatty acid biosynthetic enzymes are given by Stumpf (391, 392), Mazliak (243), and Hitchcock & Nichols (158). Microsomal cytochrome b_5 but not P_{450} has been implicated as a component of the desaturating system (440).

The nonrandom distribution of fatty acids in phospholipids of membranes could arise from specific acylation during synthesis (139, 250, 312) and/or by specific redistribution of fatty acyl chains via phospholipase and acyltransferase reactions (see, however, 209). Van den Bosch et al (417) proposed that the dissimilar location of saturated and unsaturated fatty acids in phosphatides occurs during de novo synthesis of these compounds but that the selective acylation of lysophosphatides helps preserve asymmetry during turnover. In several cellular systems, fatty acids are exchanged in membrane phospholipids by mechanisms that maintain positional specificity of saturated and unsaturated fatty acids (12, 250, 279, 292). The various base exchange, acylation-deacylation, and degradative reactions of membrane phospholipids have been reviewed (418).

Sterols, Pigments, and other Non-saponifiable Lipids

The biochemical conversion from squalene to cholesterol requires a minimum of 25 steps (160, 354) facilitated by one or more carrier proteins (330, 354). Sterol carrier proteins differ from acyl carrier proteins in that the enzymes are membrane-bound and the sterols are held by hydrophobic rather than covalent bonds. Carrier proteins have been shown to facilitate biosynthesis of several steroid hormones and bile acids, vitamin D metabolism, triglyceride biosynthesis, and cholesterol ester formation (354), as well as the biosynthesis of gibberellins (262).

Plant membranes contain cholesterol and are rich in various phytosterols with the stigmasterol- and sitosterol-like phytosterols among the more widely distributed (122, 149). The biosynthetic sequence is similar to that for sterols in mammals (116, 144, 145), with appropriate enzymes at terminal steps to yield sterols typical of plants (116, 144, 145, 149, 160, 443).

The biosynthesis of chlorophylls, carotenes, xanthophylls, quinones, and other non-saponifiable lipid or pigment constituents of chloroplast membranes has been reviewed (22, 126, 197, 233, 321). Carotenoids and quinones are present not only in plastid and mitochondrial membranes but in endomembrane components as well (193, 290, 291). Their distribution in plant endomembranes has not been determined.

Proteins

The mechanism of biosynthesis of membrane proteins has not been studied in detail, although inhibitor studies suggest involvement of ribosomes, messenger RNA, and both cytoplasmic and organellar translation systems (75, 319, 320). Protein components of the inner membrane systems of both mitochondria and chloroplasts contain two classes of peptides—one derived from protein synthesis within the cytoplasm and the other from protein synthesis within the organelle (30). Synthesis of intrinsic proteins of the plasma membranes, tonoplast, and endomembranes in general is presumed to involve ribosomes and messenger RNA—either membrane-bound or free in the cytoplasm—but there is no proof of this for plants (276). The precise sites of synthesis of most membrane proteins are unknown, as detailed in the section on sites of synthesis of membrane proteins which follows. As a result, the mechanisms are unknown.

Glycoproteins and Glycolipids

Glycoprotein or glycolipid glycosyltransferases in plant and mammalian systems transfer a glycose from an appropriate nucleotide sugar to an acceptor which is usually an incomplete carbohydrate side chain of a glycoprotein or glycolipid (60, 342). The enzymes usually exhibit specificity to both acceptor and donor. An alternative pathway is through a lipid intermediate analogous to the oligosaccha-ride-polyisoprenol derivatives that function in the biosynthesis of bacterial polymers (18, 51, 188, 216, 431, 435). The glycosyl derivatives of dolichol found in mam-malian systems (18, 51, 435) participate in the glycosylation of proteins.

Little is known about glycoproteins and glycolipids in plants except that they exist as potential membrane constituents (60, 332; J. D. Clark & D. J. Morré, unpub-lished). Roberts et al (332) studied the incorporation of ^{14}C-glucosamine into glyco-protein of a variety of plant species. The biosynthetic pathway was not determined.

Sulfated compounds occur in membranes of many species of plants and animals frequently as portions of prosthetic groups of glycoproteins and/or glycolipids (117, 329). As these molecules are synthesized, sulfate is first activated by ATP in a two-step sequence requiring two separate enzymes: ATP sulfurylase catalyzes the reaction between ATP and SO_4 to give adenosine-5'-phosphosulfate (APS). In a second activating step, APS is phosphorylated by ATP and the enzyme APS-phosphokinase to form 3'-phosphoadenosine-5'-phosphosulfate (PAPS). PAPS is the starting compound for assimilatory sulfate reduction and for sulfate ester forma-tion. In the latter, the activated sulfate is transferred from PAPS to an acceptor molecule by one of a class of sulfotransferases. APS is the activated intermediate in dissimilatory sulfate reduction (329, 361).

SUBCELLULAR DISTRIBUTION OF THE ENZYME SYSTEMS OF MEMBRANE BIOGENESIS

An important aspect of membrane biogenesis is the question of where. To what extent are membranes responsible for their own formation? What cellular compo-nents are involved? One approach is to consider the subcellular location of the synthetic pathways associated with the formation of membrane constituents.

Sites of Lipid Biosynthesis

Microsomal preparations (usually material sedimenting between 10,000 and 100,000 Xg from dilute homogenates) contain most of the enzymes for synthesis of PC (CDP-choline pathway and methylation of PE), PS (by exchange), PE (from CDP-ethanolamine), and PI in both plant (76, 173, 218, 235, 261, 267, 276, 418, 422) and animal (35, 70, 74, 156, 250, 418, 423, 448) tissues. Synthesis of triglycerides is also microsomal (168, 352, 423, 448), although soluble protein factors may be required for maximal activity (230, 340).

The choline phosphotransferase which catalyzes the formation of PC from CDP-choline and α,β-diglyceride and the methylation pathway of PC biosynthesis are microsomal enzymes in plants (235). The choline kinase required for phosphoryla-tion of choline, is largely a soluble enzyme (173, 218, 276, 398; but see 368 for mitochondrial location in Cuscuta reflexa), as in the ethanolamine kinase (235).

The location of PI synthesis in plants has been attributed to either mitochondria (393) or to microsomes (261) and the synthesis of phosphatidylglycerol to either the mitochondria (81), endoplasmic reticulum (234), or both (259).

The PS carboxylase involved in PE synthesis appears to be associated with the 100,000 X g supernatant fraction in spinach leaves (235) in contrast to its location in mitochondria in animals (74).

The oxygen and pyridine nucleotide-dependent desaturation of fatty acyl CoAs is microsomal (296, 297, 313, 440). Mitochondria from potato synthesize fatty acids from acetate (244). Additionally, ^{14}C-fatty acids are transferred in vitro from microsomes to mitochondria, or reciprocally, without apparent metabolic alteration (245). Activation and elongation of medium- and long-chain fatty acids occurs in mitochondria and microsomes (see 133) as well as chloroplasts (126, 172).

The sites of synthesis of mitochondrial phospholipids have been reviewed by Dawson (70) and Wirtz (451). Mitochondria synthesize both phosphatidic acid and cardiolipin (67, 81, 165, 371, 386, 465) but apparently lack enzymes to catalyze de novo synthesis of the major glycerophosphatides: PC, PS, and PI (70, 250, 416, 418, 451; but see 390 for outer membrane). Isolated mitochondria have been reported to enhance microsomal synthesis of these phospholipids, possibly by removing inhibitory calcium ions (331). Contrary to earlier conclusions, acyl transferases for forming PC and PE from corresponding lysophosphatides and fatty acyl CoA derivatives are present in rat liver mitochondria (281, 352, 439). These enzymes couple with the phospholipase A_2 of mitochondria (436–438) to provide a fatty acid exchange cycle which can operate independently of de novo synthesis. Liver mitochondria apparently cannot replace phospholipid bases by an exchange process (416); the calcium-catalyzed exchange of bases appears to be largely microsomal in crude subcellular fractions (8, 167, 416, 418). Similarly, the incorporation of ^{14}C-choline by mitochondrial fractions isolated from liver is accounted for by contamination of the preparations by endoplasmic reticulum or a non-energy dependent exchange process (176, 248). Acylation of lysophosphatidylcholine occurs in both microsomes and plasma membranes of rat liver (352, 385, 388, 458). Spinach mitochondria are unable to synthesize PC but may have some capacity to synthesize PE via the PS decarboxylase pathway (235).

In summary, the terminal enzymes for synthesizing major phospholipids are associated with membranes, although pathways require the participation of enzymes of the soluble cytoplasm. For biosynthesis of PC in onion stem, the choline kinase is soluble, the PC-cytidyl transferase is about equally distributed between the soluble and particulate fractions, and the PC-diglyceride transferase is exclusively microsomal (276). The work of Beevers and colleagues shows that within the microsomal fraction (181, 218, 261) the endoplasmic reticulum (219) of castor bean endosperm is the major site of incorporation of CDP-choline, -inositol, and -serine into phospholipids. Minor phospholipids unique to mitochondria are synthesized by mitochondria (63, 67, 74, 81, 250, 259, 352, 465). Plastids also synthesize membrane lipids (126). Enzymes of most membrane types catalyze exchange reactions (185, 416, 418, 458) and the incorporation of free fatty acids into phospholipids (292, 384, 458).

Sites of Sterol Biosynthesis

The multienzyme complex which converts squalene to cholesterol in mammals (116, 354, 397) and plants (121, 126) is bound to membranes of the microsome fraction, but the specific cellular location of the enzymes of cholesterol assembly are not known. Biosynthetic enzymes up to squalene are soluble (383). Smooth microsomes are major sites of steroidogenesis in the steroid-producing interstitial cells of the testis and in adrenocortical tissues (116, 397). Significantly, the 17α-hydroxylase and the C_{17}–C_{20} lyase are predominantly in the smooth microsome fraction which is most likely involved in the hydroxylation and side-chain cleavage reactions related to testosterone formation (397). In other tissues such as placenta and ovarian tissue, the enzymes are distributed almost evenly between the rough and smooth microsomes (397). The 11β-hydroxylase and the cholesterol side-chain cleaving enzyme (cholesterol 20α- and 22R-hydroxylases and the C_{20}–C_{22} lyase) of testicular tissues are found predominantly in the mitochondrial fraction (254), associated with the inner membrane (397).

An enzyme critical to sterol biosynthesis, *trans*-farnesylpyrophosphate-squalene synthetase (149), is associated with neither etioplasts nor mitochondria in etiolated leaves of bean but is found in smooth microsomes and vesicles of rough endoplasmic reticulum (145). The results are interpreted to mean that sterols in mitochondria and etioplasts are not synthesized in situ but are transferred to these organelles from the endoplasmic reticulum. S-adenosylmethionine: Δ^{24} sterol methyltransferase of *Saccharomyces cerevisiae*, a terminal enzyme of ergosterol biosynthesis, is reported to be mitochondrial (403). Depending on the pathway, intermediates are viewed as transferring from endoplasmic reticulum to mitochondria and vice versa in current schemes of steroidogenesis (397).

Sites of Biosynthesis of Membrane Proteins

Morphological and other evidence (280, 337) has shown at least two distinct subclasses of ribosomes in eukaryotic cells: those free in the cytoplasm and those associated with endomembranes. The latter plus the membranes to which they attach form the endoplasmic reticulum and outer membrane of the nuclear envelope. The generally accepted thesis is that attachment of ribosomes to membranes facilitates, or even may be obligatory, for the secretion of proteins destined for export (324, 396, 400; see 373 for summary). Also widely accepted is the thesis that proteins to be retained within tissues, particularly the soluble tissue proteins, are synthesized predominantly on free ribosomes (220, 396, 400). Some reports contradict these generalizations (150, 171, 307, 314, 318, 369, 396). The early proposals that all secretory proteins, as well as membrane proteins, were made on membrane-bound polysomes, and that all cytoplasmic proteins were made on free polysomes seems an oversimplification. Free and membrane-bound polysomes are implicated (however, see 48) but, with few exceptions, the sites of synthesis of membrane proteins are unknown (45, 98, 170, 274).

Some membrane proteins are strongly hydrophobic (45), and it would be reasonable for such hydrophobic proteins to be synthesized within or near the membrane on a membrane-associated template (274). In kinetic studies in vivo, constitutive hydrophobic proteins of endoplasmic reticulum membranes are labeled within minutes after a radioactive amino acid is administered (113). Only later does label appear in smooth membranes or plasma membranes. These findings reveal rough endoplasmic reticulum as an early site of membrane assembly but do not establish membrane-bound polysomes as the ultimate site of amino acid incorporation (268, 274). For example, also from in vivo studies, Favarger et al (96, 97) found that shortly after intravenous injection of labeled precursors of fatty acids in the mouse, most of the synthesized fatty acids were found as esters in the microsomal lipids of the liver. Endoplasmic reticulum was indicated as an early site of membrane assembly, but it is probably not the site of fatty acid biogenesis since the early biosynthetic enzymes are soluble (391).

In cell-free systems of protein synthesis in vivo, newly synthesized secretory proteins have a much higher specific activity than the proteins of the membrane vesicles that contain them (251, 270, 324–326, 377), suggesting a limited capacity to synthesize membrane proteins. Yet, free and membrane-bound polysomes of rat liver when isolated by conventional procedures have approximately equal abilities to synthesize mixed hydrophobic proteins of endoplasmic reticulum membranes in vitro (84). However, approximately 20% of the membrane-bound ribosomes settle with the free polysome fraction under these conditions (351). When fragments of endoplasmic reticulum are removed completely from carefully prepared homogenates prior to preparation of free and membrane-bound polysomes, membrane-bound polysomes show preferential in vitro incorporation of amino acids into hydrophobic proteins (268). Major differences between free and bound polysomes and bound polysomes stripped from membranes in their ability to incorporate amino acids into polypeptides were reported by Bont et al (33). Translation on bound polysomes appeared to be blocked after the polysomes were removed from the membranes. These findings, taken together, suggest that association of the polysomes with the membranes may facilitate translation of at least certain classes of membrane proteins.

If membrane-bound polysomes are involved in the biosynthesis of membrane proteins, at least the fates of the completed polypeptide chains must differ from those for secretory proteins. The large ribosomal unit which contains or surrounds the nascent polypeptide chain is bound to the membrane of the endoplasmic reticulum (349). The nascent chains of secretory proteins pass through the large ribosomal subunit and through the membrane into the lumen of the endoplasmic reticulum (324–326, 348, 377). If polysomes are dissociated from the membranes and reattached to either smooth microsomes or rough microsomes stripped of their endogenous polysomes, protein synthesis continues but without significant insertion of nascent chains into the membrane lumina (34). With membrane proteins, the nascent chains must insert into the membrane rather than into the membrane lumen. This property of membrane proteins may be related to their hydrophobicity or may be related to some characteristic of the translational process per se, which is unique to polysomes synthesizing membrane proteins.

An alternative possibility for synthesis of constitutive membrane proteins by membrane-bound polysomes is their synthesis by free polysomes (45, 127, 170, 217, 220, 318). Here, proteins of endoplasmic reticulum would arise from a transient pool of soluble intermediates synthesized first on cytoplasmic polysomes and only subsequently incorporated into membranes (370). There is evidence for an involvement of free polysomes in the synthesis of certain membrane proteins (84, 170, 318). Lowe & Hallinan (220) indicate that NADPH-cytochrome c reductase is made preferentially on free polysomes of livers from rats injected with phenobarbital. Yet in vivo, pulse-labeled NADPH-cytochrome c reductase appears first in rough endoplasmic reticulum, and only later appears in the smooth endoplasmic reticulum (65, 205). The latter observations imply a preferential association of synthetic sites with rough endoplasmic reticulum regardless of whether the ribosomes are free or membrane-bound.

In summary, the question of sites of synthesis of proteins of endomembranes remains controversial and is the subject of current investigation. There is no assurance that integral or intrinsic proteins of membranes are made at the same sites as extrinsic proteins. Both free and membrane-bound polysomes appear to be involved in the biogenesis of membrane proteins; some membrane proteins may be synthesized by both classes of polysomes (170, 268).

Ribosomes are not restricted to the endomembrane system and ground cytoplasm but are located also within the nucleus, at least as precursor particles, and in mitochondria and chloroplasts (30). A biosynthetic role for the nucleus in the formation of membrane proteins has not been indicated (98, 108, 113) except for recent findings from oocytes (100). Both mitochondria (14–17, 50, 128, 146, 180, 202, 356, 413) and chloroplasts (6, 28, 30, 85) incorporate radioactive amino acids into membrane proteins. With isolated mitochondria, the labeled fractions are identical with inner membrane, and little if any label is incorporated into the outer membrane (9, 17, 257, 282).

The sites of synthesis of mitochondrial proteins are not confined to mitochondria, however. The three largest cytochrome oxidase polypeptides of mitochondria are synthesized on mitochondrial ribosomes, whereas the four smallest polypeptides of cytochrome oxidase are synthesized on cytoplasmic polysomes (236, 356, 366, 367, 413). With cytochrome b, inhibitor studies suggest that the large polypeptide is synthesized on mitochondrial ribosomes and that of the smaller on cytoplasmic polysomes (444, 445). In the biosynthesis of cytochrome b, the following two steps have been suggested. First, protoheme is bound to the nascent mitochondrial translation product, and second, this precursor cytochrome b is integrated into a more complex membrane protein (444). The apoprotein of cytochrome c_1 is synthesized on cytoplasmic polysomes (343). With cytochrome c (179), peptide synthesis and addition of the prosthetic group are completely separate processes. Studies on rutamycin-sensitive adenosine triphosphatase (412, 413) provide an additional example of a mitochondrial enzyme complex consisting of nine polypeptides, four of which are synthesized on mitochondrial ribosomes.

Clearly, chloroplast proteins also may be translated on both nuclear and chloroplast ribosomes (6, 30, 52, 83, 164, 256, 433, 447). Of the proteins synthesized by isolated, intact chloroplasts of peas, the large subunit of Fraction I protein is the only detectable soluble product and accounts for nearly 25% of the total labeled protein synthesized in vitro (28). The remaining 75% presumably sediments with the chloroplast membrane fraction. The identity of the membrane proteins synthesized within the chloroplast is unknown, but they do not include cytochrome f, the coupling factor, or proteins of photosystems I and II (83, 85). The proteins not synthesized by isolated chloroplasts are presumably translated from messenger RNAs by cytoplasmic ribosomes. Whether these cytoplasmic ribosomes are free or membrane-bound is not known.

Identification of the sites of protein synthesis in mitochondria and chloroplasts by no means identifies the source of the messenger RNA, or the code for the protein synthesizing machinery itself (294). Nuclear messenger RNA could enter mito-chondria and code intramitochondrial protein synthesis, and the coding of protein synthesis by mitochondrial messenger RNA on cytoplasmic ribosomes is also possible. Nevertheless, the established concept is one of dual control with the eventual participation of both organellar and cytoplasmic ribosomes in translation. Whether the cytoplasmic ribosomes are free or membrane-bound is unknown except for mitochondrial cytochrome c where both free polysomes and polysomes bound to rough endoplasmic reticulum seem involved (127). Additionally, chloroplasts (231, 399, and ref. cit.) and mitochondria (204) have two ribosomal populations, one free and one membrane-bound. Thus participation of free and membrane-bound poly-somes in the formation of membrane proteins may extend to organellar (6) as well as cytoplasmic synthesis.

Sites of Glycosylation of Membrane Glycoproteins and Glycolipids

Glycosyl transferases are believed to be the primary controllers of the structure and specificity of the carbohydrate portions of complex polysaccharides in general and especially those of membranes (60, 342). Studies with cell fractions from rat liver show several glycoprotein (271, 272, 355) and glycolipid (192) glycosyl transferases to be concentrated in Golgi apparatus although present in endoplasmic reticulum (51, 60, 141, 192, 211, 216, 264, 265, 272, 324, 372). Much of the argument for localization of glycoprotein and glycolipid glycosyl transferases in plasma mem-branes is based on indirect evidence (see 304 and 345 for reviews). The enzymes for the sugar nucleotide pathway have not been demonstrated in highly purified plasma membrane fractions from animal cells (192). Enzymes of the polyisoprenoid path-way are microsomal in mammalian cells (216) and associated primarily with smooth membranes (51). With plants, glycosyl transferases are distributed throughout the endomembrane system (419). None have been implicated specifically in membrane biogenesis. Golgi apparatus also emerge as important sites of sulfation of polysac-charides (23, 123, 210, 460). Sulfotransferases have been localized in Golgi ap-paratus fractions from kidney (104) and testes (467) of the rat.

KINETICS OF MEMBRANE BIOGENESIS

As summarized in the preceding section, synthesis of a biological membrane in-volves a net increase in the total amount of membrane and is thus distinguished from exchange and molecular repair processes in which the total number of molecules may remain unchanged. Rough endoplasmic reticulum (and possibly nuclear en-velope) emerges as a major site of membrane biosynthesis at the subcellular level. With other cell components, major portions of their membrane lipids are synthesized by enzyme systems terminating exclusively in endoplasmic reticulum. The situation for membrane proteins is less clear, but data show rapid incorporation of amino acids into endoplasmic reticulum with delayed or subsequent appearance in other membrane fractions. Mitochondria and chloroplasts have two sites of synthesis of membrane proteins—some cytoplasmic, others organellar. Unfortunately, because of these considerations, findings from kinetic studies may more often provide measures of the rates of transport of membranes and/or membrane components rather than a direct measure of the rates of synthesis of the membranes.

Synthesis and Transfer of Membrane Proteins

In vivo experiments with rats show that isotopically labeled proteins are rapidly incorporated without a discernable lag (less than 1 min) into membranes of rough endoplasmic reticulum (113) and reach a maximum within 10 (113) to 20 (401) min. The very rapid appearance of label into rough endoplasmic reticulum suggests that membrane proteins have a relatively short processing time (whether synthesized on membrane-bound or free polysomes) prior to insertion into the membrane. In the experiments of Franke et al (113), the fractions were extensively washed with buffered salt solutions containing 0.1% deoxycholate which removes absorbed soluble proteins, ribosomal proteins, and plasma proteins destined for secretion (201, 270). NADH-cytochrome c reductase of these membranes exhibited kinetics of biosynthesis similar to total hydrophobic membrane proteins (270). Kuriyama & Omura (205) measured the appearance of protein label in purified NADPH-cytochrome c reductase and cytochrome b_5 of rough and smooth microsomes. The label appeared first in the rough microsomes. Equilibration with smooth microsomes occurred within 2 hr. The in vivo incorporation of leucine-^{14}C after pulse labeling occurred first in the rough microsomes of livers from rats following administration of phenobarbital as well (88, 205).

Involvement of the nuclear envelope is less clear (98, 108, 113). Rough endoplasmic reticulum and the outer membrane of the nuclear envelope which contains ribosomes on its cytoplasmic surface form a morphological continuum. Yet direct transfer of membrane from nuclear envelope to endoplasmic reticulum remains to be demonstrated. Labeled amino acids are incorporated into nuclear membranes in vivo with approximately the same kinetics as into rough endoplasmic reticulum (113).

After labeling of the rough endoplasmic reticulum and the nuclear envelope, radioactive proteins appear in smooth endoplasmic reticulum (205, 270), Golgi apparatus (113, 270), and plasma membranes (92, 113, 270). The lag between labeling of rough endoplasmic reticulum and other cell components varies from a few minutes for smooth endoplasmic reticulum to more than 20 minutes for plasma membrane. The order of labeling of membrane proteins indicates that membrane proteins follow the same route through the endomembrane system as that established for secretory proteins (20).

In preliminary studies with plants (277), incorporation of ^{14}C-leucine into microsomes (endoplasmic reticulum) and nuclear fractions was linear following a lag in uptake of isotope. Incorporation into dictyosomes and plasma membrane fractions lagged 30–60 min behind incorporation into microsomes. Incorporation into a mitochondrion-etioplast fraction was intermediate in its characteristics.

Examination of the time course of labeling of mitochondria and chloroplasts with radioactive amino acids in vivo reveals a complex pattern (16, 339). With mitochondria, the insoluble proteins of the membrane acquire label more intensely at first, but the soluble proteins of the matrix become much more radioactive after a few hours (16). The lag period in labeling of the soluble proteins of the matrix has been interpreted as additional evidence for an extraorganellar site of synthesis and eventual importation of membrane precursors (14). When isolated mitochondria (17, 339) or chloroplasts (83, 85, 143) are incubated with radioactive amino acids, certain proteins of the insoluble membrane-associated type label without appreciable lag while, with the exception of Fraction I protein of chloroplasts, matrix proteins remain essentially unchanged. Outer mitochondrial membranes are not labeled in these in vitro experiments (17, 282), but are the first mitochondrial components to be labeled in vivo (14, 50). Information is lacking on the outer membranes of chloroplasts (83, 85).

Synthesis and Transfer of Membrane Lipids

Various lipid precursors are readily incorporated into all classes of cellular phospholipids in vivo (243, 267). Yet there have been few detailed studies of the kinetics of incorporation. Radioactive choline supplied to plant tissues is incorporated into membranes in the form of lecithin (181, 267). The kinetics of ^{14}C-choline incorporation show that labeled lecithin is first formed in a light membranous (181) [=endoplasmic reticulum (219)] fraction and later appears in mitochondria and glyoxysomes (181). In earlier studies (267), the relative specific activity of ^{14}C-choline in lecithin was endoplasmic reticulum (microsomes) > dictyosomes > mitochondria-proplastids > smooth membranes (plasma membrane and/or tonoplast). The results of Marshall & Kates (235) suggest that mitochondrial PC and PE are not synthesized in situ by mitochondria but may be derived from microsomal-synthesized lipid by direct transport or exchange between microsomal membranes and mitochondria as has been demonstrated in potato or cauliflower tissues (2; see also 245) and in animals (3, 11, 29, 70, 249, 451, 452).

There is virtually universal agreement that after injection of labeled precursors in rats, the microsomal fraction obtained from the liver is labeled more rapidly than the mitochondrial fraction (418) for PC, PE, and PI. The reverse is true for phosphatidic acid (249). Subfractionation of mitochondria into outer and inner membranes provides a specific radioactivity sequence for phospholipids (other than phosphatidic acid) of microsomes > outer mitochondrial membrane > inner mitochondrial membrane (29, 50, 249). Within endomembranes, the specific activity sequence (other than sphingomyelin and PI) is rough endoplasmic reticulum > smooth endoplasmic reticulum > Golgi apparatus > plasma membrane (see 270 for liver, 267 for plants).

Synthesis and Transfer of Glycoproteins and Glycolipids

In glycoprotein biosynthesis, attachment of sugars occurs as a late event after completion of the peptide chain (38, 60, 92) with synthesis continuing throughout the cell cycle (39). Glycolipids are synthesized nearer to the plasma membrane (192) or are transferred to sites of accumulation more readily than are glycoproteins (92). Their synthesis may also be geared more closely to specific stages in the cell cycle. With murine lymphomas, glycolipid synthesis was confined almost exclusively to G_2 and M (39), while in synchronized KB cells galactose incorporation into glycolipids was maximal during M and G_1 (53). The kinetics of incorporation of glycolipids or glycoproteins into plant membranes has not been studied.

TURNOVER AND DEGRADATION OF MEMBRANE CONSTITUENTS

Turnover is defined as the flux of constituents through or within a particular membrane, i.e. protein turnover is the study of the flux of amino acids through proteins (169). As determined by usual isotopic procedures (169, 401), turnover is a measure of the dynamic state of membrane constituents and of the rate at which they are subject to degradation by normal intracellular degradative processes. Turnover does not necessarily indicate net synthesis of new membrane. A membrane constituent or component which is turning over rapidly will lose a greater relative amount of its incorporated radioactivity than one which is turning over more slowly. A constituent which turns over rapidly will also incorporate more of the isotope label initially only if synthesis and degradation are approximately equivalent.

Investigation of the degradation of membrane proteins has provided considerable insight into the dynamic state of membrane constituents (9, 73, 206, 293, 375, 376, 378, 401). Early studies by Omura et al (205, 206, 293), Schimke et al (7, 26, 73, 401), and others showed that membrane proteins of animal cells were degraded in a manner random with time and showed that turnover characteristics varied among different proteins. Membrane proteins, like cytoplasmic proteins, yielded characteristic exponential decay curves from which half-lives of 2–4 days were determined for proteins of endoplasmic reticulum or microsomal membranes (see Table 3.6 of ref. 274 for summary of literature). The turnover of plant proteins has been reviewed (40, 169), but little work has specifically involved membrane proteins.

In general, different proteins of a membrane may be degraded at different rates (376). A positive correlation between molecular size of a protein [except for cytochrome P_{450} (73)] and its rate of degradation in vivo characterizes cytoplasmic and membrane-bound proteins (401). However, the relationship for cytoplasmic proteins where rate of degradation of the protein in vivo is correlated with susceptibility to proteolytic attack by pronase in vitro is not shared by membrane proteins (401). Newly synthesized NADH-cytochrome c reductase appears more susceptible to degradation than old, membrane-bound enzymes, presumably because the newly synthesized enzyme is more accessible to degradative enzymes (205).

Turnover of membrane constituents can be shown to be coordinated (9, 237, 302, 316, 442) or independent (7, 9, 11, 73, 206, 213, 214, 293, 378, 401), depending on the experimental conditions (see Siekevitz 375, 376 for a review and discussion). Information relevant to turnover of microbodies is reviewed by de Duve (71); that for mitochondria is reviewed by Ashwell & Work (9).

The mechanisms of degradation of membrane phospholipids may involve one or more phospholipases in microsomes (27, 360), mitochondria (344, 359, 436–438), lysosomes (72, 77, 390, 437), plasma membranes (284, 427), or cytosol (315, 416). Individual classes of phospholipids of membranes may show different rates of turnover (11, 213) since different parts of the membrane, or even different parts of a phospholipid molecule, are removed or replaced at different rates (293, 378). Even different molecular species of the same phospholipid class may be degraded differently, and the rate of labeling of the backbone of a phosphoglyceride often bears no relationship to the turnover of the fatty acids (378). Acyl moieties esterified at the 1 position of PC and PE may turn over faster than the acyl moieties esterified at the 2 position (213). Although turnover estimates using isotopes are always influenced by labeling conditions (267, 376), glycerol incorporation is considered to indicate synthesis of new lipid molecules (12); there is no evidence of glycerol exchange.

Data with acetate ^{14}C supplied to stem explants of onion show turnover of microsomes and dictyosomes to be approximately equal, i.e. a half-time of 1.2 hr (0.5 hr of label) to 5 hr (2 hr of label) (267). The longer (2 hr) period of labeling would be expected to permit spreading of label into constituents such as proteins that are less readily exchanged than the fatty acid residues of membrane lipids preferentially labeled by the shorter (0.5 hr) labeling time. Since different parts of the membrane or even different parts of a phospholipid molecule are replaced at different rates through exchange and degradation (378), these differences in turnover with length of incubation time are easily reconciled. With ^{14}C-choline, labeling (267) resulted in an apparent half-life time of 43 hr, a rate much slower than with acetate but of the same order or shorter than for membrane constituents of rat liver (274; see also 243).

In general, two types of models have been proposed to account for membrane degradation. In one, membrane constituents continually associate and dissociate from the membranes, and only in the dissociated state are they subject to degradation by normal intracellular degradative processes (401). In the other model, organelles or fragments of membranes are internalized by autophagic vacuoles (32, 49, 61, 72, 77, 239). Lysosomal enzymes are added, and membrane breakdown is completed within the confines of the resulting digestive vacuole or autophagosome (72).

Both mechanisms might be expected to contribute to membrane degradation and renewal in plant and animal cells. The products would be available for utilization in redirected synthesis. Studies by Bolender & Weibel (32) provide convincing evidence for involvement of autophagic vacuoles in removing excess smooth endoplasmic reticulum membranes induced in liver cells following administration of phenobarbital. Evidence relevant to autophagy of mitochondria and other organelles from both plant (49, 61, 239) and animal (72, 77) cells has been reviewed.

Comparisons of turnover of different classes of membranes provide little information concerning endomembrane interactions. As discussed for rat liver (113), if all endoplasmic reticulum membranes tend to equilibrate, even slowly, and Golgi apparatus membranes are derived from endoplasmic reticulum, then Golgi apparatus turnover will be determined principally by the rate of turnover of endoplasmic reticulum (113, 274). To the extent that Golgi apparatus and endoplasmic reticulum contribute directly to the formation of plasma membrane, apparent rates of turnover of plasma membrane will also be influenced by rates of turnover of internal membranes (113, 274). Thus it is not surprising that different cellular membranes frequently exhibit very similar characteristics of turnover within a given cell population (11, 113, 138, 214, 274, 293, 302, 376, 401, 442).

MEMBRANE ASSEMBLY

Some variation of directed self-assembly is usually envisioned (45, 60, 86, 200, 221, 249, 335, 454). In a "single step" mechanism (59, 60), all components of the membrane, protein, lipid, and carbohydrate are assembled simultaneously. Alternatively, a "multistep" mechanism would involve primary synthesis of a basic membrane of lipid and intrinsic protein to which additional components, e.g. enzymes and sugar moieties, are added sequentially. Although evidence is largely indirect, the multistep mechanism is indicated at least for plasma membranes (60, 66, 192, 274, 277, 342, 446). Certainly, glycoproteins and glycolipids are synthesized in such a stepwise manner. All evidence indicates that polypeptides synthesized on polysomes migrate to or are inserted into newly forming membranes such as those of the endoplasmic reticulum where initial glycosylation may take place. Additional saccharide residues are added successively to the growing oligosaccharide prosthetic group in different compartments of the transitional endomembranes, especially the Golgi apparatus. The multistep or assembly line mechanism is also applicable to the biosynthesis of glycolipids (192, 342).

Selective lipid binding capacity of membrane proteins may help determine those lipids which interact directly with the proteins of the membrane (200, 462). Studies of liver, erythrocytes, and other mammalian tissues have revealed that phospholipid (249, 454, 457) and protein (86, 335) molecules may be added to existing membranes in vitro often in considerable excess over the amounts normally present (see also 30 for studies with mitochondria and chloroplasts). Whether this type of incorporation is equivalent to in vivo assembly of membranes has not been established but may be considered unlikely in part due to a lack of specificity. Plasma membranes of rat liver bind glycoproteins (459). Binding may be irreversible (425) and may represent the first of several steps in a degradative pathway ending with the lysosomal digestion of the incorporated glycoprotein (131). Many important questions concerning sites of membrane biogenesis and transport mechanisms must be answered before questions of membrane assembly can be even properly asked.

TRANSFER AND TRANSPORT MECHANISMS IN MEMBRANE BIOGENESIS

The concept of an endomembrane system (274, 275) was proposed to explain the functional continuum within which processes of membrane differentiation and physical transfer of membrane from one cell compartment to another (membrane flow) could account for the biogenesis of the cell's internal membranes. Documented continuity and potential ontogenetic relationships between each of the separate components led to the view of transfer and transformation of membranes along chains of cell components in subcellular developmental pathways (274). The alternative is that parts of the cell function in membrane biogenesis as independent entities with regulated movement of free membrane precursors through the cytosol, a view favored by some (24, 251–253).

Once incorporated into the membrane, lipid molecules are unable to move rapidly from one face to the other in the lipid bilayer, but they do diffuse laterally in the plane of the lipid layer with considerable facility (45, 247), as do antigens or proteins of the plasma membrane (115, 379, 394, 402). Additionally, phospholipids may move rapidly between different intracellular membranes by a process catalyzed or facilitated by specific proteins of the soluble cytoplasm (249, 451–456, 466). These cytoplasmic exchange proteins have been suggested to catalyze the exchange of intact phospholipid molecules (452) between different membranes within the cell (177, 249) and between isolated subcellular fractions in suspension (3, 15, 178, 182, 249, 453–457). Probably all subcellular membranes are potential phospholipid donors or acceptors in the exchange process (182, 451).

Even though continuities between endoplasmic reticulum and mitochondria are found (43, 112, 273, 347), it remains doubtful that flow mechanisms account for the transfer of membrane components from endoplasmic reticulum or cytosol to the inner membrane systems of mitochondria or plastids, nor has such transfer been demonstrated or claimed. Here, cytosol migration and specific cytoplasmic carriers are implicated (3, 163, 177, 182, 454).

However, in the origins of Golgi apparatus from endoplasmic reticulum (270, 274) and in the contributions of Golgi apparatus and endoplasmic reticulum to plasma membranes (60, 66, 157, 275, 277, 336), initial assembly of membrane in rough endoplasmic reticulum, followed by migration and progressive modification of membrane components along well-established secretory routes, are extensively documented. Alternative explanations are difficult to reconcile with the large body of morphological and biochemical evidence available, especially from plants and rat liver (270, 274, 275).

REGULATION OF MEMBRANE BIOSYNTHESIS, ASSEMBLY, TRANSPORT, AND DEGRADATION

The constancy of cellular membranes implies that their formation and degradation are carefully regulated. Details are virtually unknown.

Some of the earliest biochemical changes following effector stimulation or virus infection have been found in the metabolism of membrane phospholipids (102, 162, 189, 278, 328). Treatment of aleurone layers of barley with gibberellic acid (GA$_3$) enhances both ^{32}-phosphate incorporation into phospholipids (199) and the activity of phosphorylcholine cytidyl (173) and phosphorylcholine glyceride transferases (21, 173) as well as formation of polysomes and increased membrane synthesis (93, 430).

Synthesis of lipids and proteins of membranes are somehow coordinated so that for any given species, the composition, including individual phospholipids, is maintained within relatively narrow limits (156, 158, 350, 416). Similarly, during growth and differentiation, there is evidence for considerable coordination of the various events of membrane synthesis (129, 213, 303, 381, 400). With mitochondria, evidence that different subunits of the ATPase and cytochrome oxidase originate from mitochondrial and cytoplasmic protein synthesis raises significant questions of how synthesis of the two classes of proteins is integrated in time and space (413). The more intriguing of these questions are to what extent is the build-up of quaternary structure directed by enzyme-mediated events, or to what extent does it involve spontaneous interaction of the polypeptide units?

Related to the problem of regulation of membrane biogenesis is the question of the extent to which membranes are self-perpetuating or control their own formation (327). At one extreme, extrachromosomal inheritance occurs when self-duplicating parts of the cytoplasm are passed on to daughter cells. At the other extreme is de novo construction in successive generations from information carried by nuclear genes. Among eukaryotic membranes, there are no clear examples of either extreme. Mitochondria (9, 202, 356) and chloroplasts (30, 447) are only semiautonomous in their biogenesis and subject to regulation by nuclear genes. Yet, endoplasmic reticulum or even Golgi apparatus seem at times necessary for their own formation (275). Examples of regulation of the biosynthesis of specific cell components are cited in the sections which follow.

ORIGIN AND BIOGENESIS OF SPECIFIC MEMBRANE COMPONENTS

The preceding sections of this review underscore how little of what is known about membrane biogenesis is generally applicable to all types of membranes. The remainder of the discussion will focus on specific cell components and membranous organelles.

Nuclear envelope

Barer et al (13), in their studies on the origin of the nuclear membrane, discussed organization of newly formed nuclear envelope along the chromosomes until a complete double membrane was formed around the nucleus and "the fully formed membrane appears to be lifted off the surface of the chromosomes." Chromosomes are attached to the nuclear membrane at multiple sites (see 108, 147) throughout the cell cycle (155). At least some of the attachments of fragments of the nuclear membrane to chromatin (109) seem to persist through mitosis and may serve as foci for the formation of new nuclear membrane at telophase (58). Direct contribution of endoplasmic reticulum to nuclear envelope has been demonstrated by Flickinger (105) with *Amoeba* where nuclei were damaged microsurgically.

There is increasing evidence that the 8 + 1 subunits of the apparatus of the nuclear pores are rich in RNA (111, 357). Nuclear pore density doubles during interphase in mammalian cells (241, 242) and a threefold increase was noted during "activation" of dormant carrot discs (175), but only fragmentary information is available on their mode of formation (98, 108). A rapid rate of formation of 8 pores per second has been estimated by Scheer (358) for nuclei of *Xenopus laevis* oocytes during the midlampbrush stage of development.

In the much-enlarged nucleus of the maturing egg of *Aedes,* a karyosphere containing chromosomes is segregated from the bulk of the nuclear material by several layers of membrane formed well beneath and distant from the existing nuclear envelope (100). Prior to formation of this membrane, ring structures arise at the peripheries of modified synaptonemal complexes and then appear to dissociate and reassociate at the rim of the karyosphere to form an "annulated pseudomembrane" in which annulae are connected by fine fibers. At later stages of development, the annulated pseudomembranes seem to "gain substance" and eventually resemble nuclear envelope. The model proposed by Fiil & Moens (100) assumes that the regions which appear to give rise to annulae correspond to nuclear envelope-chromosome attachment sites and that the sheets of transverse filaments which connect the annulae are descendent (or derived in a manner similar to) constituents of the nuclear membrane. More important perhaps, the findings provide evidence for direct involvement of chromosomes or chromosomal derivatives in membrane biogenesis or organization.

Endoplasmic Reticulum

The endoplasmic reticulum (ER) is a major site for membrane biogenesis (89), and its origin is central to the derivation of other membrane systems (274). Rough ER may be responsible for aspects of its own formation and an important source of membrane and membrane constituents for the formation and replacing of other endomembranes during cellular ontogeny. As emphasized in the preceding sections, the ER plays a substantial role in phospholipid metabolism (219, 387, 418), cholesterol biosynthesis (116, 145), and the biosynthesis of membrane proteins (268). Additionally, ER may function in the biosynthesis of glycoproteins (60, 141, 211, 264, 265, 324, 372) and glycolipids (51, 192, 216). One cannot exclude the possibility that among endomembranes, synthesis of certain membrane proteins and nitrogen-containing phosphoglycerides resides predominantly or exclusively in ER (60, 219, 268).

Rough ER is continuous with the outer membrane of the nuclear envelope (305, 310) suggesting its possible derivation from the nuclear envelope or vice versa. In some cells, membranes of the Golgi apparatus appear to originate from specific ribosome-free sites along the outer membrane of the nuclear envelope (cf 275). These vesicles fuse to form flattened cisternae in a manner similar to derivation of Golgi apparatus membranes from rough ER (275). Although neither ER nor Golgi apparatus appear to depend solely upon the nuclear envelope for their biogenesis, derivation as an extension of the nuclear envelope during early development or to facilitate cellular differentiation is a definite possibility (275, 305). Comparing a variety of hormonal and environmentally induced conditions of treatment, Tata and co-workers (400) found a temporally coordinated enhancement of microsomal phospholipid formation and of microsomal RNA and protein synthesis. Interestingly, newly formed ER is thought to be localized in the perinuclear region during the initial stages of induction (400).

Systems of Transition Elements

Generation of the remainder of the cell's endomembrane system seems to involve a variety of "smooth" membranes and vesicles lacking ribosomes. Because these membranes have characteristics intermediate between rough endoplasmic reticulum at one extreme and tonoplast or plasma membrane at the other extreme, and because they may facilitate the conversion of one type of membrane to another, they are referred to as transition elements (274).

SMOOTH ENDOPLASMIC RETICULUM   The most general form of a transition element is smooth ER (298, 310) which is directly continuous with rough ER (298, 310). In this category of smooth ER is included smooth-surfaced portions of rough ER cisternae. Evidence from a variety of experiments (mostly applied to rat liver) that smooth ER is synthesized first as rough ER has been reviewed (88, 89, 162, 205, 274, 293). Evidence is insufficient to decide if smooth ER is derived from rough ER through loss of ribosomes or, alternatively, if smooth ER arises from the outgrowths of ribosome-free regions continuous with rough ER. According to the latter, ribosomes are never attached to smooth membranes. In in vitro experiments, membranes from smooth ER fractions are reported to bind ribosomes with the same affinity as rough ER membranes but have one-fifth the number of binding sites (338). Induction of drug-metabolizing enzymes in the liver is accompanied by a proliferation of the smooth ER (130). Induction of microsomal oleyl-coenzyme A desaturase has been reported during "aging" of potato tuber slices (1); parallel induction of stearyl coenzyme A desaturase activity and NADH cytochrome c reductase occurs in rat liver of fasted and refed rats (296, 297). The synthesis of new enzymes appears geared to the formation of new membranes by a mechanism that may also involve outgrowth and budding of smooth ER profiles from the rough ER at or near sites where the synthesis of the enzymes takes place (89, 295). Simultaneously, there is a decrease in the rate of phospholipid catabolism (161) which contributes to the increased phospholipid level (130). As to phospholipid biosynthesis per se, there is uncertainty as to whether any differences exist in the rate of synthesis between the rough and smooth ER in liver. Schneider (363) came to the general conclusion that little difference existed. Getz (118) found no difference for PC in adult rat liver but a higher rate in the rough ER of newborn animals. Holtzman et al (162) showed that rough ER incorporated more ^{32}P from $\gamma^{32}P$-ATP into PE in vivo than did smooth ER, while Williamson & Morré (450) found rough and smooth ER fractions to be nearly equal in the biosynthesis of PI.

Both rough and smooth segments of the ER exhibit similar relative enzymatic compositions (64, 88, 89), similar protein gel patterns (73, 268, 362), the same soluble antigens [except for a highly basic component present only in extracts of rough membranes (467)], and only small differences in lipid composition (120, 213). Despite these compositional similarities and continuity with rough membranes, the possibility of smooth ER biogenesis by assembly from components of a cytoplasmic pool may prove difficult to exclude completely. For example, newly appearing glucose-6-phosphatase activity of ER in the newborn rat (the result of de novo synthesis) is localized completely in rough microsomes. During a rapid increase in this activity, which becomes apparent a few hours after birth, activity appears in the smooth fraction, but several days are required before the two fractions display the equal activity characteristic of the mature animal (88). Although a common origin is indicated, Stetten & Ghosh (389) reported different pH optima for glucose-6-phosphatases in rough and smooth microsomes. Membranes of smooth ER differ from those of rough ER in important functional respects. Most significant is the property of smooth ER (including smooth-surfaced portions of rough ER cisternae) to give rise to small blebs or vesicles which are suggested to migrate away from the reticulum and fuse to form other systems of transition elements such as Golgi apparatus cisternae (see below).

GOLGI APPARATUS   Based on evidence from plants (275) and rat liver (270) it has been assumed that Golgi apparatus membranes are synthesized in the rough ER (see, however, 263). Membranes are thought to separate from rough ER, transform into smooth membranes and give rise to Golgi apparatus cisternae either by fusion of small ER-derived vesicles or through direct continuities with smooth ER. A similar origin has been suggested from smooth (lacking ribosomes) portions of the outer membrane of the nuclear envelope (275). As the ER-derived vesicles fuse to form the flattened cisternae, a balance is maintained by a simultaneous transformation of other membranes of the Golgi apparatus into membranes of secretory vesicles. Evidence for membrane differentiation within systems of transition elements, especially the Golgi apparatus, is considerable (78, 134, 195, 274, 275, 336, 426, and ref. cit.). Cisternae at one face of the Golgi apparatus resemble ER, whereas cisternae at the opposite face are more like plasma membrane. Intercalary cisternae are intermediate in appearance between these two extremes. Biochemical studies comparing ER, Golgi apparatus, and plasma membrane from rat liver provide one important source of information upon which the concept of membrane differentiation in Golgi apparatus is based. These findings suggest that the intermediate character of transition elements is exemplified by the enzymatic activities, lipid composition, and progressive biochemical modification (e.g. glyosylation) of the lipids and proteins of the membrane (20, 268–270, 274).

In addition to undergoing membrane differentiation, Golgi apparatus are dynamic. Calculations of Bowles & Northcote (41) suggest a turnover rate of one cisternal stack every 20 sec for root tips of maize, although alternative explanations of their findings are possible. This is very rapid when compared with values estimated for turnover in other plant and animal systems: 20 min per stack and individual cisternae released every 1–4 min (270, 275). However, work of Brown (46) indicates that the complex surface scales of *Pleurochrysis* are synthesized and discharged along with an entire Golgi apparatus cisternae in less than 2 min. It should be emphasized that these estimates of turnover are based on rates of formation and loss of cisternae through vesiculation and not from turnover through degradation of membrane constituents. The latter occurs much too slowly [$T_{1/2}$ of 37.5 hr for liver (113)] to account for any significant utilization of Golgi apparatus membranes during active secretion (turnover time of 20 min or less).

In summary, evidence for electron microscopy and compositional and kinetic analyses support a concept of Golgi apparatus biogenesis where cisternae are produced at one pole of the stacked cisternae from membrane materials derived from endoplasmic reticulum or nuclear envelope. A constant number of cisternae per dictyosome is maintained by loss of cisternae at the opposite pole as secretory vesicles are produced.

SECRETORY VESICLES   Some systems of transition elements such as Golgi apparatus give rise to secretory vesicles which detach from their sites of origin and migrate to the plasma membrane (66, 275, 336, 421, 426, 446). As the membranes of the secretory vesicles fuse with the plasma membrane, the vesicle membranes are incorporated into the plasma membrane. Secretory vesicles derived from the Golgi apparatus provide a structural basis for one pathway (113, 135, 255, 336, 421, 426) while vesicles from ER (113, 270, 275, 277, 430) or other transition elements (42) provide alternative pathways. The secretory vesicle is the most extensively documented example of a cell component derived from one membrane structure (the Golgi apparatus or other transition element) which migrates to and coalesces with another membrane structure (the plasma membrane) to effect the physical transfer of membrane, i.e. membrane flow. An essential point is that endomembrane-derived structures such as secretory vesicles do contribute membrane constituents to the plasma membrane and, in certain cell types such as fungal hyphae (42, 135) and elongating pollen tubes (277, 420, 421) represent a major, if not the sole, source of plasma membrane.

OTHER TRANSITION ELEMENTS   The cytoplasmic vesicle or *primary vesicle* which provides structural and functional continuity between systems of transition elements also provides a morphological link between smooth ER and the Golgi apparatus. The bounding membranes of these vesicles are sometimes coated by an electron-dense material organized in an alveolate or honeycomb pattern (283). Other systems of transition elements may give rise to vacuoles or other cellular structures. Fungi richly abound in a variety of transition elements, some not easily categorized along traditional systems of membrane classification (42). Other examples are given by Morré & Mollenhauer (274) and Franke et al (110).

Mitochondria

Structural details of mitochondrial genesis, including the theory that mitochondria multiply by increase in size followed by division (317, 353), remain unclear. Experiments of Luck (222 and ref. cit.), using autoradiography and density centrifugation, established a semiconservative pattern for replication of mitochondria from a choline-requiring mutant of *Neurospora crassa*. The validity of these findings has survived the decade, but our structural concepts of the mitochondrion have altered radically. Hoffmann & Avers (159) presented evidence for a single mitochondrion per cell in yeast. Using serial sectioning techniques, they showed that all the mitochondrial profiles were sections through a single branching tubular structure which upon cell division was split between the two daughter cells. *Chlorella* cells also possess a single mitochondrion (10). Giant mitochondria and/or a mitochondrial reticulum of one to several extensively branched mitochondria have been reported for a variety of algal genera (see ref. 10) and for rat liver (44), with mitochondrial organization approximating the sense of the early chondriome concept. The extensive ramification of a mitochondrial reticulum brings parts of the chondriome into association with all of the cell organelles; concepts of growth and division of numerous small mitochondria differ from those based on continuous growth and division of a reticular system at cytokinesis (159). Additionally, concepts of mitochondrial genesis based on presumptive promitochondrial structures in mature rat liver (353), for example, may require reevaluation.

The genetic information of mitochondria has for its expression a transcription and translation system which appears to have no components in common with its nucleocytoplasmic counterpart (9, 36, 37, 395). The properties of the mitochondrial system have been reviewed extensively in the proceedings of a recent conference (202) and differ somewhat among different plants and animals on the basis of sensitivity to antibiotics (132, 406). Biochemical studies indicate a modest contribution of mitochondrial DNA to the synthesis of mitochondria, mostly to synthesis of very hydrophobic proteins of the inner membrane system (257). The great majority of the mitochondrial proteins are specified by nuclear genes and synthesized on cytoplasmic ribosomes. Significant incorporation of labeled amino acids by isolated mitochondria is into proteolipid (180).

As more than 90% of the mitochondrial proteins may be synthesized outside mitochondria (146), the question arises as to which class of cytoplasmic polyribosomes are involved. Recent studies by Gonzáles-Cadavid & de Córdova (127) demonstrate that both free and membrane-bound polysomes of rat liver can effect the synthesis of cytochrome *c* but conclude that in vivo, cytochrome *c* is probably synthesized predominantly by rough endoplasmic reticulum since this fraction accounts for more than 70% of the total polysomes in this tissue.

Isolated mammalian mitochondria synthesize a number of acidic phospholipides (74) usually present in low concentrations in the mitochondria such as lysophosphatidic acid (63), phosphatidic acid (81, 249, 352, 465; see, however, 68), phosphatidylglycerol (67, 81, 259), and cardiolipin (67, 81, 165), as well as ubiquinone-9 (409). The requirement for these lipids may be satisfied by endogenous biosynthetic reactions in the mitochondrion (74). In contrast, negligible net synthesis of PC, PE, or PI, the phospholipids which constitute the bulk of mitochondrial lipid, could be shown with mitochondria in vivo (248, 249, 451). A mechanism to supply these lipids is provided by exchange or transfer from endoplasmic reticulum (2, 11, 29,

70, 176, 177, 182, 208, 249, 300, 301, 352, 387, 451–457). The exchange in vitro can be in either direction and is facilitated by cytoplasmic proteins (182, 249, 452, 453, 456).

Pulse-chase experiments (177) and autoradiography (387) provide support for this exchange, as do turnover studies in which the half-life times for each class of phospholipids of endoplasmic reticulum and mitochondria are similar. Fatty acids esterified into both mitochondrial and microsomal phospholipids also have similar half-life times (302).

When microsomes of potato tubers or cauliflower inflorescence, labeled with ^{32}P or ^{14}C-glycerol or acetate, were incubated with unlabeled mitochondria, there was a flow of labeled phospholipid, mainly PC and PE, to the mitochondria (2). The exchange was reversible and depended on the presence of a supernatant fraction whose phospholipids became heavily labeled in the process.

Transfer of lipid from microsomes to outer mitochondrial membranes has therefore been suggested as a general phenomenon to be followed by redistribution between outer and inner mitochondrial membranes (29). This concept is supported by findings from rat liver where the total fatty acid composition of PC and of other phospholipids are very similar in whole mitochondria, outer and inner mitochondrial membranes, and microsomes (55, 119, 224, 390). In contrast, outer and inner membranes of mitochondria from buds of cauliflower (*Brassica oleracea*) differ from each other and from microsomes (266). Even if the major phospholipids were assumed to originate from a common pool, the authors suggest that some local rearrangements in fatty acid composition are necessary to explain the specificity of fatty acid distribution among the different membrane fractions.

Deacylation (281, 436) and acylation (281, 352, 411, 439) reactions described for mammalian mitochondria could yield molecular classes of phospholipids specific for mitochondrial requirements in the absence of de novo synthesis (416, 418). The magnitude of independent acylating activity in mitochondria is usually considered small relative to the larger bulk transfer of newly synthesized phospholipid from endoplasmic reticulum (300, 301)

The proteins synthesized by isolated mitochondria are highly insoluble and hydrophobic (413). Early studies suggested that mitochondria synthesized a structural protein of the inner membranes, but this popular idea was abandoned when the fraction was shown to be a collection of many different proteins including those of the ATPase complex (413). The inner membrane of mammalian mitochondria contains at least 23 different protein components on polyacrylamide gel electrophoresis, and at least 12 different proteins are found in the outer membrane; smooth and rough microsomal membranes contain 15 different components (362). At least three protein species appear similar in outer mitochondrial membrane and smooth microsomes (362). In addition, the outer mitochondrial membrane is similar to the microsomal membrane in that both contain NADH-cytochrome c reductase and cytochrome b_5. However, monamine oxidase and kynurenine hydroxylase are found only in the outer mitochondrial membrane, while a number of enzymes such as glucose-6-phosphatase are found only in microsomal membranes of liver (see 9, 362). Thus, although the proteins of endoplasmic reticulum and outer mitochondrial membrane may have common origins, extensive transformation would be required for direct derivation of one membrane from the other.

Plastids

Like mitochondria, plastids contain basic components necessary for autonomy (40, 126, 196). These components are thought to be replicated in a process closely coupled with cell division (31, 311). The ribosomes and protein-synthesizing machinery resemble those of bacteria and mitochondria more closely than they do their nucleocytoplasmic counterparts (196, 256). Yet protein synthesizing systems of mitochondria and chloroplasts may differ in important respects including distinct elongation factors (54).

Three broad categories of methods have been used to elucidate the genes encoded in chloroplast DNA and the identities of the proteins synthesized by chloroplast ribosomes (197, 447). They include: (a) use of inhibitors; (b) analyses of mutants; and (c) identification of protein and RNA molecules synthesized by isolated chloroplasts. All three methods suggest a model for cooperation between nuclear and chloroplast genomes in the biogenesis of chloroplast membranes and catalytic proteins (6, 30, 52, 164, 256, 380, 433, 447) but have failed to disclose a specific role for the chloroplast genome (447). Many of the genes which affect the plastid are nuclear genes (197), including the DNA which codes for the photosystem II chlorophyll-protein complex (467). Yet at least some lamellar proteins appear to be synthesized on the 70S ribosomes of the chloroplast (28, 226).

The following steps are now recognized in the assembly of photosynthetic membranes in plants grown in the dark and then exposed to light (333, 433). The leaf cells of angiosperms contain proplastids which develop into etioplasts with a characteristic three-dimensional lattice of tubules—the prolamellar body or crystalline center. In etioplasts, carotenoids and protochlorophyllide holochromes (137) are associated with the tubular membranes of the prolamellar body. These membranes contain several peptides found in chloroplasts (106) and are presumed to serve as precursors to photosynthetic lamellae. Protochlorophyllide holochrome subunits with an apparent molecular weight of 63,000 and statistically carrying a single chromophore per protein particle have been obtained from the membranes of prolamellar bodies with the aid of saponin (152). Upon exposure to light, the photoreduction of protochlorophyllide to chlorophyllide (285, 433) triggers a series of morphological changes in the tubular membranes of the prolamellar body. It appears that in pea, bean, oat, and barley, a light-induced reorganization of the internal etioplast membranes together with a small amount of chlorophyll formation suffice to organize photosynthetic units containing both photosystem I and II activities (151, 333, 408).

Changes in the absorption and circular dichroism spectra after photoreduction of the protochlorophyllide holochrome subunits in vitro (107) and a decrease of their apparent molecular weight from 63,000 to 29,000 suggest a dissociation of the subunits and conformational relaxation of the holochrome protein as early events in the reorganization of the prolamellar body membranes (153, 207, 365). Esterification of the newly formed chlorophyllide with phytol (433) is linked with the Shibata shift in the red absorption maximum of chlorophyllide in vivo (153).

Several investigators (for references see 4, 151) have reported that photosystem I activity appears before that of photosystem II. More recently, Alberte et al (4) suggest from studies with greening leaves of jack bean that the P_{700}-chlorophyll-protein molecules are first inserted into developing lamellae, and that the photosystem I chlorophyll-protein molecules which function solely in a light-harvesting capacity are added later.

The process of etioplast-chloroplast transformation is complex with interrelated steps subject to precise control by light intensity and quality (408). The light responsive steps are apparently within the etioplast. Etioplasts isolated from *Zea mays* respond to light by increased synthesis of lamellar proteins, a "turning off" of some proteins synthesized in the dark, and the initiation of synthesis of others (148).

During normal development, as meristematic cells of the shoot develop into a green leaf, proplastids transform into chloroplasts without intermediate etioplast stages. Grana are formed directly from vesicles or tubular extensions of membranes that arise by invagination of the inner membrane of the proplastid. Formation of chlorophyll and other constituents of the chloroplast and differentiation of photosynthetic lamellae occur simultaneously (197), but details of the morphogenesis of photosynthetic lamellae during normal greening are lacking.

During degreening and regreening of *Chlamydomonas* both the electron transfer chain and the chlorophyll of the active center of photosystem I change continuously; variations in efficiency of light utilization are thought to be due both to changes in chlorophyll and in membrane composition (94) and organization (95).

Other than protein, chloroplast membranes contain predominantly mono- and digalactosyl diglycerides along with sulfoquinovosyldiglyceride and phosphatidylglycerol (187). The phospholipids represent less than 20% of the chloroplast lipids (158). The galactolipids apparently are unique to chloroplast membranes [mitochondria from etiolated mung bean hypocotyls or potato tubers contain no galactosyl diglycerides (246); see, however, (364)] and have α-linolenic acids as the predominant polyunsaturated fatty acids (392). Chlorophylls are the most important pigments. These too are confined to plastids apparently associated with the photosynthetic lamellae. Additionally, chloroplasts contain a variety of carotenoids, xanthophylls, and quinones (197). Phycobilins are found in plastids of certain algae (126, 197).

Plastids may express considerable autonomy for biosynthesis of pigments and lipids (126). They seem to carry out the complete synthesis of chlorophyll, carotenoids, xanthophylls, quinones, and a variety of lipids including the galactolipids and sulfolipids (126, 172, 187, 196, 197, 279, 299, 391, 392). On the other hand, the phosphatidylglycerol, a major component of chloroplast lamellae (158, 243), has been suggested to originate in endoplasmic reticulum in leaves (234), and evidence has been presented for extrachloroplastic biosynthesis of mono— and digalactosyl glycerides (424) and sterols (126) with subsequent transfer to plastids. Chloroplast envelopes have a lipid composition intermediate between that of chloroplast lamellae and microsomal fractions (227) as expected if both endoplasmic reticulum and plastidal biosynthesis contributed to their formation. However, with etiolated peas, a 2 hr illumination enhanced incorporation of acetate by isolated etioplasts but did not enhance microsomal lipid synthesis (299).

Plastids undergo differentiation into a variety of forms. Proplastids differentiate into any one of several different plastid types required of specialized cells (5, 197). In fruits, the ripening process results in the conversion of chloroplasts to chromoplasts (197) as one example. Biosynthetic events associated with membrane changes during plastid differentiation have been little studied.

Microbodies (Peroxisomes, Glyoxysomes)

In general, microbodies are thought to arise by dilation of the endoplasmic reticulum (71, 136, 166, 286, 289, 322, 428, 429) in agreement with early findings of Novikoff & Shin (288) and of Essner (90) and Tsukada et al (410) on the origin of microbodies in fetal rat livers. An origin of microbodies from endoplasmic reticulum in liver is supported from observations of numerous connections between these two cell components from mice treated with ethyl-α-p-chlorophenoxyisobutyrate (322, 323). Reddy & Svoboda (323) suggest that microbodies are not discrete organelles capable of independent growth and maturation but exist rather as extensions of endoplasmic reticulum during growth and development. Novikoff & Novikoff (289) observed multiple peroxisome-endoplasmic reticulum attachments in absorptive cells of mammalian small intestine which were interpreted as localized dilations of smooth endoplasmic reticulum. The class of small microbodies continuous with smooth endoplasmic reticulum has been termed microperoxisomes (289) and are suggested to function as peroxisomal progenitors (287). Continuities between microbodies and membranes of endoplasmic reticulum are rare or absent in adult rat liver (215) and elongating plant cells.

Few alternatives to derivation from endoplasmic reticulum have been suggested for cytoplasmic microbodies. A mechanism based on proliferation by fragmentation or budding from preexisting microbodies is favored by Legg & Wood (215). Membranes or microbody-like structures found in certain proteoplasts may originate from internal plastidal membranes (W. J. Hurkman and G. S. Kennedy, unpublished).

Unlike mitochondria and chloroplasts, microbodies of castor bean (glyoxysomes) lack DNA (82) and are unlikely to be even semiautonomous. Chlorella cells have been reported to contain a single microbody which has not been observed to divide autonomously prior to cytokinesis (10). Atkinson et al (10) suggest that microbodies are formed de novo in each daughter cell.

In attempts to investigate the synthesis of rat liver microbodies, Poole et al (308) separated these organelles according to size by zonal centrifugation and showed that the specific radioactivity of catalase was independent of the size of the isolated particles and that there was a homogeneity of peroxisomal enzyme distribution in particles of different sizes. Based on turnover studies, they could not decide if microbodies exist as individuals, each with a life history independent of the others, or if the microbody proteins form a common pool from which materials were readily exchanged (see 71). At least one significant protein of the interior of microbodies, catalase, is synthesized outside; addition of heme and completion of the molecule takes place within the microbody (212).

Beevers and colleagues (218, 219) provide evidence that endoplasmic reticulum is the principal site of synthesis of at least three major glycerophosphatides which are major phospholipid components of glyoxysomal membranes (80). If microbodies are indeed continuous with smooth endoplasmic reticulum during their ontogeny, specific transport proteins might not be required to effect the transfer of these lipids during microbody genesis. Except for PI, the fatty acid composition of the major phospholipids of endoplasmic reticulum and microbody membranes are very similar (R. P. Donaldson and H. Beevers, unpublished; 79).

Spherosomes and Lysosomes

A derivation of spherosomes from endoplasmic reticulum has been suggested (114) as has the reverse pathway (434). In mammalian cells, lysosomes appear to originate through cooperative action of endoplasmic reticulum and Golgi apparatus in a specialized region of the cytoplasm abbreviated GERL (288); the process probably is analogous to the formation of secretory vesicles (125). In plants distinctions between lysosomes and vacuoles are somewhat arbitrary. One concept is that vacuoles or vacuolar derivatives may function as lysosomal equivalents (49, 61, 238, 239).

Vacuole Membrane (Tonoplast)

Plant vacuoles have been suggested to arise from endoplasmic reticulum (25, 42, 49, 61, 99, 238–240, 334), Golgi apparatus-derived vesicles or cisternae (61, 232, 240, 414), or from plasma membrane (228). None of these origins are mutually exclusive, and there is little definitive evidence to substantiate or refute their generality. In general, the bounding membranes of vacuoles seem to derive from parts of preexisting membrane systems of the cytoplasm (49). The latter may be derived ultimately from endoplasmic reticulum. Direct continuities (25, 274) and close associations (101) of plant vacuole membranes with rough endoplasmic reticulum or Golgi apparatus have been reported, but more usual is the intermediate participation of transition elements or provacuolar structures derived from endoplasmic reticulum which then coalesce or enlarge to form the large central vacuole which characterizes mature plant cells (228, 274). In some fungi, membranes of young vacuoles are continuous with rough endoplasmic reticulum, suggesting a ubiquitous role of rough endoplasmic reticulum in generating tonoplast membrane (42).

Plasma Membrane

Most cells double in size during the cell cycle; if spherical, the surface area increases by a factor of $2^{2/3}$ or approximately 1.6 (129). The absolute rate of plasma membrane formation in rat liver cells has been calculated by Franke et al (113) from turnover data as 7.1 micropicograms/min/hepatocyte. In rapidly elongating fungal hyphae, the rate of increase is calculated to be 32 μ^2/min (C. E. Bracker, unpublished). Although well-studied examples are few, it appears that in dividing or elongating cells, new membrane material is incorporated at the cell surface where it accumulates (129). However, in the absence of cell division or growth, the production of new membrane is compensated for by degradation (441, 442). The latter may relate to modification or replacement of informational molecules (antigens, hormone receptors) of the cell surface during differentiation.

On the basis of experiments on the incorporation of radioactive amino acids into membrane proteins of rat liver, Ray et al (320) and Franke et al (113) concluded that newly formed proteins are not added directly to the cell surface. Ray et al (320) provided evidence for a pool of precursors that continued to be incorporated into the surface membrane even after protein synthesis was inhibited by cycloheximide.

In the synthesis and turnover studies of Franke et al (113), two phases of incorporation of ^{14}C-arginine were observed for plasma membrane of rat liver in vivo. The second phase of incorporation (between 10 and 20 min) was equated with the fusion of secretory vesicles derived from Golgi apparatus with the plasma membrane. The first phase of incorporation (2 to 10 min) was ascribed to a more direct route of incorporation involving endoplasmic reticulum but bypassing the Golgi apparatus (113). An ultrastructural basis for interpretation of the initial phase of incorporation involves secretory vesicles derived directly from smooth endoplasmic reticulum in liver (270, 274). If this interpretation is correct, the relative contributions of the two pathways would be approximately equal with about half the plasma membrane arising by fusion of secretory vesicles derived from the Golgi apparatus and an equal contribution bypassing the Golgi apparatus.

Preliminary experiments with onion stem in which the kinetics of labeling with ^{14}C-leucine were studied agree with the findings from rat liver (92, 113, 320) in that peaks of radioactivity first occurred in endoplasmic reticulum and later in plasma membranes (277). Similarly, the glucan synthetase activities of Golgi apparatus and endoplasmic reticulum fractions have been interpreted as activity enroute to its site of primary action at the cell surface (374, 419). Relative contributions of Golgi apparatus- or endoplasmic reticulum-derived vesicles to plasma membrane formation in plants (277, 430) are unknown.

Plasma membranes perhaps should be regarded as potentially the least autonomous of the cellular membranes, although there is no proof of this. Plasma membranes do not possess the enzymes required for the de novo synthesis of nitrogen-containing phospholipids such as PC or PI (268, 427, 450). The de novo synthesis of these phospholipids may occur exclusively in the endoplasmic reticulum (29, 70, 219, 249, 268, 418, 423, 451, 452) and Golgi apparatus (268, 271, 276; see, however, 423). Thus the supply of phospholipid components of the plasma membrane would seem to be highly dependent on the endomembrane system. The exchange of phospholipids between plasma membranes and microsomes via the intermediation of a soluble lipoprotein of the cytoplasm has been demonstrated for liver (455) but may not operate in brain [between microsomes and myelin (258)]. There is no information on this point from plants.

With glycoproteins, smooth membranes are labeled first in pulse-chase experiments and the labeled material then appears progressively in the plasma membrane (20, 38). Different rates of synthesis of membrane glycoproteins and glycolipids may indicate that glycolipids are incorporated into the plasma membrane independently of newly synthesized proteins (60, 92). The localization of glycolipid biosynthetic enzymes in Golgi apparatus (60, 193) is consistent with this observation. For plants, the very interesting subject area of the origin of glycolipids and glycoproteins characteristic of plasma membranes (60, 140) is largely unexplored.

CONCLUDING COMMENTS

The lack of rapid progress in understanding the biogenesis of plant membranes may be due in part to the lack of ready availability of purified and well-characterized endomembrane fractions in useful quantity from plant sources. There are also difficulties inherent in separating plastids and mitochondria and in separating the tonoplast from the plasma membrane. The absence of plastids and vacuoles enhances the probability of obtaining purified fractions from animal sources. Based on the comparative studies available for those cell components common to both types of organisms, patterns of membrane biogenesis appear similar in plants and animals.

Three points are stressed in summary. 1. No membranous cell component of eukaryotic cells is either truly autonomous or without some degree of semiautonomy in its biogenesis. The interdependence of cytoplasm and organelle is clearest for mitochondria and chloroplasts, but it now appears that some proteins of endoplasmic reticulum membranes may be synthesized on cytoplasmic polysomes while microbodies, vacuoles, and even secretory vesicles seem to exert some degree of control over their own destiny. 2. The endoplasmic reticulum is a cell component of considerable importance to membrane biogenesis in eukaryotic cells. This cell component contains the most complete array of lipid, protein, and carbohydrate biosynthetic machinery. Perhaps no cellular membrane is independent of the endoplasmic reticulum in its biogenesis. Some, like Golgi apparatus, microbodies, vacuoles, and plasma membrane may be highly dependent on endoplasmic reticulum for their formation as are mitochondria and possibly plastids. 3. Because of the high degree of interaction among various cellular components during their biogenesis, the field of regulation of membrane biogenesis—the mechanisms whereby synthesis of membrane components are initiated, coordinated, and eventually terminated when synthesis is complete—emerges as one of the most challenging and potentially rewarding study areas of contemporary plant physiology.

Literature Cited

1. Abdelkader, A. B., Cherif, A., Demandre, C., Mazliak, P. 1973. *Eur. J. Biochem.* 32:155–65
2. Abdelkader, A. B., Mazliak, P. 1970. *Eur. J. Biochem.* 15:250–62
3. Akiyama, M., Sakagami, T. 1969. *Biochim. Biophys. Acta* 187:105–12
4. Alberte, R. S., Thornber, J. P., Naylor, A. W. 1973. *Proc. Nat. Acad. Sci. USA* 70:134–37
5. Anstis, P. J. P., Northcote, D. H. 1973. *New Phytol.* 72:449–63
6. Apel, K., Schweiger, H. G. 1973. *Eur. J. Biochem.* 38:373–83
7. Arias, I. M., Doyle, D., Schimke, R. T. 1969. *J. Biol. Chem.* 244:3303–15
8. Arienti, G., Pirotta, M., Giorgini, D., Porcellati, G. 1970. *Biochem. J.* 118:3P
9. Ashwell, M., Work, T. S. 1970. *Ann. Rev. Biochem.* 39:251–90
10. Atkinson, A. W., John, P. C. L., Gunning, B. E. S. 1974. *Protoplasma* 81:77–109
11. Bailey, E., Taylor, C. B., Bartley, W. 1967. *Biochem. J.* 104:1026–32
12. Baker, R. R., Thompson, W. 1972. *Biochim. Biophys. Acta* 270:489–503
13. Barer, R., Joseph, S., Meek, G. A. 1959. *Exp. Cell Res.* 18:179–82
14. Beattie, D. S. 1969. *Biochem. Biophys. Res. Commun.* 35:67–74
15. Beattie, D. S. 1969. *J. Membrane Biol.* 1:383–401
16. Beattie, D. S., Basford, R. E., Koritz, S. B. 1966. *Biochemistry* 5:926–30
17. Ibid 1967. 6:3099–3110
18. Behrens, N. H., Carminatti, H., Staneloni, R. J., Leloir, L. F., Cantarella, A. I. 1973. *Proc. Nat. Acad. Sci. USA* 70:3390–94

MEMBRANE BIOGENESIS 473

19. Benedetti, E. L., Emmelot, P. 1967. *J. Cell Sci.* 2:499–512
20. Bennett, G., Leblond, C. P., Haddad, A. 1974. *J. Cell Biol.* 60:258–84
21. Ben-Tal, Y., Varner, J. E. 1974. *Plant Physiol.* 53(Suppl.):54 (Abstr.)
22. Bentley, R. 1970. See Ref. 440, 481–563
23. Berg, N. B., Young, R. W. 1971. *J. Cell Biol.* 50:469–83
24. Bergeron, J. J. M., Ehrenreich, J. H., Siekevitz, P., Palade, G. E. 1973. *J. Cell Biol.* 59:73–88
25. Berjak, P. 1972. *Ann. Bot.* 36:73–81
26. Bjerve, K. S. 1973. *Biochim. Biophys. Acta* 296:549–62
27. Bjørnstad, P. 1966. *Biochim. Biophys. Acta* 116:500–10
28. Blair, G. E., Ellis, R. J. 1973. *Biochim. Biophys. Acta* 319:223–34
29. Blok, M. C., Wirtz, K. W. A., Scherphof, G. L. 1971. *Biochim. Biophys. Acta* 233:61–75
30. Boardman, N. K., Linnane, A. W., Smillie, R. M., Eds. 1971. *Autonomy and Biogenesis of Mitochondria and Chloroplasts.* Amsterdam/London: North-Holland
31. Boasson, R., Gibbs, S. P. 1973. *Planta* 115:125–34
32. Bolender, R. P., Weibel, E. R. 1973. *J. Cell Biol.* 56:746–61
33. Bont, W. S., Geels, J., Rezelman, G. 1969. *Int. J. Protein Res.* 1:193–97
34. Borgese, N., Kreibich, G., Sabatini, D. 1972. *J. Cell Biol.* 55:24a (Abstr.)
35. Borkenhagen, L. F., Kennedy, E. P., Fielding, L. 1961. *J. Biol. Chem.* 236:PC28–30
36. Borst, P. 1972. *Ann. Rev. Biochem.* 41:333–76
37. Borst, P. 1974. *Biochem. Soc. Trans.* 2:182–85
38. Bosmann, H. B., Hagopian, A., Eylar, E. H. 1969. *Arch. Biochem. Biophys.* 130:573–83
39. Bosmann, H. B., Winston, R. A. 1970. *J. Cell Biol.* 45:23–33
40. Boulter, D., Ellis, R. J., Yarwood, A. 1972. *Biol. Rev.* 47:113–75
41. Bowles, D. J., Northcote, D. H. 1974. *Biochem. J.* 142:139–44
42. Bracker, C. E. 1974. *Proc. 8th Int. Congr. Electron Microsc. Canberra,* 2:558–59
43. Bracker, C. E., Grove, S. N. 1971. *Protoplasma* 73:15–34
44. Brandt, J. T., Martin, A. P., Lucas, F. V., Vorbeck, M. L. 1974. *Biochem. Biophys. Res. Commun.* 59:1097–1103
45. Bretscher, M. S. 1973. *Science* 181:622–29

46. Brown, R. M. 1969. *J. Cell Biol.* 41:109–23
47. Burton, D. N., Haavik, A. G., Porter, J. W. 1968. *Arch. Biochem. Biophys.* 126:141–54
48. Busby, W. F., Hele, P., Chang, M. C. 1974. *Biochim. Biophys. Acta* 335:246–59
49. Buvat, R. 1971. See Ref. 327, 127–57
50. Bygrave, F. L. 1969. *J. Biol. Chem.* 244:4768–72
51. Caccam, J. F., Jackson, J. J., Eylar, E. H. 1969. *Biochem. Biophys. Res. Commun.* 35:505–11
52. Chan, P. H., Wildman, S. G. 1972. *Biochim. Biophys. Acta* 277:677–80
53. Chatterjee, S., Sweeley, C. C., Velicer, L. F. 1973. *Biochem. Biophys. Res. Commun.* 54:585–92
54. Ciferri, O., Tiboni, O. 1973. *Nature New Biol.* 245:209–11
55. Colbeau, A., Nachbaur, J., Vignais, P. M. 1971. *Biochim. Biophys. Acta* 249:462–92
56. Coleman, R. 1973. *Biochim. Biophys. Acta* 300:1–30
57. Coleman, R., Finean, J. B. 1968. *Compr. Biochem.* 23:99–126
58. Comings, D. E., Okada, T. A. 1970. *Exp. Cell Res.* 62:293–302
59. Cook, G. M. W. 1971. *Ann. Rev. Plant Physiol.* 22:97–120
60. Cook, G. M. W., Stoddart, R. W. 1973. *Surface Carbohydrates of the Eukaryotic Cell.* New York/London: Academic
61. Coulomb, P., Coulomb, C., Coulton, J. 1972. *J. Microsc.* 13:263–80
62. Cuatrecasas, P. 1974. *Ann. Rev. Biochem.* 43:169–214
63. Daae, L. N. W. 1972. *Biochim. Biophys. Acta* 270:23–31
64. Dallner, G., Ernster, L. 1968. *J. Histochem. Cytochem.* 16:611–32
65. Dallner, G., Siekevitz, P., Palade, G. E. 1966. *J. Cell Biol.* 30:97–117
66. Dauwalder, M., Whaley, W. G., Kephart, J. E. 1972. *Sub-Cell. Biochem.* 1:225–75
67. Davidson, J. B., Stanacev, N. Z. 1971. *Biochem. Biophys. Res. Commun.* 42:1191–99
68. Davidson, J. B., Stanacev, N. Z. 1972. *Can. J. Biochem.* 50:936–48
69. Dawson, R. M. C. 1966. In *Essays in Biochemistry,* ed. P. N. Campbell, G. D. Greville, 2:69–115. New York: Academic
70. Dawson, R. M. C. 1973. *Sub-Cell. Biochem.* 2:69–89

71. de Duve, C. 1973. *J. Histochem. Cytochem.* 21:941–48
72. de Duve, C., Wattiaux, R. 1966. *Ann. Rev. Physiol.* 28:435–92
73. Dehlinger, P. J., Schimke, R. T. 1972. *J. Biol. Chem.* 247:1257–64
74. Dennis, E. A., Kennedy, E. P. 1972. *J. Lipid Res.* 13:263–67
75. Deutsch, J., Blumenfeld, O. O. 1974. *Biochem. Biophys. Res. Commun.* 58:454–59
76. Devor, K. A., Mudd, J. B. 1971. *J. Lipid Res.* 12:403–11
77. Dingle, J. T., Fel, H. B. 1969. *Lysosomes in Biology and Pathology.* Amsterdam: North Holland
78. Dobberstein, B., Kiermayer, O. 1972. *Protoplasma* 75:185–94
79. Donaldson, R. P., Beevers, H. 1974. *Plant Physiol.* 53 (Suppl.) :40 (Abstr.)
80. Donaldson, R. P., Tolbert, N. E., Schnarrenberger, C. 1972. *Arch. Biochem. Biophys.* 152:199–215
81. Douce, R., Dupont, J. 1969. *C. R. Acad. Sci. Paris Ser. D* 268:1657–60
82. Douglass, S. A., Criddle, R. S., Breidenbach, R. W. 1973. *Plant Physiol.* 51:902–6
83. Eaglesham, A. R. J., Ellis, R. J. 1974. *Biochim. Biophys. Acta* 335:396–407
84. Elder, J. H., Morré, D. J. 1974. *J. Cell Biol.* 63:92a (Abstr.)
85. Ellis, R. J. 1974. *Biochem. Soc. Trans.* 2:179–82
86. Enomoto, K. I., Sato, R. 1973. *Biochem. Biophys. Res. Commun.* 51:1–7
87. Epstein, E. 1973. *Int. Rev. Cytol.* 34:123–68
88. Eriksson, L., Svensson, H., Bergstrand, A., Dallner, G. 1972. In *Role of Membranes in Secretory Processes,* ed. L. Bolis, R. D. Keynes, W. Wilbrandt, 3–23. Amsterdam: North Holland
89. Ernster, L., Orrenius, S. 1973. *Drug Metab. Dispos.* 1:66–73
90. Essner, E. 1967. *Lab. Invest.* 17:71–87
91. Evans, W. H. 1970. *Biochim. Biophys. Acta* 211:578–81
92. Evans, W. H., Gurd, J. W. 1971. *Biochem. J.* 125:615–24
93. Evins, W. H., Varner, J. E. 1971. *Proc. Nat. Acad. Sci. USA* 68:1631–33
94. Eytan, G., Ohad, I. 1972. *J. Biol. Chem.* 247:122–29
95. Eytan, G., Jennings, R. C., Forti, G., Ohad, I. 1974. *J. Biol. Chem.* 249:738–44
96. Favarger, P., Gerlach, J. 1971. *FEBS Lett.* 13:285–89
97. Favarger, P., Gerlach, J., Rous, S. 1968. *FEBS Lett.* 2:289–92
98. Feldherr, C. M. 1972. *Advan. Cell Mol. Biol.* 2:273–307
99. Figier, J. 1973. *Le Botaniste* 55:311–38
100. Fiil, A., Moens, P. B. 1973. *Chromosoma* 41:37–62
101. Fineran, B. A. 1973. *J. Ultrastruct. Res.* 43:75–87
102. Fisher, D. B., Mueller, G. C. 1968. *Proc. Nat. Acad. Sci. USA* 60:1396–1402
103. Fleischer, B., Fleischer, S. 1970. *Biochim. Biophys. Acta* 219:301–19
104. Fleischer, B., Zambrano, F. 1973. *Biochem. Biophys. Res. Commun.* 52:951–58
105. Flickinger, C. J. 1974. *J. Cell Sci.* 14:421–37
106. Forger, J. M., Bogorad, L. 1973. *Plant Physiol.* 52:491–97
107. Foster, R. J. et al 1971. *Proc. 1st Eur. Biophys. Congr.* 4:137–49
108. Franke, W. W. 1974. *Phil. Trans. Roy. Soc. London Ser. B* 268:67–93
109. Franke, W. W., Deumling, B., Zentgraf, H., Falk, H., Rae, P. M. M. 1973. *Exp. Cell Res.* 81:365–92
110. Franke, W. W., Eckert, W. A., Krien, S. 1971. *Z. Zellforsch. Mikrosk. Anat.* 119:577–604
111. Franke, W. W., Falk, H. 1970. *Histochemie* 24:266–78
112. Franke, W. W., Kartenbeck, J. 1971. *Protoplasma* 73:35–41
113. Franke, W. W. et al 1971. *Z. Naturforsch.* 26b:1031–39
114. Frey-Wyssling, A., Grieshaber, E., Mühlethaler, K. 1963. *J. Ultrastruct. Res.* 8:506–16
115. Frye, L. D., Edidin, M. 1970. *J. Cell Sci.* 7:319–35
116. Gaylor, J. L. 1972. *Advan. Lipid Res.* 10:89–141
117. George, E., Singh, M., Bachhawat, B. K. 1970. *J. Neurochem.* 17:189–200
118. Getz, G. S. 1970. *Advan. Lipid Res.* 8:175–223
119. Getz, G. S., Bartley, W., Stirpe, F., Notton, B. M., Renshaw, A. 1962. *Biochem. J.* 83:181–91
120. Glaumann, H., Dallner, G. 1968. *J. Lipid Res.* 9:720–29
121. Goad, L. J. 1970. In *Natural Substances Formed Biologically from Mevalonic Acid,* ed. T. W. Goodwin, 45–77. New York:Academic
122. Goad, L. J., Goodwin, T. W. 1972. *Progr. Phytochem.* 3:113–98
123. Godman, G. C., Lane, N. 1964. *J. Cell Biol.* 21:353–66
124. Goldberg, I. H. 1961. *J. Lipid Res.* 2:103–9

125. Goldstone, A., Koenig, H. 1974. *Biochem. J.* 132:267–82
126. Goodwin, T. W. 1971. In *Structure and Function of Chloroplasts*, ed. M. Gibbs, 215–76. New York:Springer
127. Gonzáles-Cadavid, N. F., Sáez de Córdova, C. 1974. *Biochem. J.* 140:157–67
128. Goswami, B. B., Chakrabarti, S., Dube, D. K., Roy, S. C. 1973. *Biochem. J.* 134:815–16
129. Graham, J. M., Sumner, M. C. B., Curtis, D. H., Pasternak, C. A. 1973. *Nature* 246:291–95
130. Gram, T. E., Gillette, J. R. 1970. In *Fundamentals of Biochemical Pharmacology*, ed. Z. M. Bacq, 571–609. Oxford:Permagon
131. Gregoriadis, G., Morell, A. G., Sternlieb, I., Scheinberg, I. H. 1970. *J. Biol. Chem.* 245:5833–37
132. Grivell, L. A., Netter, P., Borst, P., Slonimski, P. P. 1973. *Biochim. Biophys. Acta* 312:358–67
133. Groot, P. H. E., Van Loon, C. M. I., Hülsmann, W. C. 1974. *Biochim. Biophys. Acta* 337:1–12
134. Grove, S. N., Bracker, C. E., Morré, D. J. 1968. *Science* 161:171–73
135. Grove, S. N., Bracker, C. E., Morré, D. J. 1970. *Am. J. Bot.* 57:245–66
136. Gruber, P. J., Becker, W. M., Newcomb, E. H. 1973. *J. Cell Biol.* 56:500–18
137. Guignery, G., Luzzati, A., Duranton, J. 1974. *Planta* 115:227–43
138. Gurd, J. W., Evans, W. H. 1973. *Eur. J. Biochem.* 36:273–79
139. Hajra, A. K. 1968. *Biochem. Biophys. Res. Commun.* 33:929–35
140. Hakomori, S. 1971. In *The Dynamic Structure of Cell Membranes*, ed. D. F. H. Wallach, H. Fischer, 71–96. Berlin: Springer
141. Hallinan, T., Murty, C. N., Grant, J. H. 1968. *Arch. Biochem. Biophys.* 125:715–20
142. Hardin, J. W., Cherry, J. H., Morré, D. J., Lembi, C. A. 1972. *Proc. Nat. Acad. Sci. USA* 69:3146–50
143. Harris, E. H., Preston, J. F., Eisenstadt, J. M. 1973. *Biochemistry* 12:1227–33
144. Hartmann, M. A., Benveniste, P., Durst, F. 1972. *Phytochemistry* 11:3003–5
145. Hartmann, M. A., Ferne, M., Gigot, C., Brandt, R., Benveniste, P. 1973. *Physiol. Veg.* 11:209–30
146. Hawley, E. S., Greenawalt, J. W. 1970. *J. Biol. Chem.* 245:3574–83
147. Haynes, M. E., Davies, H. G. 1973. *J. Cell Sci.* 13:139–71

148. Hearing, V. J. 1973. *Phytochemistry* 12:277–82
149. Heftmann, E. 1973. In *Phytochemistry*, ed. L. P. Miller, 2:171–226. New York: Van Nostrand/Reinhold
150. Hendler, R. W. 1974. *Biomembranes* 5:147–211
151. Henningsen, K. W., Boardman, N. K. 1973. *Plant Physiol.* 51:1117–26
152. Henningsen, K. W., Kahn, A. 1971. *Plant Physiol.* 47:685–90
153. Henningsen, K. W., Thorne, S. W., Boardman, N. K. 1974. *Plant Physiol.* 53:419–25
154. Herzog, V., Miller, F. 1972. *J. Cell Biol.* 53:662–80
155. Hildebrand, C. E., Tobey, R. A. 1973. *Biochim. Biophys. Acta* 331:165–80
156. Hill, E. E., Lands, W. E. M. 1970. See Ref. 440, 185–277
157. Hirano, H., Parkhouse, B., Nicolson, G. L., Lennox, E. S., Singer, S. J. 1973. *Proc. Nat. Acad. Sci. USA* 69:2945–49
158. Hitchcock, C., Nichols, B. W. 1971. *Plant Lipid Biochemistry*. New York: Academic
159. Hoffmann, H. P., Avers, C. J. 1973. *Science* 181:749–51
160. Holloway, P. W. 1970. See Ref. 440, 371–429
161. Holtzman, J. L., Gillette, J. R. 1968. *J. Biol. Chem.* 243:3020–28
162. Holtzman, J. L., Gram, T. E., Gillette, J. R. 1970. *Arch. Biochem. Biophys.* 138:199–207
163. Hoober, K. 1972. *J. Cell Biol.* 52:84–96
164. Hoober, K., Stegeman, W. J. 1973. *J. Cell Biol.* 56:1–12
165. Hostetler, K. Y., Van den Bosch, H., Van Deenen, L. L. M. 1971. *Biochim. Biophys. Acta* 239:113–19
166. Hruban, Z., Recheigl, M. 1969. *Int. Rev. Cytol. Suppl.* 1:1–269
167. Hübscher, G. 1962. *Biochim. Biophys. Acta* 57:555–61
168. Hübscher, G. 1970. See Ref. 440, 279–370
169. Huffaker, R. C., Peterson, L. W. 1974. *Ann. Rev. Plant Physiol.* 25:363–92
170. Hughes, R. C. 1974. *Nature* 249:414
171. Ikehara, Y., Pitot, H. C. 1973. *J. Cell Biol.* 59:28–44
172. Jacobson, B. S., Kannangara, C. G., Stumpf, P. K. 1973. *Biochem. Biophys. Res. Commun.* 52:1190–98
173. Johnson, K. D., Kende, H. 1971. *Proc. Nat. Acad. Sci. USA* 68:2674–77
174. Johnston, J. M., Paultauf, F., Schiller, C. M., Schultz, L. D. 1970. *Biochim. Biophys. Acta* 218:124–33

175. Jordan, E. G., Chapman, J. M. 1973. *J. Exp. Bot.* 24:197–209
176. Jungalwala, F. B., Dawson, R. M. C. 1970. *Eur. J. Biochem.* 12:399–402
177. Jungalwala, F. B., Dawson, R. M. C. 1970. *Biochem. J.* 117:481–90
178. Jungalwala, F. B., Freinkel, N., Dawson, R. M. C. 1971. *Biochem. J.* 123:19–33
179. Kadenbach, B. 1970. *Eur. J. Biochem.* 12:392–98
180. Kadenbach, B., Hadváry, P. 1973. *Eur. J. Biochem.* 32:343–49
181. Kagawa, T., Lord, J. M., Beevers, H. 1972. *Plant Physiol.* 51:61–65
182. Kamath, S. A., Rubin, E. 1973. *Arch. Biochem. Biophys.* 158:312–22
183. Kanfer, J. N. 1972. *J. Lipid Res.* 13:468–76
184. Kanfer, J. N., Kennedy, E. P. 1964. *J. Biol. Chem.* 239:1720–26
185. Kanoh, H., Ohno, K. 1973. *Biochim. Biophys. Acta* 326:17–25
186. Kasper, C. B. 1971. *J. Biol. Chem.* 246:577–81
187. Kates, M. 1970. *Advan. Lipid Res.* 8:225–65
188. Kauss, H. 1969. *FEBS Lett.* 5:81–84
189. Kay, J. E. 1968. *Nature* 219:172–73
190. Keenan, T. W., Leonard, R. T., Hodges, T. K. 1973. *Cytobios* 7:103–12
191. Keenan, T. W., Morré, D. J. 1970. *Biochemistry* 9:19–25
192. Keenan, T. W., Morré, D. J., Basu, S. 1974. *J. Biol. Chem.* 249:310–15
193. Keenan, T. W., Morré, D. J., Huang, C. M. 1974. In *Lactation: A Comprehensive Treatise,* ed. B. L. Larson, V. R. Smith, 191–233. New York:Academic
194. Kennedy, E. P. 1961. *Fed. Proc.* 20:934–40
195. Kiermayer, O., Dobberstein, B. 1973. *Protoplasma* 77:437–51
196. Kirk, J. T. O. 1970. *Ann. Rev. Plant Physiol.* 21:11–42
197. Kirk, J. T. O., Tilney-Bassett, R. A. E. 1967. *The Plastids.* London:Freeman
198. Kobata, A., Grollman, E. F., Torain, B. F., Ginsburg, V. 1970. In *Blood and Tissue Antigens,* ed. D. Aminoff, 497–504. New York:Academic
199. Koehler, D. E., Varner, J. E. 1973. *Plant Physiol.* 52:208–14
200. Kramer, R., Schlatter, C., Zahler, P. 1972. *Biochim. Biophys. Acta* 282:146–62
201. Kreibich, G., Sabatini, D. 1973. *Fed. Proc.* 32:2133–38
202. Kroon, A. M., Saccone, C., Eds. 1974. *The Biogenesis of Mitochondria.* New York:Academic
203. Kuksis, A., Marai, L. 1967. *Lipids* 2:217–24
204. Kuriyama, Y., Luck, D. J. L. 1973. *J. Cell Biol.* 59:776–84
205. Kuriyama, Y., Omura, T. 1971. *J. Biochem.* 69:659–69
206. Kuriyama, Y., Omura, T., Siekevitz, P., Palade, G. E. 1969. *J. Biol. Chem.* 244:2017–26
207. Laflèche, D., Bové, J. M., Duranton, J. 1972. *J. Ultrastruct. Res.* 40:205–14
208. Landriscina, C., Marra, E. 1973. *Life Sci.* 13:1373–81
209. Lands, W. E. M., Hart, P. 1964. *J. Lipid Res.* 5:81–87
210. Lane, N., Caro, L., Otero-Vilardébo, L. R., Godman, G. C. 1964. *J. Cell Biol.* 21:339–51
211. Lawford, G. R., Schachter, H. 1966. *J. Biol. Chem.* 241:5408–18
212. Lazarow, P. B., de Duve, C. 1973. *J. Cell Biol.* 59:507–24
213. Lee, T. C., Snyder, F. 1973. *Biochim. Biophys. Acta* 291:71–82
214. Lee, T. C., Stephens, N., Moehl, A., Snyder, F. 1973. *Biochim. Biophys. Acta* 291:86–92
215. Legg, P. G., Wood, R. L. 1970. *J. Cell Biol.* 45:118–29
216. Lennarz, W. J., Scher, M. G. 1972. *Biochim. Biophys. Acta* 265:417–41
217. Lodish, H. F. 1973. *Proc. Nat. Acad. Sci. USA* 70:1526–30
218. Lord, J. M., Kagawa, T., Beevers, H. 1972. *Proc. Nat. Acad. Sci. USA* 69:2429–32
219. Lord, J. M., Kagawa, T., Moore, T. S., Beevers, H. 1973. *J. Cell Biol.* 57:659–67
220. Lowe, D., Hallinan, T. 1973. *Biochem. J.* 136:825–28
221. Lu, A. Y. H., Levin, W. 1974. *Biochim. Biophys. Acta* 344:205–40
222. Luck, D. J. L. 1965. *J. Cell Biol.* 24:461–70
223. Lynen, F. 1967. *Biochem. J.* 102:381–400
224. MacFarlane, M., Gray, G. M., Wheeldon, L. W. 1960. *Biochem. J.* 77:626–31
225. Macher, B. A., Mudd, J. B. 1974. *Plant Physiol.* 53:171–75
226. Machold, O., Aurich, O. 1972. *Biochim. Biophys. Acta* 281:103–12
227. Mackender, R. O., Leech, R. M. 1974. *Plant Physiol.* 53:496–502
228. Mahlberg, P. 1972. *Am. J. Bot.* 59:172–79
229. Majerus, P. W., Vagelos, P. R. 1967. *Advan. Lipid Res.* 5:1–30
230. Manley, E. R., Skrdlant, H. B. 1974. *Fed. Proc.* 33:1525(Abstr.)

231. Margulies, M. M., Michaels, A. 1974. *J. Cell Biol.* 60:65–77
232. Marinos, N. G. 1963. *J. Ultrastruct. Res.* 9:177–85
233. Marks, G. S. 1969. *Heme and Chlorophyll.* London:Van Nostrand
234. Marshall, M. O., Kates, M. 1972. *Biochim. Biophys. Acta* 260:558–70
235. Marshall, M. O., Kates, M. 1974. *Can. J. Biochem.* 52:469–82
236. Mason, T. L., Schatz, G. 1973. *J. Biol. Chem.* 248:1355–60
237. Mastro, A. M., Beer, C. T., Mueller, G. C. 1974. *Biochim. Biophys. Acta* 352:38–51
238. Matile, P. 1968. *Planta* 79:181–96
239. Matile, P. 1974. In *Dynamic Aspects of Plant Ultrastructure,* ed. A. W. Robards, 178–218. London:McGraw-Hill
240. Matile, P., Moor, H. 1968. *Planta* 80:159–75
241. Maul, G. G., Price, J. W., Lieberman, M. W. 1971. *J. Cell Biol.* 51:405–18
242. Maul, H. M., Wessing, A. 1972. *J. Cell Biol.* 55:167a(Abstr.)
243. Mazliak, P. 1973. *Ann. Rev. Plant Physiol.* 24:287–310
244. Mazliak, P., Oursel, A., Ben Abdelkader, A., Grosbois, M. 1972. *Eur. J. Biochem.* 28:399–411
245. Mazliak, P., Ben Abdelkader, A. 1971. *Phytochemistry* 10:2879–90
246. McCarty, R. E., Douce, R., Benson, A. A. 1973. *Biochim. Biophys. Acta* 316:266–70
247. McConnell, H. M., McFarland, B. G. 1970. *Quart. Rev. Biophys.* 3:91–136
248. McMurray, W. C. 1974. *Biochem. Biophys. Res. Commun.* 58:467–81
249. McMurray, W. C., Dawson, R. M. C. 1969. *Biochem. J.* 112:91–108
250. McMurray, W. C., Magee, W. L. 1972. *Ann. Rev. Biochem.* 41:129–60
251. Meldolesi, J. 1974. *J. Cell Biol.* 61:1–13
252. Meldolesi, J., Cova, D. 1971. *Biochem. Biophys. Res. Commun.* 44:139–43
253. Meldolesi, J., Cova, D. 1972. *J. Cell Biol.* 55:1–18
254. Menon, K. M. J., Dorfman, R. I., Forchielli, E. 1967. *Biochim. Biophys. Acta* 148:486–94
255. Merritt, W. D., Morré, D. J. 1973. *Biochim. Biophys. Acta* 304:397–407
256. Mets, L., Bogorad, L. 1972. *Proc. Nat. Acad. Sci. USA* 69:3779–83
257. Michel, R., Neupert, W. 1973. *Eur. J. Biochem.* 36:53–67
258. Miller, E. K., Dawson, R. M. 1972. *Biochem. J.* 126:823–35
259. Moore, T. S. 1974. *Plant Physiol.* 54:164–68
260. Moore, T. S. 1974. *Plant Physiol.* 53(Suppl.):40(Abstr.)
261. Moore, T. S., Lord, J. M., Kagawa, T., Beevers, H. 1973. *Plant Physiol.* 52:50–53
262. Moore, T. S., Barlow, S. A., Coolbaugh, R. C. 1972. *Phytochemistry* 11:3225–33
263. Mollenhauer, H. H., Morré, D. J. 1974. *Protoplasma* 79:333–36
264. Molnar, J., Robinson, G. B., Winzler, R. J. 1965. *J. Biol. Chem.* 240:1882–88
265. Molnar, J., Tetas, M., Chao, H. 1969. *Biochem. Biophys. Res. Commun.* 37:684–90
266. Moreau, F., Dupont, J., Lance, C. 1974. *Biochim. Biophys. Acta* 345:294–304
267. Morré, D. J. 1970. *Plant Physiol.* 45:791–99
268. Morré, D. J., Elder, J. H., Jelsema, C., Merritt, W. D. 1975. In press
269. Morré, D. J., Franke, W. W., Deumling, B., Nyquist, S. E., Ovtracht, L. 1971. *Biomembranes* 2:95–104
270. Morré, D. J., Keenan, T. W., Huang, C. M. 1974. *Advan. Cytopharmacol.* 2:107–25
271. Morré, D. J., Keenan, T. W., Mollenhauer, H. H. 1971. *Advan. Cytopharmacol.* 1:159–82
272. Morré, D. J., Merlin, L. M., Keenan, T. W. 1969. *Biochem. Biophys. Res. Commun.* 37:813–19
273. Morré, D. J., Merritt, W. D., Lembi, C. A. 1971. *Protoplasma* 73:43–49
274. Morré, D. J., Mollenhauer, H. H. 1974. See Ref. 239, 84–137
275. Morré, D. J., Mollenhauer, H. H., Bracker, C. E. 1971. See Ref. 327, 82–126
276. Morré, D. J., Nyquist, S., Rivera, E. 1970. *Plant Physiol.* 45:800–4
277. Morré, D. J., Van der Woude, W. J. 1974. In *Macromolecules Regulating Growth and Development,* ed. E. D. Hay, T. J. King, J. Papaconstantinou, 81–111. New York: Academic
278. Mosser, A. G., Caliguiri, L. A., Tamm, I. 1972. *Virology* 47:39–47
279. Mudd, J. B., Van Vliet, H.H.D.M., Van Deenen, L.L.M. 1969. *J. Lipid Res.* 10:623–30
280. Murty, C. N., Sidransky, H. 1974. *Biochim. Biophys. Acta* 335:226–35
281. Nachbaur, J., Colbeau, A., Vignais, P. M. 1969. *FEBS Lett.* 3:121–24
282. Neupert, W., Brdiczka, D., Bücher, T. 1967. *Biochem. Biophys. Res. Commun.* 27:488–93
283. Newcomb, E. H. 1967. *J. Cell Biol.* 35:C17–22

284. Newkirk, J. D., Waite, M. 1971. *Biochim. Biophys. Acta* 225:224–33
285. Nielsen, O. F., Kahn, A. 1973. *Biochim. Biophys. Acta* 292:117–29
286. Novikoff, A. B., Allen, J. M. 1973. *J. Histochem. Cytochem.* 21:941–77
287. Novikoff, A. B., Novikoff, P. M. 1973. *J. Histochem. Cytochem.* 21:963–66
288. Novikoff, A. B., Shin, W. Y. 1964. *J. Microsc.* 3:187–206
289. Novikoff, P. M., Novikoff, A. B. 1972. *J. Cell Biol.* 53:532–60
290. Nyquist, S. E., Crane, F. L., Morré, D. J. 1971. *Science* 173:939–41
291. Nyquist, S. E., Matschiner, J. T., Morré, D. J. 1971. *Biochim. Biophys. Acta* 244:645–49
292. Oliveira, M. M., Vaughan, M. 1964. *J. Lipid Res.* 5:156–62
293. Omura, T., Siekevitz, P., Palade, G. E. 1967. *J. Biol. Chem.* 242:2389–96
294. Öpik, H. 1974. See Ref. 239, 52–83
295. Orrenius, S., Ericsson, J. L. E., Ernster, L. 1965. *J. Cell Biol.* 25:627–39
296. Oshino, N. 1972. *Arch. Biochem. Biophys.* 149:378–87
297. Oshino, N., Sato, R. 1972. *Arch. Biochem. Biophys.* 149:369–77
298. Palade, G. E., Porter, K. R. 1954. *J. Exp. Med.* 100:641–56
299. Panter, R. A., Boardman, N. K. 1973. *J. Lipid Res.* 14:664–71
300. Parkes, J. G., Thompson, W. 1974. *Biochim. Biophys. Acta* 306:403–11
301. Parkes, J. G., Thompson, W. 1973. *J. Biol. Chem.* 248:6655–62
302. Pascaud, M. 1964. *Biochim. Biophys. Acta* 84:528–37
303. Pasternak, C. A. 1973. *Develop. Biol.* 30:403–10
304. Patt, L. M., Grimes, W. J. 1974. *J. Biol. Chem.* 249:4157–65
305. Peel, M. C., Lucas, I. A. N., Duckett, J. G., Greenwood, A. D. 1973. *Z. Zellforsch. Mikrosk. Anat.* 147:59–74
306. Pinto Da Silva, P., Douglas, S. D., Branton, D. 1971. *Nature* 232:194–96
307. Pitot, H. C., Jost, J. P. 1968. In *Regulatory Mechanisms for Protein Synthesis in Mammalian Cells,* ed. A. San Pietro, M. R. Lamborg, F. T. Kenney, 283–98. New York: Academic
308. Poole, B., Higashi, T., de Duve, C. 1970. *J. Cell Biol.* 45:408–15
309. Porcellati, G., Arienti, G., Pirotta, M., Giorgini, D. 1971. *J. Neurochem.* 18:1395–1417
310. Porter, K. R. 1961. In *The Cell,* ed. J. Brachet, A. E. Mirsky, 2:621–75. New York: Academic
311. Possingham, J. V. 1973. *J. Exp. Bot.* 24:1247–60
312. Possmayer, F., Scherphof, G. L., Dubbelman, T. M. A. R., Van Golde, L. M. G., Van Deenen, L. L. M. 1969. *Biochim. Biophys. Acta* 176:95–110
313. Pugh, E. L., Kates, M. 1973. *Biochim. Biophys. Acta* 316:305–16
314. Puro, D. G., Richter, G. W. 1971. *Proc. Soc. Exp. Biol. Med.* 138:399–403
315. Quarles, R. H., Dawson, R. M. C. 1969. *Biochem. J.* 112:787–94
316. Quirk, S. J., Byrne, J., Robinson, G. B. 1973. *Biochem. J.* 132:501–8
317. Rabinowitz, M., Swift, H. 1970. *Physiol. Rev.* 50:376–427
318. Ragnotti, G., Lawford, G. R., Campbell, P. N. 1969. *Biochem. J.* 112:139–47
319. Ramirez, G. 1973. *Biochem. Biophys. Res. Commun.* 50:452–58
320. Ray, T. K., Lieberman, I., Lansing, A. I. 1968. *Biochem. Biophys. Res. Commun.* 31:54–58
321. Rebeiz, C. A., Castelfranco, P. A. 1973. *Ann. Rev. Plant Physiol.* 24:129–72
322. Reddy, J. K., 1973. *J. Histochem. Cytochem.* 21:967–71
323. Reddy, J., Svoboda, D. 1973. *Am. J. Pathol.* 70:421–38
324. Redman, C. M., Cherian, M. G. 1972. *J. Cell Biol.* 52:231–45
325. Redman, C. M., Sabatini, D. D. 1966. *Proc. Nat. Acad. Sci. USA* 56:608–15
326. Redman, C. M., Siekevitz, P., Palade, G. E. 1966. *J. Biol. Chem.* 241:1150–58
327. Reinert, J., Ursprung, H., Eds. 1971. *Results and Problems in Cell Differentiation. II. Origin and Continuity of Cell Organelles.* Berlin:Springer
328. Resch, K., Ferber, E., Odenthal, J., Fischer, H. 1971. *Eur. J. Immunol.* 1:162–65
329. Richmond, D. V. 1973. See Ref. 149, 3:41–73
330. Ritter, M. C., Dempsey, M. E. 1973. *Proc. Nat. Acad. Sci. USA* 70:265–69
331. Roberts, J. B., Bygrave, F. L. 1973. *Biochem. J.* 136:467–75
332. Roberts, R. M., Connor, A. B., Cetorelli, J. J. 1971. *Biochem. J.* 125:999–1008
333. Robertson, D., Laetsch, W. M. 1974. *Plant Physiol.* 54:148–59
334. Robinson, P. M., Park, D., McClure, W. K. 1969. *Trans. Brit. Mycol. Soc.* 52:447–50
335. Rogers, M. J., Strittmatter, P. 1974. *J. Biol. Chem.* 249:895–900
336. Roland, J. C., Vian, B. 1971. *Protoplasma* 73:121–37

337. Rolleston, F. S. 1974. *Sub-Cell. Biochem.* 3:91–117
338. Rolleston, F. S., Lam, T. Y. 1974. *Biochem. Biophys. Res. Commun.* 59:467–73
339. Roodyn, D. B., Wilkie, D. 1968. *The Biogenesis of Mitochondria.* London: Methuen
340. Roncari, D. A. K. 1974. *Fed. Proc.* 33:1525(Abstr.)
341. Roncari, D. A. K. 1974. *Can. J. Biochem.* 52:221–30
342. Roseman, S. 1971. *Chem. Phys. Lipids.* 5:270–97
343. Ross, E. et al 1974. See Ref. 202, 477–89
344. Rossi, C. R., Sartorelli, L., Tatò, L., Baretta, L., Siliprandi, N. 1965. *Biochim. Biophys. Acta* 98:207–9
345. Roth, S. 1973. *Quart. Rev. Biol.* 48:541–63
346. Rothfield, L. I. 1971. *Structure and Function of Biological Membranes.* New York:Academic
347. Ruby, J. R., Dyer, R. F., Skalko, R. G. 1969. *Z. Zellforsch. Mikrosk. Anat.* 97:30–37
348. Sabatini, D. D., Blobel, G. 1970. *J. Cell Biol.* 45:146–57
349. Sabatini, D. D., Tashiro, Y., Palade, G. E. 1966. *J. Mol. Biol.* 19:503–24
350. Salerno, D. M., Beeler, D. A. 1973. *Biochim. Biophys. Acta* 326:325–38
351. Sarma, D. S. R., Reid, I. M., Verney, E., Sidransky, H. 1972. *Lab Invest.* 27:39–47
352. Sarzala, M. G., Van Golde, L. M. G., De Kruyff, B., Van Deenen, L. L. M. 1970. *Biochim. Biophys. Acta* 202:106–19
353. Satav, J. G. et al 1973. *Biochem. J.* 134:687–95
354. Scallen, T. J., Srikantaiah, M. V., Seetharam, B., Hansbury, E., Gavey, K. L. 1974. *Fed. Proc.* 33:1733–46
355. Schachter, H. et al 1970. *J. Biol. Chem.* 245:1090–1100
356. Schatz, G., Mason, T. L. 1974. *Ann. Rev. Biochem.* 43:51–87
357. Scheer, U. 1972. *Z. Zellforsch. Mikrosk. Anat.* 127:127–48
358. Scheer, U. 1973. *Develop. Biol.* 30:13–28
359. Scherphof, G. L., Van Deenen, L. L. M. 1965. *Biochim. Biophys. Acta* 98:204–6
360. Scherphof, G. L., Waite, M., Van Deenen, L. L. M. 1966. *Biochim. Biophys. Acta* 125:406–9
361. Schiff, J. A., Hodson, R. C. 1973. *Ann. Rev. Plant Physiol.* 24:381–414
362. Schnaitman, C. A. 1969. *Proc. Nat. Acad. Sci. USA* 63:412–19
363. Schneider, W. C. 1963. *J. Biol. Chem.* 238:3572–78
364. Schwertner, H. A., Biale, J. B. 1973. *J. Lipid Res.* 14:235–42
365. Schultz, A., Sauer, K. 1972. *Biochim. Biophys. Acta* 267:320–40
366. Sebald, W., Machleidt, W., Otto, J. 1973. *Eur. J. Biochem.* 38:311–24
367. Sebald, W., Weiss, H., Jackl, G. 1972. *Eur. J. Biochem.* 30:413–17
368. Setty, P. N. , Krishnan, P. S. 1972. *Biochem. J.* 126:313–24
369. Shafritz, D. A. 1974. *J. Biol. Chem.* 249:89–93
370. Shannon, C. F., Hill, J. M. 1971. *Biochemistry* 10:3021–29
371. Shephard, E. H., Hübscher, G. 1969. *Biochem. J.* 113:429–40
372. Sherr, C. J., Uhr, J. W. 1970. *Proc. Nat. Acad. Sci. USA* 66:1183–89
373. Shires, T. K., Pitot, H. C. , Kauffmann, S. A. 1974. *Biomembranes* 5:81–145
374. Shore, G., Maclachlan, G. 1974. *Plant Physiol.* 53(Suppl.):16(Abstr.)
375. Siekevitz, P. 1972. *J. Theor. Biol.* 37:321–34
376. Siekevitz, P. 1972. *Ann. Rev. Physiol.* 34:117–40
377. Siekevitz, P., Palade, G. E. 1960. *J. Biophys. Biochem. Cytol.* 7:619–30
378. Siekevitz, P., Palade, G. E., Dallner, G., Ohad, I., Omura, T. 1967. In *Organizational Biosynthesis,* ed. H. J. Vogel, J. O. Lampen, V. Bryson, 331–62. New York:Academic
379. Singer, S. J. 1974. *Ann. Rev. Biochem.* 43:805–33
380. Sirevag, R., Levine, R. P. 1972. *J. Biol. Chem.* 247:2586–91
381. Skurdal, D. N., Cornatzer, W. E. 1974. *Fed. Proc.* 33:1470(Abstr.)
382. Spatz, L., Strittmatter, P. 1971. *Proc. Nat. Acad. Sci. USA* 68:1042–46
383. Staby, G. L., Hackett, W. P., De Hertogh, A. A. 1973. *Plant Physiol.* 52:416–21
384. Stadler, J., Franke, W. W. 1973. *Biochim. Biophys. Acta* 311:205–13
385. Stahl, W. L., Trams, E. G. 1968. *Biochim. Biophys. Acta* 163:459–71
386. Stanacev, N. Z., Stuhne-Sekalec, L., Brookes, K. B., Davidson, J. B. 1969. *Biochim. Biophys. Acta* 176:650–53
387. Stein, O., Stein, Y. 1969. *J. Cell Biol.* 40:461–83
388. Stein, Y., Widnell, C., Stein, O. 1968. *J. Cell Biol.* 39:185–92
389. Stetten, M. R., Ghosh, S. B. 1971. *Biochim. Biophys. Acta* 233:163–75
390. Stoffel, W., Schiefer, H. G. 1968.

Hoppe-Seyler's Z. Physiol. Chem. 349:1017–26

391. Stumpf, P. K. 1970. See Ref. 440, 79–106

392. Stumpf, P. K. 1970. *Compr. Biochem.* 18:265–92

393. Sumida, S., Mudd, J. B. 1970. *Plant Physiol.* 45:712–18

394. Sundquist, K. G. 1972. *Nature New Biol.* 239:147–49

395. Swanson, R. F. 1973. *Biochemistry* 12:2142–46

396. Takagi, M., Tanaka, T., Ogata, K. 1970. *Biochim. Biophys. Acta* 217:148–58

397. Tamaoki, B. I. 1973. *J. Steroid Biochem.* 4:89–118

398. Tanaka, K., Tolbert, N. E., Gohlke, A. F. 1966. *Plant Physiol.* 41:307–12

399. Tao, K. J., Jagendorf, A. T. 1973. *Biochim. Biophys. Acta* 324:518–22

400. Tata, J. R. 1973. *Acta Endocrinol.* 180:192–224

401. Taylor, J. M., Dehlinger, P. J., Dice, J. F., Schimke, R. T. 1973. *Drug Metab. Dispos.* 1:84–91

402. Taylor, R. B., Duffus, W. P. H., Raff, M. C., De Petris, S. 1971. *Nature New Biol.* 233:225–29

403. Thompson, E. D., Bailey, R. B., Parks, L. W. 1974. *Biochim. Biophys. Acta* 334:116–26

404. Tillack, T. W., Boland, R., Martonosi, A. 1974. *J. Biol. Chem.* 249:624–33

405. Tillack, T. W., Scott, R. E., Marchesi, V. T. 1972. *J. Exp. Med.* 135:1209–27

406. Towers, N. R. 1974. *Life Sci.* 14:2037–43

407. Trams, E. G., Lauter, C. J. 1974. *Biochim. Biophys. Acta* 345:180–97

408. Treffry, T. 1973. *J. Exp. Bot.* 24:185–95

409. Trumpower, B. L., Houser, R. M., Olson, R. E. 1974. *J. Biol. Chem.* 249:3041–48

410. Tsukada, H., Mochizuki, Y., Konishi, T. 1968. *J. Cell Biol.* 37:231–43

411. Turkki, P. R., Glenn, J. L. 1968. *Biochim. Biophys. Acta* 152:104–13

412. Tzagoloff, A., Meagher, P. 1972. *J. Biol. Chem.* 247:594–603

413. Tzagoloff, A., Rubin, M. S., Sierra, M. F. 1973. *Biochim. Biophys. Acta* 301:71–104

414. Ueda, K. 1966. *Cytologia* 31:461–72

415. Vagelos, P. R. 1971. *Curr. Top. Cell. Regul.* 4:119–66

416. Van den Bosch, H. 1974. *Ann. Rev. Biochem.* 43:243–77

417. Van den Bosch, H., Van Golde, L. M. G., Slotboom, A. J., Van Deenen, L. L.

1968. *Biochim. Biophys. Acta* 152:694–703

418. Van den Bosch, H., Van Golde, L. M. G., Van Deenen, L. L. M. 1972. *Ergeb. Physiol. Biol. Chem. Exp. Pharmakol.* 66:13–145

419. Van der Woude, W. J., Lembi, C. A., Morré, D. J., Kindinger, J. I., Ordin, L. 1974. *Plant Physiol.* 54:333–40

420. Van der Woude, W. J., Morré, D. J., Bracker, C. E. 1969. *Proc. 11th Int. Bot. Congr.*, 226 (Abstr.)

421. Van der Woude, W. J., Morré, D. J., Bracker, C. E. 1971. *J. Cell Sci.* 8:331–51

422. Vandor, S. L., Richardson, K. E. 1968. *Can. J. Biochem.* 46:1309–15

423. Van Golde, L. M. G., Fleischer, B., Fleischer, S. 1971. *Biochim. Biophys. Acta* 249:318–30

424. Van Hummel, H. C. 1974. *Z. Pflanzenphysiol.* 71:228–41

425. Van Lenten, L., Ashwell, G. 1972. *J. Biol. Chem.* 247:4633–40

426. Vian, B., Roland, J. C. 1972. *J. Microsc.* 13:119–36

427. Victoria, E. J., Van Golde, L. M. G., Hostetler, K. Y., Scherphof, G. L., Van Deenen, L. L. M. 1971. *Biochim. Biophys. Acta* 239:443–57

428. Vigil, E. L. 1970. *J. Cell Biol.* 46:435–54

429. Vigil, E. L. 1973. *Sub-Cell. Biochem.* 2:237–85

430. Vigil, E. L., Ruddat, M. 1973. *Plant Physiol.* 51:549–58

431. Villemez, C. L. 1970. *Biochem. Biophys. Res. Commun.* 40:636–41

432. Volpe, J. J., Vagelos, P. R. 1973. *Ann. Rev. Biochem.* 42:21–60

433. Von Wettstein, D. 1974. *Biochem. Soc. Trans.* 2:176–79

434. Vredevoogd, C., Morré, D. J., Ruddat, M. *Isolation and Characterization of Spherosomes from Wheat Aleurone Layers.* In preparation

435. Waechter, C. J., Lucas, J. J., Lennarz, W. J. 1973. *J. Biol. Chem.* 248:7570–79

436. Waite, M. 1969. *Biochemistry* 8:2536–42

437. Waite, M., Scherphof, G. L., Boshouwers, F. M. G., Van Deenen, L. L. M. 1969. *J. Lipid Res.* 10:411–20

438. Waite, M., Sisson, P. 1971. *Biochemistry* 10:2377–83

439. Waite, M., Sisson, P., Blackwell, E. 1970. *Biochemistry* 9:746–53

440. Wakil, S. J., Ed. 1970. *Lipid Metabolism,* 1–48. New York: Academic

441. Warren, L. 1969. *Curr. Top. Develop. Biol.* 4:197–222

442. Warren, L., Glick, M. C. 1968. *J. Cell Biol.* 37:729–46
443. Weete, J. D. 1973. *Phytochemistry* 12:1843–64
444. Weiss, H. 1974. *Biochem. Soc. Trans.* 2:185–88
445. Weiss, H., Ziganke, B. 1974. *Eur. J. Biochem.* 41:63–71
446. Whaley, W. G., Dauwalder, M., Kephart, J. E. 1972. *Science* 175:596–99
447. Wildman, S. G. et al 1973. In *The Biochemistry of Gene Expression in Higher Organisms*, ed. J. K. Pollak, J. W. Lee, 443–56. Sydney:Aust. N.Z. Book
448. Wilgram, G. F., Kennedy, E. P. 1963. *J. Biol. Chem.* 238:2615–19
449. Willemot, C., Boll, W. G. 1967. *Can. J. Bot.* 45:1863–76
450. Williamson, F. A., Morré, D. J. *Distribution of Enzymes of Phosphatidylinositol Biosynthesis in Rat Liver.* In preparation
451. Wirtz, K. W. A. 1974. *Biochim. Biophys. Acta* 344:95–117
452. Wirtz, K. W. A., Kamp, H. H. 1972. See Ref. 88, 52–61
453. Wirtz, K. W. A., Kamp, H. H., Van Deenen, L. L. M. 1972. *Biochim. Biophys. Acta* 274:606–17

Added in Proof
467. Knap, A., Kornblatt, M. J., Schachter, H., Murray, R. K. 1973. *Biochem. Biophys. Res. Commun.* 55:179–86
468. Kung, S. D., Thornber, J. P., Wildman, S. G. 1972. *FEBS Lett.* 24:185–88

454. Wirtz, K. W. A., Zilversmit, D. B. 1968. *J. Biol. Chem.* 243:3596–3602
455. Wirtz, K. W. A., Zilversmit, D. B. 1969. *Biochim. Biophys. Acta* 193:105–16
456. Wirtz, K. W. A., Zilversmit, D. B. 1970. *FEBS Lett.* 7:44–46
457. Wojtczak, L., Baranska, J., Zborowski, J., Drahota, Z. 1971. *Biochim. Biophys. Acta* 249:41–52
458. Wright, J. D., Green, C. 1971. *Biochem. J.* 123:837–44
459. Yarrison, G., Choules, G. L. 1973. *Biochem. Biophys. Res. Commun.* 52:57–63
460. Young, R. W. 1973. *J. Cell Biol.* 57:175–89
461. Yunghans, W. N., Morré, D. J., Cherry, J. H. 1973. *Proc. Indiana Acad. Sci.* 82:134–36
462. Zahler, P. 1972. *Biomembranes* 3:193–95
463. Zahler, W. L., Fleischer, B., Fleischer, S. 1970. *Biochim. Biophys. Acta* 203:283–90
464. Zalik, S., Jones, B. L. 1973. *Ann. Rev. Plant Physiol.* 24:47–68
465. Zborowski, J., Wojtczak, L. 1969. *Biochim. Biophys. Acta* 187:73–84
466. Zilversmit, D. B. 1971. *J. Lipid Res.* 12:36–42

469. Lundquist, U., Perlman, P. 1966. *Science* 152:780–82
470. Yunghans, W. N., Keenan, T. W., Morré, D. J. 1970. *Exp. Mol. Pathol.* 12:36–45

AUTHOR INDEX

SUBJECT INDEX

CUMULATIVE INDEXES

CONTRIBUTING AUTHORS VOLUMES 22-26

CHAPTER TITLES VOLUMES 22-26